ACCELERAT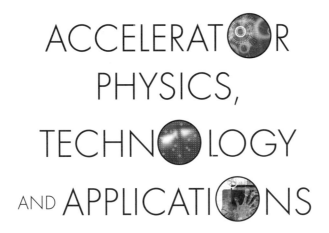R PHYSICS, TECHN LOGY AND APPLICATI NS

Selected Lectures of
OCPA International Accelerator School 2002
Singapore

ACCELERAT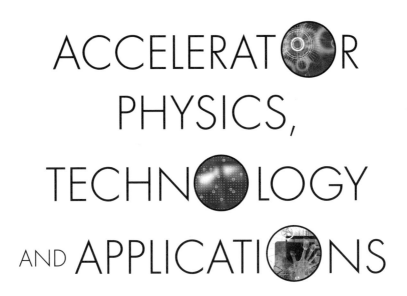R PHYSICS, TECHN LOGY AND APPLICATI NS

Selected Lectures of
OCPA International Accelerator School 2002
Singapore

Editors

Alexander Wu Chao
Stanford Linear Accelerator Center, USA

Herbert O. Moser
Singapore Synchrotron Light Source
National University of Singapore

Zhentang Zhao
Shanghai Synchrotron Radiation Facility, China

 **World Scientific**

NEW JERSEY • LONDON • SINGAPORE • SHANGHAI • HONG KONG • TAIPEI • CHENNAI

Published by

World Scientific Publishing Co. Pte. Ltd.

5 Toh Tuck Link, Singapore 596224

USA office: Suite 202, 1060 Main Street, River Edge, NJ 07661

UK office: 57 Shelton Street, Covent Garden, London WC2H 9HE

British Library Cataloguing-in-Publication Data
A catalogue record for this book is available from the British Library.

ISBN 981-238-794-3

Printed in Singapore by World Scientific Printers (S) Pte Ltd

PREFACE

This book is a result of the 3rd International Accelerator School held in 2002 under the auspices of the Overseas Chinese Physics Association (OCPA) in Singapore. Preceding schools were held in Hsinchu, Taiwan, in 1998, and Yellow Mountain, China, in 2000. The purpose of these schools is to provide students, particularly from Asia, a basic training on modern accelerators.

The past two decades have seen a rapid growth in the science and technology of accelerators as well as the development of a wealth of their applications. Accelerators, the art to build them, and the science to understand their function have become a very exciting field of research. The OCPA school series aims on conveying this excitement from the experts in the field to the students such as to fuel the manpower needs of the ongoing accelerator R&D work.

Reaching a wider audience than the participants of the school is the purpose of this book. Its contents rely mostly on the work of the school's curriculum committee and are biased towards synchrotron radiation as the local host was the Singapore Synchrotron Light Source.

The editors gratefully acknowledge the dedicated contribution of manuscripts by an expert group of scientists as well as the crucial support by K K Phua and World Scientific Publishing in producing this book. May it help many readers to become initiated into the fascinating field of accelerators.

Alex Wu Chao, SLAC
Herbert O. Moser, SSLS
Zhentang Zhao, SSRC

Organizers

Singapore Synchrotron Light Source, National University of Singapore
Department of Physics, National University of Singapore
South East Asia Theoretical Physics Association, Singapore
Division of Beam Physics, Overseas Chinese Physics Association

School Secretariat

Meeting Matters International, Ms Cheng-Hoon KHOO, General Manager

Sponsors

Singapore Lee Foundation
Singapore Synchrotron Light Source, National University of Singapore
National Natural Science Foundation of China
Synchrotron Radiation Research Center, Hsinchu
Singapore Buddhist Lodge
M & W Zander, Munich, and ACCEL, Bergisch Gladbach
South East Asia Theoretical Physics Association, Singapore
World Scientific Publishing Co Ltd
Overseas Chinese Physics Association

Organizing Committee

Alex CHAO (SLAC) — Chair
Chien-Te CHEN (SRRC)
Jiaer CHEN (NSFC)
Shouxian FANG (IHEP)
Herbert O. MOSER (SSLS)
C.H. OH (NUS)
K.K. PHUA (SEATPA)
Lee TENG (ANL)

Local Committee

Herbert O. MOSER (SSLS) — Chair
K.K. PHUA (SEATPA)
C.H. OH (NUS)
Andrew WEE (NUS)
Edward TEO (NUS)
Vince KEMPSON (SSLS)
Robert ROSSMANITH (SSLS)
Frank WATT (NUS)
Khing Ling CHONG (SSLS)

Curriculum Committee

Zhentang ZHAO (SSRC) — Chair
June-Rong CHEN (SRRC)
Shyh-Yuan LEE (U. Indiana)
Zhiyuan GUO (IHEP)
Jiaxun FANG (Peking U.)
Robert ROSSMANITH (SSLS)

CONTENTS

ACCELERATOR PHYSICS, TECHNOLOGY AND APPLICATIONS

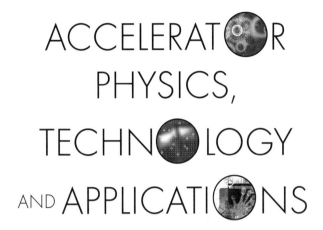

Selected Lectures of
OCPA International Accelerator School 2002
Singapore

PARTICLE ACCELERATORS: AN INTRODUCTION

Historical Evolution, Research Frontiers, Innovative Ideas and Perspectives of Accelerator Development

ZHANG CHUANG

Institute of High Energy Physics, Chinese Academy of Sciences
P.O.Box 918, Beijing 100039, China
E-mail: zhangc@mail.ihep.ac.cn

The human curiosity about the micro-world is the driving force behind the development of particle accelerators. Traced to its three roots, the history of accelerators is a continuous upgrade towards higher energy and better performance. Higher energy and higher luminosity are two frontiers of accelerator development for high energy physics. As platforms of multidisciplinary research, synchrotron radiation sources, free electron lasers, and spallation sources, are vigorously growing. Varieties of low- and medium-energy accelerators are widely applied in our society. Innovative ideas, new methods, and new technologies emerge endlessly. In this chapter, the historical evolution of particle accelerators is reviewed and their prospects are described.

1. Historical Evolution of Particle Accelerators

1.1. *Study of Matter and Particle Accelerators*

The world is made of matter. The aims of science are discovery, insight and understanding of the natural environment and the laws that govern it. Figure 1 demonstrates the nature and the science sketched by Prof. A.Glashow in a lecture.

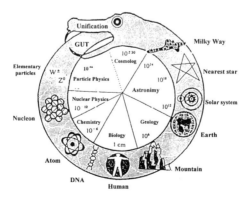

Figure 1. The Glashow snake, the nature and the science

As a part of basic science, particle physics studies "elementary particles", the bricks which the huge building of matter consists of. Nuclear physics and high energy physics play an essential role in basic science for they try to answer the following fundamental questions: What are the primal constituents of all matter and energy in the universe? What are the laws governing the behavior of those constituents that let them combine and form matter we observe?

It is known from the development of science that matter is made of atoms which are composites of a nucleus and electrons while nuclei are composites of protons and neutrons. The latter consist of even smaller particles called "quarks". As a microscope for the micro-world, the "resolution" of accelerators is limited by the de Broglie wavelength λ

$$\lambda = \frac{h}{p} = \frac{hc}{E\beta} \tag{1}$$

where $h=3.507\times10^{-15}$ eV·s is Planck's constant, p, E and β are momentum, energy, and relativistic speed of the beam used by the microscope. It can be seen from Eq. (1) that the smaller the observed substance the higher a beam energy is needed. The methods to investigate the micro-world are indicated in Table 1.

Table 1. The methods to investigate the micro-world

Observed Substance	Size (cm)	Beam Energy $E = hc/\lambda\beta$	Method
Cell/Bacteria = Aggregate of molecules	10^{-3}~10^{-5}	0.1~10 eV	Optical microscope
Molecule = Aggregate of atoms	~10^{-7}	~1 keV	Electron microscope
Atom = Nucleus + Electrons	~10^{-8}	~ 10 keV	Synchrotron radiation
Nucleus e.g., Oxygen = 8p+8n	~10^{-12}	>100 MeV	Low-energy electron or proton accelerators
Hadron = Aggregate of quarks e.g., p=u+u+d, J/ψ=c+$\bar{c}$	~10^{-13}	>1 GeV	High-energy proton accelerators
Quark, Lepton ... (u,d) (s,c) (b,t) $(e,\nu_e)(\mu,\nu_\mu)(\tau,\nu_\tau)$	<10^{-16}	>1TeV	High-energy electron or proton colliders
......	......	......	......

1.2. *Historical Roots and Development*

The history and development of particle accelerators is intimately connected to the discoveries of physics frontiers. It shows a continuous path towards higher energy. This has been successful by repeated use of the cycle of "new principle - improved technology - saturation". There are many types of accelerators in the path: Cockcroft-Walton accelerators, electrostatic accelerators, cyclotrons, betatrons, synchrocyclotrons, linear accelerators, weak-focusing synchrotrons, and strong-focusing synchrotrons, while the colliders may produce higher "effective" energy than fixed target machines. Figure 2, the Livingston chart, explains how the energy grows along with the new technologies in the history of particle accelerators.

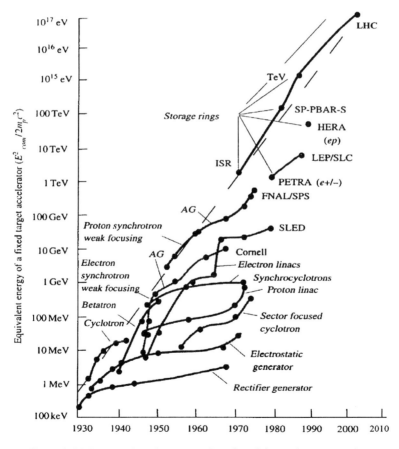

Figure 2. Livingston chart, beam energies of particle accelerators vs. time

The early history of accelerators can be traced to three roots based on their mechanism for acceleration: direct voltage, resonant, and betatron acceleration motivated by nuclear physics research in the 1920's.

The first root is based on the direct voltage acceleration, which was a consequence of the high voltage technology development. The main events on this root are listed below.

1919 E. Rutherford induced a nuclear reaction with natural α particles.

1920 H.Greinacher built the first cascade generator.

1928 G. Gamov predicted tunneling. As a result, particle energies of 500 keV might be sufficient to split a nucleus.

1928 Encouraged by Rutherford, Cockcroft & Walton started designing an 800 keV generator.

1931 Van de Graaff built the first high voltage electrostatic generator.

1932 The rectifier generator reached 700 kV and Cockcroft & Walton used the generator to accelerate protons to 400 keV to initiate the first artificial nuclear reaction Li + p $\rightarrow$ 2 He.

The beam energy of this type of accelerators was limited by the voltage that could be reached with high-voltage generators. The next question was whether particle beams could be accelerated by means of alternating voltage? The second root is based on the resonant acceleration technology, which was due to the early development of the radio frequency technology. In circular accelerators, the beam is accelerated in gaps to which the RF voltage is applied in resonance with the beams. The main events on this root are listed below.

1924 Ising proposed time-varying fields across a drift tube.

1928 Wideröe demonstrated Ising's principle with a 1 MHz-25 kV oscillator to accelerate potassium ions to 50 keV.

1929 Lawrence, inspired by Wideröe and Ising, conceived the cyclotron.

1931 Livingston demonstrated the cyclotron by accelerating protons to 80 keV.

1932 Lawrence's cyclotron produced 1.25 MeV protons and achieved nuclear fission just a few weeks after Cockcroft & Walton.

The third root, namely, betatron acceleration, is based on Faraday's Law which tells that a time-varying magnetic field can generate an electric field perpendicular to the direction of change of the magnetic field. Why can't we apply the magnetic field to bend the particle beams and the electric field to accelerate them? Wideröe, then a young Norwegian student, invented the "ray transformer" type accelerator, a precursor of the betatron, and worked at it. The main events on this root are listed below:

1923 Wideröe sketched the "ray transformer" and derived the 2-1 rule. Two years later he added the condition for radial stability.

1927 Wideröe made a model betatron, but it did not work. Discouraged he changed course and built a linear accelerator.

1940 Kerst re-invented the betatron and built the first working machine for 2.2 MeV electron beams.

1950 Kerst built the world's largest betatron with 300 MeV.

By the 1940's, the above-mentioned three acceleration principles were demonstrated and well developed. The research led to the discovery of "elementary particles", and the physics discoveries pushed up the accelerators to higher energy and higher performance. Here are some major events of the further development of particle accelerators.

1941 Touschek and Wideröe formulated the principle of a storage ring.

1944 Ivanenko, Pomeranchuk and Schwinger independently found an energy limit in circular electron accelerators due to synchrotron radiation.

1944 E. Mcmillan & V.Veksler independently discovered the principle of phase stability and invented the synchrotron.

1944 Veksler proposed an idea of a microtron.

1945 Energy loss due to synchrotron radiation was measured in a betatron.

1946 F. Goward & D. Barnes made a synchrotron work.

1946 First proton linear accelerator of 32 MeV was built at Berkeley.

1946 First electron linear accelerators were studied at Stanford and MIT.

1950 Christofilos formulated concept of strong focusing.

1952 BNL built 3 GeV Cosmotron.

1952 E. Courant et al. proposed the principle of strong focusing.

1959 CERN built 28 GeV CPS; BNL built 33 GeV AGS (1960).

1960's Several synchrotron radiation facilities were set up on rings initially for high energy physics.

1962 First single-ring e^+-e^- collider AdA of 2×250 MeV was built at Frascati.

1970 Radio-Frequency Quadrupole was suggested by I. Kapchinski.

1972 First double-ring proton collider ISR 2×28 GeV was built at CERN.

1989 First linear collider SLC of 2×50 GeV was built at SLAC.

Particle accelerators blossomed rapidly after world war II due to the rapid development of high power radio frequency and other technologies. The increased funding support from governments is another reason for the development. Along with the increase of the accelerator size, the international collaboration has become a significant characteristic of modern accelerator projects.

1.3. *Evolution of Acceleration Mechanisms*

The development of particle accelerators goes with the continuous invention of acceleration mechanisms. Figure 3 illustrates the evolution of acceleration mechanisms.

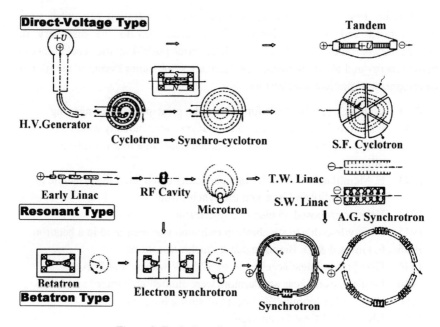

Figure 3. Evolution of acceleration mechanism

Starting from the three roots, new acceleration mechanisms were proposed and tested in order to overcome the difficulties of the existing accelerators to reach higher energy and better performance.

In the direct-voltage accelerators, the particle energy gain is proportional to the voltage applied to electrodes.

$$E = ZeV \ , \tag{2}$$

where E is the beam energy, V is the voltage applied on the acceleration electrodes, e and Z are the electron charge and the charge number of the accelerated particles, respectively. The tandem accelerator was to use the accelerating voltage twice: a negatively charged ion beam is accelerated to the high-voltage terminal where it passes through a thin foil which strips electrons from the negative ions and converts them to positive ions; they are then accelerated a second time back to earth potential

$$E = (1 + Z)eV \tag{3}$$

The beam energy in the direct-voltage accelerators is limited by sparking of the high voltage. This resulted in a development of the resonant type accelerators.

There were two types of resonant accelerators in the early stage, namely, linear accelerators and cyclotrons. In resonant accelerators, the co-existing varying magnetic and electric fields act on the particle beams. The relationship between the magnetic and the electric field is described by Faraday's Law

$$\nabla \times \vec{E} = -\frac{\partial}{\partial t} \vec{B} \ , \tag{4}$$

that correlates the electric field $\vec{E}$ with the variation rate of the magnetic field $\vec{B}$.

In linear accelerators, a series of conducting drift tubes is connected to the same RF generator. The RF frequency of the generator is adjusted in such a way that a particle traversing a gap sees an electric field in the direction of its motion and is inside the drift tube while the field is reverse. As the particle gains energy and speed, the length of the drift tubes must increase in order to maintain the resonant acceleration. This mechanism of linear acceleration was demonstrated in 1928. However, as the length of the drift tubes increases inconveniently with the velocity and as RF technology in the MHz frequency range was not available at that time, the early linear accelerators did not really find applications.

The mechanism of resonant acceleration was used in circular accelerators, where a magnetic field is applied to guide the particles moving in a circular orbit

$$\rho = \frac{p}{ZeB} \ , \tag{5}$$

where ρ is the curvature radius of the orbit and p the particle momentum.

In conventional cyclotrons, the relativistic effect acts against the increase of beam energy which led to the invention of the sector-focused cyclotron, the synchro-cyclotron and the electron cyclotron.

In the sector-focused cyclotrons, the RF frequency remains constant, while the distribution of the magnetic field is designed in order to keep synchronism

$$\overline{B}(r) = B_0 / \sqrt{1 - \omega_c^2 r^2 / c^2} \ , \quad f_c(t) = const., \quad f_{rf}(t) = const. \ , \tag{6}$$

where $\bar{B}(r)$ is the average magnetic field along the radius r, B_0 is the field at $r=0$, and $\omega_c = 2\pi \cdot f_c$ the circular frequency.

The synchrocyclotron is also called synchro-phase-cyclotron, in which the RF frequency is synchronized with the circulating frequency f_c

$$B(t) = const. \; , \quad f_{rf}(t) = h \cdot f_c \; , \tag{7}$$

where h is the RF harmonic number. The particle circulating with the frequency f_c is called synchronous particle while other particles in the beam oscillate around this particle.

For its small rest mass, the electron can easily be accelerated to a speed close to that of light. The electron cyclotron, or microtron, is designed in such a way that the difference of the successive revolution periods equals a multiple of the RF period, so that the synchronism can be maintained with a constant RF frequency

$$f_{rf}(t) = const. = h_n \cdot f_{c,n} \; , \quad B(t) = const. \quad . \tag{8}$$

The betatron is also an example of the application of Faraday's Law. Equation (4) tells us that the periodic variation of the guiding magnetic field induces an electric field on the beam axis. The betatron acceleration can be realized when the so called 2-to-1 rule is satisfied

$$\frac{dB_0(t)}{dt} = \frac{1}{2}\frac{d\bar{B}(t)}{dt} \tag{9}$$

and the transverse focusing is determined by the geometry of the guiding magnet.

In the cyclotrons and betatrons, the guiding field covers the full circular area and this makes the size of the magnets very big when the accelerated beam energy gets high. The synchrotron was invented to solve this problem. In synchrotrons, both, the magnetic field and the RF frequency, vary synchronously with the beam energy so that particles can move in circular vacuum chambers

$$B(t) = \frac{p(t)c}{Ze\rho} \quad . \tag{10}$$

In order to maintain the focusing on both horizontal and vertical planes, the focusing applied to beams is rather weak in the above-mentioned accelerators

$$0 < n = \frac{\rho}{B_0}\frac{\partial B}{\partial r} < 1 \tag{11}$$

which results in a large beam size and correspondingly large vacuum chambers and magnet apertures. The idea of the strong focusing synchrotron was proposed to reduce the transverse beam size. In a strong focusing synchrotron, high gradient focusing and defocusing (focusing on another plane) quadrupoles with magnetic index $|n| \gg 1$ are alternatively placed so that the global focusing on both horizontal and vertical planes is obtained

$$\frac{1}{F} = \frac{1}{f_1} + \frac{1}{f_2} - \frac{d}{f_1 f_2} \tag{12}$$

where f_1 and f_2 are focusing length of two successive quadrupole lenses, d is the distance between them. If the lenses have equal but opposite focal length, $f_1 = -f_2 = f$, and the overall focusing length $F = f^2/d$ which is positive, i.e., global focusing is achieved.

It is shown in Table 1 that the smaller the substance under study the higher is the center-of-mass energy $E_{c.m.}$ required. In fixed target experiments the $E_{c.m.}$ is written as

$$E_{c.m.} \approx \sqrt{2E_0 E} \tag{13}$$

The center –of- mass energy for a collision of two particles with energy $E_1 = E_2 = E$ is expressed as

$$E_{c.m.} = 2E \tag{14}$$

It is clear that $E_{cm, collider} \gg E_{cm,fixed\ target}$ when $E \gg E_0$, which is the case of high energy physics experiments.

The idea of colliders is to produce collisions between oppositely directed beams in order to generate high center-of-mass energy interactions between them. This is why almost all modern accelerators for high energy physics are collider type machines.

2. High Energy Frontier

There are four main directions in the high energy frontier: hadron colliders, electron-positron colliders, electron-proton colliders and μ^+-μ^- colliders.

2.1. Hadron Colliders

In 1954, E. Fermi proposed an accelerator with a center-of-mass energy of 3 TeV (1TeV=1000 GeV=10^{12} eV). At that time, there was no idea of "collider". Using Eq. (13) one obtained the required E =5000 TeV for E_{cm} of 3 TeV. The radius of the 5000 TeV synchrotron is about 8000 km for the magnetic field of 2 Tesla, i.e., a ring with the circumference of about the earth equator. The cost was estimated as 170 Billion USD and it would need 40 years to be built. Clearly, this was only a dream. Colliders will make Fermi's dream true. It only needs

$E = 1.5$ TeV for E_{cm} of 3 TeV. Fermilab has already reached E_{cm} of 2 TeV with an energy of 1 TeV of each beam with an updated luminosity of $2 \times 10^{31} cm^{-2} s^{-1}$ in its proton-antiproton collider Tevatron [1] (see Figure 4).

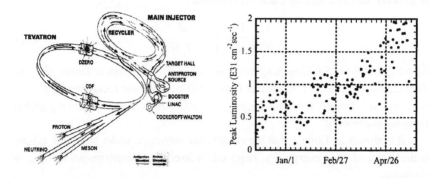

Figure 4. Layout of the Tevatron and its performance

The Relativistic Heavy Ion Collider (RHIC) [2] of BNL, with its circumference of 3834 m, can accelerate a gold beam up to 100 GeV/u, and its luminosity has reached its design value of 5×10^{26} cm^{-2}s^{-1}, shown in Figure 5.

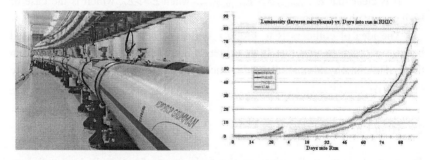

Figure 5. The Relativistic Heavy Ion Collider

The Large Hadron Collider (LHC) [3] under construction at CERN is planned to reach E_{cm} of 14 TeV with proton and antiproton beam energy of 7 TeV. This will allow scientists to penetrate still further into the structure of matter and recreate the conditions prevailing in the early universe, just after the "Big Bang". Figure 6 shows the two-in-one superconducting magnets and other devices in the LHC tunnel together with its main parameters. There are many theoretical and technological challenges in LHC in order to reach high energy with high luminosity, such as beam-beam interaction, instability control, magnificent magnets, large scale cryogenics, beam lifetime and particle loss issues, synchrotron radiation and others.

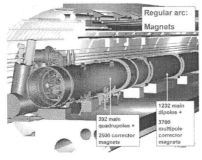

Beam energy	7 TeV
Injection energy	0.45 TeV
Magnetic field	8.33 T
Beam current	0.56 A
Circumference	27 km
Luminosity	1×10^{34} cm^{-2}s^{-1}
Completed time	2007

Figure 6. The Large Hadron Collider

Although the LHC is still under construction, the next generation of hadron collider VLHC (Very Large Hadron Collider) has been intensively studied. Table 2 lists its main parameters [4].

Table 2. Basic parameters of VLHC

	Phase 1	Phase 2
Central mass energy（TeV）	40	200
Ring circumference (km)	233	
Magnetic field（T）	2	11.2
Synchrotron radiation power（kW/beam）	70	1328
Total AC power（MW）	25	100
Luminosity 10^{34}cm^{-2}s^{-1}	1	2
Cost（10^9\$）	4	?

2.2. Hadron-Lepton Colliders

The HERA of DESY is a unique machine in the world for hadron (proton) and lepton (electron) collision [5]. Completed in 1992, HERA surpassed its design performance in 1997 and has accumulated a luminosity of 185 pb^{-1} by the year 2000. This has resulted in a number of fundamental physics observations. The kinematics range of lepton proton collisions in HERA made a large fraction of phase space accessible for measurements and provided a new understanding of strong interactions and the nature of the proton. In particular, it confirmed the point-like nature of the constituents of the proton on a scale of 10^{-18} m [6].

The 820 GeV proton and 30 GeV electron beams are brought into collision in their interaction points. Its recent upgrade of the new interaction regions by positioning the low-β quadrupoles much closer to the IP to increase the luminosity by a factor of 3.5 to 7 $\times 10^{31}$ cm^{-2}s^{-1} is in progress. Figure 7 shows the HERA tunnel and the time-integrated luminosity.

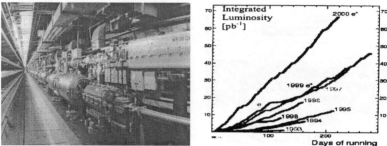

Figure 7. HERA tunnel and its time-integrated luminosity

As a future plan of a hadron-lepton collider, THERA [5], denotes TESLA-HERA-Electron-Proton-Collider, is proposed. TESLA, the 500 GeV c.m. electron-positron linear collider, presently under study, will provide electron beams up to an energy of 800 GeV. The TESLA tunnel has been aligned to allow collisions between electrons and protons stored in the HERA proton ring. The main parameters of THERA are given in Table 3.

Table 3. Main parameters of THERA

	Proton beam	Electron Beam
Energy (GeV)	1000	250
Particle/bunch	10^{11}	2×10^{10}
$\beta^*_{x,y}$ (cm)	10	50
Crossing angle (mrad)	0.05	
Luminosity (10^{30} cm^{-2}s^{-1})	4.1	

2.3. Electron-Positron Colliders

The maximum beam energy in electron-positron colliders has reached 100 GeV in LEP, the Large Electron Positron collider in CERN. The synchrotron radiation prohibits the circular colliders of this type to reach a higher energy. Now, the LEP has been dismounted in order to build LHC in the same tunnel.

Linear colliders can get rid of synchrotron radiation. It is agreed by high energy physics community that an electron-positron linear collider will be the next frontier accelerator after LHC. A starting energy of 0.5 TeV c.m. with capability of extension to about 1 TeV or more is indicated with a luminosity of the order of 10^{34} cm^{-2}s^{-1}. In terms of the first linear collider, the SLC at SLAC, the energy will be 5 to 10 (or more) times greater, the luminosity 10 times higher and the beam size at the collision point 100 times smaller.

Several approaches to meeting these requirements have been identified over the past years. Currently there are three distinct technologies being pursued actively. Japan and America are jointly pursuing a machine based on X-band linacs. Japan is also pursuing a C-band scheme. The TESLA collaboration, centered at DESY, has developed an L-band superconducting technology. For the next generation a collaboration centered at CERN is pursuing a Ka-band (30 GHz) approach with novel power sources for $E_{c.m.}$ of 3 TeV. Figure 8 gives the schematic layout of JLC/NLC and Figure 9 is the tunnel layout of TESLA.

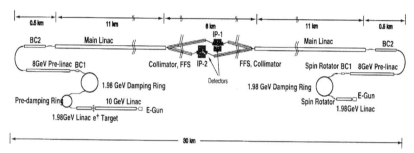

Figure 8. Layout of JLC/NLC

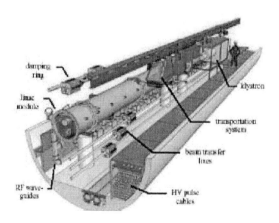

Figure 9. Tunnel layout of TESLA

Table 4 gives main parameters for the various linear colliders. KEK/SLAC collaboration adopts normal conducting, higher frequency approach to reach the high gradients as 50 MV/m. The TESLA collaboration has applied an L-band superconducting technology to reach high performance. Aimed at the next generation of linear collider when energies well above 1 TeV will be needed, the CLIC machine utilizes 30 GHz generated in a novel two-beam structure.

Table 4. Main parameters of linear colliders [7]

Collider	TESLA		JLC/NLC		CLIC	
C.M energy (TeV)	0.5	0.8	0.5	1.0	0.5	3.0
RF frequency (GHz)	1.3		11.4		30	
Luminosity (10^{33}cm^{-2}s^{-1})	34	58	25	25	14.2	10.3
Repetition rate (Hz)	5	4	150	100	200	100
Unloaded/loaded gradient (MV/m)	23.4	35	70/54		172/150	
Beam power/beam (MW)	11.3	17	8.6	11.5	4.9	14.8
Total two-linac length (km)	30	30	12.6	25.8	5.0	27.5
Proposed site length (km)	33		32		40	
Tunnel configuration	Single		Separate		Two-Beam	

The linear colliders are confronted with many challenges for the required high energy (~0.5 TeV) and high luminosity (~10^{34}cm^{-2} s^{-1}) at acceptable cost, such as multi-bunch effects and wakefield, high gradient and high repetition rate, damping rings, positron source, alignment and ground-vibration suppressors, feedback, final focusing, background and collimation. A number of test facilities and R&D work for linear colliders are pursued towards the real machines.

As an add-on to an electron-positron linear collider, a γ–γ collider is proposed [8]. In a γ-γ collider, the γ-rays are produced by Compton back scattering of laser photons from intense, high energy electron and positron beams. Figure 10 sketches the schematic layout of a γ–γ collider.

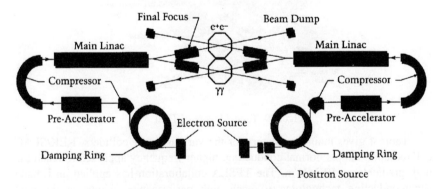

Figure 10. Schematic layout of a γ–γ collider

2.4. μ^+- μ^+ Colliders

It seems a crazy idea when one hears about the μ^+- μ^+ collider for the first time, as the muon is so unstable that its rest lifetime $\tau_{\mu 0}$ is only 2μs. However, it is an intriguing solution for a next generation machine at the high energy frontier [9].

Due to the rest mass of m_μ =105 MeV that is about 200 times larger than the rest mass of an electron, the synchrotron radiation is negligible so that the muon colliders can be circular. While there is beamstrahlung when muon bunches cross each other, the use of larger muon bunches is allowed and the energy spread of interactions is reduced. The s-channel Higgs production is enhanced by a factor of $(m_\mu / m_e)^2 = 40000$. This combined with the lower energy spreads allows more precise determination of Higgs mass, width and branch ratio that gives chances for new physics discovery.

The relativity shows that lifetime of a particle increases with its relativistic energy, i. e. $\tau_\mu = \gamma\tau_{\mu 0}$. At the muon energy of 2 TeV, γ =20000, $\tau_\mu = 40$ms, and the number of interactions can be N = $\tau_\mu / T_{rev} \approx 1500$, for $T_{rev} \sim 25$ μs. Figure 11 gives the layout of a 4 TeV μ^+- μ^+ collider.

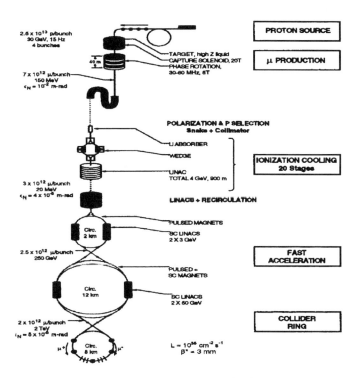

Figure 11. Overview of a 4 TeV μ^+- μ^+ collider

The muons are produced from the decay of pions, made by high intensity proton beams bombarding a liquid metal target. The large emittance of the produced muons must be quickly reduced by ion cooling before they decay. For the short lifetime of muons, recirculating linacs or pulsed synchrotrons are applied for the quick acceleration. As muons decay when stored in a circular collider, the shielding and background are significant issues and the radiation hazard is an issue in the muon colliders.

3. High Luminosity Frontier

After discoveries made with pioneer accelerators, more detailed and accurate investigations need to be made in order to extend the discovery with higher statistics. This requires higher luminosity machines, called "particle factories".

Higher luminosity calls for higher performance colliders with higher current, better final focusing, and usually double-ring structure. This brings challenges to accelerator physics and technology, such as beam instabilities and impedance, injection, vacuum, accurate bunch monitoring, interaction region, background and others.

There are B-factories, τ-charm factory, Φ-Factory in the c.m. energy regions of B-mesons from the decay of Υ (4S) resonance at 10.6GeV, τ-lepton and charm-particles (3-5 GeV) and K-mesons the Φ resonance at 1.02 GeV. Neutrino-factory is a relatively new idea derived from muon colliders.

3.1. B-Factory

Since 1999 two asymmetric double-ring B-factories, PEP-II and KEKB, have been operating at Υ(4S). Table 5 gives their main parameters [10].

Table 5. Main parameters of PEP-II and KEKB

Parameters	PEP-II		KEKB	
	LER	HER	LER	HER
Beam energy (GeV)	3.1	9.0	3.5	8.0
Circumference (m)	2200		3016	
Beam current (A)	1.78	1.06	1.37	0.92
Bunch number	800		1224	
Cross. Angle (mrad)	0		2×11	
β_x^*/β_x^* (cm)	35/0.9	50/1.25	59/0.62	61/0.7
Design/Reached Lum. (10^{33}cm^{-2}s^{-1})	4.60		7.35	
$\int L \cdot dt$ (nb^{-1}/30 days)	6666		8783	

Figure 12 displays the development of the peak and the integrated luminosities of PEP-II and KEKB since 1999. The figures show the success of both B-factories.

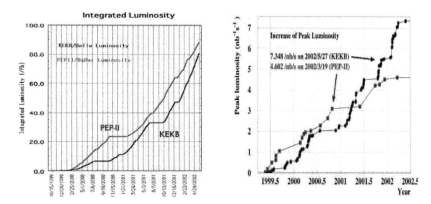

Figure 12. The increase of the integrated (left) and peak (right) luminosities of PEP – II and KEKB since 1999

The success of PEP-II and KEKB has encouraged the proposals of super-B schemes with a luminosity of 10^{35}-10^{36} cm^{-2}s^{-1} in order to compete with LHC-B experiment in the B-physics. Table 6 lists their parameters.

Table 6. Some parameters of Super-B

Parameters	Super - PEP-II		Super - KEKB	
	LER	HER	LER	HER
Beam energy (GeV)	3.5	8.0	3.5	8.0
Circumference (m)	7700		3016	
Beam current (A)	10.3	23.5	9.4	4.1
Bunch number	800		5018	
β_x^*/β_x^* (cm)	15/0.15	15/0.15	15/0.3	15/0.3
Design Luminosity	10^{36}cm^{-2}s^{-1}		10^{35}cm^{-2}s^{-1}	

The way towards 1~2 orders of magnitude higher luminosity is rather conventional: more bunches, higher beam current and smaller β-values. These mean more cost and bigger challenge in super-B factories.

3.2. DAΦNE and Φ-Factory

DAΦNE in Frascati is a high luminosity e⁺e⁻ collider designed as a Φ-factory for the production of a high rate of K-mesons from the decay of the Φ-resonance at c.m. energy of 1.02 GeV. It consists of two rings in the horizontal plane, crossing at an angle of 2×12.5 mrad in two interaction regions (IRs). Up to 120 bunches can be stored in each ring. The first IR hosts a magnetic detector (KLOE) mainly aimed at the study of CP violation. In the second IR the DEAR experiment studies the properties of kaonic atoms, namely atoms where kaons are captured in the inner shells in place of electrons [11]. Figure 13 shows the layout of DAΦNE and Figure 14 gives its luminosity and main parameters.

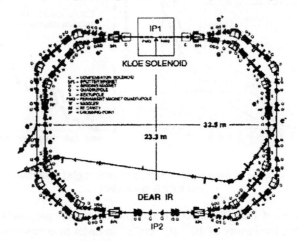

Figure 13. The Layout of DAΦNE

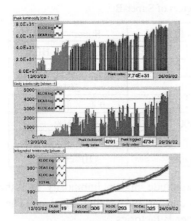

Beam Energy	510 MeV
Bunch number	47
Current	1 A
$\beta_x^* \lhd \beta_y^*$	4 m /4 cm
Emittance	1 mm·mrad
Lum Lifetime	30 min
Luminosity	$5.1 \times 10^{31} \text{cm}^{-2}\text{s}^{-1}$
$\int L \cdot dt$	2.5 pb^{-1}/day

Figure 14 The luminosity performance and parameters of DAΦNE

The short term planning at DAΦNE is to further improve luminosity performance and to increase signal to-background ratio in order to observe and measure the properties of kaonic hydrogen in DEAR. As the long term plan, the luminosity goal of $5*10^{33} - 10^{34}$ of the DAΦNE upgrade is being studied.

3.3. *BEPC and BEPCII*

The Beijing Electron-Positron Collider (BEPC) was constructed for both high energy physics (HEP) and synchrotron radiation (SR) research. The BEPC consists of a 202 m long electron-positron linac injector, a storage ring with a circumference of 240.4 m, and transport lines with the total length of 210 m. There are two interaction points in the storage ring. A general-purpose detector, the Beijing Spectrometer (BES), is installed in the south interaction region. The Beijing Synchrotron Radiation Facility (BSRF), equipped with 14 beamlines and experimental stations, is flanking the east and west of the southern area of the storage ring. Figure 15 illustrates the layout of the BEPC.

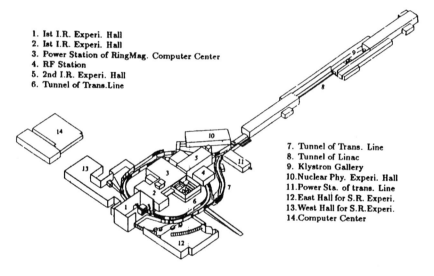

1. Ist I.R. Experi. Hall
2. Ist I.R. Experi. Hall
3. Power Station of RingMag. Computer Center
4. RF Station
5. 2nd I.R. Experi. Hall
6. Tunnel of Trans.Line

7. Tunnel of Trans. Line
8. Tunnel of Linac
9. Klystron Gallery
10. Nuclear Phy. Experi. Hall
11. Power Sta. of trans. Line
12. East Hall for S.R. Experi.
13. West Hall for S.R.Experi.
14. Computer Center

Figure 15. Layout of the BEPC

As a unique e^+-e^- collider operating in the τ-charm region and the first SR source in China, the machine has been well operated for over 13 years since it was put into operation in 1989. The success of the BEPC and the exciting physics in the τ-charm region call for a higher luminosity. BEPCII is a double-ring collider based on the BEPC. Table 7 gives the main parameters of the BEPCII in comparison with the present BEPC [12].

Table 7. The main parameters of the BEPCII in comparison with BEPC

Parameters		Unit	BEPCII	BEPC
Operation energy (E)		GeV	1.0–2.0	1.0–2.5
Circumference (C)		m	237.5	240.4
β^*-function at IP (β_x^*/β_y^*)		cm	100/1.5	120/5
Bunch number (N_b)			93	1
Bunch current (I_b)		mA	9.8 @1.89 GeV	35 @1.89 GeV
Beam current	Colliding	mA	910 @1.89 GeV	2×35 @1.89 GeV
	SR		250@25GeV	130
Luminosity@1.89 GeV		10^{31}cm^{-2}s^{-1}	100	1

3.4. Neutrino-Factory

Recent results from Super-K indicate the existence of neutrino oscillations and, therefore, motivate the building of a 20-50 GeV muon storage ring that can produce a directed beam of intense neutrinos (10^{20} -10^{21} per year) for both domestic and intercontinental experiments. A significant effort has been made in the past years by the worldwide accelerator laboratories (BNL, CERN, FNAL, KEK, LANL, RAL and others) to provide high-energy neutrino beams for particle physics experiments [13]. A schematic layout of a neutrino factory is shown in Figure 13.

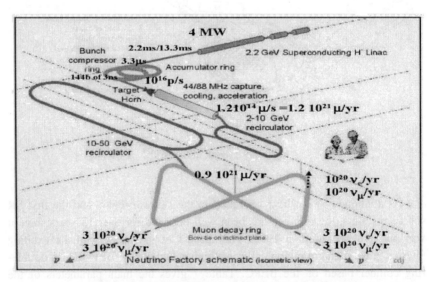

Figure 16. A schematic layout of a neutrino factory.

Comparing Figure 16 to Figure 11, it is noticed that the neutrino factory is similar to a muon collider, at least in the parts for muon generation and acceleration. A neutrino factory consists of a high intensity proton driver, a target for pion production, pion and muon capture region, phase rotators, cooling devices, a linac and linac-recirculators for rapid muon accelerating, and a muon storage ring. The storage ring is equipped with long straight sections in order to produce intense and well collimated neutrino beam in different directions.

The accelerator-based neutrino beams are an interesting challenge for accelerator physics and technology. The physics issues are high proton beam power, short proton bunches, transverse emittance cooling and design of beam channels and accelerators with large admittance. The technology issues are the components of the proton drivers, targets standing the thermal shocks, magnets and RF in high radiation environment, absorbers and cavities for the transverse cooling, cavities with high gradients for fast acceleration and a number of operational issues such as reliability and maintenance.

4.   Multidisciplinary Platforms and Application

4.1. *Synchrotron Radiation Sources*

As multidisciplinary research platforms, synchrotron radiation facilities are widely applied in the fields of basic and applied research, such as physics, chemistry, materials science, molecular biology, geology, medical diagnostics and therapy, lithography, micro-fabrication and other fields.

Relativistic charged particles traveling along a curved orbit radiate electromagnetic waves that are termed "synchrotron radiation (SR)". Synchrotron radiation is a harmful effect in terms of limiting the maximum achievable energy in the electron-positron colliders. For a growing SR user community, it has become a major type of light sources for the atomic-molecular-scale experiments. In the early 1960's, several synchrotron radiation sources were set up on the storage rings built initially for high energy physics research, in the US and Europe. These are first generation SR sources. The second generation or SR sources were dedicated facilities built in the 1980s in China, France, Germany, Japan, the UK, and the US and more countries like Brazil, India, Italy, Korea, Russia, Switzerland, Thailand and Singapore in the 1990s. Several of them were more advanced machines with intense electron beams and many straight sections for insertion devices to make even brighter radiation. This is the third generation of SR sources. Nowadays, about 50 SR light sources have been in operation serving thousands of users. They operate in three energy ranges, lower-energy (below 2 GeV, for VUV and soft X-ray), intermediate energy (2–4 GeV, for X-ray and hard X-ray) and high energy (6–8 GeV, for hard X-ray and ultimate hard X-ray). Table 8 lists the SR sources in China.

Table 8. Synchrotron radiation sources in China

	BEPC	NSRL	SRRC	SSRL
Location	Beijing	Hefei	Hsinchu	Shanghai
Completed Time	1988	1989	1993	R&D
Generation	1-st	2-nd	3-rd	3-rd
Beam Energy (GeV)	2.2	0.8	1.5	3.5
Circumference	240.4	66.1	120	396
Emittance (nm·rad)	80	27	25	4.6
Beam Current (mA)	140	200	200	300
Beam Lifetime (hrs.)	>20	>10	>9	>20

The SR sources mentioned above are all storage ring type where quantum emission of radiation from the bending magnets affects the electron beam to produce energy spread and growth in emittance. The emittance increases quadratically with electron energy. These fundamental limits on electron beam brightness and pulse length in storage rings can be overcome in linac-based light sources. There are two types of linac-based SR sources, the Free Electron Laser (FEL) as described in the next sub-section, and the energy recovery linac (ERL) [14].

ERL is a novel idea in which the energy of the used beam is recovered so that operating costs can be brought to a reasonable level. The principle has been tested at low energy in the Jefferson Laboratory. There are many challenges and opportunities provided by a higher energy ERL, delivering X rays with higher brightness and shorter pulses than any storage rings. Figure 17 gives the layout of an ERL light source. 10 MeV electrons from the injector are deflected by a weak bending magnet into a superconducting linac that is a few hundred meters long. They are brought to full energy in a single pass. The electrons are then guided around a one-turn orbit imposed by bending magnets and travel through undulators to produce the X-rays. The length of the electron orbit is set such that electrons return with a phase opposite to the linac. Thus, their energy is recovered before they are steered into a dump at an energy of about 10 MeV by another weak field magnet.

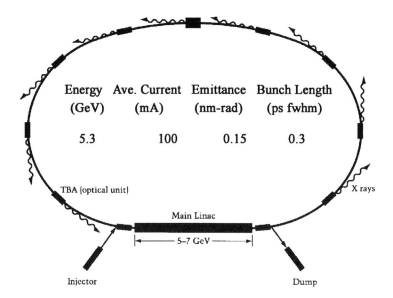

Figure 17 The layout of an ERL by Cornell

4.2. *Free Electron Lasers*

Free Electron Lasers are powerful sources of coherent electromagnetic radiation with high brightness and large peak power. They differ from conventional lasers in using a relativistic electron beam as the lasing medium, as opposed to bound atomic or molecular states, hence the term free-electron. FELs are a powerful and challenging combination of particle accelerator and laser physics and technology. An FEL consists of a high brightness electron accelerator, an undulator magnet, and optical beam lines. Storage rings, linacs, and electrostatic accelerators have been used to provide the high quality electron beams. An FEL can be operated as an amplifier of an external laser signal or of its own undulator spontaneous radiation (SASE, Self Amplified Spontaneous Emission), or as an oscillator by means of an optical cavity.

A number of FELs have been operated in the wavelength from microwave to ultraviolet with pulse length from micro-second to pico-second. There are two directions of development, i.e. higher average power in the infrared to ultraviolet region and very short wavelength (hard X-ray). The SASE X-ray FELs can satisfy the requirements of future experiments which need X-ray pulses with a large number of photons focused on a sample as small as a molecule, squeezed in a time a thousand times shorter (femtosecond range), to study the dynamics of atomic and molecular processes. Figure 18 exhibits the layout of the Linac Coherent Light Source (LCLS) at SLAC. The LCLS is a multi-institutional proposal for a single-pass X-ray FEL operating in the 1-15 Å wavelength region, using electron beams with peak current of 3.4 kA and emittance of 0.05 nm·rad from the SLAC linac at energies up to 15 GeV [15].

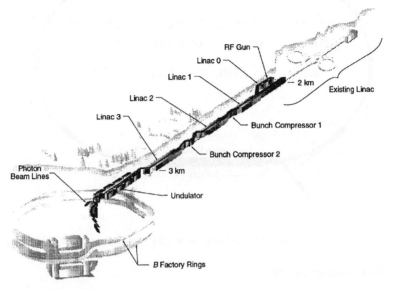

Figure 18. The layout of the Linac Coherent Light Source at SLAC

4.3. *Spallation Sources*

Slow neutrons have such unique properties as a wavelength of the order of inter-atomic distances, energies comparable to collective excitations in solids, non-zero magnetic moment in spite of zero charge. They are an ideal tool to study both, the structure and the dynamics of condensed matter. Neutron scattering has a large range of applications in the materials science, molecular biology, medicine, and many other fields. Neutron beams complement synchrotron radiation photon beams as probes for the study of condensed matter and molecular structure. Spallation Neutron Sources (SNS) where neutrons are generated by spallation are powerful platforms for multidisciplinary research, like SR sources and FELs. In contrast to reactor-based neutron sources which are operated in cw mode, spallation neutron sources can be pulsed with high peak neutron flux.

In spallation neutron sources, the neutrons are produced by protons impinging on a target. The rate of neutrons produced is proportional to the energy and intensity of the primary protons. A spallation neutron source usually consists of a high power proton linac, an accumulator ring or a rapid cycling synchrotron, and a target system. Table 9 lists the existing and planned spallation neutron sources in the world [16].

Table 9. Existing and planned spallation neutron sources

Name	Status	Accelerator energy	Average beam power (MW)	Rep. Rate (Hz)
IPNS, ANL	Operation 1981	50 MeV linac 500 MeV SCR	0.0075	30
ISIS, RAL	Operation 1985	70 MeV linac 800 MeV SCR	0.16	50
SINQ, PSI	Operation 1996	590 MeV cyclotron	≤ 0.9	C.W.
LANSCE LANL	Operation 1977	800 MeV linac	0.08	20
LANSCE II LANL	Planned	800 MeV linac	1	30
JPARC Japan	Construction	400 MeV linac 3 GeV RCS	1	25
ESS Europe	Planned	1.33 GeV linac 2 accumulators	5	50
NSNS US	Construction	1 GeV linac accumulator	1 (5)	60
CSNS China	Planned	70 MeV linac 1.6 GeV SCR	0.1 (0.4)	25

4.4. *Application of Accelerators*

Although accelerators were developed for nuclear and particle physics many thousands of them are put to practical uses in other branches of scientific research as well as in industry, agriculture, medicine, and many other fields. Besides the application in nuclear and particle physics, accelerators are widely used for analysis of materials, particle-induced X-ray emission, nuclear reaction analysis, elastic recoil detection, charged particle activation analysis, accelerator mass spectroscopy, and extended X-ray absorption fine structure study and other researches. The majority of accelerators in application are low energy linear accelerators used in factories to polymerize plastics, treat waste and sterilize food, electron beam irradiation, and in hospitals for therapy. In the field of medicine we also find cyclotrons at work producing isotopes to supply the hospitals of the world with modified biological chemicals whose location can be traced in the body by the particles they emit. Some of them, by their biochemical nature, can even select certain sites in the body for diagnosis or treatment, e.g., Positron Emission Tomography (PET).

Accelerators are widely used in the industrial inspection. High energy intensive X-rays produced by MeV range electron beams serve as a probe to detect samples, from small to very large. Recently, container inspection machines are developed by Tsinghua University as pictured in Figure 19.

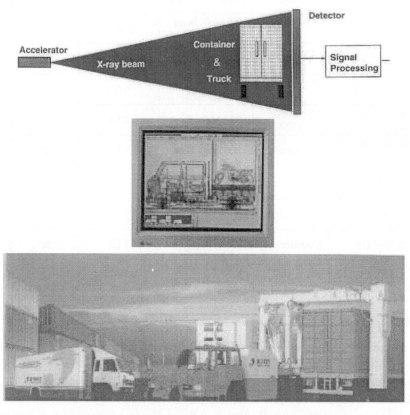

Figure 19. Tsinghua produced container inspection device

Proton and ion accelerators of a few hundred MeV have been developed in order to destroy deep tumours. Protons and ions deposit most of their energy near the end of their path causing minimum damage to surface tissue and sparing delicate organs around the target zone with millimeter precision.

High power proton linacs have been proposed for many applications. Accelerator Driven Sub-critical (ADS) reactor is one of the most promising potential uses of particle accelerators to produce clean, safe and almost inexhaustible energy. By using a particle accelerator to produce the neutrons which provoke nuclear fission, the commonly occurring and easily extracted element thorium can be used as reactor fuel instead of uranium. The reaction is not self sustaining, so there is no chance of Chernobyl-like runaway. Unlike a conventional fission reactor, ADS machine needs energy to keep it going. The amount of energy produced would be several times as much as that put in, leading to the name of "Energy Amplifier" for this device (See Figure 20).

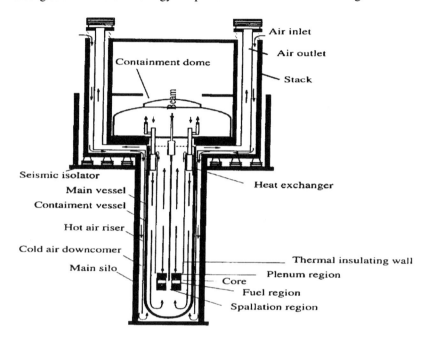

The concept of the energy amplifier can be extended to the transmutation of long-lived nuclear waste. The idea is to mix nuclear plutonium waste with the thorium fuel so that it also undergoes fission and breaks up into harmless elements. High power proton linacs can also be applied for tritium production.

Heavy ion beams are widely used to implant atoms in the surfaces of semiconductors to "print" the circuit of modern computer chips. Another industrial use is to harden metal surfaces for bearings and to etch shapes in silicon with micron precision. Heavy ion beams from rapid cycling accelerators can be used to compress and ignite deuterium-tritium pellets to drive inertial fusion. This is considered as one of the promising ways towards nuclear fusion power production. The heavy-ion fusion driver is illustrated in Figure 21.

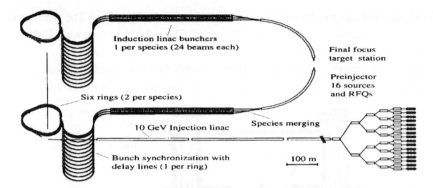

Figure 21. The heavy-ion fusion driver.

5. Novel Acceleration Methods

In Section 1 we traced the history of particle accelerators to the three roots, namely direct-voltage acceleration, resonant acceleration, and betatron acceleration. The principles of the present-day accelerators follow these conventional ways. As we have seen in the previous sections, the beam energy in circular accelerators is limited by synchrotron radiation and by keeping the circumference of the rings realistic. In linear accelerators, on the other hand, the maximum accelerating gradient is limited to lower than 100 MeV/m which makes the TeV-scale machines more than 10 km long. The practical limitation that high-energy accelerators meet today is the cost per unit beam energy. The goal of research is to develop an advanced technology that provides many more GeV per cost unit. If there is no new and more effective method for acceleration, mankind's march towards the higher energy frontier may have to be slowed down.

Intense particle beams, lasers, and plasma waves can offer much more powerful electromagnetic fields. Most of the new ideas proposed so far are based on these technologies and try to find a practical way to apply these fields.

5.1. Direct Acceleration with Laser

The maximum electric field in a laser can be as high as $10^5 \sim 10^9$ MV/m while the key issue is how to obtain the E_z component along the beam direction.

The E vector of plane waves in free space is perpendicular to the direction of laser propagation. On the other hand, the wave travels with a phase velocity of light which is always faster than the motion of the particles. The basic issues for the direct acceleration with lasers are how to turn the E vector in the right direction and to keep the particles in phase with the laser waves.

A number of schemes, such as surface wave accelerator, inverse Cherenkov effect accelerator, inverse FEL accelerator and others were proposed to solve these problems.

In contrast to a conventional linac, the surface field in the grating accelerator is set up by a laser wave, instead of a microwave, at grazing incidence on the face of a metal grating that serves as a medium to generate the parallel component of the electric field near the surface and to slow down the phase velocity of the waves.

A SLAC group proposed an idea of an "accelerator on a chip"[17]. The goal is to lithographically produce the power source, power transmission system, accelerator structures and beam diagnostics on a single substrate by semiconductor process. The use of a laser with micron wavelength as power source implies accelerating structures with millimeter dimensions. The high available peak power and high available wall-plug-to-photon efficiencies (20-40%) of modern solid-state lasers are encouraging. The arrangement of these microstructures in accelerator arrays could optimize efficiency and available beam current. An experiment in SLAC will further address this concept of laser acceleration in vacuum, aiming at an accelerating gradient of 0.3 GV/m [18].

The Cherenkov effect denotes that a particle traveling through a medium of refractive index n will emit a cone of light if its velocity is greater than c/n, c being speed of light. Conversely, a cone of laser light directed on a particle by an axial prism can accelerate the particle. The principle of inverse Cerenkov accelerator was demonstrated over a few keV.

In an Inverse Free Electron Laser (IFEL) accelerator, the laser gives energy to particles by means of an undulator which serves as a medium to turn the E vector and make the synchronization between particles and laser waves. The principle of IFEL has been used at the STELLA staging experiment to produce 1 µm long micro-bunches and to accelerate these bunches by 6 MeV/m. This important experiment demonstrated femto-second accuracy in the laser-beam synchronization. The acceleration goal for future experiments is 90 MeV/m. However, due to synchrotron radiation, the use of IFEL's is limited to beam energy below 200 GeV and would therefore be restricted to the low energy part of the accelerator [18].

5.2. *Two-Beam Accelerators*

In a general sense, most RF extracted concepts, such as klystrons, gyrotrons, wakefield and FEL's, can be configured into the two-beam acceleration.

There was an idea of the electron-ring accelerator or "smokatron" proposed in the late 1960s when much attention was attracted. Figure 22 explains its mechanism.

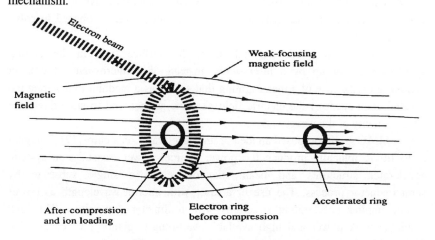

Figure 22. The electron-ring accelerator

In the smokatron, an electron ring is formed by injecting an intense current at a few MeV into a specially shaped axial field. Protons resulting from the ionization of background gas are collected in the potential well within the ring. The ring is first compressed by increasing the field and then accelerated along its axis in a tapered solenoid field. Unfortunately, this scheme was tested with a disappointing result. It was concluded that in order to accelerate a proton beam to high energy electron-ring currents beyond the limit of coupled instability between the two beams would be required.

The more recent Two-Beam Accelerators (TBAs) have been proposed as a scheme for high-gradient electron-positron colliders since the 1980s. In the TBAs, a high current, low energy drive beam is used to generate RF power and a low current beam is accelerated to high energy. The TBAs have two advantages, namely, a high efficiency for power conversion from the driving beam to RF power and a high accelerating gradient ($\geq$100MeV/m) with high frequency ($\geq$11.4 GHz). There are two main lines of TBA research: to accelerate the driving beam with an RF structure and to use induction modules. The first approach is being studied at CERN by the CLIC (Compact Linear Collider Project), and the second at LBNL and LLNL [19]. Figure 23 gives the layout of CLIC. The accelerating gradient of 150 MV/m is demonstrated with a CLIC test facility with short RF pulses, showing the TBA scheme as the most efficient and cost effective method for high energy linear colliders [20].

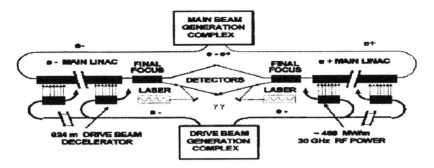

Figure 23. The Layout of CLIC

5.3. *Wakefield Accelerators*

The Wakefield Accelerator (WFA) is an old idea. Electromagnetic waves are driven in a medium by the wake of a short and dense beam that precedes the particles to be accelerated. The drive beam can be either a particle beam or a laser beam. There are two types of medium for the waves to be generated, plasma and slow-wave structure.

Plasma based accelerators replace the metallic walls of conventional RF structures with a plasma. The damage problems faced in high-gradient metallic structures are therefore no longer an issue. Laser beams (Laser Wakefield Accelerator LWFA) or charged particle beams (Plasma Wakefield Accelerator PWFA) are used to excite space-charge oscillations in the plasma. The resulting longitudinal fields can be used for particle acceleration. Accelerating gradients of up to 160 GV/m have been demonstrated in experiments and plasma structures have been built from the millimeter to the meter scale [18].

In the slow-wave structure type WFA, the wakefield is excited in metallic or dielectric structures by drive beams. This field has the phase velocity of the drive beam pulse and can be used to accelerate the less intense particle beams to high energy. Accelerating fields in this type of WFA are smaller than in the plasma-based WFA while the stability problems of plasmas are avoided. Several slow-wave structures have been proposed, including iris loaded wave guides, dielectric wave guides, concentric dielectric tubes, and as a variation, Coupled Wake Tube Accelerator (CWTA).

Table 10 summarizes characteristics of various approaches to WFAs and the status of their development [21].

Table 10. Various approaches of wakefield accelerators

WF Device	Max. Gradient	Advantages	Disadvantages	Experimental Status
LWFA	Multi-GV/m	High-gradient	Requires powerful short pulse laser. Poor efficiency.	Preliminary experiments are underway. To date electrons captured from background have been accelerated
PWFA	≈ GV/m	High-gradient	Requires difficult drive beam, alignments. Same focus problem as with LWFA. Alignments are critical.	Acceleration of injected beam demonstrated at 5-7MeV/m using few nC drive beam
Iris loaded WFA	50MV/m	Simple and well understood	Low gradient, requires good beam-beam alignment.	Similar to PWFA
DWFA	100MV/m	Very simple; deflection modes can be damped	Requires difficult drive beam. Extending length requires complicated "staging".	Similar to PWFA
CWTA	500MV/m	Stepped up gradients. Transverse beam-beam effects small. Simple extension of acceleration possible.	Requires efficient coupling of RF power. Drive beam less difficult than that of DWFA, but still not easy.	Experiment under construction

More recent experiments have demonstrated in LWFA acceleration with gradients of up to 160 GV/m, although only over millimeter lengths. It is hoped that an acceleration of 20 GeV beam can be achieved in about 1 meter length. PWFA experiments are presently constrained to lower plasma densities and have tested acceleration in the 150 MV/m range over 1.4 m. A future experiment will use shorter electron bunches and higher plasma densities to test accelerating gradients of up to 20 GV/m over 30 cm. A 2×15 m long plasma afterburner for the Stanford Linear Collider (SLC) could offer the first possibility to use plasma acceleration for particle physics. For large plasma-based linacs, both laser- and beam-driven, important problems of emittance preservation are expected and remain to be studied [18].

The novel acceleration concepts still sound exotic, like the traditional methods did in the 1930s. There is still a long way to go before the novel concepts come to real accelerators with ultra high accelerating gradient. The concepts need to be consummated, while more experiments need to be carried out. We know, from the history of accelerators from Lawrence, van de Graaff and Wideröe to the present day, these are promising.

Particle accelerators have a bright future. The R&D and construction of particle accelerators will be pursued, the rapid advances of the technology are expected, the applications of accelerators will be promoted, the dream for ultra high accelerating gradient with exotic concepts will come true, and mankind's march towards higher energy and high performance frontiers will continue.

References

1. M. Church, Tevatron Run II Performance and Plans, Proc.EPAC02 (2002).
2. F. Pilat, RHIC Status and Plans, Proc.EPAC02 (2002).
3. R. Schmidt, Status of the LHC, Proc.EPAC02 (2002).
4. P.Limon, Very Large Hadron Collider, ICFA seminar on Future Perspectives in High Energy Physics (2002).
5. F. Willeke, HERA Status and Perspectives of Future Lepton-Hadron Colliders, Proc.EPAC02 (2002).
6. U. Schneekloth, (ed) The Luminosity Upgrade, DESY-HERA Internal Report (1998).
7. M. Tigner, Review of Linear Colliders design and Path to the Future, Proc.EPAC02 (2002).
8. K-J. Kim and A. Sessler, Gamma-Gamma Colliders, Beam Line Vol.26, No.1 (1996).
9. R.Palmer, μ^+-μ^- Colliders, Handbook of Accelerator Physics and Engineering, World Scientific, p33-34 (1998).
10. K. Oide, Operation Experience and Performance Limitations in e+e- Factories, Proc.EPAC02 (2002).
11. S. Bertolucci, LNF Status and Outlook, ICFA seminar on Future Perspectives in High Energy Physics (2002).
12. BEPCII Design Team, BEPCII Design Report, IHEP-BEPCII-SB-01 & IHEP-AC-Report/2002-01 (2002).
13. K. Hübner, Accelerator Based Neutrino Beams, Proc.EPAC02 (2002).
14. K-J. Kim and A. Sessler, Gamma-Gamma Colliders, Beam Line Vol.26, No.1 (1996).
15. The LCSL Design Study Group, Linac Coherent Light Source (LCLS) Design Study Report, SLAC-R-521 (1998).
16. H.Lengeler, Spallation Sources, Handbook of Accelerator Physics and Engineering, World Scientific, p40-42, (1998).
17. C.D. Barnes et al., SLAC Proposal E-163 (2001).
18. R. W. Aβmann, Review of Ultra High Gradient Acceleration Schemes, Proc.EPAC02 (2002).
19. A.Sessler, Two-Beam Accelerators, Handbook of Accelerator Physics and Engineering, World Scientific, p44-46, (1998).
20. J.P.Delahaye, The Compact Linear Collider Study, ICFA seminar on Future Perspectives in High Energy Physics, (2002).

21. J.Simpson, Wakefield Accelerators, Handbook of Accelerator Physics and Engineering, World Scientific, p46-48, (1998).

A GUIDED SURVEY OF SYNCHROTRON RADIATION SOURCES

H.O. MOSER

Singapore Synchrotron Light Source
National University of Singapore
5 Research Link
Singapore 117603
E-mail: moser@nus.edu.sg

Synchrotron radiation sources arguably enable today's most multidisciplinary large research facilities because of the wide spectral range, the flux and the brilliance of their radiation. They have undergone, during the past three decades, an impressive evolution with regard to their number worldwide, now being about 70, their performance in terms of quality of photon beams, and the scope of the applications of the integral facility. Their continued development will lead to X-ray Free Electron Lasers that will open up to synchrotron radiation completely new regimes of brilliance, time resolution, and coherence in the foreseeable future.

1. Introduction

Light has been and continues to be the probe of choice for studying the properties of matter, non-destructively and without mechanical contact. Synchrotron radiation is probably the most versatile tool that provides for the broadest spectral range of any source, for the highest intensity over that range except for the few narrow regions where lasers exist, and is even going to incorporate the laser principle, thereby expanding the range of useful fully tunable laser radiation from the UV to hard X-rays. While synchrotron radiation is well known for its hard and soft X-ray spectral range it provides also for superior flux and brilliance from near to far infrared.

Since the discovery of synchrotron radiation in 1947 and the pursuit of more dedicated development and exploitation in the early seventies, more than 70 synchrotron radiation facilities have been set up worldwide. Today, synchrotron radiation sources enable the most multidisciplinary R&D facilities whose range of applications is covering the life sciences and environmental science as well as engineering and materials science, physics and chemistry as well as micro and nano manufacturing. The field of synchrotron radiation based science and manufacturing is as well mature as it is vigorously growing.

In this survey, we will outline some of the essential contours of the landscape of synchrotron radiation sources, including the history of the experimental discovery, the theoretical treatment of the generation of synchrotron radiation from highly relativistic electrons, the development of synchrotron radiation facilities all over the world characterised by three distinct generations, and the outlook to what may become the fourth generation of sources in the future. Applications of synchrotron radiation and the required equipment such as beamlines and experimental stations will not be considered here, but in the last lecture of this school.

2. Historical

The experimental discovery of synchrotron radiation came after the theoretical prediction. Starting from Maxwell's theory Wiechert (1896), Liénard (1897), and Schott (1907, 1912) developed the basic theoretical concepts, and later, when accelerators came up, Schiff, Ivanenko and Pomeranchuk (1944) predicted this kind of radiation. The 70 MeV synchrotron of General Electric at Schenectady, New York, was the place where technicians Gerald Knowlton or Floyd Haber working under physicists Elder, Gurevich, Langmuir, and Pollock, first observed the visible part of synchrotron radiation on 24 April 1947. The full theory was then worked out by Schwinger (1946, 1949) and Ivanenko, Sokolov, Ternov (1948). A more detailed account of the history may be found in the Handbook on Synchrotron Radiation[1]. Figure 1 shows a picture of the visible part of synchrotron radiation as produced in the Helios 2 storage ring at SSLS.

Figure 1. The visible part of synchrotron radiation generated in the Helios 2 superconducting compact storage ring at SSLS can be seen as a bright racetrack-shaped area within the glass window.

In the early 1970s the scientific exploitation of synchrotron radiation took off leading eventually to more than 70 facilities that are under operation, construction, or projected. A comprehensive list of the synchrotron radiation facilities all over the world is being maintained by H. Winick and H. Nuhn[2]. From that web site and the links provided there the interested reader may go to any facility and extract information in great detail. More broadly, the present survey tries to guide the reader through the wealth of information by highlighting common essential features.

Table 1 shows the list of facilities and Figure 2 their geographic distribution.

Table 1: Synchrotron radiation facilities worldwide (operating, under construction, planned)

Location	Name	Energy/GeV	Location	Name	Energy/GeV
Aarhus	ASTRID	0.6	Melbourne	Aust. SR	3
Aarhus	ASTRID II	1.4	Moscow	Siberia I	0.45
Allaan	SESAME	1	Moscow	Siberia II	2.5
Argonne	APS	7	Nakhon R.	SIAM	1
Barcelona	LLS	2.5	Nishi Harima	NewSUB.	1-1.5
Baton Rouge	CAMD	1.3-1.5	Nishi Harima	NIJI III	0.6
Beijing	BLS	2.2-2.5	Nishi Harima	SPring-8	8
Beijing	BSRF	1.5-2.8	Novosibirsk	Siberia-SM	0.8
Berkeley	ALS	1.5-1.9	Novosibirsk	VEPP-2M	0.7
Berlin	BESSY II	1.7-1.9	Novosibirsk	VEPP-3	2.2
Bonn	ELSA	1.5-3.5	Novosibirsk	VEPP-4M	5-7
Campinas	LNLS-1	1.35	Okasaki	UVSOR	0.75
Campinas	LNLS-2	2	Okasaki	UVSOR-II	1
Daresbury	SRS	2	Orsay	DCI	1.8
Dortmund	DELTA	1.5	Orsay	SOLEIL	2.5-2.75
Dubna	DELSY	1.2	Orsay	SuperACO	0.8
Durham	DFELL	1-1.3	Oxfordshire	DIAMOND	3
Frascati	DAFNE	0.51	Pohang	PLS	2
Gaithersburg	SURF III	0.39	Raleigh	NC STAR	2.5
Grenoble	ESRF	6	Saga	Saga LS	1.4
Hamburg	DORIS III	4.5	Saskatoon	CLS	2.9
Hamburg	PETRA II	7-14	Sendai	TLS	1.5
Hefei	NSRL	0.8	Seoul	CESS	0.1
Hiroshima	HISOR	0.7	Shanghai	SSRF	3.5
Hsinchu	SRRC	1.3-1.5	Singapore	SSLS	0.7
Ichihara	NANO-H	1.5-2	Stanford	SPEAR2	3
Indore	INDUS-I	0.45	Stanford	SPEAR3	3
Indore	INDUS-II	2	Stoughton	Aladdin	0.8-1
Ithaca	CESR	5.5	Trieste	ELETTRA	2-2.4
Karlsruhe	ANKA	2.5	Tsukuba	Accum. R.	6.5
Kashiwa	VSX	1-1.6	Tsukuba	NIJI II	0.6
Kharkov	PS / SR	0.75-2	Tsukuba	NIJI IV	0.5
Kiev	ISI-800	0.7-1.0	Tsukuba	PF	2.5
Kusatsu	Rits SR	0.58	Tsukuba	TERAS	0.8
Kyoto	KSR	0.3	Upton	NSLS I	0.8
Lund	MAX I	0.55	Upton	NSLS II	2.5-2.8
Lund	MAX II	1.5	Villigen	SLS	2.4
Lund	MAX III	0.7	Yerevan	Candle	3.2
Lund	MAX IV	1.5-3			

All existing facilities are distributed over the northern hemisphere, SSLS being the southernmost, except for the LNLS in Campinas, Brazil. The highest concentration of sources is found in Japan. New construction projects which all belong to the class of intermediate energy light sources (2.5 to 4 GeV) are being pursued in Australia, Canada, China, France, Spain, UK. The Australian Light Source will be the second built in the southern hemisphere. For a recent review on intermediate energy light sources see[3].

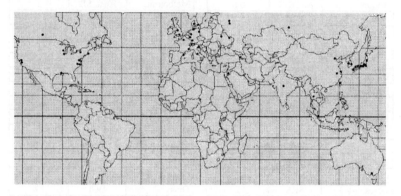

Figure 2: Geographic distribution of synchrotron radiation facilities

3. Main features of synchrotron radiation

To explain why synchrotron radiation is so much sought for, its attractive features are outlined reviewing the derivation of the most important formulae. SI units are used throughout. A more comprehensive treatment of the theory is given in lecture D3.1 by Lee Teng and in reference[4].

The most prominent features are the
- high values of flux, brightness, and brilliance, the
- broad spectral range, and the
- polarization.

The theoretical tools for calculating the properties of synchrotron radiation are Maxwell's equations together with the Liénard-Wiechert potentials for moving charges including the notion of evaluating these potentials at the "retarded" time. Figure 3 shows the reference frame and some vectors used to describe the motion of an electric charge on a circular orbit and the radiation measured by an observer.

The scalar and vector potentials of this moving charge are called its Liénard-Wiechert potentials and read (Eq. (1), Eq. (2))

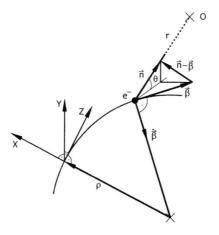

Figure 3. Reference frame for a charge moving on a circular orbit and emitting radiation.

$$V(t) = \frac{1}{4\pi\varepsilon_0} \frac{q}{\left[r\left(1 - \vec{n}\vec{\beta}\right)\right]_{ret}} \tag{1}$$

$$\vec{A}(t) = \frac{\mu_0}{4\pi} \left[\frac{q\vec{v}}{r\left(1 - \vec{n}\vec{\beta}\right)}\right]_{ret} \tag{2}$$

where $[\ ]_{ret}$ means that the expression in square brackets is evaluated at the time

$$t' = t - r(t')/c.$$

3.1. Spatial distribution of radiated power

With the expressions for the electric and magnetic field (Eq. (3)) and the Poynting vector (Eq. (4)) as derived from Maxwell's equations

$$\vec{E}(t) = -\nabla V - \frac{\partial \vec{A}}{\partial t}$$

$$\vec{B}(t) = \nabla \times \vec{A} \tag{3}$$

$$\vec{S}(t) = \frac{1}{\mu_0} \vec{E} \times \vec{B} = \frac{\vec{E}^2}{\mu_0 c} \vec{n} \tag{4}$$

we obtain Eq. (5) for the spatial power distribution emitted by the moving electric charge

$$\frac{dP(t')}{d\Omega} = \frac{1}{\mu_0 c^3} \frac{q^2}{(4\pi\varepsilon_0)^2} \frac{\left[\vec{n}\times((\vec{n}-\vec{\beta})\times\dot{\vec{\beta}})\right]^2}{(1-\vec{n}\vec{\beta})^5}$$

(5)

The spatial power distribution is determined by the unit vector from the charge to the observer and the normalized velocity and acceleration. For highly relativistic particles with $\gamma = 1/\sqrt{1-\beta^2} \gg 1$ the denominator leads to a strongly peaked distribution in the direction of the velocity (forward peak) which accounts for the small divergence angle of synchrotron radiation from an individual electron. The full spatial distribution is depicted in Figure 4 for three cases of the normalized velocity pointing in the x-direction. While this picture shows the general tendency it cannot convey the more needle-like forward peak when the particle is highly relativistic. However, it is this strong forward peak that accounts for the high brightness of synchrotron radiation. Furthermore, as the cross section of electron beams can be made rather small up-to-date – down to an order of 1000 μm^2 – it also accounts for the high brilliance.

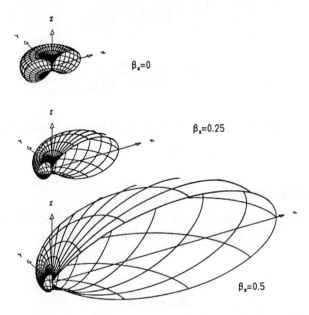

Figure 4. Spatial distribution of power density radiated by an electrical charge accelerated in z direction and moving in x direction. Three different velocities are illustrated as described by the value of β_z (Brandt, Dahmen, *Elektrodynamik*, © Springer-Verlag Berlin).

3.2. *Spectral distribution of radiated power*

From the time-dependent spatial radiated power distribution we obtain the spectral power distribution basically by Fourier transformation. Writing the power distribution as

$$\frac{dP(t)}{d\Omega} = |\vec{F}(t)|^2 \tag{6}$$

with

$$\vec{F}(t) = \frac{1}{\sqrt{\mu_0 c}} [r\vec{E}]_{ret} \tag{7}$$

we define the spectral power distribution $d^2I / d\omega d\Omega$ by the expressions (Eq. (8))

$$\int_0^\infty \frac{d^2I}{d\omega d\Omega} d\omega = \int_{-\infty}^\infty \frac{dP}{d\Omega} dt = \int_{-\infty}^\infty |\vec{F}(t)|^2 dt \tag{8}$$

Using Parsefal's theorem we write (Eq. (9))

$$\int_{-\infty}^\infty |\vec{F}(t)|^2 dt = \int_{-\infty}^\infty |\vec{G}(\omega)|^2 d\omega = 2 \int_0^\infty |\vec{G}(\omega)|^2 d\omega \tag{9}$$

with the Fourier transform (Eq. (10))

$$\vec{G}(\omega) = \frac{1}{\sqrt{2\pi}} \int_{-\infty}^\infty \vec{F}(t) \exp(i\omega t) dt \tag{10}$$

and its property $\vec{G}(-\omega) = \vec{G}^*(\omega)$ to finally obtain (Eq. (11))

$$\frac{d^2I}{d\omega d\Omega} = 2|\vec{G}(\omega)|^2 \tag{11}$$

Explicit evaluation in the above coordinate system leads to (Eq. (12))

$$\frac{d^2I}{d\omega d\Omega} = \frac{q^2}{12\pi^3 \varepsilon_0 c} \left(\frac{\omega\rho}{c}\right)^2 \left(\frac{1}{\gamma^2} + \vartheta^2\right)^2 \left[K_{2/3}^2(\xi) + \frac{\vartheta^2}{(1/\gamma^2) + \vartheta^2} K_{1/3}^2(\xi)\right] \tag{12}$$

Eq. (12) is composed of two contributions that represent the polarization of the electric field parallel and perpendicular to the midplane. Integrating over the solid angle and either frequency or time we finally obtain the total power emitted

$$P(t) = \frac{2}{3} r_0 m_0 c \dot{\beta}^2 \gamma^4 \tag{13}$$

which is the so-called Larmor formula that shows impressively how much the power radiated is increased when the particles are highly relativistic. Use is made of the following symbols

$\beta = v/c$	Normalized velocity of particle
c	Speed of light *in vacuo*
γ	Ratio of particle total energy to rest energy
$\vec{n}$	Unit vector from particle to observer (versor)
ϑ	Angle between midplane and $\vec{n}$ (latitude)
ρ	Instantaneous orbit curvature radius
ω	Angular frequency
$K_{2/3}(\xi), K_{1/3}(\xi)$	Modified Bessel functions
$\xi = (\omega\rho/3c)(1/\gamma^2 + \vartheta^2)^{3/2}$	
Ω	Solid angle
q	Electric charge of particle
r	Distance particle-observer
$r_0 = e^2/(4\pi\varepsilon_0 m_0 c^2)$	Classical electron radius
m_0	Electron rest mass
e	Elementary charge

The spectral distribution of radiation from a bending magnet is depicted in Figure 5 for the case of Helios 2, the superconducting compact electron storage ring at SSLS.

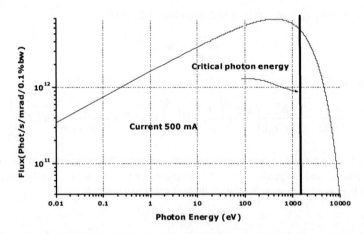

Figure 5. Spectral flux per milliradian of horizontal angle of Helios 2 compact superconducting storage ring.

4. Generation of highly relativistic electron beams as sources for synchrotron radiation

Basically, an accelerator system is needed to produce beams of highly relativistic electrons[5], and magnetic field devices for the electrons to travel through in order to generate the light.

4.1. *Accelerator systems*

Most of present-day synchrotron radiation sources are based on an electron storage ring accelerator that will be shortly discussed in the following. The use of linear accelerators is the characteristic feature of future 4th generation sources.

The very basic elements of an electron storage ring are, as depicted schematically in Figure 6,

☐ dipole magnets to bend the beam, enabling a closed orbit,
☐ quadrupole magnets for focusing, defining the location of the closed orbit,
☐ sextupole magnets for correcting the energy-dependence of the focusing (chromaticity),
☐ a topologically toroidal vacuum chamber under ultrahigh vacuum (UHV) conditions (in most cases a ring consisting of alternating curved and straight sections), to minimize the rate of collisions between electrons and residual gas molecules and atoms, and a
☐ radiofrequency (RF) system to accelerate the electrons and keep their energy.

Furthermore, the electron storage ring also needs an injector which is another accelerator producing an electron beam from the extraction out of a

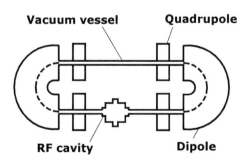

Figure 6. Schematic of a storage ring showing dipoles, quadrupoles, an RF cavity, and the vacuum vessel. In the case of Helios 2 the dipoles have a field gradient and a sextupolar field integrated.

cathode and accelerating it to some injection energy that is less or equal to the final energy in the storage ring. Very often, a chain of accelerators is used as an injector comprising an electrostatic gun, a microtron or a linear accelerator, and a booster synchrotron.

4.2. Magnetic field device

In order to be transversally accelerated and to emit synchrotron radiation the electron has to travel across a magnetic field. The force on the electron is given by the Lorentz force formula

$$\vec{F} = q\vec{v} \times \vec{B} \tag{14}$$

Magnetic flux density $\vec{B}$ and electron velocity $\vec{v}$ have to have mutually perpendicular components to yield a non-vanishing Lorentz force. This is the case for the dipole (or bending) magnets of a storage ring as well as for the so-called insertion devices. The dipoles approximate a uniform magnetic field for which case the calculations of the spatial and spectral power density were performed, above.

Insertion devices may be defined as magnetic field configurations that change the particles' design orbit only inside themselves when switched on. There is no effect on the design orbit outside. The operation of the storage ring is almost independent of their function.

Insertion devices include

☐ wavelength shifters (mostly superconducting) in which the particle undergoes only one comparably strong beam bump,

☐ wigglers (mostly superconducting) in which the particle follows an orbit with a certain number of strongly curved alternating equidistant short arcs, and, finally,

☐ undulators in which there is a larger number of weakly curved alternating equidistant short arcs.

Initially, the majority of undulators were built from permanent magnets[6,7], more recently, superconducting technology is also used, in particular, for superconducting miniundulators[8,9]. In wavelength shifters and wigglers, the orbit is strongly bent and the spectrum is a bending spectrum, eventually made harder by the stronger magnetic field in a superconducting magnet and more intense when light emitted from several wiggles is superposed. In undulators, the orbit is only weakly bent and, as a result of regularly repeated undulation, the spectrum becomes a line spectrum.

Wigglers and undulators are distinguished using the K parameter which describes the ratio between the maximum inclination angle between the orbit and the straight forward direction and the aperture angle of the synchrotron radiation

$$K = \alpha \cdot \gamma = \frac{qB\lambda_u}{2\pi \cdot m_0 c} \approx 0.934 B[T]\lambda_u[cm] \qquad (15)$$

In practice, undulators are considered to have $K \leq 2$, for wigglers $K > 2$. For a more thorough discussion of insertion devices see lecture D8.2 by C.S. Hwang and reference[10].

5. Generations of synchrotron light sources

5.1. *Some statistics on existing sources*

The higher the energy of the electrons the harder are the most energetic photons, the bigger is the machine, the broader the application range, and the more expensive the facility. Figure 7 shows the distribution of existing sources versus their energy.

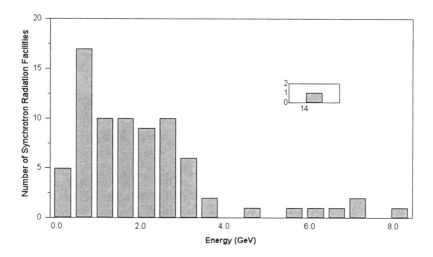

Figure 7. Number of synchrotron radiation sources versus electron energy.

Figure 8 gives an impression of the huge range of sizes among the synchrotron radiation facilities. The arcs of the smallest ones, AURORA 1, and HELIOS 2, are compared with SSRF and SPring-8 on the same scale.

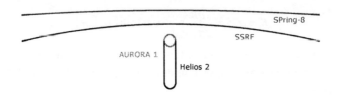

Figure 8. Comparing arc sizes of the smallest and largest storage rings at synchrotron radiation facilities. AURORA 1 is at Rits SR and Helios 2 at SSLS.

5.2. Criteria to distinguish generations

We now discuss the meaning of generations of synchrotron light sources referring to the following table.

Table 2: Main distinctive features of various generations of synchrotron radiation facilities

Criteria	Generation		
	1st	2nd	3rd
Original purpose of facility	High energy physics	Synchrotron radiation	Synchrotron radiation
Access for users	Parasitic	Dedicated	Dedicated
Machine parameters influencing light	Energy, magnetic field	Energy, magnetic field	Energy
Source	Bending magnet	Bending magnet, insertion device	Insertion device, bending magnet
Source characteristics	Flux	Flux, brightness	Flux, brightness, brilliance
Emittance (nmrad)	1000	100	3.7 [11]; 4.8 [12]
Examples	DORIS, SPEAR	SRS, NSLS, DORIS, SPEAR	SSRF, SPring-8, SRRC, SPEAR III

The 1st generation of synchrotron radiation sources includes machines such as colliders that were built for high energy physics purposes, and not for producing synchrotron radiation. Accordingly, users had only parasitic access when there was no interference with the high energy physics experiment. As the synchrotron radiation was taken from bending magnets its spectrum was determined by the electron energy and the magnetic field of the bending magnets as required by high energy physics and machine physics considerations.

The 2nd generation of synchrotron radiation sources still had this coupling between machine physics parameters and the spectrum, but they were built as sources dedicated to the production and use of synchrotron radiation.

Besides improving access to and availability of synchrotron radiation the underlying theme for developing the various generations was and is to improve the quality of the synchrotron radiation produced. This quality is described in terms of brilliance, brightness, and flux of the photon beam and is related to the emittance and energy of the electron beam as well as to the magnetic field devices used for producing the radiation. It has led to the 3rd generation of sources characterized by electron beam emittances around or below 10 nmrad and by the extensive use of insertion devices like undulators, wigglers, and wavelength-shifters for producing high-quality light. The art of building 3rd generation sources is mature, yet, new projects are vigorously pursued in several countries such as Australia, Canada, China, France, Spain, and the UK.

5.3. *Emittance and brilliance*

Here, the basic ideas of emittance and brilliance are briefly touched. The goal is to produce photon beams that can be either focused to very small spots and not too large a divergence angle in the interest of spatial resolution or made nearly parallel with a rather small cross section to provide high angular resolution in any scattering experiment. Such a property of a photon beam is called brilliance. A highly brilliant photon beam needs an electron beam with a small emittance.

Any particle beam can be characterized by giving the particles transverse coordinates x and y and momenta p_x and p_y at any point in a certain cross-sectional control plane. The space subtended by x, p_x, y, p_y is called phase space. Usually, two-dimensional subspaces are used for illustration such as x, p_x and y, p_y.

For an electron orbiting in a storage ring the points (x, p_x) or (y, p_y) as observed in any cross-sectional control plane lie on so-called phase ellipses. These are centered with respect to the origin of the x, p_x or y, p_y reference frame. In general, their principal axes do not co-incide with the frame axes except for waists or bulges. Different electrons describe different ellipses for which the length of the semiaxes depends on the betatron oscillation amplitude of the individual electron. All the electrons of a beam fill some ellipse the area of which is called the emittance of the beam. Each transverse coordinate x and y has an emittance value associated with. Care has to be taken when numerical emittance values are compared. Some use the full area of the ellipse πab for the value, others include π in the unit, frequently calling it $\pi \cdot nm \cdot rad$.

The phase space filled by the photon beam results from the superposition of the angular distribution of the emitted radiation to the electron beam phase space distribution. This is usually approximated by stretching out the ellipse describing the electron beam emittance in the angular direction such that there is quadratic addition of semiaxes.

Finally, the spectral brilliance B of a photon beam is calculated from its spectral flux Φ divided by the phase-space volume filled.

$$B = d^2\Phi/dAd\Omega \approx$$

$$\Phi/(\sigma_x \sigma_y)(\sqrt{\sigma_{p_x}^2 + \sigma_R^2}\sqrt{\sigma_{p_y}^2 + \sigma_R^2}) = \tag{16}$$

$$\Phi/(\sigma_x\sqrt{\sigma_{p_x}^2 + \sigma_R^2})(\sigma_y\sqrt{\sigma_{p_y}^2 + \sigma_R^2}) = \Phi/\varepsilon_x \cdot \varepsilon_y$$

Here, σ_x, σ_y, σ_{p_x}, σ_{p_y} are the variances of the transverse spatial and angular distributions of the electrons, σ_R is the angular distribution of the radiation, A the beam cross sectional area, Ω the solid angle, and ε_x, ε_y the transverse photon beam emittances.

For detailed discussions of phase space concepts and ensuing definitions of emittance and brilliance the reader is referred to lectures D1.3 and D2.2 by Q. Qin and D5.4 by L. Teng as well as the books by S.Y. Lee and by F. Ciocci et al.

5.4. 4th generation source concepts

As the electron beam emittance in storage rings is determined by the balance between radiation excitation and acceleration damping it becomes very costly to significantly decrease the emittance of 3rd generation storage ring sources below present values of a few nmrad. This can be immediately understood by considering that the emittance is proportional to the cube of the bending angle per dipole in a storage ring. For example, the minimum emittance in a double bend achromat (DBA) lattice is given by

$$\varepsilon_{MEDBA} = \frac{C_q \gamma^2 \varphi^3}{4\sqrt{15}J_x} \tag{17}$$

where φ is the bending angle per dipole, J_x the horizontal damping partition number, and $C_q = 3.83 \cdot 10^{-13} m$ the quantum fluctuation coefficient for electrons (see S Y Lee). Reduction of bending angle is concurrent with adding more quadrupole focusing and increasing the circumference of the storage ring.

4th generation sources are expected to exploit a radically different way of beam production, namely using a linear accelerator in which, for the same electron energy in the 2 – 3 GeV range, the emittance can be about 1-2 orders of magnitude smaller than the natural emittance in a storage ring. The obvious drawback of this approach, namely, the small average electron current delivered by linear accelerators as compared to storage rings, may be partly remedied by the use of superconducting linacs because they enable the energy recovery (ERL energy recovery linac) or power saving operation that has a significant and experimentally proven potential to increase the beam current in the sc linac[13,14]. Furthermore, the high pulsed current possible with the linac enables Free Electron Laser (FEL) operation.

The expected trends for the brilliance of the radiation are not only determined, as usually, by the emittance and the beam current, but also by the pulse length and the spectral bandwidth both of which are very much reduced by bunch compression and by coherent emission. As the product of horizontal and vertical emittance does not differ much between storage ring and FEL/ERL the increase in brilliance is mostly due to a reduction of bunch length and spectral bandwidth (Table 3).

Table 3: Typical values of emittance, beam current, and maximum brilliance in 3rd and 4th generation light sources. Note that values of average and peak current of ERLs have a high degree of uncertainty (partly using data from[15]).

Parameter \ Source type		3rd gen	4th gen	
		Storage ring	FEL	ERL
ε_x	nmrad	5	0.5	0.5
ε_y	nmrad	0.05	0.5	0.5
$\varepsilon_x \varepsilon_y$	$(\text{nmrad})^2$	0.25	0.25	0.25
I_{av}	mA	300	1	10 - 100
I_{peak}	A	1	>1000	100 - 1000
B_{av}	photons/s/mm^2/mrad^2/0.1% bw	10^{19}	10^{23}	$3 \cdot 10^{17}$ - $3 \cdot 10^{18}$
B_{peak}	photons/s/mm^2/mrad^2/0.1% bw	10^{22}	10^{32}	10^{25} - 10^{26}

On top of the brilliance issue, X-ray FELs promise to deliver photon beam pulses with durations on the fs scale. On this time scale, the influence of the motion of the nuclei on the electronic structure in a molecule can be studied. This would allow extending the so-called femtochemistry into hitherto inaccessible soft and hard X-ray spectral ranges.

6. Conclusion

After 66 years since the discovery, the field of synchrotron radiation sources is as well mature as it has bright perspectives for future development into new source concepts. Drastic performance improvements can be expected, especially from X-ray Free Electron Lasers and Energy Recovery Linac schemes, spawning a great wealth of new scientific applications.

Acknowledgement

Work supported by NUS Core Support 2002/03. The author is grateful to Messrs. Chew Eh Piew and Zheng Hongwei for help with the figures.

References

[1] E.E. Koch, D.E. Eastman, Y. Farge, *Synchrotron Radiation – A Powerful Tool in Science,* Handbook on Synchrotron Radiation, vol. 1, E.E. Koch, ed., North-Holland Publishing Company, 1983

[2] http://www-ssrl.slac.stanford.edu/sr_sources.html

[3] Z.T. Zhao, Nuclear Science and Techniques 14(1)(2003)1-8

[4] S.Y. Lee, Accelerator Physics, World Scientific, 1999, pp 395

[5] Positrons can also be used. They are advantageous with respect to ion beam trapping, but their production is more costly so that only a very few accelerator labs have used positrons.

[6] K. Halbach, J. de Physique 44(1983)C1-211; S. Krinsky, IEEE Trans. Nucl. Sci. NS-30(1983)3078; K.J. Kim, *Characteristics of Synchrotron Radiation,* X-Ray Data Booklet, D. Vaughan, ed., LBL PUB-490, 1986

[7] T. Hara, T. Tanaka, T. Tanabe, X.-M. Marechal, S. Okada, H. Kitamura, J. Synchrotron Radiation 5 (1998) 403

[8] I. Ben-Zvi, Z.Y. Jiang, G. Ingold, L.H. Yu, W.B. Sampson, Nucl. Instrum. Meth. A297(1990)301

[9] H.O. Moser, B. Krevet, G. Holzapfel, German patent P 41 01 094.9-33, Jan. 16, 1991; T. Hezel, M. Homscheidt, H.O. Moser, R. Rossmanith, et al., *First beam test of a superconductive in-vacuo mini-undulator for future x-ray lasers and storage rings,* in Free Electron Lasers 1999, J. Feldhaus, H. Weise, eds., pp. II-103, North Holland, 2000.

[10] F. Ciocci, G. Dattoli, A. Torre, A. Renieri, Insertion Devices for Synchrotron Radiation and Free Electron Laser, World Scientific, 2000

[11] SOLEIL 3, 2.75 GeV, Saclay, France; http://www.synchrotron-soleil.fr

[12] SSRF, 3.5 GeV, Shanghai

[13] M. Tigner, Nuovo Cimento 37(1965)1228

[14] G. Neill et al., Phys. Rev. Lett. 84(4)(2000)662
[15] Visions of Science: The BESSY SASE-FEL in Berlin-Adlershof, BESSY GmbH Berlin-Adlershof 2001

TRANSVERSE BEAM DYNAMICS: LINEAR OPTICS*

Q. QIN

Institute of High Energy Physics,
Beijing, 100039, P.R. China
E-mail: qinq@mail.ihep.ac.cn

The transverse motion of charged particles in a synchrotron facility is our main subject, along with descriptions of basic magnets, solutions of Hill's equation, Liouville's theorem, elementary Hamiltonian treatment, etc. Errors from magnets, energy deviations are discussed together with their effects on beam dynamics. Transverse linear coupling is treated with the example of BEPC storage ring. Methods of measuring the basic lattice parameters are briefly introduced.

1   Introduction

In this fundamental course on accelerators, we focus on the storage ring of a synchrotron radiation facility. In such a machine, dipoles, or bending magnets, keep the revolution of the charged particle beams within a vacuum chamber. This confines them in a closed orbit. The strong focusing principle needs quadrupoles to make beams oscillate around the closed orbit, i.e., the betatron motion. Such an arrangement of dipoles and quadrupoles for a certain purpose is called the accelerator "lattice".

Charged particle beams are stored in high energy rings with transverse and longitudinal motions. The transverse particle motion can be divided into two parts: a closed orbit and a small-amplitude oscillation around the closed orbit.

In the following sections, we will introduce the basic knowledge of the transverse motion of charged beams in circular accelerators.

2   Linear Betatron Motion

2.1   Coordinate System

The bending field produced by dipoles is usually vertically directed, which makes the charged particle follow a curved path in the horizontal plane. Figure 1

---

* Work supported by National Natural Science Foundation of China (10275079)

simply shows a curvilinear coordinate system for particle motion in a synchrotron.

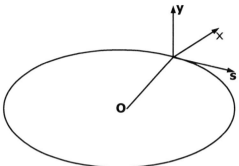

Figure 1 Curvilinear coordinate system for charged particles

Under such a coordinate system, the force acting on the charged particle in the horizontal direction is

$$\mathbf{F} = e\,\mathbf{v} \times \mathbf{B} \ , \tag{1}$$

where $\mathbf{v}$ is the velocity of the charged particle in the direction tangential to its path, $\mathbf{B}$ the magnetic guide field, and e the charge of the electron.

2.2 Displacement and Divergence

In the above curvilinear coordinate system, the transverse displacements for a charged particle in an accelerator are defined as x and y, in the horizontal and vertical planes respectively, and the divergence angles, x' and y', are

$$x' = dx/ds, \qquad y' = dy/ds \ . \tag{2}$$

Particles will leave the ideal orbit when divergences exist. In the meantime, quadrupole magnets will provide the restoring fields in the lattice to make particles oscillate about the ideal orbit.

2.3 Dipoles and Magnetic Rigidity

We assume a particle traveling perpendicularly to a guide field $\mathbf{B}$ with a relativistic momentum $\mathbf{p}$. Following a curved path of radius ρ and length ds after a duration dt, the particle's momentum becomes $p + dp$. From Eq. (1), we have

$$e\,\mathbf{v} \times \mathbf{B} = \frac{d\mathbf{p}}{dt} \ . \tag{3}$$

Since the magnitude of the force can be written as

$$e\left|\mathbf{v} \times \mathbf{B}\right| = e\left|\mathbf{B}\right|\frac{ds}{dt} \ , \tag{4}$$

where s is the longitudinal displacement and

$$\frac{d\mathbf{p}}{dt} = |\mathbf{p}|\frac{d\theta}{dt}\hat{x} = \frac{|\mathbf{p}|}{\rho}\frac{ds}{dt}\hat{x}, \tag{5}$$

in which θ is the azimuthal angle. With simple evaluations, we find the quantity magnetic rigidity:

$$(B\rho) = \frac{p}{e}. \tag{6}$$

Conveniently, the magnetic rigidity is expressed in particle dynamics with electron-volts as the units of pc, that is

$$(B\rho)[\text{T}\cdot\text{m}] = \frac{pc[\text{eV}]}{c[\text{m/s}]} = 3.3356\,(pc). \tag{7}$$

2.4   Quadrupoles

Quadrupoles are used in modern accelerators for focusing beams. The poles of quadrupoles are truncated rectangular hyperbolae and alternate in polarity.

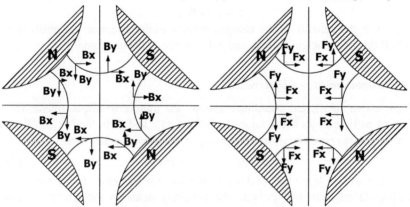

Figure 2 Magnetic components and force in a quadrupole. (Negatively charged particles enter the plane of the paper with paths perpendicular.)

Figure 2 shows the shape of quadrupole poles. Force analysis tells us that the strength increases linearly with the distance from an axis, while remaining zero on the axes. In Figure 2, the quadrupole would focus negative particles going into the plane of the paper in the horizontal plane. It is easy to see that, in the meantime, the negative particles are defocused in the vertical direction.

The strength of a quadrupole can be expressed by its gradient dB_y/dx normalized with respect to magnetic rigidity:

$$k = \frac{1}{(B\rho)}\frac{dB_y}{dx}. \tag{8}$$

When a particle goes through a short quadrupole of length l and strength k, at a distance x the angular deflection is

$$\Delta x' = lkx \ . \tag{9}$$

Like a focusing lens, a focusing quadrupole has the focal length f, so that

$$\Delta x' = -x / f \ . \tag{10}$$

Thus,

$$f = -1 / (kl) \ . \tag{11}$$

2.5 Alternating Gradient Focusing

The most important milestone in modern accelerator development is the discovery of alternating gradient focusing. Under this theorem, a system of quadrupoles can be used to focus beams in both planes although an individual quadrupole is focusing only in one plane and defocusing in the other. In this way, a much stronger focusing system is obtained.

The principle of the alternating gradient system is depicted as an optical system in Figure 3, where each lens is concave in one plane and convex in the other plane. If the spacing of the lenses is kept at a certain value, even with lenses of equal strength, it is still possible to find a ray always on axis at the defocusing lenses in the horizontal plane. The same case is true in the vertical plane.

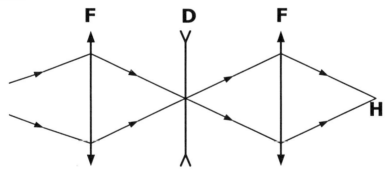

Figure 3 Alternating gradient system in horizontal plane.

For quadrupole lenses, the idea of alternating gradient works when the F lenses and D lenses are arranged along the beam line alternately. Figure 4 shows that the particle trajectories tend to be closer to the axis in D lenses than in F lenses.

The envelope of all such trajectories is described by the betatron function $\beta(s)$, which expresses the biggest excursion of a beam. With suitable strengths and spacing of lenses, $\beta(s)$ can be made periodic and kept large at all F quadrupoles and small at all D's. The same procedure can be done in the other plane.

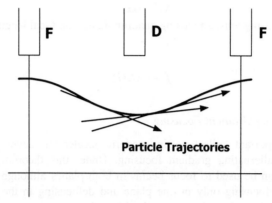

Figure 4   Particle trajectories and the envelope of betatron motion in a FODO lattice

2.6   *Equation of Transverse Motion*

Under the coordinate system shown in Figure 1, which gives the unit vectors of $\hat{x}, \hat{y}, \hat{s}$, for **x**, **y**, **s** directions, we can derive the equation of transverse motion from the very beginning.

Suppose a charged particle moves along the ideal orbit in a storage ring. The radius of the orbit is $\rho(s)$ and its inverse, the curvature, is expressed as $G(s)$,

$$G(s) = \frac{1}{\rho(s)} .$$

(12)

The displacement from the origin to the charged particle, r, and the magnetic field, B, can be written as

$$\mathbf{r} = \mathbf{r_0} + x\hat{x} + y\hat{y}$$

(13)

and

$$\mathbf{B} = B_x \hat{x} + B_y \hat{y} + B_s \hat{s} .$$

(14)

From Eqs. (3) and (13) we have

$$\frac{d}{dt}(m\frac{d\mathbf{r}}{dt}) = e \cdot [\frac{d\mathbf{r}}{dt} \times \mathbf{B(r)}]$$

(15)

and

$$\frac{d\mathbf{r}}{d\theta} = \frac{d\mathbf{r}}{dt}\frac{dt}{d\theta} = \frac{e}{G_0}[(1+Gx)\hat{x} + G_0 x' \hat{y} + G_0 z' \hat{s}] .$$

(16)

Since $B_x = B_s = 0$ at the mid-plane, after some manipulations, we get

$$\begin{cases} \dfrac{d^2x}{ds^2} + (\dfrac{1}{\rho^2(s)} + \dfrac{1}{B\rho}\dfrac{\partial B_y}{\partial x})x = \dfrac{1}{\rho(s)} \cdot \dfrac{\Delta p}{p_0} \\ \dfrac{d^2y}{ds^2} - \dfrac{1}{B\rho}\dfrac{\partial B_y}{\partial x}y = 0 \ . \end{cases} \quad (17)$$

Here $\Delta p = p - p_0$ is the energy deviation of the charged particle. After defining

$$\begin{cases} K_x(s) = \dfrac{1}{\rho^2(s)} + \dfrac{1}{B\rho} \cdot \dfrac{\partial B_y}{\partial x} \\ K_y(s) = -\dfrac{1}{B\rho} \cdot \dfrac{\partial B_y}{\partial x} \ , \end{cases} \quad (18)$$

we can finally get the equation of transverse motion as

$$\begin{cases} \dfrac{d^2x}{ds^2} + K_x(s)x = \dfrac{1}{\rho(s)} \cdot \dfrac{\Delta p}{p_0} \\ \dfrac{d^2y}{ds^2} + K_y(s)y = 0 . \end{cases} \quad (19)$$

2.7 Solution of Hill's Equation

When $\Delta p/p_0 = 0$, eq. (19) becomes Hill's equation,

$$\frac{d^2z}{ds^2} + K(s)z = 0 \ , \quad (20)$$

where z stands for both x and y. $K(s)$ is periodic in the circumference, C, i.e.,

$$K(s) = K(s+C) . \quad (21)$$

Equation (20) is reminiscent of simple harmonic motion but with a restoring constant K (s) which varies around the storage ring. The solution of Hill's equation can be assumed to have a form similar to the simple harmonic motion:

$$z(s) = a\omega(s)e^{\pm i\psi(s)} . \quad (22)$$

Here, a is a constant, but ω (s) is a function of s, different from the amplitude of simple harmonic oscillation. Note that the phase advance in Eq. (22) is also a function of s, not of time t.

Furthermore, we have

$$z'(s) = a[\omega'(s)e^{\pm i\psi(s)} \pm i\omega(s)\psi'(s)e^{\pm i\psi(s)}] . \quad (23)$$

and

$$z''(s) = a[\omega''(s) \pm 2i\omega'(s)\psi'(s) \pm i\omega(s)\psi''(s) \mp \omega(s)\psi'^2(s)]e^{\pm i\psi(s)} . \quad (24)$$

With some manipulations, we can get

$$\begin{cases} \omega'' + K(s)\omega \mp \omega\psi'^2 = 0 \\ 2\omega'\psi' + \omega\psi'' = 0 \end{cases}$$

(25)

After defining $\mu = \psi'$, we have

$$\psi' = \mu = \frac{1}{\omega^2} \quad \text{or} \quad \psi = \int \frac{1}{\omega^2} ds .$$

(26)

Then, defining $\beta = \omega^2$, we finally get the famous formula

$$2\beta\beta'' - \beta'^2 + 4K\beta^2 - 4 = 0 .$$

(27)

Now we have to define the so-called Twiss parameters, $\alpha(s)$, $\beta(s)$ and $\gamma(s)$ as

$$\begin{cases} \alpha(s) = -\omega(s)\omega'(s) = -\dfrac{\beta'(s)}{2} \\ \beta(s) = \omega^2(s) \\ \gamma(s) = \dfrac{1+[\omega(s)\omega'(s)]^2}{\omega^2(s)} = \dfrac{1+\alpha^2(s)}{\beta(s)} \end{cases}$$

(28)

so that the solution of Hill's equation can be written as

$$\begin{cases} z(s) = a\sqrt{\beta(s)} \cos[\psi(s)+\psi_0] \\ z'(s) = -a\sqrt{\beta(s)}\alpha(s) \cos[\psi(s)+\psi_0] - \dfrac{a}{\sqrt{\beta(s)}} \sin[\psi(s)+\psi_0] \end{cases}$$

(29)

with ψ_0 the initial phase advance.

2.8 Matrix Description

The solution of Hill's equation can also be expressed with matrices, which represent the transport of the beam in storage rings. From one point, s_1, to another point, s_2, the 2×2 transport matrix can be expressed as

$$\begin{pmatrix} z(s_2) \\ z'(s_2) \end{pmatrix} = \begin{pmatrix} a & b \\ c & d \end{pmatrix} \begin{pmatrix} z(s_1) \\ z'(s_1) \end{pmatrix} = M_{12} \begin{pmatrix} z(s_1) \\ z'(s_1) \end{pmatrix}$$

(30)

As the displacement and divergence of a particle should have the form of Eq. (29), we first suppose $\psi_0 = 0$, then choose $\psi(s_1) = 0$ and $\psi(s_2) = \psi$. In this way, we obtain four simultaneous equations which can be solved for a, b, c, d in terms of ω_1, ω_2 and ψ. The result is the most general form of the transport matrix:

$$M_{12} = \begin{pmatrix} \dfrac{\omega_2}{\omega_1}\cos\psi - \omega_2\omega_1'\sin\psi, & \omega_1\omega_2\sin\psi \\[4mm] -\dfrac{1+\omega_1\omega_1'\omega_2\omega_2'}{\omega_1\omega_2}\sin\psi - (\dfrac{\omega_1}{\omega_2} - \dfrac{\omega_2}{\omega_1})\cos\psi, & \dfrac{\omega_1}{\omega_2}\cos\psi + \omega_1\omega_2'\sin\psi \end{pmatrix} .$$

(31)

To simplify the above matrix, we can restrict M_{12} to be between two identical points in successive turns or cells of a periodic structure. This causes $\omega_1 = \omega_2$, $\omega'_1 = \omega'_2$ and ψ becomes the phase advance per cell, μ. Thus

$$M = \begin{pmatrix} \cos\mu - \omega\omega'\sin\mu, & \omega^2\sin\mu \\ -\dfrac{1+(\omega\omega')^2}{\omega^2}\sin\mu, & \cos\mu + \omega\omega'\sin\mu \end{pmatrix}. \tag{32}$$

Combining this with Eq. (28), the matrix now becomes even simpler — the Twiss matrix:

$$M = \begin{pmatrix} a & b \\ c & d \end{pmatrix} = \begin{pmatrix} \cos\mu + \alpha\sin\mu, & \beta\sin\mu \\ -\gamma\sin\mu, & \cos\mu - \alpha\sin\mu \end{pmatrix}. \tag{33}$$

Suppose

$$M(s) = \begin{pmatrix} m_{11} & m_{12} \\ m_{21} & m_{22} \end{pmatrix}, \tag{34}$$

the Twiss parameters can have other expressions

$$\begin{cases} \alpha = \dfrac{m_{11} - m_{22}}{2\sin\mu}, & \beta = \dfrac{m_{12}}{\sin\mu}, & \gamma = -\dfrac{m_{21}}{\sin\mu}, \\[2mm] \cos\mu = \dfrac{m_{11} + m_{22}}{2} \end{cases} \tag{35}$$

where μ is the phase advance per cell,

$$\mu = \int_c \frac{ds}{\beta(s)} . \tag{36}$$

2.9   Stability of Transverse Motion

When particles move in a storage ring without any energy deviation, say, $\Delta p = 0$, Eq. (19) has periodic solutions, which are also the solutions for an ideal machine. Thus, the particles will go around the ring on an Equilibrium Orbit.

For a real machine, due to errors of K_x and K_y, and the energy deviation, particles with different energies have different closed orbits.

In the alternating gradient focusing lattice, particles would remain focused only if they were close to the axis when passing through defocusing quadrupoles. With the matrix description, the stability condition can be achieved when the product $\prod_{Nk}\{M(s)\}^{Nk}$ does not diverge after N periods, which consist of one turn, and k turns, which define the real stable life of a beam. The conditions for this to be so are (1) automatically satisfied by a linear transport matrix, det $|M| = 1$, and (2) the matrix has real non-vanishing eigenvalues.

If (z, z') is depicted as the vector $\mathbf{Z}$, this implies

$$M\mathbf{Z} = \lambda\,\mathbf{Z} \tag{37}$$

and

$$\det|M - \lambda I| = 0 \quad . \tag{38}$$

With the Eq. (33), which is the form of M for a period, we have

$$\cos \mu = \frac{a+d}{2} = \frac{1}{2} \mathrm{Tr}\, M \quad . \tag{39}$$

Solving for λ, we can get the eigenvalues

$$\lambda = \cos \mu \pm i \sin \mu = e^{\pm i\mu} \tag{40}$$

so that the stability condition for real values of λ is obtained provided that

$$|\cos \mu| = \frac{|a+d|}{2} = \frac{1}{2} |\mathrm{Tr}\, M| < 1 \quad . \tag{41}$$

Then we can easily see that M has the characteristics of an exponential:

$$M^K = (I \cos \mu + J \sin \mu)^K = I \cos K\mu + J \sin K\mu \quad . \tag{42}$$

Equation (42) can be compared with

$$(e^{i\mu})^K = e^{iK\mu} \quad . \tag{43}$$

Thus, M^K will remain stable only if μ is real.

By decomposing M into two matrices, one can have another demonstration of the stability condition:

$$M = \begin{pmatrix} \cos \mu + \alpha \sin \mu, & \beta \sin \mu \\ -\gamma \sin \mu, & \cos \mu - \alpha \sin \mu \end{pmatrix} = I \cos \mu + J \sin \mu \tag{44}$$

where I is the unit matrix and J is described by

$$J = \begin{pmatrix} \alpha, & \beta \\ -\gamma, & \alpha \end{pmatrix} . \tag{45}$$

2.10   Betatron Tunes and Envelope Functions

In the above section one of the most important parameters in a circular machine was shown to be the Betatron Tune, i.e., the betatron wave number. If, in a constant gradient machine, a particle with the largest amplitude in the beam, $a\beta^{1/2}$, starts off with phase ψ_0, after one turn its phase has increased by

$$\Delta \psi = \oint ds / \beta = 2\pi R / \beta \tag{46}$$

where R is the radius of the ring. The quantity v is defined as the number of betatron oscillations per turn:

$$v = \frac{\Delta \psi}{2\pi} = \frac{R}{\beta} \quad \text{or} \quad \beta = \frac{R}{v} \quad . \tag{47}$$

In a normal machine, this is also approximately true. It is often written as

$$\overline{\beta} = R / v \quad . \tag{48}$$

It is important that v must not be a simple integer or vulgar fraction. Otherwise, particles will be in the resonance of $nv = p$ (n, p are integers) and lost with this dangerous condition.

In a regular storage ring, the betatron tunes, v_x and v_y in the horizontal and vertical planes respectively, are also called the working points.

Another important parameter in machine design and commissioning is the β function, which describes the envelopes of betatron oscillations made by particles and is therefore also called the Envelope Function.

In the equation $2\beta\beta'' - \beta'^2 + 4K\beta^2 - 4 = 0$, if K (s)=0, the solution will be

$$\beta(s) = \beta_0 + \frac{(s-s_0)^2}{\beta_0} ,$$ (49)

where β_0 and s_0 are the initial values with K (s) = 0. For example, at the interaction point of a collider, K (s) = 0, we can calculate the beta function of the location of the first quadrupole from Eq. (49). The β function and other lattice functions define the size of the aperture the stored beam needs, while the tunes of a machine determine the stability.

The Twiss functions, α, β, γ, and the tune, μ, are local and apply to the point chosen in the period as a starting and finishing point. Each individual element, such as drift space, dipole, quadrupole, sextupole in the ring has its own transport matrix. One can first choose the starting point, s_0, where the Twiss parameters are assumed. Starting from that point and multiplying the element matrices for one turn, one can find a, b, c and d in the matrices numerically and then apply the Eq. (33) to find the Twiss matrix.

2.11 Transport Matrices for Individual Components in a Ring

Drift spaces, dipoles, and quadrupoles comprise most of the elements of a storage ring. The simplest among these is the drift space. The divergence of the trajectory and the angle of the ray have the following relation:

$$\theta = \arctan(x') .$$ (50)

It is easy to see that in phase space, a horizontal drift length is a translation from (x, x') to ($x + l\,x'$, x'). Thus the matrix for a drift space can be written as

$$\begin{pmatrix} x_2 \\ x_2' \end{pmatrix} = \begin{pmatrix} 1 & l \\ 0 & 1 \end{pmatrix}\begin{pmatrix} x_1 \\ x_1' \end{pmatrix}.$$ (51)

A thin quadrupole, with a very small length but a finite integrated gradient of

$$lk = \frac{l}{B\rho} \cdot \frac{\partial B_y}{\partial x} ,$$ (52)

could be compared with a focusing lens. A ray diverging from the focal point passes through the lens with a displacement x, becoming parallel by a deflection

$$\theta = \frac{1}{f} \cdot x \ . \tag{53}$$

Thus, the behavior of the ray can be described by a simple matrix

$$\begin{pmatrix} x_2 \\ x_2' \end{pmatrix} = \begin{pmatrix} 1, & 0 \\ -1/f, & 1 \end{pmatrix} \begin{pmatrix} x_1 \\ x_1' \end{pmatrix} \ . \tag{54}$$

For a quadrupole, moving a displacement x, a particle sees an integrated field:

$$\Delta(Bl) = l \cdot \frac{\partial B_y}{\partial x} \cdot x \ . \tag{55}$$

Then the small deflection θ is

$$\theta = \frac{\Delta(Bl)}{(B\rho)} = \frac{l}{(B\rho)} \cdot \frac{\partial B_y x}{\partial x} = lkx \ . \tag{56}$$

One can see at once that $l\,k = 1/f$. Thus the matrix of a thin quadrupole is

$$\begin{pmatrix} x_2 \\ x_2' \end{pmatrix} = \begin{pmatrix} 1, & 0 \\ -kl, & 1 \end{pmatrix} \begin{pmatrix} x_1 \\ x_1' \end{pmatrix} \ . \tag{57}$$

Normally, the quadrupoles in a synchrotron are not short compared to their focal length. Then the matrices for long quadrupoles are

$$M_F = \begin{pmatrix} \cos(l\sqrt{k}), & \dfrac{1}{\sqrt{k}}\sin(l\sqrt{k}) \\ -\sqrt{k}\sin(l\sqrt{k}), & \cos(l\sqrt{k}) \end{pmatrix} \tag{58}$$

and

$$M_D = \begin{pmatrix} \cosh(l\sqrt{k}), & \dfrac{1}{\sqrt{k}}\sinh(l\sqrt{k}) \\ -\sqrt{k}\sinh(l\sqrt{k}), & \cosh(l\sqrt{k}) \end{pmatrix} . \tag{59}$$

The dipole magnets, which bend the beam in the accelerator, can be thought of as drift spaces as a first approximation. But in a real bending magnet, one must consider the focusing effect of their ends. In a pure sector dipole, the particles that pass at a displacement x away from the center of curvature will have longer trajectories in the magnets. This is just like a quadrupole, focusing horizontally but not vertically. The matrices for a sector dipole are

$$M_x = \begin{pmatrix} \cos\theta, & \rho\sin\theta \\ -\dfrac{1}{\rho}\sin\theta, & \cos\theta \end{pmatrix}, \quad M_y = \begin{pmatrix} 1, & \rho\theta \\ 0, & 1 \end{pmatrix} . \tag{60}$$

Most dipoles are not sector magnets, but still have parallel end faces because of easier lamination in manufacturing.

2.12 Regular FODO Lattice

Arranging the quadrupoles as shown in Figure 5, we get a regular FODO lattice. The matrix for one period between mid-planes of focusing quadrupoles is

$$
M = \begin{pmatrix} 1 & 0 \\ \mp 1/(2f) & 1 \end{pmatrix} \begin{pmatrix} 1 & L \\ 0 & 1 \end{pmatrix} \begin{pmatrix} 1 & 0 \\ \pm 1/f & 1 \end{pmatrix} \begin{pmatrix} 1 & L \\ 0 & 1 \end{pmatrix} \begin{pmatrix} 1 & 0 \\ \mp 1/(2f) & 1 \end{pmatrix}
$$

$$
= \begin{pmatrix} 1 - L^2/2f^2 & 2L(1 \pm L/2f) \\ -L/2f^2(1 \mp L/2f) & 1 - L^2/2f^2 \end{pmatrix} = \begin{pmatrix} \cos\mu + \alpha\sin\mu & \beta\sin\mu \\ -\gamma\sin\mu & \cos\mu - \alpha\sin\mu \end{pmatrix}.
$$

(61)

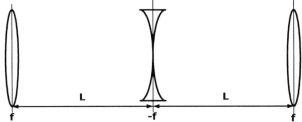

Figure 5   Regular FODO cell composed of thin lenses

Multiplying out the product of M and using Eq. (33), we obtain

$$
\begin{cases} \cos\mu = 1 - \dfrac{L^2}{2f^2} \\ \alpha_{x,y} = 0 \\ \beta = 2L\dfrac{1 \pm \sin(\mu/2)}{\sin\mu} \end{cases}
$$

(62)

where the plus sign indicates the matrix between mid-planes of focusing quadrupoles and the minus sign that of defocusing quadrupoles. The α, which means the slope of the beta function, is zero at the planes of symmetry. The ratio of the maximum and minimum β functions in one plane will be

$$
\frac{\hat{\beta}}{\check{\beta}} = \frac{1 + \sin(\mu/2)}{1 - \sin(\mu/2)}.
$$

(63)

If, more generally, we solve in the horizontal plane the matrix product for an F lens with a focal length f_1 and a D lens with a focal length f_2, we can draw a graph for the stable area of f_1 and f_2 when $\sin\mu < 1$ (plotted in Figure 6).

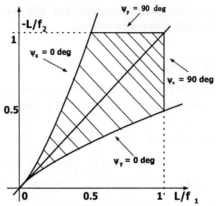

Figure 6 Stability diagram (cross hatched) for a FODO lattice. ψ is the phase advance per cell

2.13 Liouville's Theorem and Emittance

Liouville's theorem is a conservation law of phase space obeyed by particle dynamics. It is stated as (also depicted as Figure 7):

"In the vicinity of a particle, the particle density in phase space is constant if the particles move in an external magnetic field or in a general field in which the forces do not depend upon velocity."

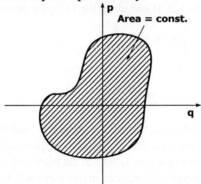

Figure 7   Illustration of Liouville's theorem

As a direct consequence of Liouville's Theorem, it is shown that, though the beam's cross section may have different shapes over the ring, the area of its phase space will not change (Figure 8).

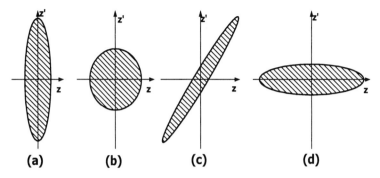

Figure 8 Development of a constant emittance beam in phase space. (a) narrow waist; (b) broad waist or maximum in beta; (c) diverging beam; (d) broad maximum at the center of an F lens

In regions where the β function is not a maximum or minimum, the ellipse of the beam phase space diagram will be tilted, as in Figure 8 (c). In Figure 9, we see that the sizes of the ellipse are related to the Twiss parameters. From Eq. (29), we can get the equation of the ellipse,

$$\beta \, z'^2 + 2\alpha \, z \, z' + \gamma \, z^2 = a^2 = \varepsilon \ , \tag{64}$$

which is often called the Courant-Snyder invariant.

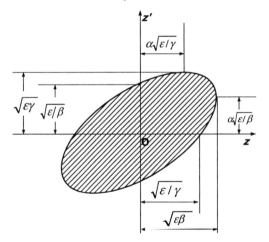

Figure 8   Parameters of a phase-space ellipse with an emittance ε at a point in the lattice between quadrupoles.

Usually, a beam of particles can be represented as a cloud of points within a closed contour, normally an ellipse in the phase space diagram. The area within the ellipse is proportional to the "emittance" of the beam. When energy is fixed, this area can be expressed as $\varepsilon = \int z' dz$ in units of $\pi \cdot$mm$\cdot$mrad. Figure 8 shows the contour at the place where the β function is at a maximum or minimum and where the major and minor axes of an upright ellipse are $\sqrt{\varepsilon\beta}$ and $\sqrt{\varepsilon/\beta}$ .The emittance is conserved, no matter how the bending and focusing magnets operate on the beam. This is a direct consequence of Liouville's theorem.

2.14 Hamiltonian in an accelerator

A dynamic system is often described with a Hamiltonian. Generally, the Hamiltonian is the sum of the kinetic energy, T, and the potential energy, V. Both T and V can be written as a function of the coordinates and canonical momenta. If q represents the coordinates, p the canonical momentum, t the independent variable or time, and $H(q, p, t)$ the Hamiltonian, the equations of motions now are

$$\frac{dq_i}{dt} = \frac{\partial H}{\partial p_i}, \quad \frac{dp_i}{dt} = -\frac{\partial H}{\partial q_i} . \tag{65}$$

With the Hamilton method applied in beam dynamics, or a relativistic charged particle in an electromagnetic field, the Hamiltonian is given by

$$H = c[m^2c^2 + (\mathbf{p} - 2\mathbf{A})^2]^{1/2} + e\varphi , \tag{66}$$

where m is the rest mass of a charged particle, $\mathbf{p}$ the momentum, c the velocity of light, e the electron charge, φ the electromagnetic potential, and $\mathbf{A}$ the magnetic vector potential.

In our curvilinear coordinate system, the Hamiltonian becomes

$$H = c[m^2c^2 + (p_s - eA_s)^2(1+\frac{x}{\rho})^{-2} + (p_x - eA_x)^2 + (p_y - eA_y)^2]^{1/2} + e\varphi , \tag{67}$$

where p_x and p_y are the projections of $\mathbf{p}$ onto the x and y directions and

$$p_s = (\mathbf{p} \cdot \hat{s})(1+\frac{x}{\rho}). \tag{68}$$

Like these canonical momenta, the canonical vector potentials A_x, A_y and A_s are also defined analogously, especially

$$A_s = (\mathbf{A} \cdot \hat{s})(1+\frac{x}{\rho}) . \tag{69}$$

As H and t are conjugate variables, a new Hamiltonian can be obtained by exchanging s and t as independent variables:

$$H = -eA_s \mp (1+\frac{x}{\rho})[(\frac{p_0 + e\varphi}{c})^2 - m^2c^2 - (p_x - eA_x)^2 - (p_y - eA_y)^2]^{1/2} , \tag{70}$$

where $p_0 = \sqrt{\mathbf{p}^2 - m^2c^4}$.

After some simplification and manipulation, we have (in the vertical plane)

$$H = -\frac{eA_s}{p} - (1 + p_y^2)^{1/2} \approx -\frac{eA_s}{p_0} + \frac{p_y^2}{2} = \frac{p_y^2}{2} + \frac{k(s)y^2}{2} \ , \tag{71}$$

where $k\,(s)$ is the normalized gradient in a quadrupole.

Applying Hamilton's equations

$$\frac{dy}{ds} = \frac{\partial H}{\partial p_y}, \qquad \frac{dp_y}{ds} = -\frac{\partial H}{\partial y} \ , \tag{72}$$

we get

$$y' = p_y \ , \qquad p_y' = -k(s)y \ . \tag{73}$$

Finally,

$$y'' + k(s)y = 0 \ , \tag{74}$$

which is the Hill's equation in the vertical plane.

3   Effect of Linear Magnet Imperfections

In a real machine, we have to deal with the errors due to the magnets' manufacture and alignment, which influence the beam performance in a storage ring. Energy differences among charged particles cause chromatic effects, which need to be corrected in the operation of a synchrotron. Extra apparatuses, such as detectors, bring in coupling in a collider and lead to luminosity degradation. In this section, we discuss mainly the effects due to imperfections in the magnets.

3.1   Errors from dipoles

In an ideal machine, the dipoles around the ring bend the beam exactly 2π in one turn. Such a beam path is called the central (or synchrotron) momentum closed orbit. But in reality, since even the best synchrotron magnets cannot be manufactured to be absolutely identical and to have perfect fields, the central orbit of the beam is distorted by the imperfections of the bending magnets. Therefore, the ideal or design orbit is not possible, but an orbit with a small distortion is possible and is called the distorted orbit. Table 1 lists the main imperfections in a synchrotron that cause the closed-orbit distortion.

In addition, other machine imperfections such as survey errors, resolution and ripple of power supplies, movement or oscillation of the ground, etc., also influence the central beam orbit; all of these are assumed to be randomly distributed around the ring.

Table 1. Main sources of closed-orbit distortion

Element	Kick source	RMS value	$\langle \Delta Bl/(B\rho)\rangle_{rms}$	Direction
Drift space ($l = d$)	Stray field	$\langle \Delta B_s\rangle$	$d\ \langle\Delta B_s\rangle/(B\rho)$	x, y
Dipole (angle = θ)	Field error	$\langle \Delta B/B\rangle$	$\theta\ \langle\Delta B/B\rangle$	x
Dipole	Tilt	$\langle \Delta \rangle$	$\theta\langle\Delta\rangle$	y
Quadrupole ($K\,l$)	Displacement	$\langle \Delta x, y\rangle$	$K\,l\,\langle\Delta x, y\rangle$	x, y

Suppose the perturbation is $f(s)$, which can be added on the right side of Hill's equation as

$$\frac{d^2z}{ds^2} + K(s)z = f(s) \ , \tag{75}$$

the orbit distortion will be

$$z(s) = \frac{\sqrt{\beta(s)}}{2\sin\pi v}\oint \sqrt{\beta(\bar{s})}f(\bar{s})\cos[\psi(s) - \psi(\bar{s}) - \pi v]d\bar{s} \ , \tag{76}$$

where $\psi(s) = \int_0^s \frac{ds}{v\beta}$. From Eq. (76), we see that the tune v must avoid integers, i.e., $v \neq n$, $n \in J$. Here we omit the procedure of getting the Eq. (76).

3.2 Gradient errors

Like dipoles, quadrupoles also can have many errors in manufacturing, surveying, etc. The gradient errors change the focusing strength $K(s)$, causing the variation of beta functions and transverse tunes.

Suppose there is a gradient error $k(s)$ at $s = 0$ with a small length Δs. When an electron passes the point $s = 0$, it will feel a kick of $\Delta x' = k(s)\cdot\Delta s \cdot x$. With the perturbation, Hill's equation then becomes

$$\frac{d^2z}{ds^2} + [K(s) + k(s)]z = 0 \ . \tag{77}$$

If the matrix form of the perturbation is written as $\begin{pmatrix} 1 & 0 \\ \delta & 1 \end{pmatrix}$ where $\delta = k(s)\Delta s, |k(s)\Delta s| \ll 1$, and the transport matrix for one turn with no perturbation is

$$M_0 = \begin{pmatrix} \cos\mu_0 + \alpha_0\sin\mu_0 & \beta_0\sin\mu_0 \\ -\gamma_0\sin\mu_0 & \cos\mu_0 - \alpha_0\sin\mu_0 \end{pmatrix} \ , \tag{78}$$

then the matrix for one turn with perturbation can be described as

$$M = \begin{pmatrix} 1 & 0 \\ \delta & 1 \end{pmatrix} \begin{pmatrix} \cos\mu_0 + \alpha_0 \sin\mu_0 & \beta_0 \sin\mu_0 \\ -\gamma_0 \sin\mu_0 & \cos\mu_0 - \alpha_0 \sin\mu_0 \end{pmatrix}$$
$$= \begin{pmatrix} \cos\mu + \alpha\sin\mu & \beta\sin\mu \\ -\gamma\sin\mu & \cos\mu - \alpha\sin\mu \end{pmatrix}. \tag{79}$$

Thus we can get

$$\frac{1}{2}\mathrm{Tr}\,(M) = \cos\mu = \frac{1}{2}\beta_0\delta\sin\mu_0 + \cos\mu_0 . \tag{80}$$

As $\cos\mu = \cos(\mu_0 + \Delta\mu) \approx \cos\mu_0 - \Delta\mu\sin\mu_0$, and $\Delta\mu = 2\pi\Delta\nu$, we have

$$\Delta\nu = -\frac{1}{4\pi}\beta_0\delta = -\frac{1}{4\pi}\beta_0 k(s)\Delta s . \tag{81}$$

Integrating for one turn, the tune shift due to gradient errors can be expressed as

$$\Delta\nu = -\frac{1}{4\pi}\oint \beta(s)k(s)ds . \tag{82}$$

Next, we will consider the variation of beta function caused by the gradient error $k(s)$. If a particle starts from s_2, and after passing s_1, it returns to s_2 for a revolution, then the whole path, which the particle covers, can be seen as $s_2 \to s_1$ (B) and $s_1 \to s_2$ (A). If there is no error, the matrix is

$$M_0 = AB = \begin{pmatrix} a_{11} & a_{12} \\ a_{21} & a_{22} \end{pmatrix} \begin{pmatrix} b_{11} & b_{12} \\ b_{21} & b_{22} \end{pmatrix} = \begin{pmatrix} a_{11}b_{11} + a_{12}b_{21} & a_{11}b_{12} + a_{12}b_{22} \\ a_{21}b_{11} + a_{22}b_{21} & a_{21}b_{12} + a_{22}b_{22} \end{pmatrix}. \tag{83}$$

If there is a perturbation δ whose matrix is $M_\delta = \begin{pmatrix} 1 & 0 \\ \delta & 1 \end{pmatrix}$ with $|\delta| \ll 1$,

then the matrix for the whole ring will be

$$M = \begin{pmatrix} m_{11} & m_{12} \\ m_{21} & m_{22} \end{pmatrix} = \begin{pmatrix} a_{11} & a_{12} \\ a_{21} & a_{22} \end{pmatrix} \begin{pmatrix} 1 & 0 \\ \delta & 1 \end{pmatrix} \begin{pmatrix} b_{11} & b_{12} \\ b_{21} & b_{22} \end{pmatrix}$$
$$= \begin{pmatrix} a_{11}b_{11} + a_{12}b_{11}\delta + a_{12}b_{21} & a_{11}b_{12} + a_{12}b_{12}\delta + a_{12}b_{22} \\ a_{21}b_{11} + a_{22}b_{11}\delta + a_{22}b_{21} & a_{21}b_{12} + a_{22}b_{12}\delta + a_{22}b_{22} \end{pmatrix} \tag{84}$$
$$= \begin{pmatrix} m_{110} + a_{12}b_{11}\delta & m_{120} + a_{12}b_{12}\delta \\ m_{210} + a_{22}b_{11}\delta & m_{220} + a_{22}b_{12}\delta \end{pmatrix},$$

if $M_0 = \begin{pmatrix} m_{110} & m_{120} \\ m_{210} & m_{220} \end{pmatrix}$ and $\begin{cases} m_{11} = m_{110} + \Delta m_{11}, & m_{12} = m_{120} + \Delta m_{12} \\ m_{21} = m_{210} + \Delta m_{21}, & m_{22} = m_{220} + \Delta m_{22} \end{cases}$. Thus we

get

$$\Delta m_{12} = m_{12} - m_{120} = a_{12}b_{12}\delta . \tag{85}$$

As the matrix from s_1 to s_2 without any perturbation can be described as

$$M = \begin{pmatrix} \sqrt{\beta_2/\beta_1}[\cos(\mu_2 - \mu_1) + \alpha_1\sin(\mu_2 - \mu_1)] & \sqrt{\beta_1\beta_2}\sin(\mu_2 - \mu_1) \\ \dfrac{-1}{\sqrt{\beta_1\beta_2}}[(1+\alpha_1\alpha_2)\sin(\mu_2 - \mu_1) + (\alpha_2 - \alpha_1)\cos(\mu_2 - \mu_1)] & \sqrt{\dfrac{\beta_1}{\beta_2}}[\cos(\mu_2 - \mu_1) - \alpha_2\sin(\mu_2 - \mu_1)] \end{pmatrix}$$

(86)

we can get

$$\begin{cases} a_{12} = \sqrt{\beta_1\beta_2}\,\sin(\mu_2 - \mu_1) \\ b_{12} = \sqrt{\beta_1\beta_2}\,\sin[\mu_0 - (\mu_2 - \mu_1)] \end{cases},$$

(87)

where $\mu_0 = 2\pi\nu$ is the phase advance for one revolution. Combining Eqs. (85) and (86), we have

$$\Delta m_{12} = \delta\beta_1\beta_2\sin(\mu_2 - \mu_1)\sin[\mu_0 - (\mu_2 - \mu_1)] .$$

(88)

Since $m_{12} = \beta_2\sin\mu_0$, the differentiation of m_{12} will be

$$\Delta m_{12} = \Delta(\beta_2\sin\mu_0) = \Delta\beta_2\sin\mu_0 + \beta_2\cos\mu_0\cdot\Delta\mu_0 .$$

(89)

From the equation of tune shift due to gradient errors, Eq. (82), and equating Eq. (88) to Eq. (89), we get

$$\Delta\beta_2 = \frac{1}{2\sin 2\pi\nu}\cdot\delta\beta_1\beta_2\cos[2\pi\nu - 2(\mu_2 - \mu_1)] .$$

(90)

Integrating $\Delta\beta$ along the whole ring, the final expression of the β function variation is

$$\Delta\beta(s) = \frac{\beta(s)}{2\sin 2\pi\nu}\oint\beta(\bar{s})k(\bar{s})\cos[2\pi\nu - 2(\mu(s) - \mu(\bar{s}))]d\bar{s}$$

(91)

From Eq. (91), we know that the transverse tunes must avoid half integers, otherwise $\Delta\beta$ will approach infinity. This implies the resonance lines

$$\left.\begin{array}{ll} 2\nu_x = p, & 2\nu_y = p \\ \nu_x = \nu_y = p, & \nu_x + \nu_y = p \end{array}\right\}.$$

(92)

3.3   Working diagram and multipole field

Up to now, we know that the perturbation due to dipoles is independent of the transverse displacement, and that the tunes should not cross the integer lines, which would cause resonances. The gradient errors, which have fields proportional to the displacement, will enhance the resonance when the tunes approach half integers. Thus we can deduce that the sextupole errors will excite third-integer resonances, and the transverse tunes should avoid 1/3 integers, which have the following forms:

$$3v_x = p, \qquad 3v_y = p$$
$$\left.\begin{array}{ll} 2v_x + v_y = p, & v_x + 2v_y = p \\ 2v_x - v_y = p, & v_x - 2v_y = p \end{array}\right\}, \quad p \in J. \qquad (93)$$

We may further think that the fourth-order resonance is driven by octupoles, and so on. Then, a diagram with v_x and v_y as its axes can be drawn, on which the mesh of lines marks danger zones for the particles. Such a diagram is called a "working diagram" and the transverse tunes are "working points." Generally, the working points should satisfy

$$l v_x + m v_y \neq p \qquad (94)$$

where l, m, p are integers. If Eq. (94) is not true, the resonance of $|l| + |m|$ order will be driven, while p is the azimuthal frequency. Figure 9 shows the working diagram for the BEPC storage ring up to fifth-order lines.

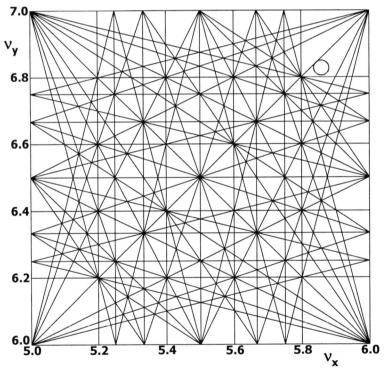

Figure 9 Working diagram of the BEPC storage ring (tunes locate in the red circle)

Summarizing the field errors, which drive the resonances, we can get the magnetic vector potential of a magnet with $2n$ poles in Cartesian coordinates as

$$A = \sum_n A_n f_n(x, y) \qquad (95)$$

72

where f_n is a homogeneous function in x and y of order n. Table 2 lists $f_n(x, y)$ for low-order multipoles with both regular and skew terms. Figure 10 shows the difference between regular and skew multipoles.

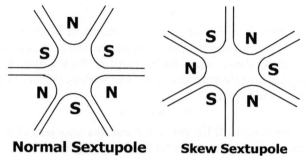

Normal Sextupole   **Skew Sextupole**

Figure 10 Normal and skew sextupoles

The function $f_n(x, y)$ can be expanded in the form of

$$f_n(x, y) = (x + iy)^n \qquad (96)$$

where the real terms stand for regular multipoles, and the imaginary parts correspond to skew ones.

Table 2 Solutions of magnetic vector potential in Cartesian coordinates.

Multipole	n	Regular f_n	Skew f_n
Quadrupole	2	$x^2 - y^2$	$2xy$
Sextupole	3	$x^3 - 3xy^2$	$3x^2y - y^3$
Octupole	4	$x^4 - 6x^2y^2 + y^4$	$4x^3y - 4xy^3$
Decapole	5	$x^5 - 10x^3y^2 + 5xy^4$	$5x^4y - 10x^2y^3 + y^5$
Dodecapole	6	$x^6 - 15x^4y^2 + 15x^2y^4 + y^6$	$6x^5y + 20x^3y^3 + 6xy^5$

Furthermore, the scalar potential $\phi(r, \theta)$ in Cartesian and polar coordinates obeys the Laplace equation

$$\frac{\partial^2 \phi}{\partial x^2} + \frac{\partial^2 \phi}{\partial y^2} = 0 \quad \text{and} \quad \frac{1}{r}\frac{\partial}{\partial r}\left(r\frac{\partial \phi}{\partial r}\right) + \frac{1}{r^2}\frac{\partial^2 \phi}{\partial \theta^2} = 0 \qquad (97)$$

which has the solution

$$\phi = \sum_{n=1}^{\infty} \phi_n r^n \sin n\theta \qquad (98)$$

in polar coordinates.

As the fields in polar coordinates have the expressions

$$B_r = \frac{\partial \phi}{\partial r}, \qquad\qquad B_\theta = \frac{1}{r}\frac{\partial \phi}{\partial \theta}$$

(99)

or $\qquad B_r = \sum_{n=1}^{\infty} \phi_n n r^{n-1} \sin n\theta, \qquad B_\theta = \sum_{n=1}^{\infty} \phi_n n r^{n-1} \cos n\theta,$

we can get the vertical field (when $y = 0$)

$$B_y = B_r \sin \theta + B_\theta \cos \theta$$

$$= \sum_{n=1}^{\infty} \phi_n n r^{n-1} [\cos \theta \cos n\theta + \sin \theta \sin n\theta] \qquad (100)$$

$$= \sum_{n=1}^{\infty} \phi_n n r^{n-1} \cos(n-1)\theta = \sum_{n=1}^{\infty} \phi_n n x^{n-1}.$$

The Taylor series of multipoles then can be expanded as

$$B_y = \phi_1 + \phi_2 \cdot 2x + \phi_3 \cdot 3x^2 + \phi_4 \cdot 4x^3 + \cdots$$

$$= B_0 + \frac{1}{1!}\frac{\partial B_y}{\partial x} x + \frac{1}{2!}\frac{\partial^2 B_y}{\partial x^2} x^2 + \frac{1}{3!}\frac{\partial^3 B_y}{\partial x^3} x^3 + \cdots . \qquad (101)$$

The first term of Eq. (101) represents dipole, the second quadrupole, the third sextupole, the fourth octupole, etc. Some multipole field shapes are drawn in Figure 11.

In general, multipoles can drive resonances of lower order. For instance, octupoles drive fourth- and second-order, sextupoles third- and first-order, etc. The nonlinear resonances are those of third-order and above, driven by nonlinear multipoles, which can reduce the dynamic aperture severely, especially in a superconducting machine.

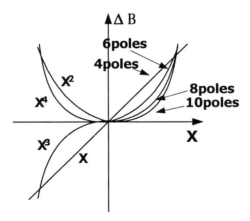

Figure 11 Multipole field shapes

4　Off Momentum Orbit

In a real machine, the energy of a particle may be deviated from the nominal value when the particle moves around the ring. Such an oscillation is called energy oscillation, or synchrotron oscillation. The particle with energy different from the nominal one is called "off-momentum particle". Normally, this oscillation is very small, affecting only the transverse motion. In this section, we will discuss the effect of the off-momentum on the particle closed orbit.

4.1　Dispersion function

Suppose the trajectory of the particle is on the horizontal plane. So the vertical motion of particles is still the betatron oscillation. In the horizontal plane, the displacement of a particle should be written as

$$x = x_\beta + x_c + x_D ,$$ (102)

where x_β is the displacement of betatron oscillation, x_c the orbit distortion, and x_D the displacement due to energy oscillation.

As the energy is $E = E_0 + \varepsilon$, where E_0 is the nominal energy and ε the deviation, we can have

$$x_D = D_x(s)\frac{\varepsilon}{E_0} ,$$ (103)

where $D_x(s)$ is defined as the dispersion function. The dispersion function means that a particle without any betatron oscillation or orbit distortion moves along a path which is proportional to ε/E_0 and the ratio is $D_x(s)$.

Consider the Hill's equation again. We add the energy deviation on the r.h.s. of the equation as (in horizontal plane)

$$x'' + K_x x = \frac{1}{\rho}\frac{\Delta p}{p_0} .$$ (104)

Neglecting the orbit distortion, since $x = x_\beta + x_D$ and $x_D = D_x\frac{\Delta p}{p_0}$, we get

$$D_x'' + K_x D_x = \frac{1}{\rho} .$$ (105)

After the particle has moved one turn from any point, the matrix equation becomes

$$\begin{pmatrix} x_D \\ x_D' \\ \Delta p/p \end{pmatrix} = \begin{pmatrix} m_{11} & m_{12} & m_{13} \\ m_{21} & m_{22} & m_{23} \\ 0 & 0 & 1 \end{pmatrix} \begin{pmatrix} x_D \\ x_D' \\ \Delta p/p_0 \end{pmatrix} .$$ (106)

From $x_D = D_x\frac{\Delta p}{p_0}$ and $x_D' = D_x'\frac{\Delta p}{p_0}$, we obtain

$$\begin{pmatrix} D_x \\ D'_x \\ 1 \end{pmatrix} = \begin{pmatrix} m_{11} & m_{12} & m_{13} \\ m_{21} & m_{22} & m_{23} \\ 0 & 0 & 1 \end{pmatrix} \begin{pmatrix} D_x \\ D'_x \\ 1 \end{pmatrix}. \tag{107}$$

It is easy to find

$$\begin{cases} D_x = m_{11}D_x + m_{12}D'_x + m_{13} \\ D'_x = m_{21}D_x + m_{22}D'_x + m_{23} \end{cases}, \tag{108}$$

and, finally, the dispersion functions are solved as

$$\begin{cases} D_x = \dfrac{m_{13}(1-m_{22}) + m_{12}m_{23}}{2(1-\cos\mu)} \\ D'_x = \dfrac{m_{23}(1-m_{11}) + m_{21}m_{13}}{2(1-\cos\mu)} \end{cases}, \tag{109}$$

where $\cos\mu = (m_{11}+m_{22})/2$.

With this formula, we can write down a matrix including the new elements m_{13} and m_{23} for all the magnetic elements of a period just as we did in the case of 2×2 matrices. Selecting a starting point, at azimuth s, it is simple to multiply all the matrices together until we move exactly one period onward.

Example:

A FODO cell shown in Figure 12 can be represented by

$$\text{CELL} = \{\frac{1}{2}\text{QF}, \ \text{B}, \ \text{QD}, \ \text{B}, \ \frac{1}{2}\text{QF}\}, \tag{110}$$

where QF and QD are focusing and defocusing quadrupoles, and B the bending magnet.

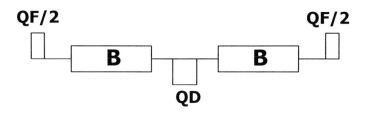

Figure 12 FODO cell with bending magnets

If we apply the thin-lens approximation, and assume L to be the length of the half cell, θ the bending angle of the half cell, and f the focal length of the quadrupoles, we obtain

$$M = \begin{pmatrix} 1 & 0 & 0 \\ -\dfrac{1}{2f} & 1 & 0 \\ 0 & 0 & 1 \end{pmatrix} \begin{pmatrix} 1 & L & \dfrac{L\theta}{2} \\ 0 & 1 & \theta \\ 0 & 0 & 1 \end{pmatrix} \begin{pmatrix} 1 & 0 & 0 \\ -\dfrac{1}{f} & 1 & 0 \\ 0 & 0 & 1 \end{pmatrix} \begin{pmatrix} 1 & L & \dfrac{L\theta}{2} \\ 0 & 1 & \theta \\ 0 & 0 & 1 \end{pmatrix} \begin{pmatrix} 1 & 0 & 0 \\ -\dfrac{1}{2f} & 1 & 0 \\ 0 & 0 & 1 \end{pmatrix}$$

$$(111)$$

Repeating the procedure of Eqs. (106) and (107), we find

$$\begin{pmatrix} D_x \\ D'_x \\ 1 \end{pmatrix} = \begin{pmatrix} 1 - \dfrac{L^2}{2f^2} & 2L(1+\dfrac{L}{2f}) & 2L\theta(1+\dfrac{L}{4f}) \\ -\dfrac{L}{2f^2}+\dfrac{L^2}{4f^3} & 1-\dfrac{L^2}{2f^2} & 2\theta(1-\dfrac{L}{4f}-\dfrac{L^2}{8f^2}) \\ 0 & 0 & 1 \end{pmatrix} \begin{pmatrix} D_x \\ D'_x \\ 1 \end{pmatrix} \quad (112)$$

Here, D_x and D'_x are the dispersion function and its derivative at the quadrupole's location. Under the thin-lens approximation, we have

$$\sin\frac{\mu}{2} = \frac{L}{2f}, \quad \beta_F = \frac{2L(1+\sin\mu/2)}{\sin\mu}, \quad \beta_D = \frac{2L(1-\sin\mu/2)}{\sin\mu}, \quad \alpha_F = \alpha_D = 0$$

$$(113)$$

Thus, we can get the dispersion and its derivative at the location of quadrupoles expressed as

$$D_F = \frac{L\theta(1+\dfrac{1}{2}\sin\dfrac{\mu}{2})}{\sin^2\dfrac{\mu}{2}}, \quad D'_F = 0$$

$$(114)$$

$$D_D = \frac{L\theta(1-\dfrac{1}{2}\sin\dfrac{\mu}{2})}{\sin^2\dfrac{\mu}{2}}, \quad D'_D = 0$$

From this example, we can conclude that in the FODO cell,

1). The dispersion function at the focusing quadrupole is bigger than that at the defocusing quadrupole, with the ratio of $\dfrac{2+\sin(\mu/2)}{2-\sin(\mu/2)}$,

2). The dispersion function is proportional to the product of length of half cell and the bending angle of half cell, i.e., $D_x \propto L\theta$,

3). When the phase advance is small, the dispersion function is proportional to the inverse quadratic power of the phase advance.

The dispersion function of the lattice of the BEPC storage ring is shown in Figure 13.

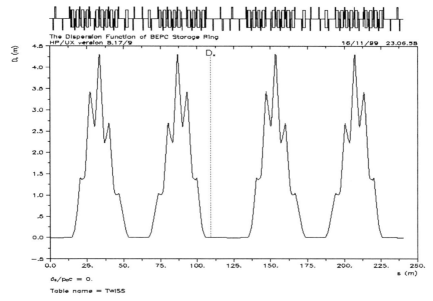

Figure 13 Dispersion function of the BEPC storage ring

Similarly to the orbit distortion equation, at point s, the dispersion function $D_x(s)$ can be deduced by replacing

$$\frac{\Delta B}{B} \rightarrow \frac{\Delta p}{p_0}\frac{1}{\rho}, \qquad z_{COD} \rightarrow D_x(s)\frac{\Delta p}{p_0} , \qquad (115)$$

then we have

$$D_x(s) = \frac{\sqrt{\beta(s)}}{2\sin\pi\nu}\oint\frac{\sqrt{\beta(\bar{s})}}{\rho(\bar{s})}\cos(|\psi(\bar{s}) - \psi(s)| - \pi\nu)d\bar{s} . \qquad (116)$$

From the above description, we know that the change in path length with momentum must depend upon the dispersion function. Thus the closed orbit will have a mean radius:

$$R = R_0 + \overline{D}_x\frac{\Delta p}{p_0} . \qquad (117)$$

The phase stability we used in previous sections argues that a particle, arriving late due to its lower energy, would see a high RF voltage from the rising waveform and, accelerated to a higher velocity, would catch up with the synchronous particle. But after the dispersion function being introduced, we find that giving the errant particle more energy will speed it up, but still may send it on an orbit of larger orbit, shown as the above equation. For the particle close to the velocity of light where acceleration can increase momentum but not velocity, the longer path length will more than cancel the small effect of velocity. The particle, instead of catching up with its synchronous partners, it will arrive even later than it did on the previous turn. This seems to violate the whole idea of phase stability.

Depending on how a synchrotron is designed and which kind of particles it accelerates, there is a certain energy where the idea of phase stability will break down. This is called the transition energy. There is also a way of ensuring stability above transition.

4.2 $\mathcal{H}$- function

Similar to the Courant-Snyder invariant, the dispersion $\mathcal{H}$-function is defined as

$$\mathcal{H}(D_x, D_x') = \gamma_x D_x^2 + 2\alpha_x D_x D_x' + \beta_x D_x'^2 = \frac{1}{\beta_x}[D_x^2 + (\beta_x D_x' + \alpha_x D_x)^2]$$

(118)

It is easy to find that the $\mathcal{H}$-function is invariant, because the dispersion function satisfies the homogeneous betatron equation of motion in the dipole-free $(1/\rho = 0)$ regions. But in the regions with dipoles, the $\mathcal{H}$-function is not invariant. In FODO cell, the dispersion $\mathcal{H}$-function is larger at the defocusing quadrupole than at the focusing quadrupole, i.e., $\mathcal{H}_F \le \mathcal{H}_D$, and

$$\mathcal{H}_D = \frac{L\theta^2 \sin\mu(1 - \frac{1}{2}\sin\frac{\mu}{2})^2}{2(1 - \sin\frac{\mu}{2})\sin^4\frac{\mu}{2}}, \qquad \mathcal{H}_F = \frac{L\theta^2 \sin\mu(1 + \frac{1}{2}\sin\frac{\mu}{2})^2}{2(1 + \sin\frac{\mu}{2})\sin^4\frac{\mu}{2}},$$

(119)

which are proportional to the inverse cubic power of the phase advance.

The dispersion $\mathcal{H}$-function is very important in the lattice matching, especially in the dedicated synchrotron radiation sources, in which the dispersion-free insertion is the most basic component in the lattice.

4.3 Momentum compaction factor

When a particle has a momentum deviation, $\Delta p/p_0 > 0$, it will pass the longer path than those particles with $\Delta p = 0$, as shown in Figure 14.

If the path length for the synchronous particle moving along the storage ring is C_0, the path length for the non-synchronous particle for one turn will be

$$C = \oint \sqrt{(1 + x/\rho)^2 + x'^2 + y'^2}\, ds \approx C_0 + \oint \frac{x}{\rho} ds + \cdots.$$

(120)

Thus, the deviation of the total path length for an off-momentum particle from that of the on-momentum closed orbit is expressed as

$$\Delta C = C - C_0 = \oint \frac{x}{\rho} ds \ . \tag{121}$$

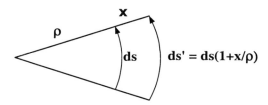

Figure 14 Paths of off-momentum and on-momentum particles

Since the difference in path lengths comes from the different momenta between off- and on-momentum particles, the x in Eq. (121) can be given by

$$x = D_x(s) \frac{\Delta p}{p_0} \ . \tag{122}$$

Then, we have

$$\frac{\Delta C}{C_0} = \frac{1}{C_0} \oint \frac{D_x(s)}{\rho(s)} ds \cdot \frac{\Delta p}{p_0} = \alpha_p \frac{\Delta p}{p_0} \ . \tag{123}$$

Here, we define the momentum compaction factor as given by

$$\alpha_p = \frac{\Delta C}{C_0} \bigg/ \frac{\Delta p}{p_0} = \frac{1}{C_0} \oint \frac{D_x(s)}{\rho(s)} ds \ . \tag{124}$$

The momentum compaction factor describes the deviation of particle trajectories due to different energy. It is proportional to the dispersion function.

Example:

For a thin-lens FODO cell, if L and θ are the length and bending angle of one half cell, the momentum compaction factor will be

$$\alpha_p \approx \frac{(D_F + D_D)\theta}{2L} \approx \frac{\theta^2}{\sin^2(\mu/2)} \ , \tag{125}$$

where μ is the phase advance of a FODO cell, and v_x the horizontal tune.

Normally, for a storage ring, there will be

$$\alpha_p \approx \frac{1}{v_x^2} \ . \tag{126}$$

4.4   Transition energy and the phase slip factor

Particles with different momenta travel along different paths in a storage ring. The revolution period of a particle is given by

$$T = C/v \ , \tag{127}$$

where C is the circumference of a storage ring and v the speed of the circulating particle. Then, the fractional change of the revolution time is

$$\frac{\Delta T}{T_0} = -\frac{\Delta f}{f_0} = \frac{\Delta C}{C_0} - \frac{\Delta v}{v} = (\alpha_p - \frac{1}{\gamma^2})\frac{\Delta p}{p_0},$$

(128)

where f_0 is the revolution frequency. The phase slip factor is defined as

$$\eta = \alpha_p - \frac{1}{\gamma^2} = \frac{1}{\gamma_t^2} - \frac{1}{\gamma^2}.$$

(129)

Here, the transition energy, γ_t is given by

$$\gamma_t \equiv \sqrt{1/\alpha_p}.$$

(130)

So, we have

$$\frac{\Delta T}{T_0} = -\frac{\Delta f}{f_0} = \eta\frac{\Delta p}{p_0} = \eta\delta.$$

(131)

For a FODO cell lattice, $\gamma_t \approx \nu_x$.

Above the transition energy, i.e., $\gamma > \gamma_t$ and $\eta > 0$, particle with higher momentum will have a longer revolution period than that of the synchronous particle. Thus, the higher energy particle will arrive at a fixed location in storage ring later than the synchronous one. Below the transition energy, i.e., $\gamma < \gamma_t$ and $\eta < 0$, the higher momentum particle will travel faster and have a shorter revolution period than the synchronous particle, and arrive at a fixed location earlier than the synchronous particle. At $\gamma = \gamma_t$, or $\eta = 0$, the revolution period is independent of the particle momentum. All particles at different momenta move around the storage ring in an equal revolution frequency. This is called isochronous condition.

Normally, high energy electron storage rings work above transition energy. Some proton and heavy ion machines operate below transition energy or with transition jump.

5   Chromaticity

5.1   Definition and source of chromaticity

In an ideal machine, particles move along the ring with identical energy, as described in Hill's equation. But normally there are momentum differences among particles, which will cause tune spread for the beam. As the tune spread due to momentum is equivalent to the chromatic aberration in an optical lens, it is called "chromaticity," which is expressed as

$$\xi = \frac{\Delta v / v}{\Delta p / p_0} \quad \text{or} \quad \xi = \frac{\Delta v}{\Delta p / p_0} \,. \tag{132}$$

Other imperfections, such as the non-uniform guide field, the sextupole component in magnetic field, $dB_y/dt \neq 0$, which causes the ripple effect in the vacuum chamber, and the saturation of bending magnets, will also introduce chromaticity into the lattice. The chromaticity due to $\Delta p/p_0$ is called natural chromaticity.

When $\Delta p/p_0 \neq 0$, Hill's equation becomes

$$z'' + Kz + \delta kz = 0 \tag{133}$$

where $K = \frac{1}{B\rho}\left(\frac{\partial B_y}{\partial x}\right)_0$ and $\delta k = \frac{\Delta p}{p_0} K$. Applying the equation derived above

in the section on gradient errors, the result of tune spread is

$$\Delta v = -\frac{1}{4\pi}\oint \beta(s)\delta k(s)ds = -\frac{1}{4\pi}\oint \beta(s)K(s)ds \cdot \frac{\Delta p}{p_0} \,. \tag{134}$$

Thus the natural chromaticity is

$$\xi = -\frac{1}{4\pi v}\oint \beta(s)K(s)ds \quad \text{or} \quad \xi = -\frac{1}{4\pi}\oint \beta(s)K(s)ds \,. \tag{135}$$

From the definition of natural chromaticity, Eq. (135), we find that the natural chromaticity remains negative and it comprises most of the chromaticity in a storage ring without any chromaticity correction.

Example:

A lattice composed of N FODO cells, the natural chromaticity is

$$\xi \approx -\frac{N}{\pi}\tan\frac{\mu}{2} \,, \tag{136}$$

where μ is the phase advance of a FODO cell.

For the BEPC storage ring, the natural chromaticities in the horizontal and vertical planes are $\xi_x \sim -10$ and $\xi_y \sim -20$, respectively in the collision mode.

In some other cases, such as injection, the momentum deviation can be as high as $\pm 2\times10^{-3}$, which may cause a large tune spread and make it difficult to inject the beam. Another effect of negative chromaticity is head-tail instability which will cause beam loss. Thus the chromaticity must be corrected to be positive.

5.2   Chromaticity correction

One way to correct the chromaticity is to introduce extra focusing which is stronger for high momentum orbits far from the central axes. Such a magnet is a sextupole, shown in Figure 15.

For a normal sextupole, the field equation is

$$\begin{cases} B_y = \lambda(x^2 - y^2) \\ B_x = 2\lambda xy \end{cases}, \tag{137}$$

where $\lambda = \dfrac{1}{2}\dfrac{\partial^2 B_y}{\partial x^2} = \dfrac{1}{2}B''$. The equation of motion of the particles passing through the sextupole can be written as

$$\begin{cases} x'' + r(x^2 - y^2) = 0 \\ y'' - 2rxy = 0 \end{cases}. \tag{138}$$

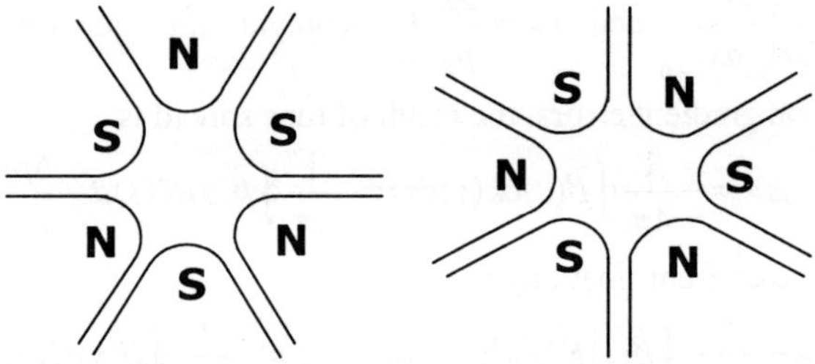

Figure 15 Sextupole magnets (Left: normal sextupole, Right: skew sextupole)

Here, $r = \dfrac{1}{|B\rho|}\lambda$. Thus, sextupoles placed in the nonzero dispersion locations will introduce a focusing

$$\delta k = -\frac{B''D_x}{B\rho} \cdot \frac{\Delta p}{p_0} \tag{139}$$

where D is the dispersion function. This perturbation can cause a tune shift of

$$\Delta \nu_x = \frac{1}{4\pi}\oint \beta_x(s) \cdot \frac{B''(s)D_x(s)}{B\rho}ds \cdot \frac{\Delta p}{p_0} = \frac{1}{4\pi}\oint \beta_x(s)2r(s)D_x(s)ds\frac{\Delta p}{p_0} \tag{140}$$

in horizontal plane, and the same expression in vertical plane. Thus, the chromaticity of a storage ring after introducing sextupoles will be

$$\xi_x = -\frac{1}{4\pi\nu_x}\oint \beta_x[K_x(s) - 2r(s)\eta_x(s)]ds$$
$$\xi_y = -\frac{1}{4\pi\nu_y}\oint \beta_y[K_y(s) + 2r(s)\eta_x(s)]ds \tag{141}$$

We can simply correct chromaticity with Eq. (141) to make it balance the natural chromaticity given by Eq. (135) and other chromaticities.

Sextupoles used for chromaticity correction are called chromatic sextupoles. Similar to quadrupoles, when a sextupole brings positive chromaticity in one direction, negative chromaticity is created in another direction, and vice versa.

Thus, we generally need two sets of sextupoles to correct horizontal and vertical chromaticities. One is near focusing quadrupoles where β_x is large, to correct the horizontal chromaticity, while the other is near defocusing quadrupoles where β_y is large to correct the vertical chromaticity.

For colliders or low emittance storage rings like synchrotron light source, chromatic sextupoles are arranged in families located in the arcs composed of FODO-cell or DBA/TBA-cell. Low β values in these lattices bring about large chromaticity so that strong sextupoles are needed to correct chromaticity. Normally, the simple chromatic correction scheme with two families of sextupoles is not able to correct the higher order chromatic effects, due to the large widths of systematic half integer stopband. We can introduce another two families of sextupoles to correct these higher order chromatic effects. More details on the chromatic correction are given in [4].

The introduction of sextupoles will cause some effects on stored beams, such as damping some instabilities, generating linear coupling, exciting third order resonance, bringing about non-linearity, etc.

6 Linear Coupling

6.1 Definition

Coupling, i.e. periodic energy exchange between two oscillators, exists widely in physics. Figure 16 shows an amplitude-coupled pendulum system, in which a pair of pendulums is linked by a weightless spring.

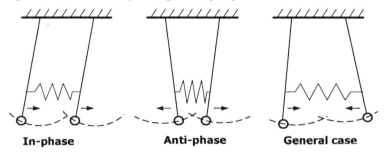

In-phase **Anti-phase** **General case**

Figure 16 Example of coupling: amplitude-coupled pendulums

In a real storage ring, the betatron motions are also coupled. The sources of coupling are mainly skew quadrupole fields and solenoids. The skew quadrupole fields arise from quadrupole rolls, fringe field of the Lambertson magnet for injection, vertical closed-orbit in sextupoles, horizontal closed-orbit in skew quadrupoles, and feed-downs from higher-order multipoles. Solenoids, which induce a longitudinal magnetic field and cause linear coupling, are used widely in the high energy detector at the interaction point of a collider, and in electron cooling storage rings.

In the operation of synchrotrons, linear betatron coupling can have both good and bad effects. On one hand, the dilation of the vertical emittance in electron rings will increase the Touschek lifetime limitation, but on the other hand, the dynamic aperture of particle motion may be reduced.

6.2 Compensation of linear coupling

In the following section, we will briefly introduce linear coupling combined with coupling compensation in the Beijing Electron Positron Collider (BEPC).

Generally, when linear coupling exists, the transverse motion equation for particles in a storage ring can be approximated to first order as

$$\begin{cases} x'' + K_x x = -(K + \dfrac{M'}{2})y - My' \\ y'' + K_y y = -(K - \dfrac{M'}{2})x + Mx' \end{cases} \tag{142}$$

where

$$K_x = \frac{1}{\rho^2} + \frac{1}{B\rho}\left(\frac{\partial B_y}{\partial x}\right)_0, \quad K_y = -\frac{1}{B\rho}\left(\frac{\partial B_x}{\partial y}\right)_0, \quad K = -\frac{1}{2(B\rho)}\left(\frac{\partial B_x}{\partial x} - \frac{\partial B_y}{\partial y}\right)_0, \quad \text{and}$$

$M = -\dfrac{1}{B\rho}B_s$ with B_s the solenoid field and $B\rho$ the rigidity of the magnet. If we apply the method of Lagrange's variation of constants, the solutions will be

$$\begin{cases} x = \left(\int_0^s \dfrac{i}{2}F_x ds + a_{x0}\right)\beta_x^{1/2}e^{i\phi_x} + c.c. \\ y = \left(\int_0^s \dfrac{i}{2}F_y ds + a_{y0}\right)\beta_y^{1/2}e^{i\phi_y} + c.c. \end{cases} \tag{143}$$

where

$$\begin{cases} F_x = \left(Ky - \dfrac{1}{2}M'y\right)\beta_x^{1/2}e^{-i\phi_x} - My(\beta_x^{1/2}e^{-i\phi_x})' \\ F_y = \left(Kx + \dfrac{1}{2}M'x\right)\beta_y^{1/2}e^{-i\phi_y} + Mx(\beta_y^{1/2}e^{-i\phi_y})' \end{cases} \tag{144}$$

and $\beta_{x,y}$ are envelope functions, $a_{x,y}$ constants of transverse amplitude, and $\phi_{x,y}$ phase advances of transverse motion.

The linear coupling of the BEPC storage ring is due to the solenoid of the detector placed at the interaction point. Other sources of coupling are minor and can be neglected. The method of compensating the linear coupling is to install skew quadrupoles at appropriate places around the ring. With the extra coupling provided by the skew quadrupoles, we can compensate the effect of the solenoid field.

From Eq. (143), we know that if $\int_0^s \frac{i}{2} F_{x,y}\,ds = 0$, the linear coupling can be compensated, and Eq. (143) becomes the solution of transverse motion without any coupling. When $F_{x,y} \neq 0$, it is impossible to satisfy the above condition on any point around the ring. But we can realize the compensation globally, that is, the effects of elements can cancel each other with no amplitude growth. In this way, we can conclude that if there are n solenoids and N pairs of skew quadrupoles, the equations for coupling compensation can be deduced as

$$
\begin{cases}
\sum_{j=1}^{N}\left[Kl\sqrt{\beta_x\beta_y}\,\sin(\phi_x-\phi_y)\right]_{ji} - \left[\frac{Ml}{4}\left(\sqrt{\frac{\beta_y}{\beta_x}}+\sqrt{\frac{\beta_x}{\beta_y}}\right)\right]_i = 0 \\[2mm]
\sum_{j=1}^{N}\left[Kl\sqrt{\beta_x\beta_y}\,\cos(\phi_x-\phi_y)\right]_{ji} = 0 \\[2mm]
\sum_{j=1}^{N}\left[Kl\sqrt{\beta_x\beta_y}\,\sin(\phi_x+\phi_y)\right]_{ji} - \left[\frac{Ml}{4}\left(\sqrt{\frac{\beta_y}{\beta_x}}-\sqrt{\frac{\beta_x}{\beta_y}}\right)\right]_i = 0 \\[2mm]
\sum_{j=1}^{N}\left[Kl\sqrt{\beta_x\beta_y}\,\cos(\phi_x+\phi_y)\right]_{ji} = 0
\end{cases}
\tag{145}
$$

$$I = 1,2,3,\ldots,m, \quad j = 1,2,3,\ldots,N\ ,$$

where M_i is the strength of the i-th solenoid and l_i its length, β_{xi}, β_{yi} are the beta functions in the mid-solenoid, ϕ_{xi}, ϕ_{yi} the phase advances, $\pm (Kl)_{ji}$ the strengths of the j-th pair of skew quadrupoles, β_{xji}, β_{yji} the beta functions of the skew quadrupoles, and $\phi_{xi}\pm\phi_{xji}$, $\phi_{zi}\pm\phi_{zji}$ the phase advances of skew quadrupoles. There is no consideration of dispersion coupling and compensation.

With Eq. (145), we can choose the phase advance of the skew quadrupoles and get the corresponding strengths. In the light of BEPC's boundary conditions, we can find that when $|\phi_x-\phi_y| = \pi/2$, $\phi_x+\phi_y = 2K\pi+\pi/2$, $K\in J$, the second and fourth equations of Eq. (145) would be satisfied. By choosing proper strengths of skew quadrupoles, the first or third equation of Eq. (145) can be satisfied too, while the left equation is approximately satisfied. Thus, the linear coupling is compensated.

The calculation in the BEPC tells us that, installing skew quadrupoles in the injection region, or $\pm$ 64.507m away from the IP, can cancel the effect of solenoid, as shown in Figure 17. The constraint to tunes caused by compensation is $|v_x-v_y| = 1$. There are three pairs of skew quadrupoles in the BEPC storage ring. Two pairs near the IP are used to compensate the coupling when beams are injected, and another pair located in the injection region is for compensation when beams collide.

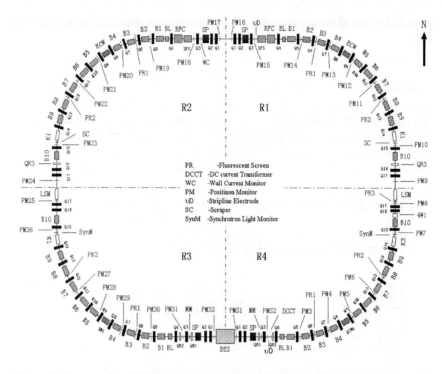

Figure 17 Layout of the BEPC storage ring (QR1 and QR2 near IP are for injection, and QR3 near injection point for collision)

The calculation of coupling shows that without skew quadrupoles the coupling reaches more than 20%, but with skew quadrupoles it drops to less than 1%. This result is consistent with the measured linear coupling of the BEPC.

7 Applications of Errors

Some beam parameters can be measured by introducing errors artificially into the storage ring, i.e., changing the magnetic field deliberately to get changes on corresponding parameters.

7.1 Measurement of β-function

As we already known, quadrupole field errors cause the variation of transverse tunes, expressed as

$$\Delta v_{x,y} = -\frac{1}{4\pi} \oint \beta(s) \Delta k(s) ds \quad ,$$

(146)

where Δk corresponds to the change of quadrupole strength. If one changes the strength of a given quadrupole, the average β-function at the location of this quadrupole can be found by measuring the variation of transverse tunes as

$$\overline{\beta} \approx -\frac{4\pi \cdot \Delta \nu}{\Delta k \cdot L_Q} .$$

(147)

If we measure the β-function of the first quadrupole near the interaction point (IP) of a collider, we can get the β-function at the IP from the solution of the following equation:

$$\beta = \beta^* + \frac{s^2}{\beta^*} .$$

(148)

Here, s is the distance between the quadrupole and the IP, and β^* the β-function at the IP. Figure 18 shows the measured and calculated horizontal β-function of the half BEPC storage ring. The error of the measurement mainly comes from the tune measurement and the resolution of power supplies for magnets.

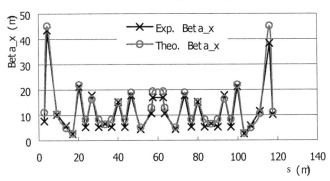

Figure 18 Comparison of measured and calculated horizontal β-function along the half BEPC storage ring. (Courtesy X. Wen, IHEP)

7.2   Measurement of dispersion function

At the locations of non-zero dispersion in a storage ring, the orbit variation caused by momentum deviation can be given by

$$\delta x = D_x(s)\frac{\Delta p}{p_0} ,$$

(149)

where D_x is the dispersion function. Combining with the definition of momentum compaction factor, i.e.,

$$\alpha_p = \frac{\Delta C}{C_0}\bigg/\frac{\Delta p}{p_0} = -\frac{\Delta f_{rf}}{f_{rf}}\bigg/\frac{\Delta p}{p_0} ,$$

(150)

where f_{rf} is the frequency of RF cavity, we will get

$$D_x = -\alpha_p \delta x \frac{f_{rf}}{\Delta f_{rf}}.$$

$$(151)$$

So after measuring the orbit as a function of RF frequency, it is easy to get the dispersion function at the locations of BPMs. The accuracy of the dispersion function measurement depends on the precision of the BPM system, and also on effects of power supply ripple. Figure 19 shows the measurement of the dispersion function at the BEPC storage ring.

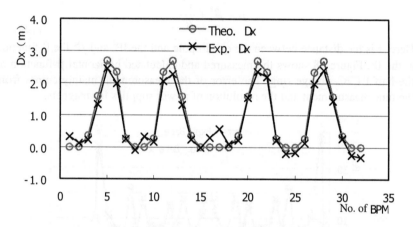

Figure 19 Comparison of measured and calculated dispersion function along the BEPC storage ring. (Courtesy X. Wen, IHEP)

7.3   Measurement of chromaticity

The natural chromaticity of a storage ring can be obtained by changing the field of dipole magnet and measuring the corresponding transverse tune variation, expressed as

$$\xi_0 = \frac{\Delta v}{\Delta B / B}.$$

$$(152)$$

After being corrected with sextupoles, the chromaticity of a storage ring can be obtained by changing the RF frequency and measuring the transverse tune variation with the expression of

$$\xi = -\alpha_p f_{rf} \frac{\Delta v}{\Delta f_{rf}}.$$

$$(153)$$

7.4   Measurement of linear coupling

By changing the strength of a certain quadrupole, and measuring the corresponding tunes, we can find the curves shown in Fig 20. The linear coupling can be calculated approximately from the following formula:

$$\kappa = \frac{(\Delta \nu)^2}{(\nu_x - \nu_y)^2 + 2(\Delta \nu)^2} \tag{154}$$

where ν_x, ν_y are the transverse tunes of the machine, and $\Delta \nu$ the distance between the peak of the lower curve and the valley of the upper curve in Fig 20.

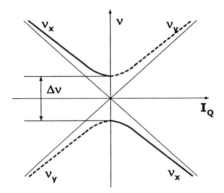

Figure 20 Method of tune splitting for coupling measurement

8   Summary

In this lecture, the transverse beam dynamics of single particle is reviewed. The transverse motion of the ideal particle in a storage ring is described as the betatron motion, together with the solution of the equation of motion, Hill's equation. The effects of magnetic imperfections are discussed for a real machine. For non-ideal particles, off-momentum orbits and chromatic effect are investigated. At last, linear coupling due to the longitudinal solenoid field and skew quadrupole component is treated, along with the compensation method. All of the contents given in the lecture are the most basic knowledge of the modern high energy storage ring. For further study, the reader is referred to the literature or the references listed at the end of this lecture note.

Acknowledgements

The lecturer would like to express his thankfulness to the organizers and sponsors of the OCPA IAS'02 for their efficient work and hospitality.

References

1. E. Courant and H. Snyder, "Theory of the Alternating Gradient Synchrotron", Ann. Phys. 3, 1 (1958).
2. M. Sands, "The Physics of Electron Storage Rings, An Introduction", SLAC-121, 1977.
3. E.J.N. Wilson, AIP Conference Proceedings 153, p.3, 1983.
4. S.Y. Lee, "Accelerator Physics", textbook of USPAS, 1998.
5. H. Wiedemann, "Particle Accelerator Physics-I, II", Springer-Verlag, 1993 and 1995.
6. S.X. Fang, N. Huang, and C. Zhang, 'The Compensation of solenoid effect on BEPC', IHEP internal report, 1990.
7. P.J. Bryant, CERN 94-01, p. 207, 1994.

TRANSVERSE BEAM DYNAMICS: CLOSED ORBIT CORRECTION AND INJECTION

CHIN-CHENG KUO

National Synchrotron Radiation Research Center
101 Hsin-Ann Road, Hsinchu Science-Based Industrial Park
Hsinchu, Taiwan
E-mail: cckuo@srrc.gov.tw

Closed orbit distortion resulting from magnetic field imperfections and misalignment of magnets is an important issue in the design of the accelerators. The closed orbit needs to be optimized and corrected during the commissioning stage and routine operations. The calculations of the orbit distortions due to errors are described. Several closed orbit correction methods are discussed. Orbit stability control in real time is of paramount importance in modern light source machines as well as in advanced colliders. The control algorithm is demonstrated. Beam orbit manipulation to form local bumps for injection, extraction, and various purposes are given. Finally, several injection schemes and extraction methods are introduced.

1. Introduction

Modern accelerators such as the third generation synchrotron light sources, linear colliders, and collider factories are aiming for their ultimate performances. For instance, the synchrotron light sources are pushing the brightness to the diffraction limit and the collider factories are trying to increase luminosity. As linear colliders are getting longer the precision of their beam trajectory becomes more demanding. Therefore, beam orbit control in these particle accelerators is important. During the design stage of the accelerators, it is essential to carry out a detailed analysis of the impacts of the magnetic field and mechanical errors in the beam orbit, emittance, and ring acceptance. Consequently, orbit distortion budget and correction schemes need to be fully investigated in order to ensure the desired performance. In this lecture, we focus on the effects and corrections of the circular accelerators.

A circular accelerator normally consists of a number of periodic magnet structure sets, called "magnet lattice." They basically include bending dipole-, quadrupole-, and sextupole magnets, and so forth. Errors always exist in the construction and alignment of these magnets. Linear field errors include dipole- and quadrupole field errors. Dipole field kicks might arise from field imperfections in the dipole bending magnets, from the roll error of the dipole magnets, from the misalignment of the quadrupole magnets, and from the feed-down of the nonlinear components such as sextupole magnets, etc. Dipole kicks create closed orbit distortions in the circular machines. Emittance coupling can be increased if the vertical closed orbit is off-center at the sextupole magnets. On the other hand, quadrupole field errors generate betatron beating of the lattice optical functions. These quadrupole field errors can be the manufacture errors or feed-down from the sextupole magnets if the beam orbit is not in the center of the sextupole magnets. Moreover, nonlinear elements and field errors can reduce the dynamic aperture substantially. Simulations provide guidelines for the tolerance budgets of the magnet field qualities.

In this lecture, we emphasize the discussions of the error sources of the closed orbit and methods to correct the closed orbit or to utilize these effects. Because orbit stability in the third generation light sources and in the modern colliders are critical, sources generating the time-dependent orbit fluctuation are discussed. Orbit feedback systems are also presented.

A high energy accelerator usually consists of a number of smaller accelerators, so the extraction from the pre-accelerator and injection into the host accelerator are necessary. The design goal of the injection and extraction schemes is to obtain efficient transfer and accumulation of the beam particles with little beam dilution after transfer. We will discuss the types of injections and extractions. Brief descriptions of the hardware structure of the injection elements are given.

The readers can find reference material in some accelerator textbooks[1], proceedings of the accelerator schools in Europe and the United States, and proceedings of the accelerator conferences PAC, EPAC, and APAC.

2. Orbit Distortion due to Dipole Field Errors

In the presence of magnetic field errors, Hill's equation can be expressed as

$$x'' + K_x(s)x = \frac{\Delta B_z}{B\rho}, \text{ and } z'' + K_z(s)z = \frac{\Delta B_x}{B\rho}, \tag{1}$$

where $K_x(s) = 1/\rho^2 + (\partial B_z/\partial x)/B\rho$, $K_z(s) = -(\partial B_z/\partial x)/B\rho$, ρ is the bending radius of the dipole magnet, and the perturbation fields ΔB_z and ΔB_x can be expressed by

$$\Delta B_z + j\Delta B_x = B_0 \sum_{n=0}^{\infty} (b_n + ja_n)(x + jz)^n. \tag{2}$$

Here j is an imaginary number, B_0 is the main dipole field, b_0 and b_1 are the dipole and quadrupole field errors, respectively, b_2 is the sextupole field term.

a_n represents the skew terms. These terms can be obtained from the following equations

$$b_n = \frac{1}{B_0 n!} \frac{\partial^n B_z}{\partial x^n} \Big|_{x=z=0}, \quad a_n = \frac{1}{B_0 n!} \frac{\partial^n B_x}{\partial x^n} \Big|_{x=z=0}. \tag{3}$$

Expanding the perturbed terms, we can write

$$\Delta B_z = B_0 [b_0 + (b_1 x - a_1 z) + (1/2)(b_2 (x^2 - z^2) - a_2 xz) + ..$$
$$\Delta B_x = B_0 [a_0 + (b_1 z + a_1 x) + (1/2)(a_2 (x^2 - z^2) + b_2 xz) + .. \tag{4}$$

It is shown that the dipole field errors can be the bending field error in the dipole magnets, which can be generated from the off-center orbit in the quadrupole- and sextupole magnets.

In Table 1, we list the main sources of the closed orbit distortion with the corresponding effective kick angles in an accelerator.

Table 1: Main sources of the closed orbit distortions.

Element type	Source of kick	Kick angle	Direction
Dipole (bend angle Φ)	Field error $\Delta B_z/B_z$	$\Phi(\Delta B_z/B_z)$	x
Dipole (bend angle Φ)	Roll φ	$\Phi \varphi$	z
Quadrupole (KL)	Displacement Δx, Δz	$KL\Delta x$, $KL\Delta z$	x, z

2.1. Existence of the Closed Orbit

In an ideal accelerator assuming a dipole error at $s = s_0$ with kick-angle $\theta = \Delta Bds / B\rho$ introduced, where ΔBds is the integrated dipole field error and $B\rho$ is the momentum rigidity of the beam, the closed orbit condition for the linear lattice can be expressed as

$$M \begin{pmatrix} y_0 \\ \dot{y}_0 \end{pmatrix} + \begin{pmatrix} 0 \\ \theta \end{pmatrix} = \begin{pmatrix} y_0 \\ \dot{y}_0 \end{pmatrix} \tag{5}$$

where M is the one-turn linear transfer matrix of the circular machine

$$M(s_0 + C | s_0) = \begin{pmatrix} \cos \pi v + \alpha_0 \sin \pi v & \beta_0 \sin \pi v \\ -\gamma_0 \sin \pi v & \cos \pi v - \alpha_0 \sin \pi v \end{pmatrix} \tag{6}$$

$$= I \cos \pi v + J \sin \pi v,$$

where

$$I = \begin{pmatrix} 1 & 0 \\ 0 & 1 \end{pmatrix}, \quad J = \begin{pmatrix} \alpha_0 & \beta_0 \\ -\gamma_0 & -\alpha_0 \end{pmatrix},$$

v is the betatron tune of the machine, and α_0, β_0 and γ_0 are the betatron amplitude functions at the kick location s_0. The closed orbit distortion and its divergence angle at $s = s_0$ are calculated,

$$\begin{pmatrix} y_0 \\ y_0' \end{pmatrix} = (I - M)^{-1} \begin{pmatrix} 0 \\ \theta \end{pmatrix}. \tag{7}$$

One can also write the resulting orbit as follows

$$y_0 = \frac{\beta_0 \theta}{2\sin \pi v} \cos \pi v, \quad y_0' = \frac{\theta}{2\sin \pi v}(\sin \pi v - \alpha_0 \cos \pi v) \tag{8}$$

The closed orbit at location s in the accelerator can be obtained from the linear transfer (response) matrix between s and s_0, that is

$$\begin{pmatrix} y(s) \\ y'(s) \end{pmatrix}_{co} = M(s \mid s_0) \begin{pmatrix} y_0 \\ y_0' \end{pmatrix}_{co}, \tag{9}$$

The transfer matrix can be written in terms of the betatron amplitude functions between two points.

$$M(s \mid s_0) = \begin{pmatrix} \frac{\sqrt{\beta}}{\sqrt{\beta_0}}(\cos(\psi-\psi_0)+\alpha_0\sin(\psi-\psi_0)) & \sqrt{\beta\beta_0}\sin(\psi-\psi_0) \\ \frac{-1}{\sqrt{\beta\beta_0}}((\alpha-\alpha_0)\cos(\psi-\psi_0)+(1+\alpha\alpha_0)\sin(\psi-\psi_0)) & \frac{\sqrt{\beta_0}}{\sqrt{\beta}}(\cos(\psi-\psi_0)-\alpha\sin(\psi-\psi_0)) \end{pmatrix}, \tag{10}$$

Substituting Eq. (8) and Eq. (10) into Eq. (9), we obtain the closed orbit as

$$y_{co}(s) = G(s,s_0)\theta(s_0), \tag{11}$$

where

$$G(s,s_0) = \frac{\sqrt{\beta(s)\beta(s_0)}}{2\sin \pi v} \cos(\pi v - \psi(s) + \psi(s_0)) \tag{12}$$

is the Green's function of the Hill's equation. The Green's function can be re-written in a more general form regardless of the sign of the phase advance difference between two locations, i.e.,

$$G(s,s_0) = \frac{\sqrt{\beta(s)\beta(s_0)}}{2\sin \pi v} \cos(\pi v - |\psi(s) - \psi(s_0)|). \tag{13}$$

Equation (13) shows that the closed orbit becomes infinite when $\sin \pi v = 0$, i.e., the betatron tune is an integer number. Hence, integer tune or the tune close to integer should be avoided in the machine design.

Let $\phi(s) = (1/\nu)\int_0^s ds'/\beta(s')$, and $\psi(s) = \nu\phi(s)$, we can write the closed orbit in a real machine with distributed dipole errors along the ring as

$$y_{co}(s) = \frac{\sqrt{\beta(s)}}{2\sin\pi\nu}\int_s^{s+c}\sqrt{\beta(s')}\frac{\Delta B(s')}{B\rho}\cos(\pi\nu - \psi(s) + \psi(s'))ds' \tag{14}$$

$$= \frac{\nu\sqrt{\beta(s)}}{2\sin\pi\nu}\int_\phi^{\phi+2\pi}\left(\beta^{3/2}(\varphi)\frac{\Delta B(\varphi)}{B\rho}\right)\cos\nu(\pi - \phi + \varphi)d\varphi. \tag{15}$$

In other words, the orbit response of the inhomogeneous Hill's equation is

$$y_{co}(s) = \int_s^{s+c}G(s,s')\frac{\Delta B(s')}{B\rho}ds', \tag{16}$$

where the Green's function is given by

$$G(s,s') = \frac{\sqrt{\beta(s)\beta(s')}}{2\sin\pi\nu}\cos(\pi\nu - |\psi(s) - \psi(s')|). \tag{17}$$

In dipole angular kick form, we can write as

$$y_{co}(s) = \frac{\sqrt{\beta(s)}}{2\sin\pi\nu}\sum_{i=1}^N\theta_i\sqrt{\beta(s_i)}\cos(\pi\nu - |\psi(s) - \psi(s_i)|), \tag{18}$$

where $\theta_i = \dfrac{\Delta B(s_i)}{B\rho}ds_i$.

Figures 1a and 1b are examples of the closed orbit responses to the kick errors in the 1.5 GeV storage ring TLS at NSRRC in Taiwan. The kick errors are in the middle of the ring circumference in the horizontal and vertical planes, respectively. It is worth mentioning that the lattice structure of the TLS is a six-fold super-symmetry type. The normalized orbit distortions $y_{co}(s)/\sqrt{\beta(s)}$ versus the phase advance $\psi(s) = 2\pi\mu$ are shown in Figures 1c and 1d.

2.2. Statistical Estimation of the Closed Orbit Errors

Equation (16) and Eq. (18) show that the errors in the real machine are mainly due to the random construction errors in the dipole magnets and misalignment in the quadrupoles. These errors are not exactly known during the design stage of the machine. From Eq. (14) one can estimate the rms closed orbit with the random distribution of the errors

$$y_{co,rms} = \frac{\overline{\beta}}{2\sqrt{2}\,|\sin\pi\nu|}\sqrt{N}\theta_{rms}, \tag{19}$$

with $\overline{\beta}$ the average of the betatron functions in kick elements, N the number of kick elements, θ_{rms} the rms kick strength for each element.

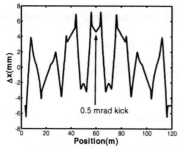

Figure 1a: Closed orbit response to a kick of 0.5 mrad at TLS ring center in the horizontal plane.

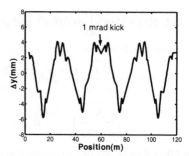

Figure 1b: Closed orbit response to a kick of 1.0 mrad at TLS ring center in the vertical plane.

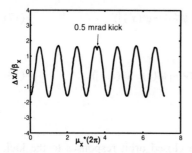

Figure 1c: Normalized closed orbit response to a kick of 0.5 mrad as a function of phase advance at TLS ring center in the horizontal plane.

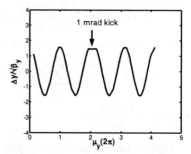

Figure 1d: Normalized closed orbit response to a kick of 1.0 mrad as a function of phase advance at TLS ring center in the vertical plane.

The kick strengths in the quadrupole magnets are mainly from the misalignment. When quadrupole magnets are misaligned by a distance Δy, the effective angular kick is

$$\theta_i = \frac{B_1 \ell}{B\rho} \Delta y = \frac{\Delta y}{f}, \tag{20}$$

where $B_1 = B_0 b_1$ in Eq. (4) for the normal gradient and $B_1 = \pm B_0 a_1$ for the skew gradient, f is the focal length. Hence the rms closed orbit due to N quadrupole misalignment Δy_{rms} can be written as

$$y_{co,rms} = \left(\frac{\bar{\beta}}{2\sqrt{2}\,\bar{f}\,|\sin \pi \nu|} \sqrt{N} \right) \Delta y_{rms}, \tag{21}$$

where $\bar{f}$ is the average focal length.

The term inside the curly brackets is called sensitivity or amplification factor. In the third generation synchrotron light sources, the machines are designed with strong focusing strengths in order to obtain small emittance, i.e., the focal length is short. Therefore the amplification factors are usually as high as 50 or more with $\sin \pi \nu \approx 0.5$. It shows that the amplification factor increases with the square root of the number of magnets, which means the sensitivity increases with the size of the accelerator. If the required rms closed orbit is a few mm, then the tolerance for the misalignment is as small as a hundred microns, which demands more precise survey and alignment.

2.3. Fourier Harmonics of the Closed Orbit

Since the term $\beta^{3/2}(\phi) \dfrac{\Delta B(\phi)}{B\rho}$ in Eq. (15) is a periodic function with period 2π, we can expand it in a Fourier series:

$$f(\phi) \equiv \beta^{3/2}(\phi)\frac{\Delta B(\phi)}{B\rho} = \sum_{k=-\infty}^{\infty} f_k e^{jk\phi}, \tag{22}$$

where the Fourier amplitude f_k is the integer stopband integral expressed as

$$f_k = \frac{1}{2\pi\nu}\oint \beta^{3/2}(\phi)\frac{\Delta B(\phi)}{B\rho}e^{-jk\phi}d\phi \tag{23}$$

$$= \frac{1}{2\pi\nu}\oint \sqrt{\beta}\frac{\Delta B}{B\rho}e^{-jk\phi}ds. \tag{24}$$

The closed orbit thus can be written as

$$y_{co}(s) = \sqrt{\beta(s)} \sum_{k=-\infty}^{\infty} \frac{\nu^2 f_k}{\nu^2 - k^2}e^{jk\phi}. \tag{25}$$

This equation proves that there are poles in the integer harmonics and the closed orbit is dominated by a few harmonics near $[\nu]$, where $[\nu]$ is the nearest integer of ν.

3. Measurement of the Closed Orbit

There are various types of sensors that can be used to measure the closed orbit. These sensors can be classified into three categories: (1) the electrostatic type which includes the split can, strip-line and button type monitors; (2) the magnetic type which includes magnetic pickup; and (3) the electromagnetic type which includes the direction coupler, etc.

The button or strip-line type beam position monitors located off the mid-plane to avoid the synchrotron light are commonly adapted in the lepton machines. In the proton machines, magnetic pickup may be chosen to avoid the secondary emission effects. In the collider ring, directional coupler can be used for two beam position detections. Some major factors on the closed orbit measurements will be briefly described.

To measure the closed orbit in a circular accelerator, a number of beam position monitors (BPMs) are required. Four position monitors per betatron wavelength are normally needed. Therefore, the number of required monitors is proportional to the machine betatron tune rather than the path length.

BPMs can be calibrated on the bench in the laboratory. Using beam-based measurement method in the real machine also reveals the gain and off-set of the BPMs. It is necessary to conduct the calibration procedure in the modern accelerators because the required precision of the orbit typically is in the order of micron or even sub-micron range.

The signal processing electronics of the button type monitors, for example, becomes rather sophisticated nowadays. Typically, difference over sum method is employed. Multiplexed signal for each button is then intensity independent and usually the beam position signal is averaged. The averaged beam position signal is good for the closed orbit correction because the resolution of the signal becomes smaller. Parallel processing electronics of the beam position monitoring system has become affordable recently. There are two other kinds of signal processing methods: (1) AM to PM conversion. This is for single turn monitoring system, but it is expensive; (2) Log ratio methods. This is good for single turn system too, and it is less expensive.

Performance requirements of the BPMs for the third generation light sources, in particular, are critically stringent. The stability and resolution should be in the order of micron with good dynamic range, say, 0.1 mA to 1 A. Other requirements for the high performance BPMs such as linearity, bandwidth, mechanical stability, electrical dependence, and so forth, need to be well considered in the design stage.

4. Closed Orbit Correction

Closed orbit correction is important during the commissioning and the routine operations of the accelerators. Usually the dipole correctors are distributed around the ring. The location and the number of correctors depend on the orbit correction methods. Normally, the number of correctors is less than that of the

BPMs, and if using too many correctors for orbit correction, they will fight each other.

It is known that if the closed orbit distortion is too large, the transverse beam acceptance becomes small, and the beam lifetime and dynamic aperture will be drastically reduced. Moreover, it is of paramount importance to correct the closed orbit to allow the acceptance of the injected beam in the host accelerator in the first-turn beam test. Simulations with error source data obtained from the survey and the magnetic field can provide some guidance for the correction settings. In addition, beam threading method allows us to steer the beam through the beam line stepwise and the closed orbit can be found iteratively. Nowadays, a well aligned accelerator does not need orbit correction during the first injection try.

Once the beam is stored in the ring, one can apply any orbit correction method to reduce the closed orbit or to change it to the desired position. Dispersion-like orbit correction is necessary in both transverse planes, either by changing the accelerating radiofrequency for the horizontal plane or by employing skew quadrupoles to correct the transverse coupling.

The orbit correction magnets produce dipole orbit kicks and have the same effects when the dipole errs. The change of the orbit at the BPM position s_j resulted from the corrector kick θ_i at s_i can be written as

$$\Delta y(s_j) = \frac{\sqrt{\beta(s_j)}}{2\sin \pi v} \sum_{i=1}^{N} \theta_i \sqrt{\beta(s_i)} \cos(\pi v - |\psi(s_j) - \psi(s_i)|) . \tag{26}$$

This can be expressed in a matrix equation

$$\Delta \vec{y}_m = A \vec{\theta}_n , \tag{27}$$

where $\Delta \vec{y}_m$ is the vector formed by the change of the orbit at m BPMs and $\vec{\theta}_n$ is the n corrector vector. A is the response matrix with

$$A_{ji} = \frac{\sqrt{\beta(s_j)}}{2\sin \pi v} \sqrt{\beta(s_i)} \cos(\pi v - |\psi(s_j) - \psi(s_i)|) . \tag{28}$$

Note that the response matrices in some machines are obtained from the measured data instead of the model calculated values. The distorted orbit can be corrected at least at the BPMs to the desired value so that $\Delta \vec{y}_m = -\Delta \vec{u}_m$, where $\Delta \vec{u}_m$ is the difference between the measured orbit and the reference orbit with the following expression

$$\vec{\theta}_n = A^{-1} \Delta \vec{y}_m . \tag{29}$$

It is obvious that these linear algebraic equations can be solved exactly for $n \geq m$ if there is no degeneracy in the equations. The solution is not practical because of the badness of BPMs, the machine modeling error, the feed-down effect from the higher-order fields, and so forth. Usually, we employ some sophisticated methods in solving the least-squares problems to avoid the unnecessary large strengths in correctors. These methods minimize, in a sense, the quadratic residual of the corrected orbit $|A\vec{\theta} + \Delta u|$, which can be expressed as follows:

$$\vec{\theta}_n = -(A^T A)^{-1} A^T \Delta \vec{u}_m , \qquad (30)$$

The numerical recipes[2] such as Gaussian elimination, LU decomposition, Householder transformation, and so forth, can be employed to solve the linear-squares problems. The regularization method can also reformulate the ill-posed problems into more well-posed ones. Another popular method is the singular value decomposition (SVD) algorithm. Detailed discussion will be given later. We will also mention some popular correction methods, such as harmonic correction and stopband correction, and beam-bump method.

4.1. Stopband and Harmonic Correction Methods

In Eq. (25) the closed orbit can be decomposed into Fourier harmonics such that

$$y_{co}(s) = \sqrt{\beta(s)} \sum_{k=-\infty}^{\infty} \frac{v^2 f_k}{v^2 - k^2} e^{jk\phi} , \qquad (25)$$

and $f_k = \frac{1}{2\pi v} \sum_i \sqrt{\beta_i} \frac{\Delta B ds_i}{B\rho} e^{-jk\phi_i} = \frac{1}{2\pi v} \sum_i \sqrt{\beta_i} \theta_i e^{-jk\phi_i} , \qquad (27)$

where θ_i and ϕ_i are the corrector angular kick strength and the normalized phase advance of the ith corrector, respectively. Placing correctors and BPMs at the locations of high betatron functions uniformly with respect to the betatron phase is an effective way to reduce the closed orbit using harmonic method. Since the most effective correction term is around tune $[v]$, the stopband can be computed as follows:

$$f_{[v]} = \frac{1}{2\pi v} \sum_i \sqrt{\beta_i} \theta_i e^{-j[v]\phi_i} , \qquad (28)$$

and we can choose two correctors with phase advance of $[v]\phi_i \approx \pi/2$ such that the strength of these two correctors can be adjusted independently to reduce the orbit stepwise.

On the other hand once we have enough well distributed correctors and monitors, we can choose a few harmonics near the tune to minimize the orbit with these correctors.[3]

The closed orbit expressed in Eq. (25) can be re-written as

$$\frac{y_i}{\sqrt{\beta_i}} = \sum_{k=0}^{\infty} (a_k \cos k\phi_i + b_k \sin k\phi_i) . \tag{29}$$

Similarly the jth corrector θ_j can be decomposed into Fourier harmonics

$$\theta_j = \frac{1}{\sqrt{\beta_j}} \sum_{k=0}^{\infty} (a_k \cos k\phi_j + b_k \sin k\phi_j) . \tag{30}$$

Using least-squares method, we can get a set of a_k and b_k from the measured orbit at distributed monitors and then with these a_k and b_k we obtain corrector strengths to cancel a few selected dangerous stopbands. The harmonic correction method can also be used for the real-time orbit feedback control.

4.2. Local Bump Method

Local bumps can be constructed using three corrector magnets with angular kicks θ_i (i=1, 2, 3) at locations 1, 2, and 3. The condition for locality is

$$\sum_{i=1}^{3} A_{ki}\theta_i = 0, \text{ and } \sum_{i=1}^{3} A_{ji}\theta_i = 0, \tag{31}$$

where j and k are the monitors inside and outside the local bump, respectively. It is easy to get the following relations between three kicks in terms of the lattice parameters

$$\frac{\theta_1 \sqrt{\beta_1}}{\sin \psi_{32}} = \frac{\theta_2 \sqrt{\beta_2}}{\sin \psi_{13}} = \frac{\theta_3 \sqrt{\beta_3}}{\sin \psi_{21}}, \text{ where } \psi_{ij} = \psi_i - \psi_j . \tag{32}$$

Hence, one needs to minimize the residual of the sum of the measured orbit and the local bumped orbit stepwise and iteratively. The orbit therefore can be flattened. In this method, the monitor errors and accuracy of lattice parameters are rather sensitive.

One also can combine two 3-magnet bumps to form a 4-magnet local bump so that both angle and position at location between the second- and third correctors inside the local bump can be controlled, especially in the local orbit feedback control.

4.3. *Micado Method*

The orbit minimization package MICADO[4] is the first orbit control algorithm. This minimization algorithm is based on the Household transformation and the pivoting method to reduce the matrix to a triangle form. In this method, one can find the single most effective corrector in the first iteration, and then the second most effective corrector in the second iteration, and so forth. One can choose any number of correctors to be used. The corrector strength can be constrained. In the iteration the corrector strengths of the previous iteration are re-calculated.

4.4. *Singular-Value Decomposition Method*

It is mentioned in the beginning of this section that the linear algebraic equations can be solved either directly with matrix inversion or by using least-squares minimization procedure. The most promising and robust method is the SVD method. An $M \times N$ matrix A ($M \geq N$) can be written as the product of an $M \times N$ column-orthogonal matrix U , an $N \times N$ diagonal matrix W (singular values) with positive or zero elements, and the transpose of an $N \times N$ orthogonal matrix V .

$$A = UWV^T , \quad U^TU = VV^T = V^TV = I , \tag{33}$$

That means that the response matrix can be decomposed into the product of the corrector-eigenvector, the orbit-eigenvector, and the singular values connecting these eigenvectors. Small w_j requires large kick strength in order to produce a required orbit response. The corrector strengths for the correction of the orbit to the desired value can be obtained using

$$\Delta \vec{\theta}_n = -A^{-1}\vec{u}_m = -VW^{-1}U^T\vec{u}_m = -V(diag(1/w_j))U^T\vec{u}_m , \tag{34}$$

if $1/w_j$ is above the threshold value, let $1/w_j = 0$. Rejection of the singular values or close to the singular values can reduce the rms value of the correctors. Therefore it is helpful to reduce the vertical dispersion functions with the resulting smaller rms values of the vertical corrector strength. With this method, one can choose the number of correctors and the threshold of w_js as needed.

SVD can be employed not only for the underdetermined linear least-squares problems but also for the over-determined cases. SVD is commonly used in the real-time orbit feedback control because it converts problems of the coupled multiple input (BPM orbits) and output (corrector strengths) into problems of many decoupled single input and output feedback loops.

5. Off-Momentum Orbit and Correction

As we have already discussed about the closed orbit due to errors for the on-momentum particle, we would now like to introduce the off-momentum orbit distortion. From the off-momentum orbit, one can derive some important beam lattice parameters such as momentum compaction factor, dispersion function, beam emittance, and so forth.

The equation of motion for the particle with momentum p, which deviates from the momentum p_0 (on-momentum) of the synchronous particle, can be written as

$$x'' - \frac{\rho + x}{\rho^2} = -\frac{B_z}{B\rho}\frac{p_0}{p}(1 + \frac{x}{\rho})^2 , \tag{35}$$

where $B_z = B_0 + B_1 x$, $\delta = (p - p_0)/p_0$. Therefore, Eq. (35) can be expressed as

$$x'' + (\frac{1-\delta}{\rho^2(1+\delta)} + \frac{K(s)}{(1+\delta)})x = \frac{\delta}{\rho(1+\delta)} . \tag{36}$$

Let $x = x_\beta + D(s)\delta$, where $D(s)\delta$ is the off-momentum orbit. To the lowest order in δ, the dispersion function is generated from an inhomogeneous equation

$$D'' + K_x(s)D = \frac{1}{\rho}, \text{ with } K_x(s) = (1/\rho^2) + B_1/B\rho . \tag{37}$$

The dispersion function thus can be obtained

$$D(s) = \frac{\sqrt{\beta_x(s)}}{2\sin \pi v_x} \int_s^{s+C} \frac{\sqrt{\beta_x(s')}}{\rho} \cos(\pi v_x - |\psi_x(s') - \psi_x(s)|)ds . \tag{38}$$

If the RF frequency remains the same, the change of circumference due to thermal expansion or tidal effect can generate the dispersion-like orbit in order to keep the single-turn particle trajectory constant and thus the particle energy is changed. This off-momentum particle orbit can be corrected by changing the RF master-clock frequency as follows:

$$\frac{\Delta f_{rf}}{f_{rf}} = -\alpha_c \frac{\Delta x}{D_x} , \tag{39}$$

where α_c is the momentum compaction factor. This correction procedure can be automated in the machine operations.

6. Real-Time Orbit Correction

The orbit error sources in an accelerator can be time-dependent and consequently the beam orbit is in motion raising the issue of real-time orbit correction[5]. The frequency of interest in these error sources ranges from 10^{-5} to 10^4 Hz. In the third generation light sources, the tolerable orbit stability can be as low as a few microns. Typically, there are two approaches to maximize orbit stability: (1) removing the sources of orbit motion, and (2) using orbit feedback to minimize the orbit motion.

Sources of orbit motion can be categorized on various time scales. Ground settlement and seasonal temperature changes result in the displacements of magnets, and surveys and realignments are therefore needed. Tidal motion, diurnal variations, underground water-tables, synchrotron radiation heating, and so forth, require slow orbit control on minute scales. Ground and mechanical vibrations, power supply ripples and electrical disturbances, insertion devices, machine working tune compensation, and so forth, need the fast orbit correction on short time scales, e.g., ranging from 10^{-1} to 10^2 Hz. Coupled-bunch instabilities usually are in hundred kHz order and need to be minimized with the precision control of the cavity temperature, the suppression of higher-order cavity modes, and the fast damping feedback systems.

Suppression of these time varying error sources is necessary, although the real-time orbit correction procedure nowadays can achieve the required orbit stability at sub-micron level. Those include, for instance, the temperature control of cooling water and air conditioning within ±0.1°C, damping pads in the girders, and reduction of power supply ripples, and so forth.

In the light sources, the variation of magnetic gaps of the insertion devices in real-time necessitates the follow-gap orbit corrections. Orbit controls with the built-in follow-up tables are used in the control systems. Sometimes machine betatron tunes are kept at some fixed point during the gap change with the help of tune correction using a quadrupole lookup table.

The orbit correction algorithms for global and local orbit feedback control are similar to those for the DC orbit corrections. SVD is most suited for orbit feedback control in the existing machines.

There are two types of feedback control: analog and digital systems. An analog system with wide bandwidth can be built at relatively lower cost than the digital system, but the digital system is more flexible in terms of the expansion of number of control channels and control software modifications.

7. Injection and Extraction

A high energy accelerator usually consists of a chain of accelerators to increase the energy of particle beam stepwise from the beginning of the particle beam source to the final design energy and to accumulate the particle beam intensity to the final design strength. Therefore, one needs to devise methods to extract the particle beam from an accelerator and to inject into another via a transfer line between these two chained accelerators. The main goal of the extraction and injection processes is to minimize beam loss and dilution while efficiently transferring the particle beam from one accelerator to another. There are some different types of extraction and injection processes.

Injection into or extraction from the linear accelerators is less complicated than from the circular accelerators. But special care is also needed even in the linac injection and extraction. For discussion purpose, we can categorize the processes of injection into and extraction from the circular accelerators as follows: (1) single-turn injection, (2) multi-turn injection, (3) charge-exchange injection, (4) resonance injection and extraction, and (5) fast extraction.

Injection and extraction processes need elements such as septum, kicker, and bumper magnets, and so forth. The septum magnets can be DC or pulsed type. The kicker and bumper magnets are the pulsed form and their switching periods are usually in the order of tens of nanoseconds to microsecond. There are a number of different types of design in these magnets, which will be discussed in this lecture.

Knowing the beam dynamic parameters such as lattice optical parameter β-function, phase advance, tune, dynamic aperture, acceptance, and so forth, is necessary in the calculation of design specifications for the injection elements.

7.1. *Single-Turn Injection*

The fast damping effects in the transverse and longitudinal planes of the electron and positron beam or the proton machine with cooling favor the single-turn injection. The single-turn injection can be either on- or off-axis injection. The injection from linac to booster synchrotron in the electron or positron machine usually employs the on-axis single-turn injection, while in the storage ring off-axis single-turn injection is normally used. However, one needs to note that on-axis single turn bucket-to-bucket transfer process in the extremely large proton or heavy ion machine is also a common method.

The injection process for the single-turn injection takes place only in one turn per injection cycle. Injection cycle usually ranges from a few seconds to tenths of a second. The damping time is usually in the millisecond range for the electron/positron rings and in the second range in the proton machine with cooling so that the previously injected beam already damps down before the next injection cycle occurs.

A schematic drawing for the single-turn on-axis and off-axis injection is shown in Figures 2a and 2b, respectively. For the on-axis injection, one kicker magnet is located downstream of the injection point. A septum magnet is usually employed so that the bend angle of the kicker magnet can be reduced. The fast rising time and decay time of the kickers usually are in the range between hundreds of nanoseconds to a few micro-seconds. The septum pulse could be as long as hundreds of micro-seconds. The strength of the kicker magnet can be calculated using the formula below.

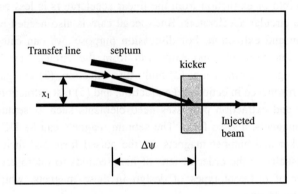

Figure 2a: Schematic of the on-axis injection scheme.

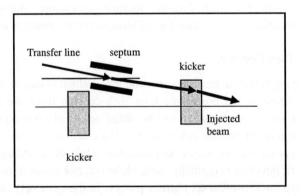

Figure 2b: Schematic of the off-axis injection scheme.

Let the transfer matrix of the beam position and angle be expressed with Courant-Snyder parameters between two points in the beam line. The on-axis injection with a septum (exit of septum denoted 1) and a kicker (denoted 2) of kick angle θ can be written as

$$x_1' = -(\alpha_1 + \cot\Delta\psi)x_1 / \beta_1 \ , \ \theta = x_1 /(\sqrt{\beta_1\beta_2} \sin\Delta\psi) \qquad (40)$$

where $\Delta\psi = \psi_2 - \psi_1$ and $x_2 = 0$.

Therefore, high betatron functions are useful to reduce the kicker strengths. The above formula can be derived from the transfer matrix given in Eq. (10). That is

$$x_2 = \frac{\sqrt{\beta_2}}{\sqrt{\beta_1}}(\cos(\psi_2 - \psi_1) + \alpha_1 \sin(\psi_2 - \psi_1))x_1 + \sqrt{\beta_2\beta_1} \sin(\psi_2 - \psi_1)x_1' \ . \qquad (41)$$

For the single-turn off-axis injection, usually we need to employ more than one kicker to create a local orbit bump in the injection region so that both the injected and stored beams are inside the ring acceptance. In this way, we can ensure the efficient beam capture.

The amplitude of the local bump for the injection is determined by the acceptance of the ring. The maximum allowable initial orbit betatron oscillation amplitude of the injected beam in the horizontal plane (usually chosen because of larger acceptance) can be determined as

$$x_1 = n\sigma_i + m\sigma_s + \eta_x[(\frac{\Delta p}{p})_i + ((\frac{\Delta p}{p})_s] + x_{co} + x_s \ , \qquad (42)$$

where $\sigma_i, \sigma_s, \eta_x, (\frac{\Delta p}{p})_i, (\frac{\Delta p}{p})_s, x_{co}, x_s$ are the injected beam size, stored beam size, dispersion function at the injection point, energy deviation of the injected and stored beam, allowable closed orbit, effective septum thickness, respectively, n and m are the integer numbers. Usually Courant-Snyder machine lattice parameters of the injected beam at the injection point are matched or identical to the host machine's lattice parameters. Figure 3 shows the transverse phase space acceptance of the stored and bumped beam of the storage ring TLS, the kicker amplitude of the stored beam, and the initial orbit amplitude of the injected beam.

The kickers can be arranged such that the maximum bump is created at the injection point. Kicker strengths of the local bump for 4 kickers with the bump amplitude and angle at injection point x_i, x_i' satisfy the local bump condition.

In the four-kicker bump, both position and angle at injection point are tunable. In the third generation light sources, a common scheme of the injection consists of four kickers, which are symmetrically located in one long straight section. A septum exit is near the middle of this section. It is a rather simple configuration which offers ample space for the installation of the injection elements. If the straight section is not long enough, we can use a three-kicker bump with a proper phase advance between the first and the third kickers which equals π. We can also use a proper phase advance between first kicker and the injection point to obtain the most efficient beam capture. Two kickers can also form a local bump, but it is seldom chosen. In some special cases, one kicker is sufficient in the injection process.

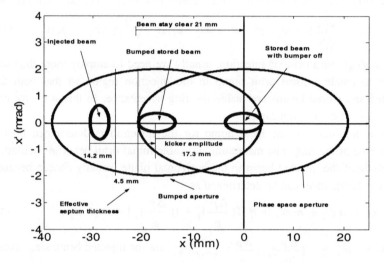

Figure 3: TLS transverse phase space acceptance defined by the septum edge, in which the stored and bumped beam acceptance in the horizontal plane is given. The injected beam is within the acceptance of the bumped stored beam. The initial orbit amplitude of the injected beam A=14.2 mm. The kicker amplitude is 17.3 mm. Four kickers are located in a long straight section symmetrically with respect to the long straight center. The dispersion function is designed to be zero.

The kicker waveform can be a half-sine wave with the base-width of a few µs. As a result, the bumped beam can be properly reduced so that the injected beam will not be scraped by the septum edge. Figures 4a-4c show the changes of the injected and stored beam inside the bumped phase space acceptances without errors. Figure 4a is at the time when the injected beam arrives and the stored beam is bumped to the maximum amplitude. Figure 4b is the beam distribution in the phase space after one turn (400 nsec later). After two turns the beam population is shown in Figure 4c.

Analysis of the impact on the injection due to the leakage field from septum, and the kicker mismatch is necessary in order to provide the field tolerance and timing jitter budget.

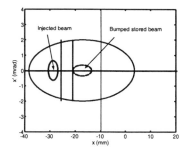

Figure 4a: The ideal injected and stored beam inside the acceptance at injection in TLS.

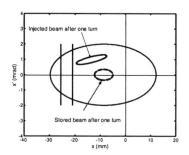

Figure 4b: The ideal injected and stored beam inside the acceptance after one turn in TLS.

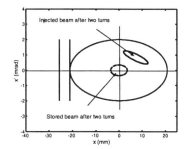

Figure 4c: The ideal injected and stored beam inside the acceptance after two turns in TLS.

7.2. *Multi-turn injection*

Multi-turn injection scheme is often used to inject low energy proton or heavy ions into the circular machine. Because of the slow damping effect, one needs to inject beams into the bumped acceptance many times in an injection cycle to accumulate beam intensity before the bumped closed orbit is removed and beam energy is ramped. Therefore, the bumped orbit and machine tune should be carefully selected so that the injected beam will not hit the septum wall. Multi-turn injection results in the painting of the injected beam in the phase space, which is shown in Figure 5.

110

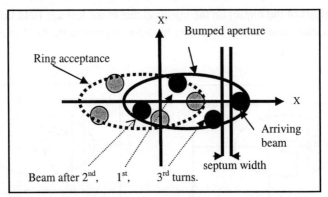

Figure 5: Evolution of the injected beam in the phase space acceptance of the multi-turn injection scheme. The machine fractional tune is slightly larger than 0.25.

7.3. *Stripping injection scheme*

The charge exchange injection can be either H^- or H_2^+ ions hitting a thin foil of a few $\mu g/cm^2$ to a few mg/cm^2 so that electrons are stripped and proton or deuteron beam can be captured. The closed orbit of the stored circulating beam is bumped onto the injection orbit of the injected H^- or H_2^+ beam by bumpers and a chicane magnet as shown in Figure 6. The painting of the injected beam can be achieved with the variation of the bumped orbit. The emittance blow-up due to scattering of the stripping foil needs to be taken care of, the number of the passages of the circulating beam through the foil is therefore limited. The injection efficiency of H^- beam stripping is high and therefore becomes a popular method.

Recently the laser stripping scheme is employed in which the injection efficiency can be increased and the beam emittance blow-up problem can be reduced.

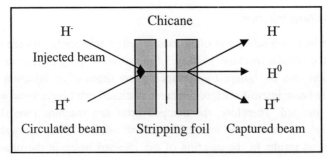

Figure 6: Charge exchange injection of H^- beam. A stripping foil is located between two dipoles.

7.4. Fast Extraction

Extraction from the circular machine can be the reversal of the injection process. A kicker, an extraction septum, and several bumpers are equipped. Figure 7 shows the extraction orbit trajectory of the NSRRC booster extraction scheme.[6] A slow orbit bump is necessary to reduce the kicker strength.

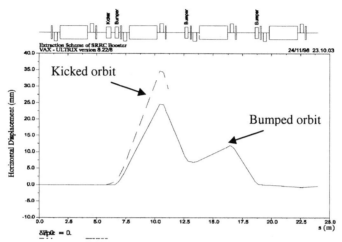

Figure 7: Beam orbit trajectory of the NSRRC extraction system. Three bumpers are excited to reduce the strength of the extraction kicker. The kicked beam orbit is directed into the extraction septum chamber and beam is transferred into the transfer line.

7.5. Resonance extraction

Slow extraction is required in high energy experiments and medical treatments. Nonlinear resonance extraction such as third-integer resonance extraction using sextupoles can be employed. Half-integer resonance extraction is also applicable but the nonlinear element of the magnetic octupoles is needed.

A schematic of the third-integer resonance extraction is shown in Figure 8. The stable region can be reduced by increasing the strength of the sextupole or adjusting the betatron tune to approach the third-integer resonance and excited beam can be extracted out of the stable region near the unstable fixed point. A thin wire septum is located near the extraction point.

A slow spill could be controlled by the above methods. A convenient method is to keep the sextupole strength constant and to steer the beam to spill beam slowly for different momentum particles.

Resonance injection is the reverse process of the resonance extraction and is applied in the small, low energy rings.

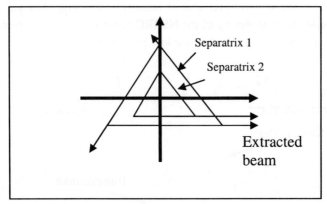

Figure 8: A third-order resonance extraction scheme. The stored beam is extracted by reducing the separatrix.

7.6. *Septum Units and Kicker*

DC or fast pulsed magnetic field generated by septa and kickers are key elements in the injection and extraction. The reliability and stability of these elements are important factors for high performance of injection and extraction efficiency.

There are several different types of septum units. Figure 9 is a schematic drawing of the typical current-sheet septum. Field distribution is shown in the drawing. Typical field leakage is less than 10^{-3} with sufficient thickness (about a few mm) of current sheet and/or high-μ mental. This type of septum is used in NSRRC. Figure 10 shows the Lambertson iron-type septum, in which the incoming and outgoing particles bend in perpendicular directions in this septum magnet. In addition, eddy current septum and electrostatic septum are also adopted.

The bend angle of particle beam is much larger for the septum magnet than for the kickers. But the rise and falling time of the pulse should be short (usually in a few hundred ns to a few μs.) Ferrite widow-frames are usually used in which the ferrite is positioned outside the ceramic chamber.

7.7. *Timing for the injection and extraction*

Timing system in the injection and extraction processes need to be well controlled and adjusted. Longitudinal phase space should be matched between two accelerators during the injection and extraction in order to position the incoming beam inside the proper RF bucket. The adjustment of the RF phase angle between two acceleration RF systems is necessary.

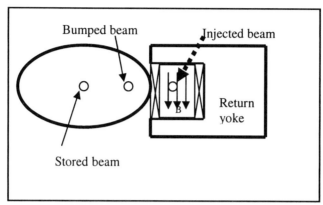

Figure 9: Schematic of a current sheet septum. The injected beam is bent into the acceptance of the stored beam by the septum or the extracted beam is directed out by the septum and into the transfer channel.

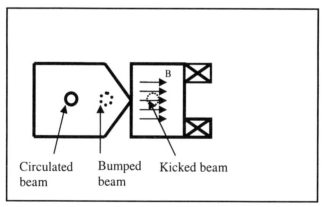

Figure 10: Schematic of a Lambertson septum. The horizontally kicked beam is bent vertically in the septum channel.

8. Summary and Acknowledgements

The closed orbit distortions resulted from the field errors and misplacements of the magnets are formulated and analyzed. Orbit measurement methods and orbit correction algorithms are discussed. Finally, injection and extraction methods and techniques are introduced.

The author would like to thank S.Y. Lee, A. Chao and H.P. Chang for the interesting discussions, and also thank H.J. Tsai, M.H. Wang, and M. Lu for their help in the preparation of this manuscript.

References

1. S.Y. Lee, *Accelerator Physics*, World Scientific, 1999.
    Helmut Wiedemann, *Particle Accelerator Physics*, Springer-Verlag, 1993.
    Alexander Wu Chao and Maury Tigner, *Handbook of Accelerator Physics and Engineering*, World Scientific, 1999.
    D.A. Edwards and M.J. Syphers, *An introduction to the Physics of high Energy Accelerators*, John Wiley & Sons, 1993.
    Herman Winick, *Synchrotron Radiation Sources - A Primer*, World Scientific, 1994.
    Philip J. Bryant and Kjell Johnson. *The Principle of Circular Accelerators and Storage Rings*, Cambridge University Press, 1993.
2. W. H. Press. S.A. Teukolsky, W.T. Vetterling, B.P. Flannery, *Numerical Recipes in C*, Cambridge University Press, 1988.
3. L.H. Yu, E. Bozoki, J. Galayda, S. Krinsky, G. Vignola, *Real Time Harmonic Closed Orbit Correction*, Nucl. Instr. Meth. A 284, 268 (1989).
4. B. Autin and Y. Marti, Closed Orbit Correction of A.G. Machines Using a Small Number of Magnets, CERN ISR-MA/7317(1973).
5. S-I. Kuakawa, S.Y. Lee, E. Perevedentsev, S. Tuner, *Beam Measurement*, Proceedings of the Joint US-Japan-Russia School on Particle Accelerators, pp.277-297, 1998.
    C.J. Bocchetta, ;, EPAC98, pp.28-32, 1998.
6. J.P. Chiou, et al, *The Upgrade of SRRC Booster Extraction*, EPAC2001, pp.145-147.

TRANSVERSE BEAM DYNAMICS: DYNAMIC APERTURE[*]

Q. QIN

Institute of High Energy Physics,
Beijing, 100039, P.R. China
E-mail: qinq@mail.ihep.ac.cn

1. Introduction

The determination of dynamic aperture is one of the most important issues that needs being considered very seriously in modern accelerator design, such as dedicated synchrotron radiation sources, high luminosity electron positron colliders or factories, and large hadron colliders. But in the early days of circular accelerators, in which dynamic aperture was larger than physical aperture in general, the study of dynamic aperture was not a crucial issue in their designs.

In modern circular accelerators, complicated non-linear forces may act on the beam particles, leading to unstable motion. The non-linearities mainly come from sextupoles, which are introduced to compensate the natural chromaticities, and the transverse multipolar fields stemming from the imperfections of magnets. The latter one is more serious in superconducting magnets. Investigating the reasons of dynamic aperture limitations and improving the dynamic aperture are main efforts for the design study of various circular machines in many laboratories. Such studies on dynamic aperture include analytical approaches, numerical simulations, or combinations of both methods.

Analytical methods try to obtain the criteria used to calculate the amplitude threshold of stable motion of particles. Numerical simulation based on the particle tracking aims at determining the dynamic aperture with powerful computer. Many computer codes were developed in different laboratories for the simulation of dynamic aperture. But both of these two methods have their difficulties to simulate a real machine, since so many complicated phenomena are combined in the non-linear dynamics. Thus, the experimental results from some real machines are important to modify the physics models in analytical approaches and the ways for numerical simulations.

---

[*]Work supported by National Natural Science Foundation of China (10275079).

In this lecture, the different approaches dealing with dynamic aperture are reviewed. The limitations of these methods are discussed. Some numerical examples on dynamic aperture determination are also given.

2. Dynamic Aperture and Some Related Concepts

2.1. *Motion of single particle*

The motion of charged particles in circular accelerators can be simply described by reference to a "design" or "ideal" trajectory. The design trajectory of charged particles contains the closed orbit in straight sections and arcs in a circular accelerator. Particles move along this orbit with the design energy, so the accelerator is an ideal or perfect machine. The so called "synchronous particle" can reach the RF accelerating cavity at the accelerating phase accurately.

But in a real machine with many kinds of imperfections in each element, the real particles move along a real trajectory. The charged particles will move no longer along the design orbit, but oscillate around the ideal trajectory. This kind of oscillation makes the length of real trajectory different from the ideal one. The energy of the charged particles will not be the design energy, but a little different from that. Thus, the velocity of the charged particles will not be the same as that of the ideal velocity. It results in a phase advance difference between the arrival of particles at the RF cavity and the field of acceleration, causing the oscillations in the energy and phase of the charged particle.

2.2. *Errors*

In a real machine, charged particles usually have some deviations of energy compared to the synchronous particle which has the design energy. The real magnets also have some errors from the ideal ones, such as field errors, misalignments, etc. Additionally, power supplies of magnets have ripples, finite resolutions in manufacture. All these can lead to the deviation of real magnetic field from the ideal field.

Some correction elements are added in accelerators to correct these deviations and keep the high quality of beams. Steering dipoles are introduced to correct the closed orbit deviation, sextupoles are for chromaticity correction, and skew quadrupoles for coupling compensation. Thus, some other errors of misalignment, manufacture, as well as non-linearities, are also introduced together with these correction elements.

As a result, all these errors affect the dynamic aperture of accelerators, especially the third generation synchrotron light source and superconducting collider, since the sextupoles with high strengths and superconducting coils with large errors induce very strong non-linearities.

2.3. *Physical aperture*

In a storage ring, a vacuum chamber is composed of various vacuum components, such as vacuum tube, valves, bellows, flanges, kickers, electro-static separators, RF cavities, diagnostic devices, etc. Charged particles move along a closed orbit in such a chamber. The transverse apertures of these vacuum components, called physical aperture, limit the beam motion along its path. It is very clear that the larger the physical aperture, the less particles loss. But on the other hand, a circular machine with a large physical aperture means a large amount of budget, as well as a big scale of the machine itself. As a result, large gaps and high fields for magnets and other elements are also required. So a suitable physical aperture is crucial to a circular machine.

2.4. *Definition of dynamic aperture*

In a very simple saying, dynamic aperture is a kind of amplitude threshold. When the amplitude of the motion of a charged particle is smaller than this threshold, the particle will not get lost as a consequence of single particle dynamics effect. When the amplitude exceeds this threshold, the betatron oscillation of the particle will not have any bounds, and the motion will become unstable. Then, the particle cannot circulate in the accelerator.

Another kind of definition of dynamic aperture is described in Fig. 1 which shows the different regions for the particle motion in circular accelerator. Within the regular and quasi-linear regions, particles survive as long as possible in vacuum chamber and their motions are stable enough. Outside the stable regions, in the weak chaotic region, there are only some stable islands, on which particles' motions keep stable. The survival time of particles are getting less and less. In the strongly chaotic region, few stable islands exist. Beyond the chaotic regions, motion of particles becomes unbounded, and the survival time approaches none. The dynamic aperture thus is given at the border between weakly chaotic and regular regions.

In the design of accelerator, it is very important to ensure the motion of the beam is bounded within a rather long time (damping times in transverse direction for electron machine), i.e., beam loss will not happen. When the initial amplitude and phase of the particle motion, (A_0, ϕ_0), are given, it is required that the amplitude at any time t should satisfy $|A(t)| < B$, where B is a finite boundary. If we can find a maximum A_0 for any phase ϕ_0 such that $A(t)$ keeps finite for a long enough time, and the betatron motion with an initial amplitude smaller than A_0 is stable, such an A_0 is called dynamic aperture[1].

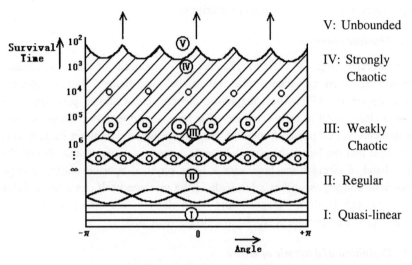

Fig. 1 Description of dynamic aperture (Courtesy W. Scandale, CERN)

The ideal case is that the dynamic aperture is equal to or larger than the physical aperture of a storage ring, which is mainly determined by the size of vacuum chamber. If the dynamic aperture is too small, people have to do some work on reducing the non-linear effects to improve the dynamic aperture. Figure 2 shows a simulation result of the dynamic aperture compared with the physical aperture of a storage ring.

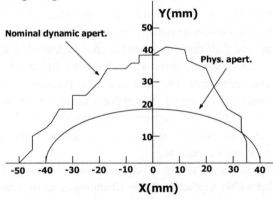

Fig. 2 An example of the simulated dynamic aperture

In Figure 2, we draw the stable horizontal amplitude as a function of the stable vertical amplitude in a finite phase space, for a given particle with a fixed momentum deviation $\delta p/p$.

2.5. Symplecticity

Let's first go back to one degree of freedom. If we write Hamilton's equation of a linear system, $H = k_x x^2/2 + p_x^2/2$, in matrix form as[2]

$$H = \frac{1}{2}(x \quad p_x)\begin{pmatrix} k_x & 0 \\ 0 & 1 \end{pmatrix}\begin{pmatrix} x \\ p_x \end{pmatrix} \equiv \frac{1}{2}\vec{X}^T\mathcal{H}\vec{X}, \tag{1}$$

where $\vec{X}^T$ is the transpose of $\vec{X}$ and $\mathcal{H} = \begin{pmatrix} k_x & 0 \\ 0 & 1 \end{pmatrix}$, we may write the equations

of motion for single particle as

$$\vec{X}' = \begin{pmatrix} 0 & 1 \\ -k_x & 0 \end{pmatrix}\vec{X} = \begin{pmatrix} 0 & 1 \\ -1 & 0 \end{pmatrix}\begin{pmatrix} k_x & 0 \\ 0 & 1 \end{pmatrix}\vec{X} \equiv S\mathcal{H}\vec{X}. \tag{2}$$

Here, we define $S \equiv \begin{pmatrix} 0 & 1 \\ -1 & 0 \end{pmatrix}$ and adopt $x' = \partial H/\partial p_x$ and $p_x' = -\partial H/\partial x$. It is easy

to find that from our definitions, we have the conditions $\mathcal{H}^T = \mathcal{H}$, $S^T S = I$, and $S^2 = -I$ with I the unit matrix.

If eq. (2) has two independent solutions, $\vec{X}_1$ and $\vec{X}_2$, it is easy to prove that the Wronskian, $\vec{X}_2^T S \vec{X}_1 = p_{x1}x_2 - x_1 p_{x2}$, is an invariant. This result is not limited only to one degree of freedom. The bilinear form, $p_{x1}x_2 - x_1 p_{x2} + p_{y1}y_2 - y_1 p_{y2} + \cdots$, extends the Wronskian to n degrees of freedom and keeps invariant. The matrix S in n degrees of freedom is

$$S \equiv \begin{pmatrix} 0 & 1 & & & & \\ -1 & 0 & & & \Large 0 & \\ & & 0 & 1 & & \\ & & -1 & 0 & & \\ & & & & \bullet & \\ \Large 0 & & & & & \bullet \end{pmatrix}. \tag{3}$$

From this invariant, we can see that if a transform matrix M propagates $\vec{X}$ between s_1 and s_2 in an accelerator, i.e.,

$$\begin{aligned} \vec{X}_1(s_2) &= M\vec{X}_1(s_1) \\ \vec{X}_2(s_2) &= M\vec{X}_2(s_1) \end{aligned}, \tag{4}$$

so

$$\vec{X}_2^T(s_2)S\vec{X}_1(s_2) = \vec{X}_2^T(s_1)M^T SM\vec{X}_1(s_1).$$ (5)

Since the Wronskian is an invariant, which means

$$\vec{X}_2^T(s_2)S\vec{X}_1(s_2) = \vec{X}_2^T(s_1)S\vec{X}_1(s_1),$$ (6)

we can easily get

$$M^T SM = S.$$ (7)

The condition expressed by eq. (7) is called symplectic condition. The matrix M, which satisfies eq. (7), is symplectic. Though we derive the symplectic condition from the linear Hamiltonian system, the matrix describing motion in the neighborhood of a particular trajectory will be symplectic if we consider the non-linear motion in a Hamiltonian system.

Given a transfer map $\mathcal{M}$, we define the Jacobi matrix $M(z^i)$ as

$$M_{ab}(z^i) = \frac{\partial z_a^r}{\partial z_b^i},$$ (8)

where z^i and z^f stand for initial and final conditions, respectively, with $a, b = 1, 6$. If the system is Hamiltonian, $M(z^i)$ is symplectic for all z^i,

$$\vec{M}^T(z^i)S\vec{M}(z^i) = S, \quad \forall z^i.$$ (9)

A transfer map $\mathcal{M}$ satisfying eq. (9) is a symplectic map. Truncating a Taylor map generally violates eq. (9).

The set of all symplectic maps forms such a group that the inverse of any symplectic map exists, and is also symplectic. The product of any two symplectic maps is again a symplectic map.

3. Determination of Dynamic Aperture

3.1. *General description*

Hamiltonian formalism is often adopted to describe the stability of particle motion and to determine the limit of the stability. Particle's motion can be given by the vector $\{x(t), p_x(t), y(t), p_y(t), z(t), p_z(t)\}$, which varies with time in a 6-dimensional phase space and $p_x(t)$, $p_y(t)$, $p_z(t)$ are canonical momenta conjugate respectively to x, y, z. Under this coordinate system, in the presence of non-linear field, the betatron motion of particle can be expressed with a Hamiltonian:

$$H = \frac{1}{2}(p_x^2 + k_x x^2 + p_y^2 + k_y y^2) + H_1,$$ (10)

where k_x and k_y are the linear focusing strengths in x and y directions, and H_1 the Hamiltonian of perturbation given as

$$H_1 = \sum_{n,m} A_{mn} x^m y^n . \tag{11}$$

If one introduces the action-angle variables (J, ψ), the new Hamiltonian of the system can be written as

$$H = v_x J_x + v_y J_y + H_1 . \tag{12}$$

Here, v_x and v_y are transverse betatron tunes of storage ring. Thus, the standard perturbation theory can be used to deal with the effects of non-linear fields.

With a symplectic transfer map $\mathcal{M}$, we can treat the Hamiltonian-generated canonical transformation which describes the evolution of the particle motion. When a particle is launched into the storage ring at initial conditions, it will return to the start plane after a single turn with transformed coordinates under the action of the map $\mathcal{M}$. Applying the transformation $\mathcal{M}$ for many times and recording the behavior of the particle, we may find the stability of the orbit of the particle. Such a procedure of iterating the transfer map to determine whether the particle motion is stable is called tracking.

For a real machine with non-linear elements such as sextupoles, the transfer map $\mathcal{M}$ is a non-linear map, leading to unstable motion.

3.2. Factors limiting dynamic aperture

1) Non-linear elements

In chromaticity dominated machines, for example, the storage ring of a third generation synchrotron radiation source, strong focusing is used to achieve low emittances and high brilliances with very strong sextupoles to correct the large chromaticities. Consequently, sextupoles with high strengths introduce various kinds of geometric and chromatic aberrations which limit the stable amplitudes, i.e., the dynamic aperture. But normally, in this kind of machine, a large dynamic aperture is necessary to accommodate the oscillations of scattered particles to obtain a long beam lifetime. So some methods are used to improve the dynamic aperture for these rings.

It is also very important to have a good dynamic aperture for electron positron colliders, since high luminosity also requires very small beam sizes in transverse direction at the interaction point, which generates large chromaticities from the quadrupoles in the mini-β insertion. Thus, strong sextupoles are introduced to correct the chromaticities with strong non-linearities, limiting the dynamic aperture of colliders. How to optimize the correction scheme and then enlarge the dynamic aperture becomes a major issue in the design of modern colliders.

Octupoles, introduced in the circular machines to provide adequate Landau damping on curing instabilities, are another kind of non-linearity source. Strategies to compensate the effect due to octupoles in linear lattice can be beneficial to the dynamic aperture.

2) Magnet errors

Errors in magnets include multipolar fields, persistent current fields and misalignments. Error dominated machines, for example, in the large hadron collider (LHC), superconducting magnets and RF cavities are applied to achieve very high energy. In such machines, the higher-order field errors of magnets are the main source of non-linearities, due to the poor field quality of superconducting magnets. Errors are composed of the systematic field imperfections which are generated by persistent current field distortions at low excitation, and the random errors from manufacturing tolerances. So the dynamic aperture is limited by these errors and needs to be improved by compensating these errors.

4. Analytical Methods to Dynamic Aperture

Though the numerical tracking of particle motion is a very powerful tool to determine the dynamic aperture of a storage ring, various analytical methods are developed to better understand the characteristics of particle motion in the presence of non-linear forces and get the amplitude limitations of stable motion. A correct analytical approach also provides good models for tracking and saves the simulation time as a result.

The analytical ways are generally restricted on low orders of the perturbing field strength. After a few order calculations, the analytical developments become difficult. In addition, another limiting factor is that, in many cases, the generalization to magnetic non-linear elements other than sextupoles dos not yet exist, or the extension to two-dimensional systems remains to be done. The convergence of the perturbation process is doubtful.

4.1. Resonance approach[3]

This method is applied in the case that a single resonance can be identified. So the transverse tune of the particle is near a rational number and a driving term acts on the particle with a frequency close to the frequency of the particle motion. Such a driving term may increase the amplitude of the particle's motion and cause it to be unstable.

We take the example of a one-dimensional point sextupole driven resonance here. In this case, the equation of motion can be expressed as

$$\frac{d^2x}{ds^2} + K(s)x = -S\delta(s)x^2,$$ (13)

where $K(s)$ represents the linear focusing of the ring, S the strength of sexupole, and the delta-function term describes the effect of the impulse (proportional to the displacement squared and S) at $s = 0$. Using the well-known "Floquet transformation" to introduce new variables u and θ defined as

$$u = \frac{x}{\sqrt{\beta}}, \qquad \theta = \int \frac{1}{\nu\beta} ds,$$ (14)

we obtain

$$\frac{d^2u}{d\theta^2} + \nu^2 u = -\nu\beta^{3/2} S\delta(\theta).$$ (15)

Here, β and ν are the linear lattice parameters. The above equation of motion is canonical with the Hamiltonian written as

$$H(u,p,\theta) = \frac{1}{2}(p^2 + \nu^2 u^2) + \frac{1}{3}\nu\beta^{3/2} Su^3 \delta(\theta),$$ (16)

where p is the momentum conjugate to u, like $p_x = dx/ds$ the momentum conjugate to x. If we take another canonical transformation in which the so called action-angle variables are introduced as

$$u = \sqrt{2J/\nu}\cos\psi, \qquad p = -\sqrt{2J\nu}\sin\psi,$$ (17)

the Hamiltonian is then

$$H(J,\psi,\theta) = \nu J + \frac{1}{3} S\nu\beta\delta(\theta)\left(\frac{2J}{\nu}\right)^{3/2}\cos^3\psi.$$ (18)

Using Fourier transforming in the driving term and neglecting the fast varying terms, we can get

$$H(J,\psi,\theta) = \nu J + \frac{1}{24\pi} S\nu\beta^{3/2}\left(\frac{2J}{\nu}\right)^{3/2}\cos(3\psi - m\theta).$$ (19)

If we make the following canonical transformation:

$$\psi \rightarrow \phi - \psi - \frac{m}{3}\theta,$$ (20)

$$J \rightarrow I = J$$

the Hamiltonian with these variables can be expressed with θ independent:

$$H(I,\phi) = (\nu - \frac{m}{3})I + \frac{1}{24\pi} S\nu\beta^{3/2}\left(\frac{2I}{\nu}\right)^{3/2}\cos 3\phi.$$ (21)

The Hamiltonian can confine the boundary of particle motion in (I, ϕ) space. By inverse transformations, one can easily return to the (x, p_x) phase space. The stability limit can be identified, together with the phase space distortions growing with amplitude or increased driving term.

Figure 3 shows the plot of phase space for an ideal machine with non-linear elements such as sextupoles. The boundary of stable amplitude is clear in the graph.

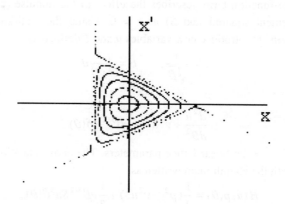

Fig. 3 Tracking result for a real machine with sextupoles

4.2. Non-resonant calculations[3]

If the tunes are not close to strong resonances, there will be no strong resonance dominating since it is possible to make the tune "irrational enough" to ensure the accelerator to operate. Thus, all the non-linear driving terms in the equation of motion need to be considered. In this case, perturbation theory can be used to deal with the non-linear behavior.

Phase space distortion calculation is one of these approaches. This method examines the "distortion" of the phase space caused by a specific nonlinear effect. For example, the phase space distortion due to sextupoles can be calculated from the equation of motion. From the Courant-Snyder invariant, a phase space ellipse will be transported in a perfect linear machine with no distortion. But a machine with non-linearities will distort this ellipse, producing a "squashed-egg" effect or other results. The distortion is attributed to a set of "distortion functions". With these distortion functions, a self-consistent result for the "shape" of a single particle phase space is developed. The analytical expressions for the first and second order distortion functions can be got, and what causes the distortion is possible to be determined.

One can find the phase space distortion through the Hamiltonian, too. Including the specific non-linear driving term, the Hamiltonian for the single particle motion can be written as[4]

$$H(u, p, \theta) = \frac{1}{2}(p^2 + v^2 u^2) + F(u, p, \theta) \qquad (22)$$

with θ the azimuthal position around the ring. In general, the non-linear form can be expressed as

$$F(u, p, \theta) = \sum_{k=3}^{\infty} A_k(\theta)u^k . \qquad (23)$$

Since the Hamiltonian is time dependent, it is not an invariant. To make H θ-independent, a constant of the motion to define the invariant curves in phase space would be provided. That the curves are closed or open means the motion is stable or not. Eliminating the θ-dependence in the Hamiltonian by a series of canonical transformations order by order, we may have the Hamiltonian with the following form after the desired number of transformations

$$\overline{H} \sim (p^2 + u^2) + B_3 u^3 + B_4 u^4 + \cdots + B_n u^n + A_{n+1}(\theta)u^{n+1} + \cdots , \qquad (24)$$

where B_n are the magnetic field coefficients due to the multipolar components of the field in the beam-line element, such as B_3 for the octupole component, B_4 for the decapole component, etc. In this equation, all terms of order n and lower are θ-independent. If we neglect terms of order $n+1$ and higher, a θ-independent invariant is obtained:

$$\overline{H} \sim (p^2 + u^2) + B_3 u^3 + B_4 u^4 + \cdots + B_n u^n . \qquad (25)$$

This invariant can be used to examine the geometry of the phase trajectories and to study the stability of single particle motion.

4.3. Other treatments

Lie transformation is used in the non-linearity analysis. With this kind of method, Dragt[5] deals with the computation of non-linear lattice functions and builds a connection between the conventional methods and the group theoretical tools. Forest[6] provides a powerful method for analysis of phase space distortion and quantification of displacements from the origin of phase space. For details, the reader is referred to the literature mentioned above.

There are some other analytical approaches, such as successive linearization method[7], secular perturbation[8], etc. Here, we will not cover all of these treatments.

5. Numerical Approaches to Dynamic Aperture

Generally speaking, analytical approaches are limited only to some simple cases. So, it is necessary for us to have a numerical modeling for the machine, including all kinds of errors and perturbations. The numerical method used to simulate the particles motion in accelerator with such a modeling is called tracking. One can get the single particle motion directly from tracking and determine the stable regions of phase space.

The process of tracking can be described as follows:
- Give certain initial conditions
- Launch a test particle at a convenient location. The particle's motion can be described with a transfer map.
- Record the position of the particle after one turn and take it as the initial condition for next turn.
- Repeat the above process for test particles with different initial conditions to identify the stable region in phase space.
- Find the maximum stable amplitude, i.e., the dynamic aperture.

Many codes have been developed to perform the above numerical iteration process for particle tracking. Here are some methods used in the tracking codes.

5.1. *Thin lens model*

The thin lens model is the simplest method applied for describing the transfer map, and also called "kick" approximation. In this model, linear elements, such as dipoles and quadrupoles, are represented with linear transfer matrices and non-linear elements, such as sextupoles and octupoles, are treated as in the impulse approximation. The transverse coordinates of the particle keep unchanged when the particle passes by these magnetic elements. The transformation can be expressed as a function of the initial conditions as follows:

$$q_f = q_i$$
$$p_f = p_i + f(q_i) \tag{26}$$

where q is the transverse coordinates x and y, and p the conjugate momentum of q. The subscripts "i" and "f" stand for the initial and final conditions.

The accelerator is then consisting of a sequence of interleaved linear matrices and non-linear impulse transformations. With the six-dimensional phase space, the energy oscillation can be dealt with and the effect of RF system can also be simulated.

The thin lens model can describe all orders of perturbations, since the transformations are easily evaluated. The more non-linear elements are in the lattice, the longer the computer time is required. This model is widely used in many accelerator designs. A large number of programs, such as BETA[9], PATRICIA[10], RACETRACK[11], TEAPOT[12], are all developed with this method.

5.2. Canonical integration method[13]

This method integrates the equation of motion numerically for particles in accelerators. An integration algorithm, which preserves Poisson brackets and is explicitly canonical, is used in this method. Each integration step is a canonical transformation.

The canonical integration can be regarded as an extension of the thin lens method, including the higher order terms in the element length. By modifying both positions and momenta by terms of linear, quadratic, or higher order in element length, the canonical integration algorithms increase the accuracy. In this method, each element (magnet and drift) is simulated using several integration steps, instead of one step, i.e., the step size is decreased and the number of steps is increased with the cost of much computing time consumption.

The program MAD[14] applies this method as an option.

5.3. Lie transformation techniques

It represents the transfer map in terms of Lie transformation, which are generated from Hamilton's equations. The Lie algebra is used in this formalism to express the Hamilton's equations as:

$$\frac{dq_i}{dt} = \frac{\partial H}{\partial p_i} = -\{H, q_i\} = -: H : q_i$$

$$\frac{dp_i}{dt} = -\frac{\partial H}{\partial q_i} = -\{H, p_i\} = -: H : p_i \qquad (27)$$

where symbol ":" is the Lie operator. The solution of the above equation can be obtained by integrating:

$$u_i(t) = e^{-t:H:} u_i(0), \qquad (28)$$

where u_i ($i = 1,2,...,6$) is any of the phase space coordinates and the exponential operator is a "Lie transformation". Equation (28) shows the exact canonical transformation of the motion in a single element. It is possible to combine several Lie transformations to get a single transformation for the transfer map through a series of elements. One can construct a canonical transformation to represent the transfer through a finite length element accurately to high order. In practice, the description is limited to third or fourth order. This formalism is applied in the program MAD and MARYLIE[15].

5.4. *Other methods*

Higher order matrix method[1], in which each component of the six-dimensional vector specifying the phase space is considered as a function of all components, is used in the program MAD and DIMAD[16]. This function is written as a Taylor's expansion and then truncated at some order. This method is restricted to second or third order, but no limitation on the length or strength for non-linear elements.

Generating function methods[1], in which the transfer map can be represented in terms of generating functions, are implemented in MAD, MARYLIE and DIMAD. This method is restricted to the description of low order non-linear terms.

Besides modeling the effects of chromaticity sextupoles or other non-linear elements in a lattice, a tracking code must deal with all non-linear effects potentially limiting the dynamic aperture including

1). any other multipole term. Systematic and random errors due to designing and manufacturing magnets are concerned.

2). closed orbits. Closed orbit distortion and its correction are important. Small apertures of magnets limit severely the allowable orbit distortions and the available dynamic aperture. In the new generation of synchrotron light sources, the lattice is so sensitive to closed orbit errors that too small a dynamic aperture might be obtained. Thus, the optimization of adequate orbit corrections becomes a crucial process in design.

5.5. *Limitations of numerical tracking*

One limitation comes from the computing power and the time needed for tracking. Different types of accelerator have different considerations. Due to the radiation damping effect, electron storage rings are short term machines and can be tracked over 10^2 to 10^4 turns only which is equivalent to a transverse damping time. But for proton rings or large hadron colliders which are long term machines, the stable operation will be for minutes at injection and hours in collision. It means that the machine should be tracked for 10^8 to 10^9 turns and, thus, the CPU time limits the tracking study.

Another limitation comes from the requirement that the motion in simulation should keep symplectic. This means the transformation used in simulation codes must be canonical. Otherwise, unphysical damping or growth of the phase space will be obtained, resulting in an incorrect estimation of the dynamic aperture.

6. Example of Particle Tracking

Here we give an example of particle tracking done on the Beijing Electron-Positron Collider (BEPC). BEPC works not only as an e^+e^- collider, but a parasitic synchrotron light source. The simulation studies have been performed for the BEPC lattice for collision beam, using the program MAD. The main source of non-linearity of BEPC storage ring is the sextupoles used to correct chromaticities. Magnetic errors, such as multipolar components and misalignment errors, influence the dynamic aperture.

In the simulation studies, one of the usual ways to represent the dynamic aperture is to make cuts in the stable volume (6-dimensional phase space) and to plot the stable horizontal amplitudes as a function of the stable vertical amplitudes for a given momentum deviation. In electron machines, practically, the vertical amplitudes are defined according to the cut $\varepsilon_y = 1/2\varepsilon_x$ (ε_x and ε_y are horizontal and vertical emittances respectively), which corresponds to full coupling.

Figure 4 shows the phase spaces for the on-momentum particles ($\delta p/p = 0$) launched at the interaction point (IP) after being tracked over 4096 turns. The innermost ellipse stands for the particle with the initial conditions of $x = 8\sigma_x$ and $y = 8\sigma_y$, the middle one with $x = 14\sigma_x$ and $y = 14\sigma_y$ initially, and the outermost ellipse for the particle with $x = 24\sigma_x$ and $y = 24\sigma_y$ initially. Figure 5 gives the dynamic aperture with different momentum deviation ($\pm 12\sigma_e$) of the BEPC storage ring. Figure 6 shows the effect of higher order magnetic fields to the dynamic aperture. Compared with the dynamic aperture without the multipolar errors it is clearly smaller than for the bare lattice.

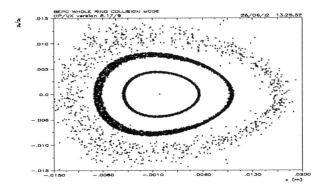

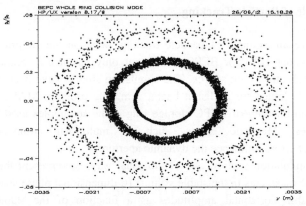

Fig. 4 Phase space portraits of three particles over 4096 turns (top: horizontal, bottom: vertical) (Courtesy Y. Wei, IHEP)

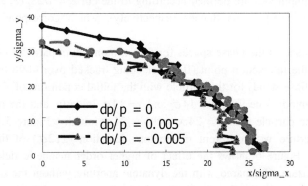

Fig. 5 Dynamic apertures with different initial condition of momentum deviation (Courtesy Y. Wei, IHEP)

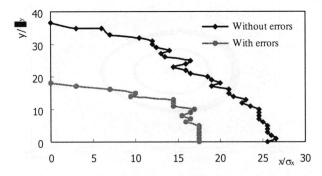

Fig. 6 Effects of magnetic multipolar errors on the dynamic aperture (Courtesy Y. Wei, IHEP)

7.   How to improve the dynamic aperture

In dynamic aperture estimation, the analytical approaches give us a better understanding on how the non-linear field influences the motion of particles in accelerators, and the numerical tracking provides a powerful tool to find the maximum stable amplitude. However, the practical problem encountered in accelerator design is how to correct the effect due to the non-linear fields and to limit the perturbation of motion. Therefore, it is important to compensate the non-linear effects and to design the scheme of correction for improving the dynamic aperture.

7.1.   *Optimization of non-linear elements*

Sextupoles are often used to correct chromaticities in storage rings. The strategy pertaining to the arrangement of sextupoles in the lattice and calculations of some physical quantities which characterize the perturbation due to non-linear fields plays an important role in the lattice design. As a result, minimization of the perturbations from non-linear fields introduced by sextupoles will enlarge the dynamic aperture. Calculations on resonance strengths, non-linear variations of transverse tunes and beta functions with momentum, tune shifts with amplitude driven by sextupoles in second order, amplitude distortions, anharmonicities will be beneficial to optimize the distribution of sextupoles.

Large scale colliders, such as LEP[17], generally consist of regular FODO cells. Besides the chromaticity sextupoles, additional sextupoles are introduced to compensate the modulation of the beta functions for off-momentum particles. The higher-order chromatic effects without generating additional non-linear resonances should be compensated with the sextupoles.

Interleaved sextupole scheme is usually adopted for such machines to provide an intrinsic cancellation of the driving terms of non-linear resonances. This scheme will work when the phase advance per cell is rational and the lattice is composed of supercells containing two identical subsections, each of which has a phase advance of $n\pi$.

Another kind of sextupole arrangement is the non-interleaved sextupole scheme[18]. Since the large chromaticities in colliders mainly come from the quadrupoles in the interaction region, local chromaticity compensation is proposed to minimize the chromatic effects with the non-interleaved sextupoles installed near these insertion quadrupoles. This sextupole compensation scheme is adopted in B-Factories, such as KEK-B[19] and PEP-II[20].

For small scale machines, which have more complex and irregular lattices than big ones and have not enough space for more sextupoles, the above compensation methods can not be applied. Compensation schemes for amplitude dependent distortions are used to maximize the dynamic aperture[21].

In synchrotron light sources, a small emittance is achieved with strong focusing, leading to high chromaticities. In some machines, harmonic sextupoles are introduced to compensate the driving terms of the most harmful harmonics and the non-linear amplitude-dependent tune shifts induced by the chromaticity sextupoles. According to the perturbation theory which utilizes harmonic expansion, the amplitude-dependent tune shifts can be written as

$$\Delta v_x = M_{11} \cdot 2J_x + M_{12} \cdot 2J_y$$
$$\Delta v_y = M_{12} \cdot 2J_x + M_{22} \cdot 2J_y \ , \tag{29}$$

where J_x, J_y are the canonical action variables. The horizontal and vertical beam trajectories can be described by

$$x = \sqrt{2J_x\beta_x} \ \cos\psi_x$$
$$y = \sqrt{2J_y\beta_y} \ \cos\psi_y \tag{30}$$

with the conjugate angle variables ψ_x and ψ_y. The coefficients M_{11}, M_{12}, and M_{22} are given by

$$M_{11} = -18\sum_m \left(\frac{A_{3m}^2}{3v_x - m} + \frac{A_{1m}^2}{v_x - m} \right)$$

$$M_{12} = 36\sum_m \left(\frac{2B_{1m}A_{1m}}{v_x - m} + \frac{B_{+m}^2}{v_x + 2v_y - m} + \frac{B_{-m}^2}{v_x - 2v_y - m} \right) , \tag{31}$$

$$M_{22} = -18\sum_m \left(\frac{4B_{1m}^2}{v_x - m} + \frac{B_{+m}^2}{v_x + 2v_y - m} + \frac{B_{-m}^2}{v_x - 2v_y - m} \right)$$

where

$$A_{1m} = \sum_k \frac{S_k}{48\pi} \beta_x^{3/2} \cos\left[\int\frac{1}{\beta_x}ds - v_x\theta + m\theta \right]$$

$$A_{3m} = \sum_k \frac{S_k}{48\pi} \beta_x^{3/2} \cos\left[3(\int\frac{1}{\beta_x}ds - v_x\theta) + m\theta \right]$$

$$B_{1m} = \sum_k \frac{S_k}{48\pi} \beta_x^{1/2} \beta_y \cos\left[\int\frac{1}{\beta_x}ds - v_x\theta + m\theta \right] \tag{32}$$

$$B_{\pm m} = \sum_k \frac{S_k}{48\pi} \beta_x^{1/2} \beta_y \cos\left[\int(\frac{1}{\beta_x} \pm \frac{2}{\beta_y})ds - (v_x \pm 2v_y\theta + m\theta \right]$$

and S_k denotes the sextupole strength. The A's, B's, and their products in eqs. (31) and (32) are resonant harmonic coefficients and drive the sextupole resonances. From the harmonic analysis of the sextupole fields in the lattice of some synchrotron radiation light sources, it is found that the main contribution to these tune shifts, expressed in eq. (29), comes from the resonance driving terms bounding the dynamic aperture. For example, in the design of SPring-8 storage ring[22], the coefficients M_{11}, M_{12} and M_{22} in eq (29) only with linear chromaticity correction are

$$M_{11} = -0.145 \times 10^6, \qquad M_{12} = -0.840 \times 10^5, \qquad M_{22} = -0.212 \times 10^5. \quad (33)$$

After introducing the harmonic sextupoles, divided into four families and located among the quadrupole triplets, the resonance driving terms and the amplitude dependent tune shifts are suppressed. The above coefficients have been reduced to

$$M_{11} = -0.503 \times 10^4, \qquad M_{12} = -0.366 \times 10^4, \qquad M_{22} = -0.359 \times 10^4. \quad (34)$$

By optimization, the dynamic aperture of the storage ring is increased by three times.

The method of harmonic correction has been successfully adopted in the design of many synchrotron light sources, such as Pohang Light Source (PLS)[23] and Shanghai Synchrotron Radiation Facility (SSRF)[24].

7.2. *Tolerance control*

Field errors in magnets are another important source of non-linear effect, especially in the superconducting machines, like the Large Hadron Collider (LHC)[25]. Therefore, tolerance control becomes dominant for such kind of accelerator. Studying the effect of the individual multipolar component of magnetic field can give designers the information of tolerance control. By optimizing the error distribution and applying the sorting technique, the dynamic aperture can be improved.

8. Summary

The dynamic aperture plays an important role in the design of modern accelerators, such as factory-like colliders and the third generation synchrotron light sources. This lecture reviews the definition of dynamic aperture, the basic treatments of dynamic aperture, such as analytical and numerical approaches, and the way to improve the dynamic aperture, together with the practical examples.

Due to the time limitation, the experimental study on dynamic aperture is not covered here. Dynamic aperture studies cover many other fields, and are progressing day by day. Numerous references to literature on dynamic aperture and related fields can be found for further studies.

Acknowledgement

The author would like to thank the organizers of the OCPA IAS'02 for their efforts on the school and the hospitality.

References

1. A. Ropert, "Dynamic Aperture", Proc. of CERN Accelerator School, CERN 90-04, 26(1990).
2. D. Edwards and M. Syphers, "An Introduction to the Physics of High Energy Accelerators", Wiley, 1993.
3. D. Douglas, "Dynamic Aperture Calculations for Circular Accelerators and Storage Rings", AIP Conference Proc. 153, 390(1988).
4. R. Ruth and W. Weng, AIP Conference Proc. 87, 1982.
5. A. Dragt, "Non-linear Lattice Functions", Proc. Of Summer Study on the Design and Utilization of the SSC, 1984.
6. E. Forest, "Normal Form Algorithm on Nonlinear Symplectic Maps", SSC Central Design Group Report, SSC-Report-29.
7. G. Guignard and H. Hagel, "Sextupole Correction and Dynamic Aperture, Numerical and Analytical Tools", CERN-LEP-Th/85-3 (1985).
8. H. Hagel and H. Moshammer, "Analytic Calculation of the Dynamic Aperture for the Two Dimensional Betatron Motion in Storage Rings", Proc. of EPAC'88, 696(1988).
9. L. Farvacque, J. Laclare, A. Ropert, "BETA User's Guide", ESRF-SR/LAT/88-08 (1987).
10. H. Wiedemann, "Chromaticity Correction in Large Storage Rings", PEP Note 220 (1976).
11. A. Wrulich, "RACETRACK, A Computer Code for the Simulation of Nonlinear Particle Motion in Accelerators", DESY 84-026 (1984).
12. L. Schachinger and R. Talman, "TEAPOT, A Thin Lens Accelerator Program for Optics and Tracking, SSC Central Design Group Note SSC-52 (1985).
13. R. Ruth, "A Canonical Integration Technique (A Symplectic Map)", LBL Note LBL-14770 (1982).
14. H. Grote and F. Iselin, "The MAD Program, User's Reference Manual", CERN/SL/90-13 (AP) (1990).

15. A. Dragt, et al, "MARYLIE, A Program for Nonlinear Analysis of Accelerators and Beam Line Lattices", IEEE Trans. Nuc. Sci., NS-32, 2311 (1983).
16. R. Servranckx, et al, "User's Guide to the Program DIMAD, SLAC Report 285, UC-28(A) (1985).
17. G. Guignard and H. Hagel, "Sextupole Correction and Dynamic Aperture —Numerical and Analytical Tools", CERN-LEP-TH/85-3 (1985).
18. K. Oide and H. Koiso, "Dynamic Aperture of Electron Storage Rings with Non-interleaved Sextupoles", Phys. Rev. E, Vol. 47, No. 3, 2010 (1993).
19. "KEKB B-Factory Design Report", KEK Report 95-7, (1995)
20. "PEP-II, An Asymmetric B Factory, CDR, LBL-PUB-5379, SLAC Report 418, (1993).
21. B. Autin, "Non-linear Betatron Oscillations", Proc. Of CERN Accelerator School, CERN 90-04 (1990).
22. "SPring-8 PROJECT, Facility Design 1991", (1991).
23. "Pohang Light Source, Conceptual Design Report", (1990).
24. "Shanghai Synchrotron Radiation Facility, Conceptual Design Report (Draft)", SSRF Report-01, (1996).
25. "LHC, The Large Hadron Collider, Conceptual Design", CERN/AC/95-05 (LHC), (1995).

LONGITUDINAL BEAM DYNAMICS—ENERGY OSCILLATION IN AN ELECTRON STORAGE RING

Y. JIN

National Synchrotron Radiation Laboratory
University of Science and Technology of China
Hefei, Anhui 230029, P.R.China
E-mail: jin@ustc.edu.cn

This lecture discusses longitudinal motion of particles in an electron storage ring. It includes the phase motion equation, small energy oscillation, large energy oscillation, and finally derives the formula of energy aperture. The phase diagram for large energy oscillation (bucket) is also given here.

1. Introduction

An electron storage ring can be considered as an electron synchrotron frozen in time. The basic function of a storage ring is that of a synchrotron, electron beams are generally not accelerated but only stored to orbit for long time of several hours, or even several ten hours [1]. The energy of the stored electrons must stay constant. But during electron beam revolution in the storage ring the energy will decrease due to synchrotron radiation, therefore there is an RF accelerating field to replenish the energy of the electrons. As in the synchrotron, an electron motion in the storage ring must meet two conditions: one is keeping orbit radius constant; another is the synchronicity condition. To preserve the synchronicity condition, the radio frequency must be an integer multiple of the revolution frequency.

Because oscillating RF fields are used, the phase of the RF fields in the accelerating cavity must reach specific values at the moment the particles arrive. At the specific values of the phase the particles can be accelerated. The synchronous particle will arrive at the same accelerating phase when it passes through the RF cavity on every revolution. This phase is called synchronous phase ψ_s. According to the principle of phase focusing, those non-synchronous particles near the synchronous phase can also be accelerated, but their phases will oscillate around the synchronous phase ψ_s. This is called phase oscillation, or phase motion. Since the particles at different phase will gain different energy, it is also called energy oscillation.

2. Phase Motion Equation

We assume E_0 (or p_0) to be the energy (or momentum) of the synchronous particle and L_0 its orbit length. With the energy difference between the non-synchronous particle and the synchronous particle ΔE (or momentum difference Δp) and the difference of their orbit length Δl we then have [2]

$$\frac{\Delta l}{L_0} = \alpha \frac{\Delta p}{p_0} = \alpha \frac{\Delta E}{E_0} \tag{1}$$

where α is the momentum compaction factor. Equation (1) describes the relation between the orbit length and the momentum (or energy). α can be expressed as

$$\alpha = \frac{\Delta l / L_0}{\Delta p / p_0} = \frac{\Delta l / L_0}{\Delta E / E_0} \tag{2}$$

Figure 1 shows the trajectories of the synchronous and the non-synchronous particle as they are moving through a dipole magnet. Let x indicate the radial

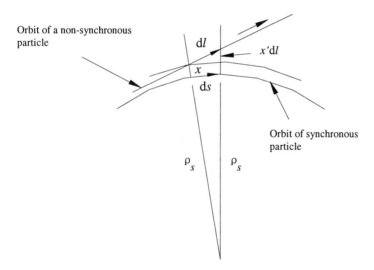

Figure 1 Electron trajectory near the design orbit

displacement of the trajectory of the non-synchronous particle with respect to the trajectory of the synchronous particle. Suppose the angle x' is small. Then, from Fig.1, we can obtain the path element dl of the trajectory at x as

$$dl = \frac{\rho_0 + x}{\rho_0} ds = \left(1 + \frac{x}{\rho_0}\right) ds = (1 + G_0 x) ds \tag{3}$$

where ρ_0 is the curvature radius of the trajectory of the synchronous particle, and G_0 is the curvature function defined as $G_0 = 1/\rho_0$. Then, the orbit length of the non-synchronous particle can be expressed by

$$L = \oint dl = \oint [1 + G(s) x(s)] ds \tag{4}$$

where $G(s)$ is a function of s, thus, formula (4) can be extended from the special case of a circular orbit to the general case of an orbit with varying curvature.

Formula (4) shows that the first term of the integral is the orbit length L_0 of the synchronous particle, the second term is the orbit length difference Δl between the non-synchronous particle and the synchronous particle. So, formula (4) can be written as

$$L = \oint [1 + G(s) x(s)] ds = L_0 + \Delta l \tag{5}$$

where Δl is

$$\begin{aligned}
\Delta l &= \oint G(s) x(s) ds \\
&= \oint G(s) \eta(s) \frac{\Delta E}{E_0} ds \\
&= \frac{\Delta E}{E_0} \oint G(s) \eta(s) ds
\end{aligned} \tag{6}$$

with $\eta(s)$ the momentum dispersion function. Substituting formula (6) into formula (2), the momentum compaction factor becomes

$$\alpha = \frac{1}{L_0} \oint G(s)\eta(s)ds \qquad (7)$$

Formula (7) shows that the momentum compaction factor is a parameter that is characteristic of the total guide field. It is a very important parameter for discussing synchrotron oscillation (energy oscillation).

If the curvature of the all dipole magnets is the same, i.e., $G(s) = G_0$ is a constant for all the bending magnets, and $G(s) = 0$ for all the straight sections, then

$$\alpha = \frac{G_0}{L_0} \int_{magnet} \eta(s)ds \qquad (8)$$

where the integral is to be taken along all of the bending magnets only. Suppose the total length of the bending magnets is l_{magnet}, and the average value of $\eta(s)$ for all bending magnets is

$$\langle \eta \rangle_{magnet} = \frac{1}{l_{magnet}} \int_{magnet} \eta(s)ds \qquad (9)$$

then we obtain

$$\alpha = \frac{\langle \eta \rangle_{magnet}}{R_0} \qquad (10)$$

Here, $R_0 = L_0/2\pi$ is the average radius of synchronous particle orbit.

As Fig. 2 shows, the synchronous electron at the center of the bunch provides a convenient reference point for the study of the longitudinal oscillation of the electrons in a bunch. Suppose an arbitrary electron $e1$ at the position A is ahead of the bunch center by the longitudinal distance y_1 at a certain azimuth position in the Fig. 1. When the synchronous electron traveled once around the design orbit to reach the same azimuth position, the electron $e1$ reached the position B, the longitudinal displacement has decreased to y_2.

140

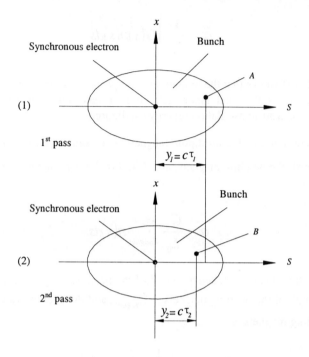

Figure 2 Longitudinal motion of an electron in a bunch

For one revolution, the synchronous electron travels a length L_0 which just equals the length of the design orbit

$$L_0 = cT_0 \tag{11}$$

where c is the speed of light, and T_0 is the revolution time of the synchronous electron. In the present discussion, we neglect the difference between the actual speed of an electron βc and the speed of light c. Since the electron $e1$ travels also at the speed c, it also has covered a path length equal to L_0. But, as $e1$ has an energy deviation ΔE from the nominal energy E_0, the path length for one complete revolution will be greater than L_0 by the amount Δl as given by

$$\Delta l = \alpha \frac{\Delta E}{E_0} L_0 \qquad (12)$$

Therefore, while the synchronous electron comes back to the same azimuth position after traveling one revolution, the electron $e1$ reaches the position B that deviates from its previous azimuth by a small distance $\delta y = -\Delta l$ which is given by

$$y_2 - y_1 = \delta y = -\Delta l = -\alpha \frac{\Delta E}{E_0} L_0 \qquad (13)$$

or

$$\frac{\delta y}{L_0} = -\alpha \frac{\Delta E}{E_0} \qquad (14)$$

From Fig. 2, we can also see that while the change in the displacement is δy, the change in time deviation from the synchronous particle after one revolution is $\delta \tau$ as given by

$$\delta \tau = \tau_2 - \tau_1 = \frac{\delta y}{c} = -\alpha \frac{\Delta E}{E_0} \frac{L_0}{c} = -\alpha \frac{\Delta E}{E_0} T_0 \qquad (15)$$

where τ_1, τ_2 are the time displacements from the bunch center that are defined by $\tau(t) = y(t)/c$. Formula (15) can be written as

$$\frac{\delta \tau}{T_0} = -\alpha \frac{\Delta E}{E_0} \qquad (16)$$

This is the time rate-of-change when the electron $e1$ travels a revolution, so we can also write it as

$$\frac{d\tau}{dt} = -\alpha \frac{\varepsilon}{E_0} \qquad (17)$$

Next, we look for the energy variation. While the electron $e1$ travels one revolution, it gains the energy $eV(\tau)$ from the RF cavity, and looses the energy $U_{rad}(\varepsilon)$ by radiation. The net change in energy during one revolution is then

$$\delta U = eV(\tau) - U_{rad}(\varepsilon) \tag{18}$$

The term of the energy radiation can be written as

$$U_{rad}(\varepsilon) = U_0 + D\varepsilon \tag{19}$$

where U_0 is defined as the energy radiated by the synchronous electron, and $D\varepsilon$ is the difference between the energy radiated by the non-synchronous and the synchronous electron. Since we shall be interested only in small energy deviations, we need to keep only the linear term. Then, the coefficient D becomes

$$D = \left(\frac{\partial U}{\partial \varepsilon}\right)_{\tau=0} \tag{20}$$

The derivative is evaluated at the nominal energy. When averaged over a complete revolution, the rate of energy change is $\delta U / T_0$, so we can write equation (18) as a differential equation

$$\frac{d\varepsilon}{dt} = \frac{eV(\tau) - U_{rad}(\varepsilon)}{T_0} \tag{21}$$

This equation is the energy oscillation equation, or phase motion equation.

3. Small Energy Oscillations

We are now ready to analyze in detail the energy oscillations of the electrons in a bunch. When we discuss the energy oscillation, we first look at the case of small oscillations for small time displacements with small energy deviations of a stored electron which we call "small energy oscillations". In this case, the variation of τ is also limited to a small interval that corresponds to an approximately linear segment of $V(\tau)$.

For small τ and ε, by equation (19), we can write equation (21) as

$$\frac{d\varepsilon}{dt} = \frac{1}{T_0}\left[eV(\tau) - U_0 - D\varepsilon\right] = \frac{1}{T_0}\left[e\dot{V}_0\tau - D\varepsilon\right] \tag{22}$$

Here we have used the linear approximation for the case that the non-synchronous electron is very near the synchronous electron (see Fig. 3).

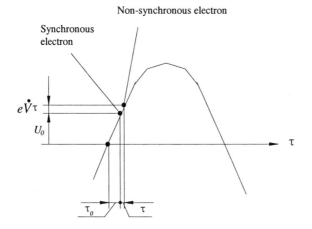

Figure 3 Small energy oscillations

From Fig. 3, we can get

$$eV(\tau) = e\dot{V}_0\tau + U_0 \tag{23}$$

Where $\dot{V}_0 = (dV/dt)_0$ stands for the derivative of the RF voltage to time at $\tau = \tau_0$.

The equation (22) can now be combined with equation (17) to give a differential equation for ε or τ. Suppose we choose τ. Taking the time derivative of the equation (17) and eliminating ε, we can get the oscillation equation for τ as following.

$$\frac{d^2\tau}{dt^2} + \frac{D}{T_0}\frac{d\tau}{dt} + \Omega^2\tau = 0 \tag{24}$$

where

$$\Omega^2 = \frac{\alpha e \dot{V}_0}{T_0 E_0} \qquad (25)$$

Here, Ω is the angular frequency of the synchrotron oscillation, and $\dot{V}_0$ is the rate of RF voltage change with time, as given by

$$\dot{V}_0 = \omega_{rf} \hat{V} \left[1 - \left(\frac{U_0}{e\hat{V}} \right)^2 \right]^{1/2} \qquad (26)$$

Equation (26) can be derived as follows: It is common for the RF voltage of a storage ring to have a sinusoidal variation with time such as

$$V(\tau) = \hat{V} \sin \omega_{rf} (\tau + \tau_0) \qquad (27)$$

where $\hat{V}$ is the peak RF voltage and $\omega_{rf} \tau_0$ is the synchronous RF phase angle. Then,

$$V_0 = \hat{V} \sin \omega_{rf} \tau_0 \qquad (28)$$

where V_0 is the value of RF voltage the synchronous electron saw when it passed through the RF cavity. Therefore for synchronous electron we would obtain

$$U_0 = eV_0 = e\hat{V} \sin \omega_{rf} \tau_0 \qquad (29)$$

Suppose the phase angles of the non-synchronous electron $e1$ and the synchronous electron are $\omega_{rf}(\tau_0 + \tau)$ and $\omega_{rf}(\tau_0)$, respectively, corresponding to the RF voltages $V(\tau)$ and V_0. So for small values of τ , the derivative of the RF voltage with respect to time is

$$\dot{V} = \frac{dV}{dt} = \frac{\Delta V}{\tau} = \frac{V(\tau) - V_0}{\tau}$$

$$= \frac{\hat{V} \sin \omega_{rf}(\tau_0 + \tau) - \hat{V} \sin \omega_{rf} \tau_0}{\tau} \approx \hat{V}\omega_{rf} \cos \omega_{rf} \tau_0$$

Therefore, when taking the derivative of the RF voltage with respect to time at the phase angle of the synchronous electron, we can write, using (29),

$$\dot{V}_0 = \left(\frac{dV}{dt}\right)_0 = \omega_{rf}\hat{V} \cos \omega_{rf} \tau_0 = \omega_{rf}\hat{V}\left[1 - \left(\frac{U_0}{e\hat{V}}\right)^2\right]^{1/2} \tag{30}$$

If we set

$$\alpha_\varepsilon = \frac{D}{2T_0} \tag{31}$$

equation (24) becomes

$$\frac{d^2\tau}{dt^2} + 2\alpha_\varepsilon \frac{d\tau}{dt} + \Omega^2\tau = 0 \tag{32}$$

The equation (32) describes a damped harmonic oscillation with the oscillation angular frequency Ω, and damping coefficient α_ε. The solution of the equation (32) can be written as

$$\tau(t) = Ae^{-\alpha_\varepsilon t} \cos(\Omega t - \theta_0) \tag{33}$$

where A and θ_0 are arbitrary constants. We can also write the solution in complex form as

$$\tau(t) = \tilde{\tau}e^{-(\alpha_\varepsilon - i\Omega)t} \tag{34}$$

where $\tilde{\tau}$ is a complex constant.

Alternatively, we can take the second order derivative of the equation (22), and, using equation (17), we will obtain

$$\frac{d^2\varepsilon}{dt^2} + \frac{D}{T_0}\frac{d\varepsilon}{dt} + \Omega^2\varepsilon = 0 \qquad (35)$$

It can be seen that the form of equation (35) for ε is the same as equation (24) for τ, therefore it has the same solution with time variable as formula (34),

$$\varepsilon(t) = \tilde{\varepsilon}e^{-(\alpha_\varepsilon - i\Omega)t} \qquad (36)$$

where $\tilde{\varepsilon}$ is also a complex constant. From equation (17), (34), (36), we derive the relation of the complex $\tilde{\varepsilon}$ and $\tilde{\tau}$ as

$$\tilde{\varepsilon} = \frac{(\alpha_\varepsilon - i\Omega)E_0}{\alpha}\tilde{\tau} \qquad (37)$$

Because $\alpha_\varepsilon \ll \Omega$ ($\alpha_\varepsilon \approx 10^2$, $\Omega \approx 10^5$), α_ε may be neglected, leading to an approximation of formula (37) such as

$$\tilde{\varepsilon} = -i\frac{\Omega E_0}{\alpha}\tilde{\tau} \qquad (38)$$

Equation (38) shows that the oscillation of ε and τ will have a phase difference of $\pi/2$.

Figure 4 shows the phase diagram of energy oscillations. The case of Fig. 4 (a) is without damping, the contour is an ellipse. The phase point moves cycling on an ellipse. The angular frequency of the phase motion is Ω. In the absence of damping, ε and τ are conjugate variables. The ratio of the major semi-axes of the ellipse is

$$\frac{\varepsilon_{max}}{\tau_{max}} = \frac{|\tilde{\varepsilon}|}{|\tilde{\tau}|} = \frac{\Omega E_0}{\alpha} \qquad (39)$$

The case of Fig. 4 (b) is with damping. The size of the ellipse decreases slowly and the trajectory of the phase motion is a slow inward spiral. The amplitude of oscillation is decreasing gradually due to damping. The damping case depends on the damping coefficient D or $dU_{rad}/d\varepsilon$. If $dU_{rad}/d\varepsilon > 0$, the electron is losing a little amount of energy while on the upper half of the ellipse. If, on the other hand, $dU_{rad}/d\varepsilon$ <0, the electron is gaining a little amount of energy while on the lower half of the ellipse.

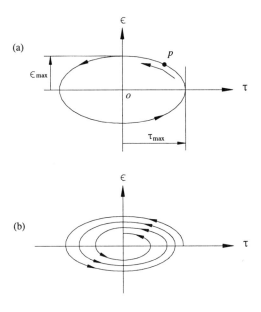

Figure 4 Phase diagram of energy oscillations, (a) without damping, (b) with damping

4. Large Energy Oscillations — Energy Aperture

We now consider the case of the oscillation amplitudes of ε and τ to be rather large, the so-called "large energy oscillations". In large oscillations, $V(\tau)$ departs significantly from a linear dependence on τ. Any storage ring guide field can usually accept particles only over a small range of energy deviations, typically only a few percent or a few thousandth of the nominal energy. If the energy deviations exceed that range, the particles will be lost. When we study the large energy oscillations, we want to determine the energy aperture, or energy acceptance, or say longitudinal acceptance. Although we shall be dealing with large energy oscillations, the maximum energy deviation will still be small such as a very small fraction of the total energy of the particles.

We begin with the two basic results of equations (21) and (19) as obtained before. Starting with equation (21), we write

$$\frac{d\varepsilon}{dt} = \frac{eV(\tau) - U_0}{T_0} - \frac{D\varepsilon}{T_0} \tag{40}$$

Taking the second order derivative of equation (17), we find

$$\frac{d^2\tau}{dt^2} = -\frac{\alpha}{E_0}\frac{d\varepsilon}{dt} \tag{41}$$

Substituting formulas (17) and (40) into equation (41), we derive

$$\frac{d^2\tau}{dt^2} = -\frac{\alpha}{E_0 T_0}[eV(\tau) - U_0] - \frac{D}{T_0}\frac{d\tau}{dt} \tag{42}$$

This equation describes the variation of τ for all oscillation amplitudes.

In order to understand equation (42), let us consider another equation

$$m\frac{d^2x}{dt^2} = F(x) - \mu\frac{dx}{dt} \tag{43}$$

This equation represents the motion of a particle of mass m in a conservative force field in which $F(x)$ is a conservative force, $\mu\frac{dx}{dt}$ is a friction force, and μ the friction factor.

Comparing equations (42) and (43), we find that they are similar in shape. Hence, we may assume that equation (42) describes the motion of a particle of unit mass in a conservative force field. The first term on the right-hand side of equation (42) looks like a conservative force

$$F(\tau) = -\frac{\alpha}{E_0 T_0}[eV(\tau) - U_0] \tag{44}$$

The second term of equation (42) is equivalent to the friction force. Since D is very small, it can be taken away in a first instance, and treated later as a perturbation term. Then, equation (42) becomes

$$\frac{d^2\tau}{dt^2} = -\frac{\alpha}{E_0 T_0}[eV(\tau) - U_0] \qquad (45)$$

Such an equation is commonly handled by defining a "potential energy" function $\Phi(\tau)$ to discuss. The function $\Phi(\tau)$ is the negative of the integral of the force. Let us define

$$\Phi(\tau) = \frac{\alpha}{E_0 T_0} \int [eV(\tau) - U_0]d\tau \qquad (46)$$

We can then analyze the motion by the principle of the law of energy conservation. According to the law of energy conservation, we may write that

$$\frac{1}{2}\left(\frac{d\tau}{dt}\right)^2 = \Phi_0 - \Phi(\tau) \qquad (47)$$

in which $1/2(d\tau/dt)^2$ is the "kinetic energy", $\Phi(\tau)$ is "potential energy", and Φ_0 is the total energy. Here we have assumed the mass of the "particle" is one unit (m=1). From equation (47), we can see that when the potential energy $\Phi(\tau)$ reaches its maximum, the kinetic energy becomes zero, and vice versa. The total energy Φ_0 is a constant in this system. The energy gain function $eV(\tau)$ and potential energy function $\Phi(\tau)$ are shown in Fig. 5 where (a) is the variation of the RF voltage, and (b) the variation of the potential energy function. There is a potential minimum at $\tau = 0$ where the electron can stay stationary to be a synchronous electron. In Fig. 5 (b), the electron at point A on the hill will slide down the hill, pass through the zero point, and coast up the other side to point B due to inertia. In the case of vanishing friction force, both A and B are at the same height $\Phi_A = \Phi_B = \Phi_0$. At τ_A and τ_B the kinetic energy will be zero. It can be seen that the motion of the electron located at point A is stable. The point C is located at the top of the potential energy curve, it is the margin point of the oscillation motion. All electrons located at $\tau > \tau_C(\tau_3)$ will be lost.

150

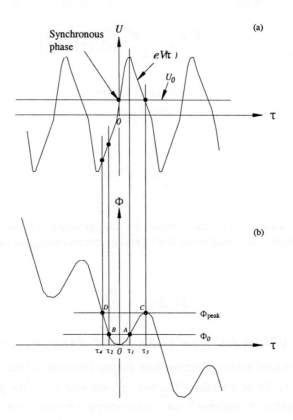

Figure 5. RF acceleration function (a) and potential energy function (b)

Point D has the same height as point C, i.e. $\Phi_D = \Phi_C$, it is also a margin point. Therefore the electrons located between $\tau_C(\tau_3)$ and $\tau_D(\tau_4)$ are stable, and all the electrons outside of this range will be unstable.

From equation (47), we can get

$$\frac{d\tau}{dt} = \pm\sqrt{2}[\Phi_0 - \Phi(\tau)]^{1/2} \tag{48}$$

where $d\tau/dt$ is equivalent to a "velocity". According to equation (17), the velocity is

$$\frac{d\tau}{dt} = -\alpha\frac{\varepsilon}{E_0} \qquad (49)$$

so the energy deviation at each τ is given by

$$\frac{\varepsilon(\tau)}{E_0} = \mp\frac{\sqrt{2}}{\alpha}[\Phi_0 - \Phi(\tau)]^{1/2} \qquad (50)$$

Using equation (50), a phase diagram for ε versus τ can be plotted as shown in Fig. 6. This phase diagram holds for large energy oscillations. The elliptical curve a is located inside the bucket, it is a stable phase trajectory. Curve b is the margin of the phase stability. Inside of the curve b, the oscillation is stable, and outside of the curve b, the oscillation is unstable. So, curve b is the separatrix.

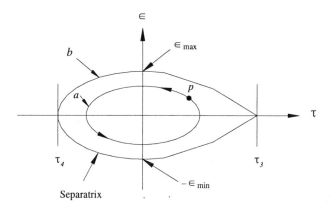

Fig. 6 Phase diagram for large energy oscillations (bucket)

We now discuss how to determine the energy aperture of a storage ring. From equation (50), we can obtain

$$\left|\frac{\varepsilon_{max}}{E_0}\right| = \frac{1}{\alpha}|2\Phi_{max}|^{1/2} \tag{51}$$

For a pure sinusoidal RF voltage, we find that

$$\Phi_{max} = \frac{\alpha U_0}{2\pi h E_0} F(q) \tag{52}$$

with

$$F(q) = 2\left[\sqrt{q^2 - 1} - \cos^{-1}\left(\frac{1}{q}\right)\right] \tag{53}$$

where $q = e\hat{V}/U_0$ is called the "over voltage coefficient". Thus the energy aperture ε_{max} for this case is then given by

$$\frac{\varepsilon_{max}}{E_0} = \left[\frac{U_0}{\pi\alpha h E_0} F(q)\right]^{1/2} \tag{54}$$

where $F(q)$ is the energy aperture function, as shown in formula (53), and h the RF harmonic number. Equation (54) describes the curve b in Fig. 6. It is the margin of phase stable region. It is also called the "RF bucket" for the region of the stable phase oscillation.

We will derive the equation (52). Setting the time displacement of the synchronous particle to τ_0, we define the time displacement of an arbitrary particle relative the synchronous particle to be τ. The RF voltage for the arbitrary particle is

$$V(\tau) = \hat{V} \sin \omega_{rf}(\tau + \tau_0) \tag{55}$$

For the synchronous particle, we have

$$U_0 = eV(\tau_0) = e\hat{V} \sin \omega_{rf}\tau_0 \tag{56}$$

Therefore

$$
\begin{cases}
\sin \omega_{rf} \tau_0 = \dfrac{U_0}{e\hat{V}} = \dfrac{1}{q} \\[4mm]
\cos \omega_{rf} \tau_0 = \sqrt{1 - \left(\dfrac{1}{q}\right)^2}
\end{cases}
\tag{57}
$$

Substituting formulas (55) and (56) into equation (46), we then get

$$
\Phi(\tau) = \frac{\alpha}{E_0 T_0}\left[\int_0^\tau e\hat{V} \sin \omega_{rf}\left(\tau + \tau_0\right)d\tau - \int_0^\tau U_0 d\tau \right]
\tag{58}
$$

From Fig. 5, we see that the potential energy is maximum for the stable particles at point C, so we integrate from 0 to $\tau_C\left(\tau_3\right)$ for the maximum potential energy $\Phi(\tau)_{\max}$. In this way, equation (58) becomes

$$
\begin{aligned}
\Phi(\tau)_{\max} &= \frac{\alpha}{E_0 T_0}\left[\int_0^{\tau_c} e\hat{V} \sin \omega_{rf}\left(\tau + \tau_0\right)d\tau - \int_0^{\tau_c} U_0 d\tau \right] \\[2mm]
&= \frac{\alpha U_0}{2\pi h E_0}\left\{ 2\left[\sqrt{q^2 - 1} - \cos^{-1}\left(\frac{1}{q}\right)\right]\right\}
\end{aligned}
\tag{59}
$$

Finally, writing

$$
\Phi(\tau)_{\max} = \frac{\alpha U_0}{2\pi h E_0} F(q)
\tag{60}
$$

we obtain the same result as in formula (52).

5.   Hamiltonian Formulation in Large Energy Oscillation

In the preceding paragraph, we have derived the phase motion equation for the large energy oscillation, and given the phase diagram, as well as the bucket. We now will derive the phase motion equation from the Hamiltonian canonical equations.

We write the equation (40) and (17) again as following

$$\begin{cases} \dfrac{d\varepsilon}{dt} = \dfrac{eV(y)-U_0}{T_0} \\[2mm] \dfrac{dy}{dt} = -c\alpha\dfrac{\varepsilon}{E_0} \end{cases} \tag{61}$$

neglecting the friction term of $D\varepsilon/T_0$, and substituting variables according to $y = c\tau$.

We may write equation (61) in the form of Hamiltonian canonical equations, then [3]

$$\begin{cases} \dfrac{d\varepsilon}{dt} = \dfrac{\partial H}{\partial y} \\[2mm] \dfrac{dy}{dt} = -\dfrac{\partial H}{\partial \varepsilon} \end{cases} \tag{62}$$

where the Hamiltonian is

$$H(\varepsilon, y) = \int\frac{\partial H}{\partial \varepsilon}d\varepsilon + \int\frac{\partial H}{\partial y}dy$$

$$= \frac{c\alpha}{2E_0}\varepsilon^2 + \frac{1}{T_0}\int_0^y[eV(y)-U_0]dy \tag{63}$$

When the variation of the RF voltage is sinusoidal, we have

$$V(y) = \hat{V}\sin\omega_{rf}\left(\frac{y+y_0}{c}\right) \tag{64}$$

where y_0 is the displacement of the synchronous particle, and y is the displacement of an arbitrary particle relative to the synchronous particle. For the synchronous particle, we have

$$\begin{cases} \sin \omega_{rf} \dfrac{y_0}{c} = \dfrac{U_0}{e\hat{V}} = \dfrac{1}{q} \\[4mm] \cos \omega_{rf} \dfrac{y_0}{c} = \sqrt{1 - \left(\dfrac{1}{q}\right)^2} \end{cases} \tag{65}$$

and

$$\omega_{rf} \frac{y}{c} = \frac{hy}{R} \tag{66}$$

where h is harmonic number, R is the average radius of the orbit. Substituting formulas (64), (65), (66) into the equation (63) we obtain after some rearrangement

$$H(\varepsilon, y) = \frac{c\alpha}{2E_0} \varepsilon^2 + \frac{ce\hat{V}}{2\pi h} \left\{ \sqrt{1 - \left(\frac{1}{q}\right)^2} \left[1 - \cos\left(\frac{yh}{R}\right)\right] + \frac{1}{q}\left[\sin\left(\frac{yh}{R}\right) - \frac{yh}{R}\right] \right\} \tag{67}$$

Using equation (67), we can calculate the phase trajectory $\varepsilon = \varepsilon\left(y|H\right)$ of the phase plane $\varepsilon - y$ as shown in Fig. 7. There are three phase trajectories for three H values representing the cases

$H < H^*$: the trajectory of the phase motion is closed which corresponds to a stable phase motion

$H > H^*$: the trajectory is not closed which corresponds to an unstable phase motion

$H = H^*$: the trajectory is a separatrix between the stable and unstable phase space regions. Its crossing point with the y axis is y^* which is given by

$$y^* = \frac{2R}{h} \cos^{-1}\left(\frac{1}{q}\right) \tag{68}$$

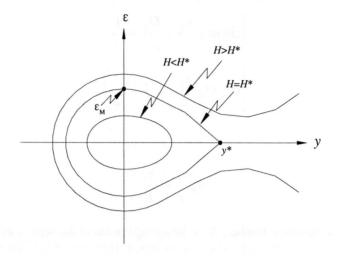

Fig. 7 Phase trajectories for longitudinal motion

and its Hamiltonian is

$$H^* = \frac{ce\hat{V}}{\pi h}\left[\sqrt{1-\left(\frac{1}{q}\right)^2} - \frac{1}{q}\cos^{-1}\left(\frac{1}{q}\right)\right] \qquad (69)$$

Thus, from equation (67), we can obtain the phase trajectory equation as following

$$\varepsilon(y) = \pm\sqrt{\frac{e\hat{V}E_0}{\pi ch}}\left\{\sqrt{1-\left(\frac{1}{q}\right)^2}\left[1+\cos\left(\frac{yh}{R}\right)\right] - \frac{1}{q}\left[\sin\left(\frac{yh}{R}\right) - \frac{yh}{R} + 2\cos^{-1}\left(\frac{1}{q}\right)\right]\right\}^{1/2}$$

$$(70)$$

Equation (70) is a boundary equation for the phase stable area. Using this equation, we can calculate the region of the phase stability. When $y = 0$, the maximum energy deviation accepted for the RF is

$$\varepsilon_{max} = E_0 \sqrt{\frac{2U_0}{\pi \alpha h E_0}\left[\sqrt{q^2-1} - \cos^{-1}\left(\frac{1}{q}\right)\right]} \tag{71}$$

In order to calculate more conveniently, we substitute the phase ψ into y in equation (70), i.e. $hy/R = \psi - \psi_s$, and ψ_s is the synchronous phase. We then obtain

$$\varepsilon(\psi) = \pm\sqrt{\frac{e\hat{V}E_0}{\pi \alpha h}}\left\{\sqrt{1-\left(\frac{1}{q}\right)^2}\left[1+\cos(\psi-\psi_s)\right] -\frac{1}{q}\left[\psi_s-\psi+\sin(\psi-\psi_s)+2\cos^{-1}\left(\frac{1}{q}\right)\right]\right\}^{1/2}$$

$$\tag{72}$$

When $\psi = \psi_s$, the RF system can accept the maximum deviation of

$$\varepsilon_{max} = \varepsilon(\psi_s) = E_0\sqrt{\frac{2U_0}{\pi \alpha h E_0}\left[\sqrt{q^2-1}-\cos^{-1}\left(\frac{1}{q}\right)\right]} \tag{73}$$

Equation (73) is called "energy aperture".

References

[1] H. Wiedemann, Particle Accelerator Physics, Springer-Verlag (1993).

[2] M. Sands, The Physics of Electron Storage Rings, An Introduction, SLAC-121 (1970).

[3] A. Renieri, Problems in Single-Particle Dynamics Specific to Electrons, CERN 77-13 (1977).

PHOTOINJECTORS

ILAN BEN-ZVI

Collider-Accelerator Department Building 817
Brookhaven National Laboratory
Upton NY 11973 USA
E-mail: benzvi@bnl.gov

Photoinjectors, or laser photocathode RF guns, are electron sources of exceptional high brightness. High brightness electron pulses are critical for certain applications such as free-electron lasers. In this article I review a bit of the development of the photoinjectors, critical issues in the physics of these devices and technological aspects.

1. Introduction

The device called "photoinjector" is a fascinating element of linear accelerators. The number of photoinjectors around the world is large and growing fast, and there is a large body of research on its many aspects. In this manuscript we will explore the science and technology of photoinjectors. This is not a survey document and does not attempt to fully describe the many variations of photoinjectors, but try to be complete in the description of the most common photoinjector, the S-band 1.6 cell device developed at Brookhaven National Laboratory (BNL). We will start by explaining why the photoinjector is relevant for a variety of applications, such as the generation of ultra-short pulses of electromagnetic radiation [Uesaka], for the generation of high-brightness photon beams and many other applications.

2. Electron bunches for high brightness

An ultra short pulse with few photons is usually not very useful. To generate ultra-short pulses of electrons, or photons that are produced by the electrons, we need high density of electrons, or many electrons in a small volume in space and time. The density in one spot is not sufficient, since we need an extended length of the interaction region in which we use the high-density electrons. Thus we need to worry about the divergence of the electrons. The combination of small divergence angles and small spatial extent lead naturally to consideration of phase-space, thus we define high brightness when an electron bunch has a high electron density in 6-dimensional phase-space. It is not sufficient to have a short bunch if its transverse dimensions are large. Tolerances alone would make such a bunch useless, since it would be impossible to make use of the small longitudinal dimension as long as the transverse size dominates. The volume occupied by a collection of particles in phase space is a conserved quantity under most circumstances.

Phase-space volume is described by emittance. Emittance is no more than a measure of the volume in phase-space. Usually one assumes independence of the two dimensional projections of the 6-dimensional phase space. Then emittance can be measured by an area in position x momentum space. A further simplification (possible at a constant energy) results from transition to configuration space of positions and angles. Thus one may talk about an emittance in units of meters x radians.

As an example, we can consider the relation between electron brightness and photon brightness: Equation 1 provides the brightness (in the usual units, photons per second per mm-mrad squared per % bandwidth) of an undulator source, where N_W is the number of periods of the undulator, K is the strength parameter, λ is the wavelength of the photons, q is the electron beam bunch charge, τ is the bunch length, and $\varepsilon x \, \varepsilon y$ are the (non-normalized) emittances in the two transverse planes.

$$B \cong \frac{2 \cdot 10^{18}}{\left(\varepsilon_x + \lambda / 4\pi\right)\left(\varepsilon_y + \lambda / 4\pi\right)} \frac{N_W K^2}{1 + K^2 / 2} \frac{q}{\tau} \tag{1}$$

The brightness is proportional to the charge divided by the 5-dimensional volume of phase space (if one neglects the photons "emittance" $\lambda/4\pi$). The expression does not contain the 6th missing phase space dimension of energy spread only because this expression assumes that the energy spread is much smaller than the undulator natural width.

Bunch compression [Uesaka, in sections 2.2.2, 2.2.3 and 2.2.4], using an energy chirp followed by an appropriate drift length (at low energy) or magnetic system, can trade off energy spread and bunch length. Therefore there is no need to generate the electron beam necessarily with a short bunch. The high-brightness is all that matters.

This leads us to the question: How does one generate the highest brightness electron bunch which still has a useful number of electrons? The answer is the photoinjector.

3. Pulsed photoinjector

The photoinjector [Fraser1, Fraser2] is a basic tool for the production of high-brightness electron beams. The term photoinjector stands for laser photocathode

RF gun. The name indicates the principles of this device. A photocathode is located inside a short microwave resonant cavity, usually followed immediately by one or more additional cavities. A high-power source of microwaves energizes the gun's cavities, thus operating as a short linear accelerator. The frequency of the microwaves (or Radio Frequency, RF) can be almost anything. Guns have been built and operated from 144 MHz to 17 GHz, but the two most common frequencies are about 1.3 GHz and about 2.9 GHz. A laser illuminates the photocathode by a short pulse (short relative to the period of the microwave) synchronized with the microwave frequency. Electrons are produced at the cathode surface by the photoelectric effect. The timing (or phase) of the laser relative to the microwave field is such that the electrons are accelerated by the electric field of the cavity. A system of solenoids provides an axial magnetic field to perform what is called emittance compensation (see later on). The magnetic field on the cathode is usually set at zero, although there are some specialized applications (flat beam or magnetized beam) that require a particular value of the field on the cathode.

Thanks to the combination of the high surface field on the cathode and the high yield of electrons possible by photo emission, a very large current density, J $\sim 10^4$ to 10^5 A/cm^2, is possible. This current density is much larger than that possible by thermionic emission (about 10 A/cm^2). The normalized thermal rms brightness is proportional to the current density. The rapid acceleration also serves to reduce the space-charge induced emittance growth. It also makes the PI a very compact accelerator.

To increase the brightness of an electron source, it is necessary to increase its peak current while keeping a very small transverse emittance. This leads to the use of high electric fields to reduce the influence of space charge forces. Since DC voltages in a gun are limited to a few hundred kilovolts, it is more appropriate to use RF fields to extract high peak current from a cathode. Following this idea, G. Westenskow and J. Madey designed and operated in 1985, the first microwave gun consisting of a thermionic cathode located in an S-band RF cavity. One of the problems of the thermionic cathode RF gun is the fact that the emission extends over a large range of the RF phase, leading to non-linear effects that limit the emittance. To obtain shorter pulses, it is natural to think about optical switching. A short laser pulse illuminating a photocathode provides an almost ideal way to produce such short pulses. It was first experimented with at Los Alamos by Fraser and Sheffield [Fraser1, Fraser2]. The combination of acceleration in a high RF field and generation of electrons by short laser pulses hitting a photocathode make a quasi perfect bright injector. Today, lasers are able to produce very short pulses (down to less then 1 ps), photocathodes can deliver high current densities (several thousands of kA/cm2) and RF cavities can sustain electric fields as high as 100 MV/m, so that RF photo-injectors can reach very high brightness.

3.1. Layout of the BNL photoinjector

A schematic diagram of a photoinjector can be seen in Figure 1. This shows the first BNL gun design [Batchelor, McDonald], later on termed "Gun I". Most guns are operated at room temperature (superconducting guns, operated at a few degrees K will be discussed later on). Such guns are constructed out of high grade OFHC copper, in some cases Hot Isostatic Pressed (HIP) copper process is used, to reduce dark currents and breakdown. In subsequent BNL guns a longer cathode cell design was adopted, beginning in Gun II [Lehrman] as well as other changes, such as a symmetric coupling and a different cathode plate in Gun III, [Palmer]. One can identify the basic elements of a typical photoinjector. There are a number of cavities, or cells. In Figure 1 there are two cavities, but there may be anything from 1 to 10 or more cells. The first one is usually shorter for two reasons. First, the electrons start at an advanced phase, in order to provide a finite field on the cathode at the time of the laser illumination. That necessitates making the cell shorter to avoid a reversal of the RF electric field before the electrons exit the cell. The other reason is that the electrons start out at a low velocity, again requiring a shorter cell to allow them to exit. The gun shown in Figure 1 has what is called 1 ½ cells, that is the first cavity is one half of the full size. The full size cavity length (period for a multi-cavity gun) is determined by mode of the structure. In the π mode the length of each unit is practically half the wavelength of the RF waveform, since the electrons are already fairly close to the speed of light.

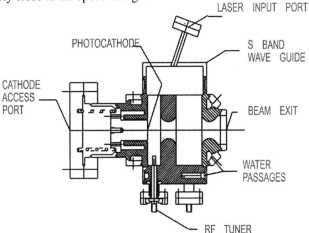

Figure 1. A schematic drawing of a photoinjector (BNL Gun I).

The BNL Gun I, for example, has a 78.75 mm long cavity, is 83.08 mm inner diameter and its beam aperture diameter is 20 mm. It has a Q of 11900 and a shunt impedance of 57 MΩ/m, which corresponds to beam energy of 4.65

MeV at a structure peak power of 6.1 MW. At this power, the peak surface electric field is 119 MV/m and the cathode field is 100 MV/m. Indeed one of the main reasons to go to the 1.6 cell gun was to reduce the peak surface electric field, which was equal to the cathode field in this variation. Another advantage is a reduction of the exit rms radial divergence of the beam caused by the exit RF kick, thanks to the somewhat higher final energy. These operating conditions can be achieved after a short period of careful RF conditioning. The non-linear RF field contribution to the emittance is minimized by providing a nearly linear dependence of the transverse fields on beam radius by a suitable cell design. To reduce the non-linear field components, the aperture can be shaped to approximate the idealized prescription [McDonald] near the aperture, given by equation 2:

$$ r = \left[a^2 - \left(4d / \pi \right)^2 \log \left(\sin \pi z / 2d \right) \right]^{1/2} \tag{2} $$

where a is the aperture radius and d is the length of the cell.

The photocathode cell has a few special features: One can identify a laser beam port and a cathode mounting flange. A beam port for injecting the laser light is common, but not absolutely necessary, since some facilities prefer to inject the laser light from the beam exit port. Such a method of injection requires placing a turning mirror for the laser light fairly close to the electron beam path. That is dangerous due to beam halo interception by the mirror. The injection at an angle into the cathode cavity has the advantage that the quantum efficiency for a 70 degrees angle is higher by about a factor of 3 [Davis]. However some of this gain is lost by the need to introduce extra optical elements whose function is to tilt the wavefront of the laser light such that emission from all parts of the cathode will take place simultaneously. The cathode mounting flange is needed for the replacement of the cavity. For semiconductor based cathodes, which are extremely sensitive to contamination, one has to introduce a load-lock chamber. This device, while complicated in construction and operation, allows one to manufacture and replace cathodes with no exposure to air. The cathode shown in Figure 1 is designed for metallic cathodes, which are robust but lower in quantum efficiency. This particular cathode has a choke-joint located at a high electric field region of the cavity. Later designs of BNL guns adopted a larger flange, sealing the back opening at a low-field, high-current part of the cavity (near the large diameter), which are less prone to breakdown and multipactoring that may take place in the choke-joint.

Both cells in Figure 1 have tuner ports and an input coupling port. Once again, such features may be different in other guns. For example, the later BNL guns have a single tuner in the full cell, relying on some tuning capability in the cathode plate and temperature tuning of the whole gun to provide the necessary tuning for both cells. Another difference would be the input coupler, which is introduced to the full cell only and is made symmetrical by the addition of an opposing port. In this case the RF power is introduced to the cathode cell through the beam iris. Water cooling is important in order to control the temperature of the gun, which has to be good to one tenth of one degree. BNL Gun IV has enough water cooling to be operated stably up to 50 Hz pulse repetition rate.

Gun construction uses electron beam welding for nearly all joints, with a few elements which are vacuum brazed to the gun (these being the beam, tuning and laser ports and in some cases a vacuum sealing surface for the cathode plate. The manufacturing tolerances are quite tight in order to maintain the exact frequency tolerances of the cells. Following welding steps the frequency is being measured and final tuning applied. Bead pulling is used to measure the field strength in each cell and tuning is applied to flatten the field profile and obtain the exact microwave frequency of the accelerator, RF and laser.

Figure 2 shows a schematic view of a gun assembly, now depicting additional features that are necessary for a working photoinjector. The photoinjector is precision mounted on a emittance compensation solenoid (a technique adopted from Gun III onwards). The gun and solenoid assembly are mounted on a support table with precision alignment adjustments. An inline vacuum valve is provided to protect the gun if downstream sections are let up to atmosphere. The gun is pumped through the waveguide connection as well as the beam pipe, usually by titanium sputter pumps. The figure also shows diagnostics chambers with various electron beam diagnostic (and optional laser injection and or visual inspection and alignment window).

3.2. Performance considerations

The following considerations affect the performance of the photoinjector:

a. The electric field of the gun, which must be as high as possible.

b. The geometry of the gun should be such that minimizes high-order terms in the electric field and maximizes the ratio of the cathode field to the peak surface electric field.

c. The geometry of the cells and RF power coupler should be such that departures from cylindrical symmetry are minimized.

d. The surface of the cavity must be smooth and clean to minimize "dark current" (field emission).

e. The gun material has to be a good electrical and thermal conductor and minimize dark current.

f. The vacuum in the gun must be very good to prolong the life of the cathode.

g. The location of the photoinjector relative to the rest of the linear accelerator is important.

h. The quantum efficiency of the cathode should be high enough that the laser would not become too difficult.

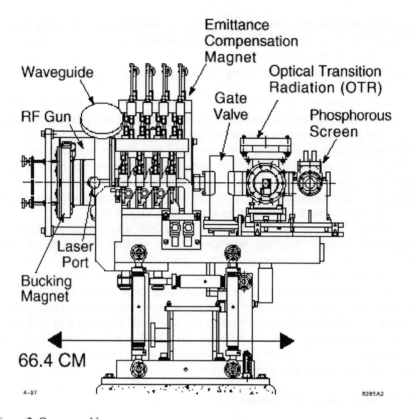

Figure 2. Gun-assembly

i. The cathode size (and laser spot size on the cathode). There is an optimum size, depending upon other parameters.

j. The cathode should produce a uniform emission.

k. The cathode should be as robust and long lived as possible (in a rather demanding environment).

l. The laser illumination should be as uniform as possible, transversely and longitudinally; and should provide a round spot on the cathode.

m. Short term and long term stability of the various systems are extremely critical: The RF power, the laser pulse energy, laser pointing, laser mode and the laser phase.

n. The solenoid producing the magnetic field must be longitudinal with a high precision to avoid breaking the cylindrical geometry.

o. Good diagnostics must be provided for the laser and electrons.

A photograph of an assembled S-band photoinjector (BNL Gun IV) is shown in Figure 3, showing the massive waveguide input coupling port and its opposing symmetrizing port, the vacuum flange for the electron beam pipe, the water cooling tubes and the stainless vacuum sealing surface between the gun body and the cathode plate.

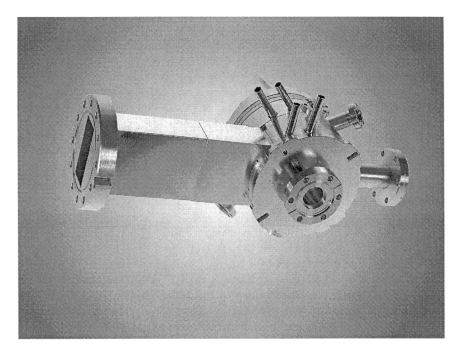

Figure 3. Photograph of an S-band photoinjector.

3.3. Laser and photocathode considerations

The mode-locked lasers that drive the photoinjector provide interesting possibilities. The pulses can be made extremely short (to sub picosecond) and intense (tens of nC at a few ps). The spatial and temporal laser power distributions can be tailored to arbitrary profiles. Particular profiles can lead to the reduction of the emittance of the photoinjector [Gallardo]. The pulse format is very flexible and pulse trains of arbitrary length and spacing can be generated.

The ideal photocathode material would have high emission efficiency (for drive laser cost containment) and high ruggedness. A study of various materials [Srinivasan1] for the photocathode has shown that certain metals have a good combination of quantum efficiency, high damage thresholds, and good mechanical and chemical stability. Copper and yttrium metal cathodes proved particularly robust. Yttrium has a work function of about 3.1 eV and quantum efficiency of up to 10^{-3} at 266 nm. Copper's work function is 4.3 eV and it has a QE of up to 10^{-4}. Magnesium is widely used in photoinjectors and has demonstrated relative high QE under modest vacuum condition [Wang]. The magnesium metal can be reliably attached to the copper back-plate by friction welding. The magnesium cathode is prepared mechanically by polishing; using three different sizes diamond polishing compounds, progressing 9 μm, to 6 μm, then to 1 μm grain size. The polished surface is rinsed with hexane and then immersed in an ultra-sound cleaning hexane bath for 20 minutes, blown with dry nitrogen, and finally placed in a high-vacuum chamber for bake-out at 150°C [Srinivasan2].

The cathode has to be laser cleaned in order to achieve its peak performance. The laser cleaning is an easy and dependable technique for improving the quantum efficiency of metallic cathodes [Srinivasan3, Wang2]. In this technique the laser intensity is increased up to a point that the vacuum in the gun increases slightly, indicating that impurities are evolved out of the cathode surface. The increase in the laser intensity is usually done by reducing its spot size on the cathode. The typical intensity is ~ 20 times lower than the damage threshold of the material: for magnesium it is about 7 mJ/cm^2 per pulse. The laser spot is scanned over the cathode surface a few times while stabilizing its energy, to avoid surface damage. The quantum efficiency that is achieved in this way can reach 2 to 3 tenths of one percent, and after a slight initial decay can be maintained at one to two tenths of one percent (at a peak field of 100 MV/m) for many months on end for vacuum levels in the 10^{-9} scale. The cleaning can be repeated an arbitrary number of times to regain the quantum efficiency after a long time or accidental poor vacuum conditions. The quantum efficiency is highly dependent on the electric field strength, in what is known as the Schottky effect [Schottky, Herring]. For electrons with energy very close to the threshold for emission (E-E$_T$<<E$_T$), the quantum efficiency is given by the following expression

$$\eta = K \left(h\nu - W + \sqrt{e/4\pi\varepsilon_0} \sqrt{\beta E} \right)^2 \tag{3}$$

Here η is the quantum efficiency, K is a material dependant constant, hν is the photon energy, W the work function, E the surface electric field at the time of photoemission and β is the field enhancement coefficient. At fields of the order of 100 MV/m the enhancement due to the term with the electric field is appreciable, 0.38 eV. The dependence of the quantum efficiency on photon energy, work function, electric field strength, field enhancement and intensity of laser cleaning has been studied [Smedley]. A number of surprising results were found. First, it was established that the field enhancement coefficient is close to one (no enhancement for the photoemission effect) even if it were much larger for field emission. Second, laser cleaning does not change the work function as it improved the quantum efficiency. Rather, it was necessary to expand the photoemission effect from a three step process [Fan, Mayer, Puff] to four steps, in which some of the electrons are intercepted near the surface by some "screen". The laser cleaning can reduce the screen value to no interception, and that recovers the full potential of the material.

Semiconductor cathodes such as Cs_3Sb or CsK_2Sb offers a much higher QE, up to several percent at 532 nm [O'Shea], but require a much better vacuum and have short lifetimes, a few hours to a few days. The longer wavelength that these cathodes require is also an asset. The laser power required increases as the wavelength becomes shorter, due to energy of the individual photons and due to the inefficiency of frequency multiplication. The cost is a factor of ~0.5, ~0.3 and ~0.2 for 2,3 and 4th harmonics of the ~1μm fundamental, respectively, in routine operation.

Another important photocathode material is cesium telluride, which is intermediate between the metallic cathodes and cathodes such as CsK_2Sb in terms of quantum efficiency and lifetime. Such cathodes find application in guns that require an intermediate average current. The Tesla Test Facility (TTF) [Schreiber] where beam currents in excess of 1 mA are produced is one good example. Quantum efficiency better than 0.5% is reported, but the required vacuum to maintain this performance is 10^{-10} torr, requiring a lock-load system.

The outstanding quantum efficiency of semiconductor photocathodes is not necessary for facilities that run at pulse repetition rates of up to about 100 Hz with charge of the order of 1 nC, since diode-pumped solid-state lasers can easily provide the necessary pulse trains for this service. However, they are essential in tow types of services: For the production of polarized electrons, where gallium arsenide cathodes must be used, or high average current photoinjectors, where the average current is over 1 mA or so.

The laser plays a significant role in the performance of the photoinjector. High power, short pulse lasers are complicated systems and require considerable attention. Fortunately the state-of-the-art of lasers has been advancing very rapidly. Diode pumped lasers such as Nd:YLF, Nd:YAG, Nd orthovanadate and other solid state lasers provide short pulses with a considerable power at reasonable costs. Other laser systems are based on a Ti:sapphire oscillator followed by various amplifiers, such as Ti:sapphire, alexandrite or excimer. While these systems are generally more complex, they provide several advantages such as higher repetition rates and the possibility of shaping the temporal intensity of the laser.

An important issue is the phase lock stability between the laser and the RF system of the gun (and linac). Sub picosecond phase lock systems are available commercially. However, one must be careful about a number of technical details, such as thermal drift due to cable and beam path length, or jitter in the energy of the high-power RF drive of the gun. The last can produce significant jitter in the flight time of the electrons, which are not totally relativistic at the gun.

3.4. *Emittance and energy spread of the photoinjector*

An early analytical model for the emittance and energy spread of a photoinjector was developed by K-J. Kim [Kim]. This model provides scaling laws that provide insight into the relationship of some of the design parameters of photoinjectors. The Kim model calculates the effects of space charge and RF fields on the emittance. The space-charge component is created by the variation of the space-charge force along the electron bunch. The RF component is due to the differential focusing applied to various parts of the bunch by the fields at the exit of the cavity. These are linear effects. Kim's model does not account for possible charge distribution changes in the bunch, to thermal emittance or to the process of emittance compensation (which does not take place in the photoinjector but in the space downstream of the photoinjector.

Using practical units, he obtained for the normalized, space-charge related emittance:

$$\varepsilon_{sc} \approx 3.8 \ 10^{3} q \left(2\sigma_{x} + \sigma_{b}\right)^{-1} \left(E_{0} Sin\phi_{0}\right)^{-1} \tag{4}$$

and for the normalized, RF related emittance

$$\varepsilon_{rf} \approx 2.7 \ 10^{-5} E_{0} f^{2} \sigma_{x}^{2} \sigma_{b}^{2} \tag{5}$$

where the emittances are expressed in π mm mrad, E_{0} is the cathode peak electric field in MV/m, f is the gun frequency in GHz, q is the charge in nC, σ_{b} is the rms bunch length in ps and σ_{x} is the rms transverse size in mm. ϕ_{0} is the launch phase, typically 30° to 60°.

For a given cathode electric field, charge and beam size, the emittance is optimized by

$$f_{opt} = 1.2 \times 10^{4} \left(\sigma_{b} \sigma_{x} E_{0} Sin\phi_{0}\right)^{-1} q^{1/2} \left(\sigma_{b} + 2\sigma_{x}\right)^{-1/2} \tag{6}$$

and then the optimized total emittance (neglecting correlations as well as thermal emittance) is

$$\varepsilon_{min} \approx \left[\varepsilon_{rf}^{2} + \varepsilon_{sc}^{2}\right]^{1/2} \approx 5.4 \times 10^{3} q \left(E_{0} Sin\phi_{0}\right)^{-1} \left(\sigma_{b} + 2\sigma_{x}\right)^{-1} \tag{7}$$

Since the minimum emittance is proportional to the charge q (and thus to the peak current), the highest brightness is not necessarily associated with the highest charge. Since we have left out the thermal emittance in these expressions, one should not conclude that the brightness is maximized for a vanishing small charge.

The minimum emittance (using the given optimized frequency) is inversely proportional to the electric field. Thus clearly for a given set of beam parameters (charge, bunch size), we would like to apply the highest possible electric field. As we increase the field (ceteris paribus) the optimal frequency is lowered. However, for a number of practical reasons the technically achievable field is smaller at lower frequencies. At some field and frequency the photoinjector will operate at the limit of breakdown or available RF power. Once we cross that limit the assumptions of the optimization break down and one can not apply these results. In addition, the thermal emittance limit may be approached. The thermal emittance of a photocathode with electron excess emission energy Δ and a cathode radius ρ is given by

$$\varepsilon_N = \frac{\rho}{4} \sqrt{\frac{2\Delta}{mc^2}} \tag{8}$$

For excess emission energy of 0.5 eV (which is believed to approximately represent metallic cathodes) the thermal normalized emittance is 0.35 microns for a cathode radius of 1 mm. The issue of excess photon energy in calculating the thermal emittance requires a bit more discussion, as the orientation of the excess photon momentum is not strictly random for electrons which have sufficient perpendicular momentum to overcome the work function. The orientation of the photon momentum is random initially, of course, but the requirement that the electrons follow a trajectory which will allow escape (step 3 in the three step model) imposes a selection on emitted electron which will forward peak their energy distribution, and thus reduce the thermal emittance.

The beam of a photoinjector has significant correlations of the longitudinal position and transverse phase space. This is the key to 'emittance correction' schemes (discussed later on). When one uses an emittance correction scheme of one sort or another, the space charge emittance is reduced. This will invalidate the conditions of the calculation presented above, pushing the optimum towards lower frequencies, lower electric fields and smaller beam size. In addition to the above discussion, the high brightness of the photoinjector will be diluted by any one of a large collection of effects: Wake fields, beam transport aberrations, dispersion, skew quadrupoles and more. For example, ambient magnetic field B on the cathode (from magnetic lenses or ion pumps), produces an emittance increase given by

$$\varepsilon_n \approx \left[\varepsilon_{n0}^2 + e^2 B^2 \sigma_r^4 m^{-2} c^{-2} \right]^{1/2} \tag{9}$$

Thus, fields of the order of 10 gauss may be detrimental. Good design practices call for a rapid acceleration of the beam to a few tens of MeV before applying dipole fields. Thus, magnetic pulse-compression is better done above, say, 70 MeV. Pulse-compression has always been part of conventional electron gun technology. Although the beam pulses of a photoinjector start out short, the brightness can be further increased by magnetic pulse-compression.

An important point that was not fully appreciated in early work is the significance of the uniformity of the emission from the cathode. A non uniformity can result from either a non-uniform laser illumination or non-uniform quantum efficiency across the area of the cathode. It turns out that the laser cleaning technique described above leads to a very uniform quantum efficiency, which may explain in part the high brightness of the BNL Accelerator Test Facility (BNL-ATF) beam, where this cleaning technique is routinely applied. In a recent experimental study [Zhou], F. Zhou et al. characterized the emittance as a function of non-uniformity (creating artificial non-uniformity by laser masks). One of the products of this study is shown in Figure 4, in which the emittance is measured for a range of rms non-uniformity values for a checkerboard pattern of non-uniformity. With a highly uniform emission from its

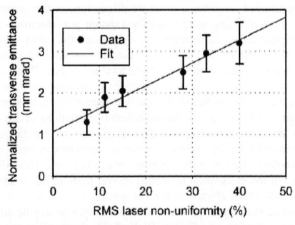

Figure 4. The effect of laser non-uniformity of photoinjector emittance.

high quantum efficiency magnesium cathode, the BNL-ATF photoinjector exhibited a record brightness, as described by Yakimenko. The result was an rms normalized emittance of 0.8 μm for a bunch charge of 0.5 nC.

3.5. *Emittance correction*

An interesting photoinjector subject is emittance correction. We define the 'slice-emittance' as the transverse emittance measured for a short longitudinal slice of the bunch. It has been observed computationally [Carlsten] that the slice-emittance is considerably smaller than the total emittance (that is integrated over the full length of the bunch). This effect is due to the variation of the space-charge force as a function of longitudinal position in the bunch within a regime of longitudinal laminarity: indeed, since electrons of a slice do not mix with electrons of other slices (lack of synchrotron motion), each slice behaves like an independent beam subject to a space charge field which varies from slice to slice (the amount of variation, in a rms sense, is minimized for a uniform distribution of the current in the bunch). Bruce Carlsten [Carlsten] proposed a simple scheme of reducing the total emittance by using the space-charge force to compensate its own effect. The method employs a lens set to produce a beam size waist with no cross-over. The electrons 'reflect' relative to the beam axis due to space-charge forces. This condition, that can restore the effects of the linear space-charge force, has been verified in an experiment [Qiu]. In this experiment, the emittance of a picosecond long slice of an electron bunch was measured. A short slice is selected out of an energy chirped beam by a slit in a dispersive region. The emittance is measured using the quadrupole scan technique. The process of emittance compensation of the beam is observed by repeating the measurement for various values of the compensating solenoid and for several slices.

The Carlsten emittance correction technique corrects linear space-charge effects. Other correction schemes have been proposed [Gallardo, Gallardo2, Serafini2] to produce the same correction by laser pulse shaping, Radio Frequency Quadrupoles and asymmetric RF cavities, respectively. However the Carlsten scheme is simple and has been tested experimentally. Other correction schemes have been proposed to correct RF time dependent effects [Serafini1] and non-linear space-charge effects [Serafini2, Gao]. Finally, a correction scheme for ultra-short, disk-like bunches using an optimized charge distribution has been proposed by Serafini [Serafini3]. Rosenzweig and Colby [Rosenzweig] provided wavelength and charge scaling laws for photoinjectors as a function of various parameters such as bunch size and electric field. One finds that, when optimum conditions are maintained for a variable wavelength, the gun electric field and solenoid magnetic field must scale inversely with the wavelength, the beam size σ (in any dimension), the charge q and the emittance scale with the wavelength, and the brightness with the wavelength to the inverse second power

$$E, B \propto \lambda^{-1} \qquad \sigma, q, \varepsilon \propto \lambda \qquad B \propto \lambda^{-2} \qquad (10)$$

If the charge is constrained to a particular value, the scaling is different (Rosenzweig2). The beam size then would be proportional to the cube root of the charge:

$$\sigma \propto q^{1/3} \tag{11}$$

Also, the normalized rms emittance is given as a function of the charge as the following:

$$\varepsilon_N = \sqrt{aq^{4/3} + bq^{8/3}} \tag{12}$$

where a and b are constants that depend on the specific photoinjector. Rosenzweig et al. (Rosenzweig2) provide the values of a, b for a Plane Wave Transformer (PWT) photoinjector as a=1.34, b=0.11 when the emittance is in microns and the charge in nC.

The emittance correction mechanism was fully explained in a beautiful work by Serafini and Rosenzweig [Serafini4]. They developed a theory of longitudinal slice-by-slice electron bunch propagation with space charge, electric and magnetic fields that describe the photoinjector and the linac that follows it. In this theory there is an invariant envelope (which is basically a beam envelope for a matched beam that will undergo optimal emittance correction): this predicts an evolution of the beam rms spot size through the booster linac following the gun that scales like the inverse of the accelerating gradient times the square root of the current divided by the beam energy. If the beam is properly matched out of the gun into the booster linac by adjusting a beam waist at the linac entrance and regulating the linac gradient in such a way to respect the invariant envelope condition, the final emittance at the linac exit would be minimized. Advanced designs of high brightness injectors for X-ray FEL, like the LCLS injector, make full profit of these theoretical criteria, achieving on simulations best emittances [Bolton].

4. CW photoinjector

Continuous wave (CW) photoinjectors are a separate class of devices with very challenging problems in quantum efficiency (for a high average current), demanding lasers and (for normal-conducting devices) extremely difficult thermal management problems. No true CW has been in operation. The device which came closest is the Boeing gun [Dowell]. The 433 MHz Boeing gun has been operated at a 25% duty factor with an average current of 135 mA during the macropulse, driven by a 13-watt laser on the photocathode. One may expect rapid progress on this subject, since such a photoinjector has a unique potential for a high average current, high-brightness electron beam. Applications include ultra-high power Free-Electron Laser and electron cooling of heavy ion, high energy storage rings. Work is in progress at Los Alamos National Laboratory, Advanced Energy Systems Inc. and Brookhaven National Laboratory.

Another direction of CW photoinjector development is the super-conducting photoinjector. This device holds the promise of continuous (100% duty factor) operation at very low RF power and very high stability. Since the RF power is the largest cost item in a photoinjector, superconducting devices also hold the promise of lower system cost. There are two main problems. One is the need for a demountable cathode plate, something that is difficult to maintain in a superconducting cavity that is supposed to achieve a high field. The other problem is that the high quantum efficiency needed for high current applications uses materials that are not compatible with the strict cleanliness demanded by superconducting RF technology. The first attempt at this challenging but most promising device was made at the University of Wuppertal [Michalke], using a CsK₂Sb cathode. A newer design has been produced by a wide collaboration with a photoinjector located at Rossendorf [Janssen]. This photoinjector has demonstrated an electron current of 80 microamperes in a recent result [Janssen2]. A different approach towards a superconducting photoinjector was taken by a collaboration of industry and Brookhaven National Laboratory. This gun is constructed of solid niobium [Cole] with no demountable cathode plate. The cathode material is niobium, cleaned by the same process described above for magnesium, including laser cleaning. The resulting quantum efficiency, tested on laboratory samples, can be as good as copper, at 0.01%. Thus this gun, shown in Figure 5 is promising for medium average electron currents, up to about one mA.

Figure 5. An all niobium superconducting photoinjector

Finally, I will mention in passing that photoinjectors are used to produce what is called magnetized electrons. Electrons that are generated with a solenoidal longitudinal magnetic field on the cathode acquire angular momentum. This shows up usually as undesired emittance as can be seen from eq. (9), but in certain applications, such as electron cooling or producing flat beams for linear colliders this property is important.

Acknowledgements

This work was done under the auspices of the United States Department of Energy.

References

[Batchelor] K. Batchelor, J. Sheehan and M. Woodle, BNL-41766 (1988), Upton NY 11973.

[Bolton] P.R. Bolton et al., Nucl. Instr. & Meth. A483, 296 (2002)

[Carlsten] B.E. Carlsten, Nucl. Instr. & Meth. A285, 313 (1989).

[Cole] M. Cole et al., Proc. 2001 Particle Accel. Conf. Chicago IL, June 18-22, 2001, p. 2272.

[Davis] P. Davis et al., Proc. 1993 Part. Accel. Conf., May 17-20, Washington DC.

[Dowell] D. H. Dowell et. al., Appl. Phys. Lett. 63 (15), 2035 (1993)

[Fan] H.Y. Fan, Phys. Rev. 68, 43 (1945)

[Fraser1] J. Fraser, R. Sheffield, E. Gray and G. Rodenz, IEEE Trans. Nucl. Sci. 32 (5), 1791 (1985);

[Fraser2] Fraser J., Sheffield R., Nucl. Instr. & Meth. A250 (1986), pp. 71-76.

[Herring] C. Herring, M. H. Nichols: Rev. Mod. Phy. 21, 185 (1949)

[Gallardo] J. C. Gallardo and R.B. palmer, Proc. Workshop Prospects for a 1Å FEL, p. 136, BNL 52273 1990, Upton NY

[Gallardo2] J.C. Gallardo and R.B. Palmer, IEEE J. of Quantum Electronics 26 No.8, 1328 (1990).

[Gao] J. Gao, Nucl. Instr. & Meth. A304, 353 (1991).

[Janssen] D. Janssen et al., Proc. 1999 Particle Accelerator Conf., NY NY, March 29th - April 2nd, 1999 p.2033

[Janssen2] D. Janssen, private communication, March 6, 2002.

[Kim] K-J. Kim, Nucl. Instr. & Meth. A275, 201 (1989).

[Lehrman] I.S. Lehrman et al., Nucl. Instr. & Meth. A318, 247 (1992).

[Mayer] H. Mayer, H. Thomas, Z. Physik 147, 419 (1957)

[McDonald] K. McDonald, IEEE Trans. Elec. Dev. ED-35 2052, 1988.

[Michalke] Michalke A. et al., Proc. European Particle Accelerator Conf., Berlin, March 24-28, 1992, p. 1014.

[O'Shea] P.G. O'Shea et. al. Nucl. Instr. & Meth. A318, 52 (1992).

[Palmer] D. T. Palmer et al., Proc. 1997 Particle Accelerator Conf. Vancouver BC May 12-16, 1997 p. 2843.

[Puff] H. Puff, Phys. Stat. Sol. 1, 636, 704 (1961)

[Qiu] X. Qiu, K. Batchelor, I. Ben-Zvi and X.J. Wang, Phys. Rev. Let. 76, 3723 (1996)

[Rosenzweig] J.B. Rosenzweig and E. Colby, AIP Conference Proceedings 335, 724 (1995).

[Rosenzweig2] J.B. Rosenzweig et al., Proc. 1999 Particle Accelerator Conf., NY NY, March 29-April 2, 1999 p. 2045.

[Schreiber] S. Schreiber et al., Proc. 2002 European Particle Accelerator Conf., Paris France, June 3-7, 2002 p. 1804

[Serafini1] L. Serafini, R. Rivolta and C. Pagani, Nucl. Instr. and Meth. A318, 301 (1992).

[Serafini2] L. Serafini, R. Rivolta, L. Terzoli and C. Pagani, Nucl. Instr. and Meth. A318, 275 (1992).

[Serafini3] L. Serafini, AIP Proc. 3rd Workshop on Advanced Accel. Concepts, Port Jefferson NY June 14-20 1992.

[Serafini4] L. Serafini and J.B. Rosenzweig, Phys. Rev. E 55 7565 (1997).

[Schottky] W. Schottky: Ann. Physik 44, 1011 (1914)

[Smedley] John Smedley, Ph.D. thesis, Stony Brook University, Stony Brook NY, 2001.

[Srinivasan1] T. Srinivasan-Rao, J. Fischer and T. Tsang, Journal Appl. Phys. 69, 3291 (1991).

[Srinivasan2] T. Srinivasan-Rao, J. Fischer and T.Tsang, J. Opt. Soc. of Am. B Vol 8, No 2, 294, (1991)

[Srinivasan3] T. Srinivasan-Rao et al., Proc. 1997 Particle Accelerator Conf. Vancouver BC May 12-16, 1997   p. 2790.

[Uesaka] M. Uesaka, Editor, Femtosecond Beam Science, Imperial College Press/World Scientific, 2003

[Wang] X.J. Wang et al, Nucl. Instr. And Meth A356 (1995) 159-166.

[Wang2] X. Wang et al., Proc. Linac'98 conference, Chicago, August 23-28, 1998, page 866.

[Yakimenko] V. Yakimenko et al., Nucl. Instr. & Meth. A483, 277 (2002)

[Zhou] F. Zhou et al., Phys. Rev. ST Accel. Beams 5, 094203 (2002)

SYNCHROTRON RADIATION*

LEE C. TENG

Argonne National Laboratory
Advanced Photon Source
9700 South Cass Avenue
Argonne, IL 60439, USA
E-mail: teng@aps.anl.gov

As early as 1900, immediately after the pioneer formulation by Liénard and Wiechert of the retarded potentials of a point charge (electron), calculations of the characteristics of the radiation from an accelerated electron or electron beam have been performed. But it was not until 1947 when John Blewett actually observed the synchrotron radiation from the beam in the 70-MeV General Electric electron synchrotron, the phenomenon was accepted as physical reality instead of some interesting mathematical deduction. It was soon recognized that the synchrotron radiation provides an extremely brilliant photon beam over a very broad frequency range from infrared to hard x-ray, and hence makes an ideal tool for use in extranuclear research. By the end of the twentieth century there were dozens of electron storage rings operated around the world at energies from a few hundred MeV to 8 GeV for experiments. Here we will study first the properties of the synchrotron radiation and then the effects of the emission of synchrotron radiation on the electron beam.

1. Properties of the Synchrotron Radiation

Classical theory is entirely adequate for the evaluation of the copious emission of photons (bosons) by electrons (fermions).

The calculation of the spectral and angular distribution of the synchrotron radiation is straightforward but laborious. We will only sketch the procedures and give the results here, and refer the students to standard textbooks for the detailed mathematics.

1.1. *Basic Electromagnetic Procedures and Formulas*

We start with the Liénard-Wiechert potential for a point-charge e (electron).

---

* This work is supported by the U.S. Department of Energy, Office of Basic Energy Sciences, under Contract No. W-31-109-ENG-38.

$$
\begin{cases}
\text{Scalar potential} \quad \phi\left(\vec{r},t\right) = e\left(\dfrac{1}{\kappa R}\right)_{t'} \\[4mm]
\text{Vector potential} \quad \vec{A}\left(\vec{r},t\right) = e\left(\dfrac{\vec{\beta}}{\kappa R}\right)_{t'}
\end{cases}
\tag{1}
$$

where

$$
R\hat{n} \equiv \vec{r} - \vec{r}_e = \text{vector from e} \left(\text{at } \vec{r}_e\right) \text{ to field point at } \vec{r},
$$

$$
\vec{\beta} \equiv \frac{1}{c}\left(\overrightarrow{\text{velocity}}\right) \quad \text{of electron,}
$$

and where the right-hand-side expressions are evaluated at the retarded time t' given by

$$
t = t' + \frac{R\left(t'\right)}{c}.
$$

Also we have denoted the retardation factor by

$$
\kappa \equiv \frac{dt}{dt'} = 1 - \hat{n}\cdot\vec{\beta}.
$$

These potentials then give the field vectors

$$
\begin{cases}
\text{Electric field} \quad \vec{E} \equiv -\vec{\nabla}\phi - \frac{1}{c}\dot{\vec{A}} \\[4mm]
\qquad = e\left(\dfrac{1}{\gamma^2}\dfrac{\hat{n}-\vec{\beta}}{\kappa^3 R^2}\right)_{t'} + \dfrac{e}{c}\left\{\dfrac{\hat{n}\times\left[\left(\hat{n}-\vec{\beta}\right)\times\dot{\vec{\beta}}\right]}{\kappa^3 R}\right\}_{t'} \\[4mm]
\text{Magnetic field} \quad \vec{B} = \vec{\nabla}\times\vec{A} = \hat{n}\times\vec{E}
\end{cases}
\tag{2}
$$

where in $\vec{E}$ the first term is dependent only on the velocity $\vec{\beta}$ and drops off rapidly as $\dfrac{1}{R^2}$. This field exists only in the neighborhood of the charge and is the Coulomb field (near field or longitudinal field). The second term depends on the acceleration $\dot{\vec{\beta}}$ and drops off slowly as $\dfrac{1}{R}$. This field extends to infinity and is the radiation field (far field or transverse field). For the last expression of the magnetic field $\vec{\beta}$ in Eq. (2), as for all later formulas, we specialize into the "radiation zone" where only the radiation field exists.

The Poynting vector (energy flux vector) is

$$\vec{S} = \frac{c}{4\pi}\vec{E}\times\vec{H} = \frac{c}{4\pi}\vec{E}\times\vec{B} = \frac{c}{4\pi}\left|\vec{E}\right|^2\hat{n} \qquad \text{(in vacuum)}, \qquad (3)$$

and the angular distribution of the radiation power P_γ or energy U is

$$\frac{dP_\gamma}{d\Omega} = \frac{d^2U}{dt\,d\Omega} = R^2\left(\vec{S}\cdot\hat{n}\right) = \frac{c}{4\pi}\left|R\vec{E}\right|^2$$

$$= \frac{e^2}{4\pi c}\left|\frac{\hat{n}\times\left[\left(\hat{n}-\vec{\beta}\right)\times\dot{\vec{\beta}}\right]}{\left(1-\hat{n}\cdot\vec{\beta}\right)^3}\right|_{t'}^2 \equiv \frac{e^2}{4\pi c}\left|\vec{f}(t')\right|^2. \qquad (4)$$

The total radiation power can be derived from Eq. (4) by integrating over solid angle Ω, or more easily by a covariant transformation of the simple expression in the electron rest-frame back to the lab-frame. This gives

$$P_\gamma = \frac{2}{3}\frac{e^2}{m^2c^3}\left(\left|\dot{\vec{p}}_\parallel\right|^2 + \gamma^2\left|\dot{\vec{p}}_\perp\right|^2\right), \qquad (5)$$

where $\vec{p}_\parallel$ and $\vec{p}_\perp$ are the component momenta parallel and perpendicular to the electron velocity $\vec{\beta}$. Equation (5) shows that perpendicular acceleration yields much more powerful radiation.

To get the spectral (frequency) distribution we use the Parseval theorem

$$\int_{-\infty}^{\infty} \left| \vec{f}(t') \right|^2 dt' = \int_{-\infty}^{\infty} \left| \vec{g}(\omega) \right|^2 d\omega, \tag{6}$$

where

$$\vec{g}(\omega) \equiv \frac{1}{\sqrt{2\pi}} \int_{-\infty}^{\infty} \vec{f}(t') e^{-i\omega t'} dt' = \text{Fourier Transform of } \vec{f}(t').$$

Applying this theorem to Eq. (4) and carrying out some simplifying procedures involving partial integration, we get

$$\frac{d^2 U}{d\omega d\Omega} = \frac{e^2 \omega^2}{4\pi^2 c} \left| \int_{-\infty}^{\infty} \left[\hat{n} \times \left(\hat{n} \times \vec{\beta} \right) \right] e^{-i\omega \left(t - \frac{\hat{n} \cdot \vec{r}_e}{c} \right)} dt \right|^2. \tag{7}$$

1.2. Radiation from a Circular Orbit

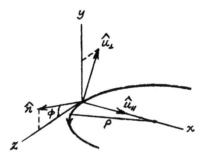

Figure 1. Coordinates for circular orbit.

We take the orbit plane to be the xz plane and orient the right handed coordinate system (x,y,z) as shown in Figure 1. Evaluating Eq. (7) in the $\left(\hat{u}_\parallel, \hat{u}_\perp, \hat{n} \right)$ system we get

$$\vec{g}(\omega) = \sqrt{\frac{3}{2\pi}} \gamma \left(1 + \gamma^2 \phi^2 \right) \left(\frac{\omega}{\omega_0} \right) \left[K_{\frac{2}{3}}(\xi) \hat{u}_\parallel - i \frac{\gamma \phi}{\left(1 + \gamma^2 \phi^2 \right)^{\frac{1}{2}}} K_{\frac{1}{3}}(\xi) \hat{u}_\perp \right] \tag{8}$$

and

$$\frac{d^2U}{d\omega d\Omega} = \frac{e^2}{2\pi c}\left|\vec{g}(\omega)\right|^2 = \frac{3e^2}{4\pi^2 c}\gamma^2\left(1+\gamma^2\phi^2\right)\left(\frac{\omega}{\omega_c}\right)^2\left[K_{\frac{2}{3}}^2(\xi) + \frac{\gamma^2\phi^2}{1+\gamma^2\phi^2}K_{\frac{1}{3}}^2(\xi)\right], (9)$$

where

$$\xi \equiv \frac{1}{2}\frac{\omega}{\omega_c}\left(1+\gamma^2\phi^2\right)^{\frac{3}{2}}, \quad \omega_c \equiv \frac{3}{2}c\frac{\gamma^3}{\rho} \equiv \text{critical frequency},$$

and where

$$K_{\frac{1}{3}}(\xi) \equiv \frac{\sqrt{3}}{2}\int_{-\infty}^{\infty}\exp\left[-i\frac{\xi}{2}\left(3x + x^3\right)\right]dx$$

and

$$K_{\frac{2}{3}}(\xi) \equiv i\frac{\sqrt{3}}{2}\int_{-\infty}^{\infty}x\exp\left[-i\frac{\xi}{2}\left(3x + x^3\right)\right]dx$$

are the fractional order modified Bessel functions. Because of the horizontal revolution of the motion, Eq. (7) applies essentially to a single pass and U is to be interpreted as the radiation energy per turn.

The features of this spectral-angular distribution, Eq. (9), are the following:

a. The vertical angle ϕ appears only in the combination $\gamma\phi$, hence scales as $\frac{1}{\gamma}$. The Bessel functions $K_{\frac{1}{3}}(\xi)$ and $K_{\frac{2}{3}}(\xi)$ fall off sharply for $\xi > 1$.

 Thus, the vertical spread angle of the radiation is roughly $\phi \sim \pm\frac{1}{\gamma}$

 which is very small at high γ.

b. Because of the orbit turning the horizontal angular spread of the radiation is not meaningful. The radiation is spread out in a horizontal sheet.

c. The frequency ω appears only in the combination $\dfrac{\omega}{\omega_c}$, hence the radiation spectrum scales with ω_c. Since $\omega_c \propto \gamma^3/\rho \propto \gamma^2 B$ one can stretch or compress the whole spectrum by adjusting the bend field B and/or the electron energy $mc^2\gamma$.

d. The $K_{\frac{2}{3}}$ term with $\vec{E}$ polarized parallel to the orbit plane $(\hat{u}_{\parallel})$ is called the σ-mode. The $K_{\frac{1}{3}}$ term with $\vec{E}$ polarized along $\hat{u}_{\perp}$, nearly perpendicular to the orbit plane, is called the π-mode.

e. In the forward direction $\phi = 0$ and

$$\left(\frac{d^2U}{d\omega d\Omega}\right)(\phi = 0) = \frac{3e^2}{4\pi^2 c}\gamma^2\left(\frac{\omega}{\omega_c}\right)^2 K_{\frac{2}{3}}^2\left(\frac{1}{2}\frac{\omega}{\omega_c}\right). \tag{10}$$

One gets a pure σ-mode.

f. Integrating Eq. (9) over the spectrum, we get the pure angular distribution

$$\frac{dU}{d\Omega} = \frac{e^2}{16}\frac{\gamma^5}{\rho}\frac{1}{\left(1+\gamma^2\phi^2\right)^{5/2}}\left(7 + 5\frac{\gamma^2\phi^2}{1+\gamma^2\phi^2}\right), \tag{11}$$

which exhibits clearly the vertical spread of $\phi \sim \pm\dfrac{1}{\gamma}$ and the γ^5/ρ dependence.

g. Integrating Eq. (9) over the solid angle, one gets the pure spectral distribution

$$\frac{dU}{d\omega} = \frac{8\pi e^2}{9c}\gamma\left[\frac{9\sqrt{3}}{8\pi}\left(\frac{\omega}{\omega_c}\right)\int_{\frac{\omega}{\omega_c}}^{\infty} K_{\frac{5}{3}}(x)dx\right] \equiv \frac{8\pi e^2}{9c}\gamma S\left(\frac{\omega}{\omega_c}\right), \tag{12}$$

where the function $S(\xi)$ is normalized $\left(\int_0^\infty S(\xi)d\xi = 1 \right)$ and is plotted in Figure 2. It has a maximum of ~0.6 at $\dfrac{\omega}{\omega_c} \sim \dfrac{1}{4}$ and a long tail toward high $\dfrac{\omega}{\omega_c}$.

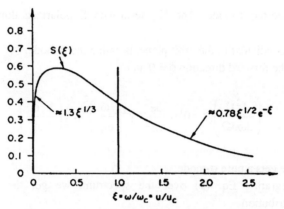

Figure 2. The spectral distribution function $S(\xi)$.

h. Integrating Eq. (12) over all ω, we get the total radiation energy per revolution

$$U = \frac{4\pi e^2}{3}\frac{\gamma^4}{\rho} = \frac{4\pi}{3}\frac{e^2}{\left(mc^2\right)^4}\frac{E^4}{\rho} \equiv C_\gamma \frac{E^4}{\rho}, \tag{13}$$

where $\mathscr{E}$ = energy of the particle and where for electrons

$$C_\gamma = \frac{4\pi}{3}\frac{r_e}{\left(mc^2\right)^3} = 8.85 \times 10^{-5} \frac{m}{(GeV)^3}. \tag{14}$$

The instantaneous radiation power is

$$P_\gamma = \frac{c}{2\pi\rho}U = \frac{c}{2\pi}C_\gamma\frac{E^4}{\rho^2} \propto B^2E^2, \tag{15}$$

where in the last expression B = bending magnetic field. The instantaneous power P_γ can also be derived directly from Eq. (5).

i. For a beam of electrons with current I, the photon flux within a frequency bandwidth $\dfrac{\Delta\omega}{\omega} = 10^{-3}$ (or per $\dfrac{\Delta\omega}{\omega}$, with $\dfrac{\Delta\omega}{\omega}$ expressed in units of 10^{-3}) is

$$\frac{\Delta n}{\Delta t}\left(\frac{\Delta\omega}{\omega} = 10^{-3}\right) = \left(\frac{10^{-3}}{\hbar\omega}\frac{dU}{d\omega}\right)\frac{I}{e} = \frac{10^{-3}}{\hbar}\frac{dU}{d\omega}\frac{I}{e}.$$

This is sometimes called the spectral flux. The "brilliance" of the radiation is defined as the spectral flux per source area per divergence solid angle. The source area is just the cross-sectional area of the e-beam. In addition, the divergence of the e-beam usually dominates over the intrinsic $\dfrac{1}{\gamma}$ divergence of emission of the radiation. Both of these distributions of the e-beam are roughly Gaussian with standard deviations $\pi\sigma_x\sigma_y$ and $\pi\sigma_{x'}\sigma_{y'}$. The product of these is also roughly the product of x- and the y-emittances $\pi^2\varepsilon_x\varepsilon_y \cong (\pi\sigma_x\sigma_{x'})(\pi\sigma_y\sigma_{y'})$. Therefore, we can write the brilliance as

$$B = \frac{1}{\pi^2\varepsilon_x\varepsilon_y}\frac{\Delta n}{\Delta t} = \frac{10^{-3}}{\pi^2\varepsilon_x\varepsilon_y}\left(\frac{1}{\hbar}\frac{dU}{d\omega}\right)\frac{I}{e}, \tag{17}$$

where ε_x and ε_y are usually given in mm-mrad units. Thus the unit of B is generally

$$bu \equiv (\text{photon no.}/\sec)\times 10^{-3}\bigg/\frac{\Delta\omega}{\omega}\bigg/mm^2\bigg/mrad^2. \tag{18}$$

j. For the APS

C = 1104 m	ρ = 39 m	T = 3.68 μs
E = 7 GeV	γ = 13700	I = 100 mA.

This gives

$$U = 5.45 \text{ MeV} \qquad P = UI = 0.545 \text{ MW}$$

$$\omega_c = \frac{3}{2} c \frac{\gamma^3}{\rho} = 2.96 \times 10^{19} \text{ s}^{-1} \qquad u_c = \hbar\omega_c = 19.5 \text{ keV}$$

$$\lambda_c = \frac{2\pi c}{\omega_c} = 0.635 \text{ Å}$$

For beam brilliance we take beam acceptance $a = \dfrac{1}{2\pi} \dfrac{\Delta x}{L} = \dfrac{1}{2\pi} \dfrac{5 \text{ mm}}{50 \text{ m}} =$

1.59×10^{-5} and e-beam emittances $\varepsilon_x = 10 \; \varepsilon_y = 8.2 \times 10^{-3}$ mm-mrad.

Then at $\omega = \omega_c$, $S\left(\dfrac{\omega}{\omega_c}\right) = S(1) = 0.4$ and we have from Eqs. (12) and

(17)

$$B = \frac{10^{-3}}{\pi^2 \varepsilon_x \varepsilon_y} \frac{8\pi a}{9} \alpha \gamma S\left(\frac{\omega}{\omega_c}\right) \frac{I}{e} = 1.67 \times 10^{16} \text{ bu.}$$

1.3. *Radiation from an Undulated Orbit*

We shall study two types of undulators. In the Planar Undulator the electron with velocity β travels on the average along the z-axis and is forced to follow in a planar (say, xz-planc) sinusoidal trajectory by an undulating y-magnetic field. The transverse velocity is then

$$\beta_x = \beta_0 \sin \frac{2\pi z}{\lambda_u}, \qquad (19)$$

where λ_u = undulator period length.

In the Helical Undulator the electron travels in a trajectory with both an x- and a y-oscillation of the same amplitude but out of phase by 90°. Thus, the trajectory is a helix and the transverse velocity $\beta_\perp = \sqrt{\beta_x^2 + \beta_y^2}$ is a constant.

In an undulated orbit an electron radiates at the frequencies of its transverse oscillation. This concentrates the radiation energy into the specific spectral lines, thereby greatly increasing the brilliance, sometimes by as much as a factor of 10^5.

1.3.1 Orbit Kinematics

<u>Helical</u> – This is the simpler case in which both the transverse velocity $\beta_\perp$ and the longitudinal velocity $\beta_z \left(= \sqrt{\beta^2 - \beta_\perp^2} \right)$ are constant.

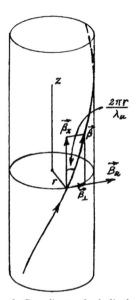

Figure 3. Coordinates for helical orbit.

The radius of the helix, r, and $\beta_\perp$ (see Figure 3) are related by

Geometry: $\qquad \dfrac{\beta_\perp}{\beta_z} = \dfrac{2\pi r}{\lambda_u}, \quad (\beta_z \cong 1) \quad$ or

$$r \cong \frac{\lambda_u}{2\pi}\beta_\perp , \tag{20}$$

Force balance: $\qquad e\beta_z B_u = mc\gamma \dfrac{\beta_\perp^2}{r} \quad$ or

$$r = \frac{mc\gamma}{eB_u}\beta_\perp^2, \tag{21}$$

where B_u is the amplitude of the undulator magnetic field.

Solving Eqs. (20) and (21) for $\beta_\perp$ and r, we get

$$\beta_\perp = \frac{\lambda_u}{2\pi}\frac{eB_u}{mc\gamma} \equiv \frac{K}{\gamma}, \qquad r = \frac{\lambda_u}{2\pi}\frac{K}{\gamma}, \tag{22}$$

where we have defined the "deflection parameter"

$$K \equiv \frac{\lambda_u}{2\pi}\frac{eB_u}{mc} = \left(93.4 \text{ m}^{-1}\text{T}^{-1}\right)\lambda_u B_u \quad \text{(for electrons)}. \tag{23}$$

The longitudinal velocity can also be expressed in terms of K as

$$\beta_z = \sqrt{\beta^2 - \beta_\perp^2} = \left(1 - \frac{1}{\gamma^2} - \frac{K^2}{\gamma^2}\right)^{\frac{1}{2}} \cong 1 - \frac{1}{2\gamma^2}\left(1 + K^2\right). \tag{24}$$

In the frame moving with velocity $c\beta_z$ (co-moving frame), the motion is a circle in the xy plane with diameter $2r \cong \frac{\lambda_u}{\gamma}\frac{K}{\pi}$. The circular motion has a single frequency and yields radiation only at the fundamental frequency. In this frame the undulator period length is Lorentz contracted to λ_u/γ. If the diameter $2r$ is smaller than the period length $\frac{\lambda_u}{\gamma}$ (K<π), the radiation source may be approximated by two mutually perpendicular dipole oscillators 90° out of phase. The angular distribution of the radiation will be proportional to $1 + \cos^2\theta$, where θ is the polar angle from the forward (z) axis. When transformed back to the lab-frame, this will give an intense peak at $\theta = 0$. This is the case of the "undulator." If the circle is large, the radiation distribution in the co-moving frame will be that of an electron in a circular orbit (shaped like a pinwheel), which will transform to a cone in the lab-frame (see Fig. 4). This is the case of the "wiggler" and is generally of less interest.

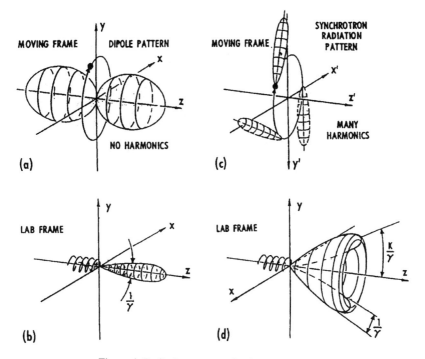

Figure 4. Radiation patterns for four different cases:
  (a)   "undulator," $K < \pi$, in co-moving frame,
  (b)   "undulator," $K < \pi$, in lab frame,
  (c)   "wiggler," $K > \pi$, in co-moving frame,
  (d)   "wiggler," $K > \pi$, in lab frame.

As an example, if we take $B_u = 1T$, $\lambda_u = 3$ cm, we get $K = 2.8$, which is close to the limit of the undulator type of operation.

<u>Planar</u> – In this case the transverse velocity is

$$\beta_\perp = \beta_x = \frac{K}{\gamma}\cos\frac{2\pi z}{\lambda_u} \; . \tag{25}$$

But since $\beta_x = \frac{1}{c}\frac{dx}{dt} \cong \frac{dx}{dz}$, Eq. (25) gives

$$x = \frac{\lambda_u}{2\pi}\frac{K}{\gamma}\sin\frac{2\pi z}{\lambda_u} \tag{26}$$

and

$$\beta_z = \left(\beta^2 - \beta_x^2\right)^{\frac{1}{2}} = \left(1 - \frac{1}{\gamma^2} - \frac{K^2}{\gamma^2}\cos^2\frac{2\pi z}{\lambda_u}\right)^{\frac{1}{2}}$$

$$= 1 - \frac{1}{2\gamma^2}\left(1 + \frac{K^2}{2}\right) - \frac{K^2}{4\gamma^2}\cos\frac{4\pi z}{\lambda_u}. \tag{27}$$

In this case, since $\beta_z \neq$ constant, we have only an average co-moving frame, namely the frame moving with velocity $c\langle\beta_z\rangle = c\left[1 - \frac{1}{2\gamma^2}\left(1 + \frac{K^2}{2}\right)\right]$. In this frame the electron moves in a figure-eight orbit standing along x in the xy plane. The figure-eight motion gives radiations at all harmonics of the fundamental frequency.

1.3.2 Radiation Wavelength

The radiation wavelength can be derived from the following simple argument. One wavelength is generated as the electron travels one undulator period length. In the forward direction, when the electron traveling at β_z has traversed a distance λ_n, the radiation traveling at $\beta = 1$ will have reached λ_u/β_z. Thus, the radiation wavelength is

$$\lambda = \lambda_n\left(\frac{1}{\beta_z} - 1\right) \cong \lambda_u\left(1 - \beta_z\right) = \frac{\lambda_u}{2\gamma^2}\begin{cases}\left(1 + K^2\right) & \text{Helical undulator} \\ \left(1 + \frac{K^2}{2}\right) & \text{Planar undulator.}\end{cases} \tag{28}$$

In a direction at angle θ from forward, when the electron arrived at $z = \lambda_u$, the wavefront has reached only $\lambda_u\cos\theta$. Whereas in the θ-direction the radiation will have reached λ_u/β_z. Therefore, the wavelength is

$$\lambda_\theta = \lambda_u\left(\frac{1}{\beta_z} - \cos\theta\right) \cong \lambda_u\left[1 - \beta_z\left(1 - \frac{\theta^2}{2}\right)\right]$$

$$= \frac{\lambda_u}{2\gamma^2} \left\{ \begin{array}{ll} \left(1 + K^2 + \gamma^2\theta^2\right) & \text{Helical undulator} \\ \left(1 + \dfrac{K^2}{2} + \gamma^2\theta^2\right) & \text{Planar undulator.} \end{array} \right. \tag{29}$$

These wavelengths could also be derived by first obtaining the radiation from the electron in its co-moving frame when it is shaken by the undulator field (transformed to the co-moving frame), then Doppler shifting it back to the lab-frame. Given in Eqs. (28) and (29) are the fundamental wavelengths. For the planar undulator one gets also the harmonic wavelengths.

The radiation wavelength is determined by the undulator only through the deflection parameter K. One can vary K through B_u by adjusting either the gap between the two sets of poles or by sliding them longitudinally (z) relative to each other. This way one can shift the wavelength as much as a factor of 3. With the harmonic radiations of the planar undulator this essentially makes the whole spectrum continually available.

1.3.3   Spectral-Angular Distribution

The procedure for deriving the spectral-angular distribution of the radiation from an electron traveling in an undulated orbit is basically identical to that for the circular orbit. One first sets up an orbit coordinate system, expresses Eq. (7) in these coordinates, and then evaluates the integral. The general expression for the distribution is, as expected, rather complex. For that the student is referred, e.g., to section 11.3 of [1]. Here we will only give the result in the forward direction ($\theta = 0$).

Helical undulation

$$\left(\frac{d^2U}{d\omega d\Omega}\right)(\theta = 0) = \frac{2e^2}{c} N_u^2 \gamma^2 \left[\frac{K^2}{\left(1 + K^2\right)^2}\right] T(\zeta), \tag{30}$$

where

$$T(\zeta) = \left(\frac{\sin\zeta}{\zeta}\right)^2 \quad with \quad \zeta \equiv \pi N_u \frac{\delta\omega}{\omega_1} = \pi N_u \frac{\omega - \omega_1}{\omega_1},$$

$$\omega_1 = \frac{2\pi c}{\lambda_u} \frac{2\gamma^2}{1 + K^2},$$

and N_u = total number of undulations.

The interference function $T(\zeta)$ gives the line shape and has the integral $\int_{-\infty}^{\infty} T(\zeta)d\zeta = 2\pi$. It has a large central peak as shown plotted in Figure 5. This is the feature of the undulator radiation that makes it much more brilliant and more desirable for experiments than the dipole radiation. For the helical undulator the radiation has only one line at the fundamental frequency ω.

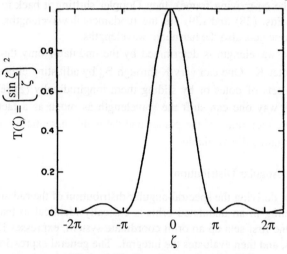

Figure 5. The interference function $T(\zeta)$.

Planar undulation

As was already discussed, the radiation from a planar undulator contains all harmonics h of the fundamental frequency ($\omega_h = h\omega_1$). The spectrum in the forward direction is

$$\left(\frac{d^2U}{d\omega d\Omega}\right)_h (\theta = 0) = \frac{2e^2}{c} N_u^2 \gamma^2 A_h(K^2) T(\zeta_h), \qquad (31)$$

where h = odd integer,

$$A_h(K^2) = h^2 \frac{\frac{K^2}{2}}{\left(1 + \frac{K^2}{2}\right)^2} \left[J_{\frac{h-1}{2}}\left(\frac{h}{2} \frac{\frac{K^2}{2}}{1 + \frac{K^2}{2}}\right) - J_{\frac{h+1}{2}}\left(\frac{h}{2} \frac{\frac{K^2}{2}}{1 + \frac{K^2}{2}}\right) \right]^2, \qquad (32)$$

the J's are the Bessel functions, and where the argument of T is now

$$\zeta_h \equiv \pi N_u \frac{\delta\omega_h}{\omega_1} = \pi N_u \frac{\omega - h\omega_1}{\omega_1} .$$

1.3.4 Features of radiation

We discuss the features of the planar undulator radiation. Most of the discussion applies also to the helical undulator radiation.

a. It is a line spectrum with line frequencies

$$h\omega_1 = h\left(\frac{2\gamma^2}{1 + \frac{K^2}{2}} \frac{2\pi c}{\lambda_u} \right) .$$

b. The N_u^2 dependence indicates that emissions coming from different undulations are coherent. But they superpose spatially to give an intense forward peaking only if $K \lesssim \pi$ (condition to distinguish an undulator from a wiggler).

c. The amplitude function $A_h(K^2)$ in Eq. (32) gives the relative intensity of the different harmonic lines. For example, with $K^2 = 2$, A_h has the following values

$$A_1(2) = 0.185, \ A_3(2) = 0.060, \ A_5(2) = 0.023,$$

$$A_7(2) = 0.009, \ A_9(2) = 0.004.$$

d. In the forward direction interference between radiations from different undulations is such as to make all even harmonics vanish. Even harmonics appear in off-forward radiations.

e. The polarization of the radiation is as follows:

  σ-mode, $\vec{E} \parallel$ undulating plane,

  π-mode, $\vec{E} \perp$ undulating plane.

  The π-mode vanishes in the undulating plane. Thus, the forward radiation given by Eq. (31) is all in the σ-mode.

f. We can integrate Eq. (31) over the spectrum to get the total forward radiation for each harmonic line. From

$$\int_0^\infty T(\zeta)d\omega = \frac{\omega_1}{\pi N_u} \int_0^\infty T(\zeta)d\zeta = \frac{\omega_1}{N_u}$$

we get

$$\left(\frac{dU}{d\Omega}\right)_h (\theta = 0) = \frac{2e^2}{c} N_u \gamma^2 \omega_1 A_h \left(K^2\right). \tag{33}$$

g. In the off-forward direction the amplitude function is dependent on the polar and the azimuthal angles θ and ϕ, and the argument of T becomes

$$\zeta_h = \pi N_u \left(\frac{\delta\omega_h}{\omega_1} + h \frac{\gamma^2 \theta^2}{1 + \frac{K^2}{2}} \right). \tag{34}$$

This shows that the "forward radiation" ($\theta = 0$) has the line-width

$$\Delta\omega = \frac{\omega_1}{N_u}. \tag{35}$$

For "on-frequency radiation" ($\omega = h\omega_1$) the angular spread is

$$\gamma^2 \theta^2 = \frac{1}{hN_u}\left(1 + \frac{K^2}{2}\right), \tag{36}$$

namely, the solid angle spread is

$$\Delta\Omega = \pi\theta^2 = \frac{\pi}{hN_u\gamma^2}\left(1 + \frac{K^2}{2}\right). \tag{37}$$

h. We can approximate the solid-angle integration for on-frequency radiation by multiplying Eq. (31)

$$\left(\frac{d^2 U}{d\omega d\Omega}\right)_h (\theta = 0, \omega = h\omega_1) = \frac{2e^2}{c} N_u^2 \gamma^2 A_h \left(K^2\right)$$

by the solid angle spread from Eq. (37). This gives

$$\left(\frac{dU}{d\omega}\right)_h (\delta\omega = 0) = \frac{2\pi e^2}{c} N_u \frac{1+\frac{K^2}{2}}{h} A_h\left(K^2\right)$$

$$= \frac{2\pi e^2}{c} N_u \frac{\frac{K^2}{2}}{1+\frac{K^2}{2}} h[J,J]^2, \tag{38}$$

where we have used $[J,J]^2$ as a shorthand for the Bessel function factor in Eq. (32).

i. We can get the total radiated energy from Eq. (5) which is, in this case,

$$U = \frac{N_u \lambda_u}{c} P_\gamma = \frac{2}{3} \frac{e^2}{m^2 c^4} N_u \lambda_u \gamma^2 \left|\ddot{\vec{p}}_\perp\right|^2. \tag{39}$$

We have, now

$$\left|\ddot{\vec{p}}_\perp\right|^2 \cong \left[c\frac{d}{dz}\left(mcK\cos\frac{2\pi z}{\lambda_u}\right)\right]^2 = \left(mc^2 K \frac{2\pi}{\lambda_u}\right)^2 \sin^2\frac{2\pi z}{\lambda_u}. \tag{40}$$

Substituting $\left\langle\sin^2\frac{2\pi z}{\lambda_u}\right\rangle = \frac{1}{2}$ we get

$$U = \frac{4\pi^2 e^2}{3} \gamma^2 \frac{N_u K^2}{\lambda_u} = \left(7.26\times10^{-11}\ \tfrac{m}{GeV}\right) E^2 \frac{N_u K^2}{\lambda_u}. \tag{41}$$

j. We have not discussed the rather complex angular distribution of the undulator radiation. Here we reproduce from [1] the computer plots of the angular distribution of the first three harmonics radiation from a planar undulator (Fig. 6).

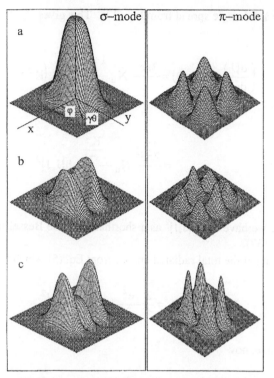

Figure 6. Angular distribution of the first three
harmonics radiation from a planar undulator.

k. For the APS undulator A we have

$N_u = 70,$ $\qquad$ $K = 2.6,$ $\qquad$ $I = 100 \text{ mA},$

$\varepsilon_x = 10 \; \varepsilon_y = 8.2 \times 10^{-3}$ mm-mrad.

The brilliance of the on-frequency radiation from this undulator is given
by Eqs. (17) and (38):

$$B = 2\pi\alpha N_u \, \frac{\dfrac{K^2}{2}}{1 + \dfrac{K^2}{2}} h[J, J]^2 \frac{I}{e} \frac{10^{-3}}{\varepsilon_x \varepsilon_y}$$

$$= \left(2.3 \times 10^{19} \, \text{bu}\right) h[J, J]^2. \tag{42}$$

This is more than 10^3 times the brilliance of the radiation from a
circular orbit.

2. Effects of Emission of Synchrotron Radiation on the Electron Beam

With the emission of synchrotron radiation the electron beam becomes an open—non-Hamiltonian—system, for which some regularities, such as the Liouville theorem, do not apply. In addition to energy loss, which is replenished by an RF acceleration system, the emission of synchrotron radiation produces two major effects on the oscillatory motions of the beam electrons in all three degrees of freedom: Damping and Excitation.

2.1. Damping of Oscillations

2.1.1   Damping of Vertical (y) Oscillation

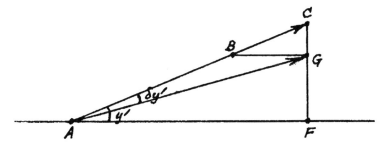

Figure 7. Vertical damping with the emission of a synchrotron radiation photon.

Let the initial momentum of the electron be $\overrightarrow{AC}$ (vector from A to C in Fig. 7). After the emission of a synchrotron radiation photon with momentum $\overrightarrow{BC}$ and energy $u = \left|\overrightarrow{BC}\right| \equiv \overline{BC}$, the momentum of the electron becomes $\overrightarrow{AB}$. The RF system restores the electron energy and adds momentum $\overrightarrow{BG}$ in the forward direction and with $\overline{BG} = \overline{BC}$. The reduction in y' is then given by

$$\frac{\delta y'}{y'} = -\frac{\overline{GC}}{\overline{FC}} = -\frac{\overline{BC}}{\overline{AC}} = -\frac{u}{E}. \tag{1}$$

Over one turn of the ring we can write

$$\left(\frac{\delta y'}{y'}\right)_{turn} = -\frac{U}{E} \quad \text{and} \quad \delta y = 0, \tag{2}$$

where U is the radiation energy per turn and E is the electron energy. With continued emission of synchrotron radiation this "RF effect" will lead to damping of the vertical oscillation amplitude $A_y^2 = \gamma y^2 + 2\alpha yy' + \beta y'^2$. With shifts δy and $\delta y'$ we have

$$A_y \delta A_y = \left(\gamma y^2 + \alpha yy'\right)\frac{\delta y}{y} + \left(\alpha yy' + \beta y'^2\right)\frac{\delta y'}{y'}. \tag{3}$$

Since synchrotron radiation is emitted uniformly over all dipoles around the storage ring, the terms in the coefficients should be averaged over all dipoles. With $\langle \alpha yy' \rangle = 0$ and $\langle \gamma y^2 \rangle = \langle \beta y'^2 \rangle = \dfrac{A_y^2}{2}$, Eq. (3) gives

$$\frac{\delta A_y}{A_y} = \frac{1}{2}\frac{\delta y}{y} + \frac{1}{2}\frac{\delta y'}{y'}. \tag{4}$$

Substituting Eq. (2) in Eq. (4), we get the vertical damping rate

$$\frac{1}{\tau_y} = -\frac{1}{A_y}\frac{dA_y}{dt} = \frac{1}{2T}\frac{U}{E} \equiv \frac{1}{\tau}J_x, \tag{5}$$

where T is the revolution period and where in the last expression we have defined

$\tau \equiv$ standard damping time

$\quad$ = time to radiate away 2E at the fixed rate of U/turn,

and

$J_x \equiv$ vertical partition number = 1.

2.1.2 Damping of Horizontal (x) Oscillation

In addition to the rf effect similar to that in the y-motion, there is dispersion (D,D') in the x-motion, which causes additional phase coordinate shifts with each emission of photon u. We have

$$\begin{cases} \delta x = D\dfrac{u}{E} \\ \delta x' = D'\dfrac{u}{E} - x'\dfrac{u}{E} \,. \end{cases} \tag{6}$$

This gives, with emission of u

$$A_x(\delta A_x)_u = a\delta x + b\delta x' = (aD + bD' - bx')\dfrac{u}{E}, \tag{7}$$

where $a = \gamma x + \alpha x'$ and $b = \alpha x + \beta x'$. With this orbit shift, integration of δA_x around the ring becomes

$$A_x(\delta A_x)_{turn} = \dfrac{1}{E} \int\limits_{turn} (aD + bD' - bx')\left(1 + \dfrac{\delta P_\gamma}{P_\gamma}\right)\left(1 + \dfrac{x}{\rho}\right)P_\gamma\dfrac{ds}{c}. \tag{8}$$

In the integral we do two things. First, from the relation $P_\gamma \propto E^2 B^2$ we get $\dfrac{\delta P_\gamma}{P_\gamma} = 2\dfrac{B'x}{B}$. Second, we average the coefficients around the ring to give

$$\begin{cases} \langle ax \rangle = \langle \gamma x^2 \rangle + \langle \alpha x'x \rangle = \dfrac{A_x^2}{2}, \quad \langle bx' \rangle = \langle \alpha xx' \rangle + \langle \beta x'^2 \rangle = \dfrac{A_x^2}{2}, \\ \langle a \rangle = \langle b \rangle = \langle bx \rangle = \langle bx'x \rangle = 0 \,. \end{cases} \tag{9}$$

Altogether we get

$$-\left(\dfrac{\delta A_x}{A_x}\right)_{turn} = \dfrac{1}{2E} \int\limits_{turn}\left[1 - D\rho\left(\dfrac{1}{\rho^2} + 2K\right)\right]P_\gamma\dfrac{ds}{c}, \tag{10}$$

where $K \equiv \dfrac{B'}{B\rho}$ is the usual field gradient index. Remembering that $U = \int P_\gamma\dfrac{ds}{c}$, we can write the horizontal damping rate as

$$\frac{1}{\tau_x} = -\frac{1}{A_x}\frac{dA_x}{dt} = \frac{U}{2TE}\left(1-D\right) \equiv \frac{1}{\tau}J_x \,, \qquad (11)$$

where, in the last expression, we have designated the horizontal partition number as $J_x \equiv 1-D$, with the dispersion index $\mathscr{D}$ given by

$$D = \frac{1}{U}\int D\rho\left(\frac{1}{\rho^2}+2K\right)P_\gamma\frac{ds}{c} = \frac{\int\frac{D}{\rho}\left(\frac{1}{\rho^2}+2K\right)ds}{\int\frac{ds}{\rho^2}}. \qquad (12)$$

To arrive at the last expression, we have used the relations $P_\gamma \propto \dfrac{1}{\rho^2}$ and

$U = \int P_\gamma\dfrac{ds}{c}$.

In a separated function lattice, $K = 0$ in the dipoles and $\rho = \infty$ in the quadrupoles. If further, ρ is the same in all dipoles (isomagnetic), we get from Eq. (12)

$$D = \frac{1}{2\pi\rho^2}\int Dds = \frac{1}{\rho}\langle D\rangle_{dipole}, \qquad (13)$$

which is generally much less than 1 giving $J_x \cong 1$.

In a combined function lattice with only focusing gradient magnet ($K > 0$), $\mathscr{D}$ can be larger than 1 or $J_x = 1 - \mathscr{D} < 0$. In such a ring we have horizontal antidamping.

2.1.3 Damping of Energy (E) Oscillation

An energy shift generally leads to an orbit shift, hence we need to start with the integral

$$U + \delta U = \int\left(1+\frac{\delta P_\gamma}{P_\gamma}\right)\left(1+\frac{D}{\rho}\frac{\delta E}{E}\right)P_\gamma\frac{ds}{c}\,. \qquad (14)$$

From $P_\gamma \propto E^2 B^2$ we get

$$\frac{\delta P_\gamma}{P_\gamma} = 2\frac{\delta E}{E} + 2\frac{\delta B}{B} = 2\left(1 + \frac{B'D}{B}\right)\frac{\delta E}{E}. \tag{15}$$

This gives

$$\delta U = \frac{\delta E}{E}\left[2U + \int D\rho\left(\frac{1}{\rho^2} + 2K\right)P_\gamma\frac{ds}{c}\right]. \tag{16}$$

The Energy damping rate is then given by

$$\frac{1}{\tau_E} = -\frac{1}{A_E}\frac{A_E}{dt} = \frac{1}{2T}\left(\frac{\delta U}{\delta E}\right)_{turn} = \frac{U}{2TE}(2+D) = \frac{1}{\tau}J_E, \tag{17}$$

showing that the energy partition number is $J_E = 2+\mathscr{D}$.

One interesting relation among the partition numbers is that their sum

$$J_x + J_y + J_E = (1-D) + 1 + (2+D) = 4$$

is independent of the lattice. This is known as the Robinson Theorem, which can be proved to hold true in general. The rather simple proof can be found, e.g., in Chapter 8.2 of [2].

2.2. Quantum Excitation

In the previous sections the synchrotron radiation was treated as though it is emitted totally smoothly. If this were true, the amplitudes of the electron oscillations would, in time, be damped to infinitesimally small values in all three degrees of freedom. This is not the case, because the synchrotron radiation is emitted in quanta of energy u. This granular emission effectively provides excitations to the oscillations.

200

2.2.1 Excitation of Vertical (y) Oscillation

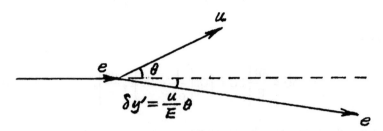

Figure 8. Excitation of vertical oscillation from the emission of a synchrotron radiation photon.

With the recoil of $\delta y'$ derived from each emission of a quantum u as shown in Figure 8, the increment in amplitude averaged around the ring is given by

$$\delta\!\left(A_y^2\right) = \beta_y \left(\delta y'\right)^2 = \beta_y \left(\frac{u}{E}\theta\right)^2 , \qquad (18)$$

where θ is the emission angle. This gives a rate of increase

$$\frac{d\!\left(A_y^2\right)}{dt} = \frac{1}{E^2}\left\langle\beta_y\theta^2\right\rangle \int_0^\infty u^2 n(u)\,du , \qquad (19)$$

where we have averaged $\beta_y\theta^2$ around the ring and where $n(u)du = n(\hbar\omega)\hbar d\omega$ is the normalized energy distribution of the synchrotron radiation quanta as given by the spectrum Eq. (12) of Section 1. The result of the integration is

$$\int_0^\infty u^2 n(u)\,du = \frac{55}{24\sqrt{3}} u_c P_\gamma , \qquad (20)$$

where

$$u_c = \hbar\omega_c = \frac{3}{2}\frac{\hbar c \gamma^3}{\rho} = \text{critical quantum energy}$$

and

$$P_\gamma = \frac{U}{T}.$$

Altogether we have

$$\frac{d\left(A_y^2\right)}{dt} = \left(\frac{55}{8\sqrt{3}}\frac{\hbar}{mc}\right)\frac{U}{2TE}\frac{\gamma^2}{\rho}\left\langle\beta_y\theta^2\right\rangle \equiv \frac{4}{\tau}Q_y, \tag{21}$$

where we have defined

$$Q_y \equiv C_q \frac{\gamma^2}{\rho}\left\langle\beta_y\theta^2\right\rangle \tag{22}$$

with $C_q \equiv \dfrac{55}{32\sqrt{3}}\dfrac{\hbar}{mc} = 3.8\times10^{-13}\,\text{m}$ for electrons. The average of $\beta_y\theta^2$ is, of course, over dipoles only and has the approximate value

$$\left\langle\beta_y\theta^2\right\rangle \cong \frac{R}{v_y}\frac{1}{\gamma^2}. \tag{23}$$

Together with the damping effect, the time development of A_y^2 is

$$\frac{d}{dt}\left(A_y^2\right) = -\frac{2}{\tau_y}\left(A_y^2\right) + \frac{4}{\tau}Q_y. \tag{24}$$

At equilibrium $\dfrac{d}{dt}\left(A_y^2\right) = 0$, and the equilibrium values of A_y^2 and the emittance $\varepsilon_y = \dfrac{1}{2}A_y^2$ are

$$\left(A_y^2\right)_{eq} = 2\frac{Q_y}{J_y}, \qquad \varepsilon_y = \frac{Q_y}{J_y}. \tag{25}$$

2.2.2 Excitation of Horizontal (x) Oscillation

Again, here, because of the dispersion, with the emission of a quantum one gets, in addition to the recoil, a shift in the orbit. This gives

$$\begin{cases} \delta x = D\dfrac{u}{E} \quad, \\ \delta x' = D'\dfrac{u}{E} + \dfrac{u}{E}\theta. \end{cases} \tag{26}$$

On averaging around the ring one obtains

$$\delta\!\left(A_x^2\right) = \frac{u^2}{E^2}\left\langle\left(\gamma_x D^2 + 2\alpha_x DD' + \beta_x D'^2\right) + \beta_x\theta^2\right\rangle, \tag{27}$$

which leads to

$$\frac{d}{dt}\!\left(A_x^2\right) = 4\frac{Q_x}{\tau} \quad \text{with} \quad Q_x \equiv C_q\frac{\gamma^2}{\rho}\left(\langle H\rangle + \left\langle\beta_x\theta^2\right\rangle\right). \tag{28}$$

The additional excitation due to the dispersion effect is

$$\langle H\rangle \equiv \left\langle\gamma_x D^2 + 2\alpha_x DD' + \beta_x D'^2\right\rangle \cong \frac{R}{v_x}\frac{1}{\gamma_t^2}, \tag{29}$$

where $\gamma_t = $ transition $-\gamma$. The equilibrium A_x^2 and emittance are then

$$\left(A_x^2\right)_{eq.} = 2\frac{Q_x}{J_x}, \qquad \varepsilon_x - \frac{Q_x}{J_x}. \tag{30}$$

We see here that because of the dispersion contribution, $\langle H\rangle$, the horizontal excitation Q_x, hence the horizontal emittance ε_x, are much bigger than the corresponding vertical quantities. Hence the electron beam in a storage ring is generally in the shape of a horizontal ribbon with $\varepsilon_x \gg \varepsilon_y$. But, of course, the relative value $\varepsilon_y/\varepsilon_x$ can be varied by adjusting the horizontal-vertical coupling using skew quadrupoles. This leaves $\varepsilon_x + \varepsilon_y$ constant.

2.2.3 Excitation of Energy (E) Oscillation

The emission of a quantum u gives directly an energy shift of $\delta E = u$. Thus we have

$$\frac{d\left(A_E^2\right)}{dt} = \frac{1}{E^2}\int u^2 n(u)dt \equiv 4\frac{Q_E}{\tau}, \tag{31}$$

where

$$Q_E = C_q\frac{\gamma^2}{\rho}. \tag{32}$$

Thus the equilibrium values are

$$\left(A_E^2\right)_{eq} = 2\frac{Q_E}{J_E}, \qquad \varepsilon_E = \left(\frac{\delta E}{E}\right)_{rms}^2 = \frac{Q_E}{J_E}. \tag{33}$$

2.3. Quantum Lifetime

The stochastic process of emission of synchrotron radiation makes the electron beam distribution Gaussian in all three degrees of freedom. A Gaussian distribution has a long tail. A finite beam pipe radius will clip the tail of the transverse distributions and impart a finite lifetime to the beam. Here we can calculate the beam lifetime as a function of the pipe size.

We denote the degree of freedom under consideration by x. For this investigation it is adequate to assume a harmonic x-oscillation. We shall use polar coordinates (A,θ) in the phase plane, which are related to (x,x') by

$$x = A\cos\theta, \qquad x' = A\sin\theta, \qquad dxdx' \Rightarrow AdAd\theta. \tag{34}$$

The equilibrium distribution is Gaussian in A and uniform in θ, namely

$$dN = \frac{N_0}{2\pi\sigma^2}e^{-\frac{A^2}{2\sigma^2}}AdAd\theta = \frac{N_0}{\sigma^2}e^{-\frac{A^2}{2\sigma^2}}AdA, \tag{35}$$

where σ = standard deviation, and where in the last expression we have integrated over θ.

At any amplitude A the excitation outward is balanced by the damping inward. But if there is a wall at A, the particles excited outward will be lost by striking the wall. The lost rate is therefore equal to the damping rate at A, given by $\frac{dA}{dt} = -\frac{1}{\tau}A$. Therefore, the loss rate with wall at A is given by

$$\frac{dN_o}{dt} = \frac{dN}{dA}\frac{dA}{dt} = -\frac{N_o}{\tau}\frac{A^2}{\sigma^2}e^{-\frac{A^2}{2\sigma^2}} \tag{36}$$

or

$$\frac{1}{\tau_q} \equiv -\frac{1}{N_o}\frac{dN_o}{dt} = \frac{1}{\tau}\frac{A^2}{\sigma^2}e^{-\frac{A^2}{2\sigma^2}} \tag{37}$$

or

$$\frac{\tau_q}{\tau} = \frac{e^\xi}{2\xi} \quad \text{with} \quad \xi \equiv \frac{A^2}{2\sigma^2}. \tag{38}$$

This is a steep function of $\frac{A}{\sigma}$. Some numerical values are:

$$\frac{\tau_q}{\tau}\left(\frac{A}{\sigma}=1\right)=1.65, \quad \frac{\tau_q}{\tau}\left(\frac{A}{\sigma}=3\right)=10.0,$$

$$\frac{\tau_q}{\tau}\left(\frac{A}{\sigma}=5\right)=1.07\times10^4, \quad \frac{\tau_q}{\tau}\left(\frac{A}{\sigma}=7\right)=8.73\times10^8.$$

Equation (38) applies to all three degrees of freedom. Since the damping time τ is generally of the order of milliseconds, A must be larger than 7σ to get a reasonable beam lifetime. Including errors and misalignment, it is safer to make $A \gtrsim 10\sigma$ in practice. When this is satisfied, however, it is only the quantum lifetime that is removed. There are still beam lifetimes due to other loss processes such as gas scattering, intrabeam scattering, Touschek effect, etc., which must be considered.

References

1. H. Wiedemann, *Particle Accelerator Physics II*, (New York: Springer-Verlag, 1995).
2. D.A. Edwards and M.J. Syphers, *An Introduction to the Physics of High Energy Accelerators*, (New York: John Wiley & Sons, Inc., 1993).

LATTICE DESIGN FOR SYNCHROTRON RADIATION SOURCE STORAGE RINGS

Y. JIN
National Synchrotron Radiation Laboratory
University of Science and Technology of China
Hefei, Anhui 230029, P. R. China
E-mail: jin@ustc.edu.cn

The modern third generation of synchrotron radiation sources is characterized by an extremely bright synchrotron radiation achieved in low emittance storage rings. At the same time, a large dynamic aperture is also required for a synchrotron radiation source storage ring. In the lattice design for synchrotron radiation source storage rings, we need to consider both problems. This lecture will discuss the minimum emittance lattice and the scaling law of lattice design for synchrotron radiation source storage rings.

1. Introduction

The magnet lattice of electron storage rings for synchrotron radiation sources requires a low beam emittance in order to obtain the highest brightness of the light. It is different from the colliding beam storage rings where high beam emittance is desired, so as to reach the maximum luminosity. Here we only discuss the lattice design for the synchrotron radiation storage rings, and ignore the colliding beam storage rings.

So far there are three generations of synchrotron radiation sources around the world [1]. The first generation of the storage ring light sources were built as a part of high-energy physics programs and initially used parasitically and then in most cases as dedicated light sources. The emittance of these storage rings is generally in the range of one hundred to a several hundred nanometer-radian. Their construction began in the 60-70's of the 20th century, such as SPEAR, ADONE etc. The second generation of storage ring light sources is designed for use as dedicated synchrotron radiation facility, the emittance generally being in the range of 40-150 nanometer-radian. Their construction started in the 70-80's of the last century, such as BESSY, NSLS etc. The third generation of storage ring light sources is also dedicated to synchrotron radiation applications, but their emittance is significantly less than 40 nanometer-radian and, thus, lower than that of the second generation of light sources. Moreover, they feature many long straight sections for installing insertion devices. Their construction began in the 1980-1990's, such as ALS, PLS, SRRC, APS, ESRF, SPring-8 etc.

Emittance is the phase space volume occupied by the beam and is essentially the product of the transverse beam size and divergence (both horizontal and vertical). The horizontal emittance of the beam in an electron storage ring is determined by a dynamic equilibrium between the radiation excitation and the radiation damping. The resultant emittance depends on the magnet lattice.

The emittance is a constant around the storage ring (Liouville's theorem), and varies quadratically with the electron energy for constant lattice optics. The vertical emittance is usually a small fraction of the horizontal emittance and is primarily determined by coupling of the horizontal to the vertical.

A low emittance lattice generally has short bending magnets, since emittance grows in the bending magnets as the third power of the angle of the bending. There are strong quadrupole magnets between the bending magnets to focus the beam tightly. But the stronger quadrupole will produce larger chromaticity. The chromaticity must be corrected in order to store high current. Correction is provided by sextupole magnets. The sextupole is a non-linear element, and its non-linear fields will cause a reduction of the dynamic aperture. Non-linear fields of wiggler and undulator magnets also reduce dynamic aperture. One of the main challenges in the design of the synchrotron radiation light source is to achieve the desired low emittance while maintaining adequate dynamic aperture.

2. Magnet Lattice Type for Synchrotron Radiation Source Storage Ring

The magnet lattice of the storage ring is characteristic for the synchrotron radiation source. At present, there are two common types throughout the world: one is the DBA (Double Bend Achromat) lattice which is also called Chasman-Green lattice, the other is the TBA (Triple Bend Achromat) lattice. The DBA lattice has two bend magnets in each cell, while the TBA has three. The cells are achromatic in the long straight section. These two lattices were suggested in the seventies of the last century. During the last decade, QBA, FBA, and SBA were considered which have four, five, and seven bend magnets in each cell, respectively. Their layouts are shown in Figure 1. In the lattice design for a synchrotron radiation source, one aims at a beam emittance as low as possible. For a given energy of the storage ring, if the amount of bend magnets is the same, the emittance of the TBA lattice is lower than that of the DBA lattice [2]. Analogously, the emittance of the QBA is lower than the TBA, the FBA is lower than the QBA, and the SBA is lower than the FBA. But the dynamic aperture also varies according to the order of DBA, TBA, QBA, FBA, and SBA in decreasing order. In designing a lattice, on one hand we require the beam emittance as low as possible, and on the other hand we want the dynamic aperture to be as large as possible. Thus the designer needs to consider the two targets, and adopt a compromising proposal. The typical lattice of ESRF and its machine functions are shown in Figure 2. It is a DBA lattice.

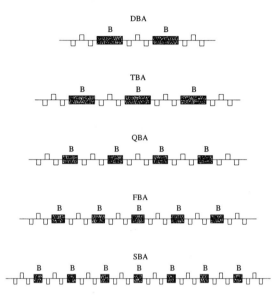

Figure 1 Lattice types for synchrotron radiation source storage rings

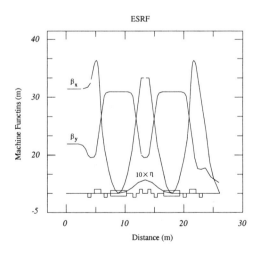

Figure 2 ESRF lattice and machine functions

3. Minimum Emittance Lattice for Synchrotron Radiation Storage Rings

In the following, we derive formulae to calculate the minimum emittance of various lattices of synchrotron radiation storage rings ([3], [4]). The natural horizontal emittance of a storage ring is

$$\varepsilon_x = \frac{C_q \gamma^2}{J_x} \cdot \frac{\langle H \rangle_{dipole}}{\rho} \tag{1}$$

where $C_q = 3.832 \times 10^{-13}$ m, γ is the energy in $m_0 c^2$ units, $\gamma = E/m_0 c^2$, J_x the horizontal damping partition number, ρ the bending radius, and H a function depending on the Courant-Snyder parameters and the dispersion function η with the average of H in the bending magnets given by

$$
\begin{aligned}
\langle H(s) \rangle_{dipole} &= \frac{1}{2\pi\rho} \int_{dipole} \frac{1}{\beta} \left[\eta^2 + \left(\beta\eta' - \frac{1}{2}\beta'\eta \right)^2 \right] ds \\
&= \frac{1}{2\pi\rho} \int_{dipole} \left(\gamma\eta^2 + 2\alpha\eta\eta' + \beta\eta'^2 \right) ds
\end{aligned}
\tag{2}
$$

Here α, β, γ are the Courant-Snyder parameters and η, η' are the dispersion function and its derivative, respectively. If we set

$$F = \frac{\rho^2}{L^3} \langle H \rangle_{dipole} \tag{3}$$

equation (1) becomes

$$\varepsilon_x = \frac{C_q \gamma^2}{J_x} \cdot F\theta^3 \tag{4}$$

where $\theta = L/\rho$ is the bending angle of each bend magnet assumed identical, and L is the length of each bend magnet. F is a numerical value function. From equation (4), we know that the beam emittance is proportional to the square of the energy and the third power of the bending angle. Once the energy of the storage ring is determined, then its beam emittance is only proportional to the third power of the bending angle. Therefore, if the number of bend magnets is larger, then the emittance is smaller. In addition, the emittance is also related to the F function. Once the energy and the number of bend magnets have been determined, minimizing the F function will make the emittance smaller. The function F is a numerical value function related to the lattice parameters of the storage ring. We will now discuss the F function.

Let us look at equation (3), in order to minimize the F function, we must minimize the H function. From formula (2), we can see that the minimization of the H function requires the β and η functions to reach their minimum in the dipole. We assume the minimum of β and η functions to occur at the center of the dipole, as shown in Figure 3.

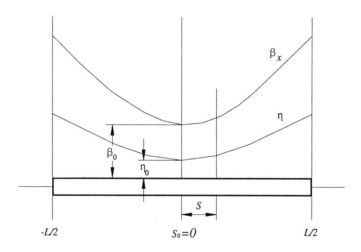

Figure 3 β, η functions in the bending magnet

We set the origin of the orbit coordinate $s_0 = 0$ at the center of the dipole, where β_0, η_0 are the minimum values of the β, η functions, respectively. Taking an arbitrary small segment s, then from $s_0 = 0$ to s, the transformation matrix is

$$M_{dB} = \begin{pmatrix} m_{11} & m_{12} & m_{13} \\ m_{21} & m_{22} & m_{23} \\ m_{31} & m_{32} & m_{33} \end{pmatrix} = \begin{pmatrix} 1 & s & s^2/2\rho \\ 0 & 1 & s/\rho \\ 0 & 0 & 1 \end{pmatrix} \qquad (5)$$

and from $s_0 = 0$ to s , using the matrix elements in formula (5), the transformation matrix for α, β, γ is

$$N_{dB} = \begin{pmatrix} m_{11}^2 & -2m_{11}m_{12} & m_{12}^2 \\ -m_{11}m_{21} & 1+m_{12}m_{21} & -m_{12}m_{22} \\ m_{21}^2 & -2m_{21}m_{22} & m_{22}^2 \end{pmatrix} = \begin{pmatrix} 1 & -2s & s^2 \\ 0 & 1 & -s \\ 0 & 0 & 1 \end{pmatrix} \quad (6)$$

According to N_{dB} matrix formula (6), from $s_0 = 0$ to s , we can get

$$\begin{pmatrix} \beta \\ \alpha \\ \gamma \end{pmatrix} = N_{dB} \begin{pmatrix} \beta_0 \\ \alpha_0 \\ \gamma_0 \end{pmatrix} = \begin{pmatrix} 1 & -2s & s^2 \\ 0 & 1 & -s \\ 0 & 0 & 1 \end{pmatrix} \begin{pmatrix} \beta_0 \\ \alpha_0 \\ \gamma_0 \end{pmatrix} \quad (7)$$

where $\alpha_0, \beta_0, \gamma_0$ are the values of α, β, γ at $s_0 = 0$. For the case in Figure 3 at $s_0 = 0$, β_0 is the minimum, and $\alpha_0 = 0$. Therefore at s , the $\alpha(s), \beta(s), \gamma(s)$ will be

$$\begin{cases} \beta = \beta_0 - 2s\alpha_0 + s^2\gamma_0 = \beta_0 + \dfrac{s^2}{\beta_0} \\[2mm] \alpha = 0 + \alpha_0 - s\gamma_0 = -\dfrac{s}{\beta_0} \\[2mm] \gamma = 0 + 0 + \gamma_0 = \dfrac{1+\alpha_0^2}{\beta_0} = \dfrac{1}{\beta_0} \end{cases} \quad (8)$$

In the same way, at s , the η, η' can be expressed by

$$\begin{pmatrix} \eta \\ \eta' \end{pmatrix} = \begin{pmatrix} m_{11} & m_{12} \\ m_{21} & m_{22} \end{pmatrix} \begin{pmatrix} \eta_0 \\ \eta_0' \end{pmatrix} + \begin{pmatrix} m_{13} \\ m_{23} \end{pmatrix} \quad (9)$$

Because at $s_0 = 0$, η_0 is the minimum, so $\eta_0' = 0$, therefore

$$\begin{cases} \eta = \eta_0 + s\eta_0' + \dfrac{s^2}{2\rho} = \eta_0 + \dfrac{s^2}{2\rho} \\ \eta' = 0 + \eta_0' + \dfrac{s}{\rho} = \dfrac{s}{\rho} \end{cases} \tag{10}$$

Thus at s, we obtain α, β, γ, and η, η' ($\eta' = d\eta/ds$) as following

$$\begin{cases} \beta = \beta_0 + \dfrac{s^2}{\beta_0} \\ \alpha = -\dfrac{s}{\beta_0} \\ \gamma = \dfrac{1}{\beta_0} \\ \eta = \eta_0 + \dfrac{s^2}{2\rho} \\ \eta' = \dfrac{s}{\rho} \end{cases} \tag{11}$$

Now we can get the H function from (11) at an arbitrary point s as follows

$$\begin{aligned} H(s) &= \gamma\eta^2 + 2\alpha\eta\eta' + \beta\eta'^2 \\ &= \dfrac{1}{\beta_0}\left(\eta_0 + \dfrac{s^2}{2\rho}\right)^2 + 2\left(-\dfrac{s}{\beta_0}\right)\left(\eta_0 + \dfrac{s^2}{2\rho}\right)\dfrac{s}{\rho} + \left(\beta_0 + \dfrac{s^2}{\beta_0}\right)\dfrac{s^2}{\rho^2} \\ &= \dfrac{\eta_0^2}{\beta_0} - \dfrac{\eta_0 s^2}{\beta_0\rho} + \dfrac{s^4}{4\beta_0\rho^2} + \dfrac{\beta_0 s^2}{\rho^2} \end{aligned} \tag{12}$$

The average value of the H function over the whole bending magnet is

$$< H(s) >_{dipole} = \dfrac{1}{L}\int_{-L/2}^{L/2}\left(\dfrac{\eta_0^2}{\beta_0} - \dfrac{\eta_0 s^2}{\beta_0\rho} + \dfrac{\beta_0 s^2}{\rho^2} + \dfrac{s^4}{4\beta_0\rho^2}\right)ds$$

$$= \frac{1}{L} \left(\frac{\eta_0^2 L}{\beta_0} - \frac{\eta_0 L^3}{12\beta_0 \rho} + \frac{\beta_0 L^3}{12\rho^2} + \frac{L^5}{320\beta_0 \rho^2} \right)$$

$$= \frac{L^3}{12\rho^2} \left(\frac{12\eta_0^2 \rho^2}{\beta_0 L^3} - \frac{\eta_0 \rho}{\beta_0 L} + \frac{\beta_0}{L} + \frac{3L}{80\beta_0} \right) \qquad (13)$$

We can demonstrate that the H function is minimum for $\eta_0 \approx \dfrac{1}{24} \dfrac{L^2}{\rho}$ (making

$\dfrac{\partial \langle H \rangle}{\partial \eta_0} = 0$). Setting $x = \dfrac{\beta_0}{L}$, taking $\eta_0 \approx \dfrac{1}{24} \dfrac{L^2}{\rho}$, and substituting into

(13) we get the average value of the H function as

$$< H(s) >_{dipole} = \frac{L^3}{12\rho^2} \left(x + \frac{1}{60x} \right) \qquad (14)$$

In order to obtain the minimum average value of the H function within the bending magnet, we need to make the value in parentheses minimum in the formula (14). Setting $z = x + \dfrac{1}{60x}$, and making $\dfrac{\partial z}{\partial x} = 0$, we obtain

$x = \dfrac{1}{2\sqrt{15}}$. Hence,

$$x = \frac{\beta_0}{L} = \frac{1}{2\sqrt{15}} \qquad (15)$$

Substituting formula (15) into formula (14), we get the minimum value of the H function as

$$\langle H(s) \rangle_{dipole} = \frac{L^3}{12\rho^2} \cdot \frac{1}{\sqrt{15}} \qquad (16)$$

Therefore the minimum of the F function is given by

$$F_{min} = \frac{\rho^2}{L^3} \cdot \langle H(s) \rangle_{dipole}$$

$$= \frac{\rho^2}{L^3} \cdot \frac{L^3}{12\rho^2} \cdot \frac{1}{\sqrt{15}}$$

$$= \frac{1}{12\sqrt{15}} \tag{17}$$

The above result is not valid in all cases as η cannot reach zero outside the bending magnet if we set the minimum of the η function at the center of the dipole. If we require $\eta = 0$ at the long straight section, we must put $\eta = \eta' = 0$ at one end of the bending magnet, which is next to the long straight section. We suppose the minimum of β function at $s = s_0$, and at one end of the bending magnet we have $s = 0$, $\eta = \eta' = 0$, which is shown in Figure 4. In this case, formula (11) becomes

$$\begin{cases} \beta = \beta_0 + \dfrac{(s - s_0)^2}{\beta_0} \\ \alpha = -\dfrac{s - s_0}{\beta_0} \\ \gamma = \dfrac{1}{\beta_0} \\ \eta = \dfrac{s^2}{2\rho} \\ \eta' = \dfrac{s}{\rho} \end{cases} \tag{18}$$

The average value of the H function within the bending magnet is

$$< H(s) >_{dipole} = \frac{1}{L} \int_0^L (\gamma\eta^2 + 2\alpha\eta\eta' + \beta\eta'^2) ds$$

$$= \frac{1}{L\rho^2} \int_0^L \left(\frac{s^4}{4\beta_0} - \frac{s_0 s^3}{\beta_0} + \beta_0 s^2 + \frac{s_0^2 s^2}{\beta_0} \right) ds$$

$$= \frac{1}{L\rho^2} \left[\frac{s^5}{5 \times 4\beta_0} - \frac{s_0 s^4}{4\beta_0} + \frac{\beta_0 s^3}{3} + \frac{s_0^2 s^3}{3\beta_0} \right]_0^L$$

$$= \frac{1}{3} \cdot \frac{L^3}{\rho^2} \left[\frac{\beta_0}{L} + \frac{L}{\beta_0} \left(\frac{s_0^2}{L^2} - \frac{3s_0}{4L} + \frac{3}{20} \right) \right] \qquad (19)$$

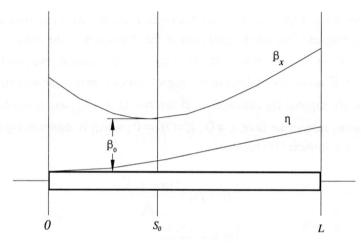

Figure 4 β, η functions for $s = 0$, $\eta = \eta' = 0$ at the end of the bending magnet

In order to minimize the average value of the H function within the bending magnet, we need to make the value in parentheses minimum in formula (19). Let $y = \dfrac{s_0}{L}$ and put

$$u = \frac{s_0^2}{L^2} - \frac{3s_0}{4L} + \frac{3}{20} = y^2 - \frac{3}{4}y + \frac{3}{20} \qquad (20)$$

Taking derivative of u with respect to y, and making $\dfrac{\partial u}{\partial y} = 2y - \dfrac{3}{4} = 0$, we get $y = \dfrac{3}{8}$, that is $\dfrac{s_0}{L} = \dfrac{3}{8}$. Hence, for $s_0 = \dfrac{3}{8}L$, the value of the parentheses is minimum. Taking $\dfrac{s_0}{L} = \dfrac{3}{8}$ and substituting into the formula (20), we obtain

$$u = \frac{3}{320}.$$

To make the value in the brackets of formula (19) a minimum, we set $x = \dfrac{\beta_0}{L}$ again, and let $f = x + \dfrac{3}{320x}$. Taking the derivative of f with respect to x, and setting it to zero, i.e.

$$\frac{\partial f}{\partial x} = 1 - \frac{3}{320x} = 0$$

we find the solution as

$$x = \sqrt{\frac{3}{320}}$$

Substituting u and x into formula (19), the minimum average value of the H function is found to be

$$\langle H(s) \rangle_{dipole} = \frac{1}{3} \cdot \frac{L^3}{\rho^2} \cdot 2\sqrt{\frac{3}{320}} \tag{21}$$

Then, substituting this minimum average value of the H function into formula (3), we finally obtain the minimum of the F function as

$$\begin{aligned} F_{min} &= \frac{\rho^2}{L^3} \langle H(s) \rangle_{dipole} \\ &= \frac{\rho^2}{L^3} \cdot \frac{1}{3} \cdot \frac{L^3}{\rho^2} \cdot 2\sqrt{\frac{3}{320}} \\ &= \frac{1}{4\sqrt{15}} \end{aligned} \tag{22}$$

Now we are ready to compose lattice cells. A bending magnet as shown in Figure 3 must be located in the middle of a cell for a symmetric lattice, such as

TBA, QBA, etc. that are shown in Figure 1. We call this sort of bending magnet a **mid-dipole**. A bending magnet as shown in Figure 4 must be located at the end of a cell for an achromatic lattice to match the dispersion free straight sections. So we call it the **matching dipole**, such as both dipoles of the DBA and the two end dipoles of the TBA, QBA, etc. From formulas (17) and (22), we can see that the contribution of the **matching dipole** to the emittance is 3 times that of a **mid-dipole**. That is the reason why the emittance of the lattice with more **mid-dipoles** is smaller than the lattice with less **mid-dipoles**. When we design the lattice for a synchrotron radiation storage ring, we require the emittance to be as small as possible as well as the dynamic aperture to be large enough. These two aspects have to be traded of.

In real synchrotron radiation storage rings, there are many long straight sections for installing special insertion devices requiring many matching dipoles. Except for the DBA lattice, all other lattices have mid-dipoles. Therefore the F_{min} function is the average of the two kinds of dipoles. For the TBA lattice that has one mid-dipole between two matching dipoles (Figure 1), the minimum F is given by

$$
\begin{aligned}
F_{min} &= \frac{1}{3}\left(2 \times \frac{1}{4\sqrt{15}} + \frac{1}{12\sqrt{15}} \right) \\
&= \frac{1}{1.29} \cdot \frac{1}{4\sqrt{15}}
\end{aligned}
\tag{23}
$$

In a similar way, we calculate the F_{min} values for QBA, FBA, SBA lattices, and list them in table 1.

Table 1. F_{min} for different lattices

Type of Lattice	F_{min}
DBA (two dipoles each cell)	$\dfrac{1}{4\sqrt{15}}$
TBA (three dipoles each cell)	$\dfrac{1}{1.29} \cdot \dfrac{1}{4\sqrt{15}}$
QBA (four dipoles each cell)	$\dfrac{1}{1.50} \cdot \dfrac{1}{4\sqrt{15}}$
FBA (five dipoles each cell)	$\dfrac{1}{1.67} \cdot \dfrac{1}{4\sqrt{15}}$
SBA (seven dipoles each cell)	$\dfrac{1}{1.91} \cdot \dfrac{1}{4\sqrt{15}}$

In summary, we can calculate the minimum emittance for any lattice by inserting F_{min} values of table 1 as well as the energy E and the bending angle θ of a storage ring into formula (4).

For example, in the case of the Hefei Light Source (HLS) storage ring, the energy is 800MeV, the lattice TBA with four cells and 12 bend magnets, accordingly, and an angle of 30 degree for each dipole. With the resulting $\gamma = 800/0.511 \approx 1566$, $\theta = 30/57.3 \approx 0.52356$, and setting $J_x \approx 1$ the minimum emittance of the HLS storage ring is found to be

$$
\varepsilon_x = \frac{C_q \gamma^2}{J_x} \cdot F\theta^3
$$
$$
= \frac{3.832 \times 10^{-13} \times (1566)^2}{1} \cdot \frac{1}{1.29} \cdot \frac{1}{4\sqrt{15}} \cdot (0.52356)^3 \, m \cdot rad
$$
$$
= 6.7 \times 10^{-9} \, m \cdot rad = 6.7nm \cdot rad
$$

Because the optimal condition is not too easy to obtain, this minimum emittance value is only a theoretical value. The real minimum emittance is about twice as large, for HLS we have achieved a minimum emittance of $13.4nm \cdot rad$ for the HBLS configuration.

4. Scaling Law for Lattice Design of a Storage Ring

In the lattice design for synchrotron radiation source storage ring, sometimes we need to change the geometry size of the storage ring, enlarge it or reduce it, for various reasons. But, at the same time, we want to keep the working point (tune) $\nu(\nu_x, \nu_y)$ and the emittance $\varepsilon(\varepsilon_x, \varepsilon_y)$ constant because looking for a good working point we have to spend much time, and getting the optimal emittance is not too easy either. The scaling law will tell you how to change the parameters of the elements of a storage ring to keep $\nu(\nu_x, \nu_y)$ and $\varepsilon(\varepsilon_x, \varepsilon_y)$ constant [5], [6].

(1) Scaling Law

We assume a storage ring with energy E_0, circumference C_0, working point (tune) $\nu_0(\nu_{x0}, \nu_{y0})$, and emittance $\varepsilon_0(\varepsilon_{x0}, \varepsilon_{y0})$.

If we change the circumference from C_0 **to** C **(** $C = qC_0$ **where** q **is an arbitrary positive real number), or change the length of all the elements (including dipoles, quadrupoles, drift sections) by** q **, and put the bending radius** $\rho = q\rho_0$ **(or** $B = B_0/q$ **), the focusing strength of the quadrupoles** $K = K_0/q^2$ **, then** $\nu_0(\nu_{x0}, \nu_{y0})$ **and** $\varepsilon_0(\varepsilon_{x0}, \varepsilon_{y0})$ **will stay constant. This is the scaling law.** In the following, this will be described in detail.

We assume the parameters of the designed original storage ring with energy E_0 to be the

circumference C_0 , the lengths of drift sections L_{D0i} , the lengths of bend magnets L_{B0i} , the bending radius ρ_{0i} , the bending angle θ_0 for each dipole, the lengths of quadrupoles L_{Q0i} , the focusing strength K_{0i} , the tune $\nu_0(\nu_{x0}, \nu_{y0})$, the emittance $\varepsilon_0(\varepsilon_{x0}, \varepsilon_{y0})$, the Courant-Snyder parameters $\alpha_0(s)$, $\beta_0(s)$, $\gamma_0(s)$, the dispersion function $\eta_0(s)$, and the chromaticity ξ_0 .

Let the new parameters of the storage ring with energy E_0 be

$$C = qC_0, L_{Di} = qL_{D0i}, L_{Bi} = qL_{B0i}, L_{Qi} = qL_{Q0i},$$
$$\rho_i = q\rho_{0i}(orB_i = B_{0i}/q), K_i = K_{0i}/q^2, \theta = \theta_0.$$

Then the new working point and emittance of the storage ring will be

$$\nu(\nu_x, \nu_y) = \nu_0(\nu_{x0}, \nu_{y0}) \text{ and } \varepsilon(\varepsilon_x, \varepsilon_y) = \varepsilon_0(\varepsilon_{x0}, \varepsilon_{y0}).$$

In addition, Courant-Snyder parameters are

$$\alpha(s) = \alpha_0(s), \beta(s) = q\beta_0(s), \gamma(s) = \gamma_0(s)/q,$$

and the dispersion function and the chromaticity are

$$\eta(s) = q\eta_0(s), \ \xi = \xi_0.$$

(2) Demonstration for Scaling Law

Setting the transformation matrix of a cell for the original storage ring as

$$M_0 = \begin{pmatrix} m_{11} & m_{12} \\ m_{21} & m_{22} \end{pmatrix} \tag{24}$$

Then the Courant-Snyder parameters are [7]

$$\begin{cases} \alpha_0 = \dfrac{m_{11} - m_{22}}{2\sin\mu_0} \\[2mm] \beta_0 = \dfrac{m_{12}}{\sin\mu_0} \\[2mm] \gamma_0 = -\dfrac{m_{21}}{\sin\mu_0} \end{cases} \tag{25}$$

where

$$\cos\mu_0 = \frac{m_{11} + m_{22}}{2} \tag{26}$$

and

$$\nu_0 = \frac{N\mu_0}{2\pi} \tag{27}$$

Here N is the number of periods (cells).

While the original storage ring is changed to the new one, the length of each element will vary from L_{D0i}, L_{B0i}, L_{Q0i} to qL_{D0i}, qL_{B0i}, qL_{Q0i}, the bending radius from ρ_0 to $q\rho_0$, and the focusing strength from K_{0i} to K_{0i}/q^2. Then the transformation matrix of a period of the new storage ring is

$$M = \begin{pmatrix} m_{11} & qm_{12} \\ m_{21}/q & m_{22} \end{pmatrix} \tag{28}$$

where m_{11}, m_{12}, m_{21}, m_{22} are the matrix elements of the original storage ring (The proof of formula (28) will be given in the Appendix]. The Courant-Snyder parameters of the new storage ring are

$$\begin{cases} \alpha = \dfrac{m_{11} - m_{22}}{2\sin\mu_0} = \alpha_0 \\[2mm] \beta = \dfrac{qm_{12}}{\sin\mu_0} = q\beta_0 \\[2mm] \gamma = -\dfrac{m_{21}}{q\sin\mu_0} = \dfrac{\gamma_0}{q} \end{cases} \qquad (29)$$

where

$$\cos\mu = \frac{m_{11} + m_{22}}{2} = \cos\mu_0 \qquad (30)$$

and

$$\nu = \frac{N\mu}{2\pi} = \frac{N\mu_0}{2\pi} = \nu_0 \qquad (31)$$

Formula (31) shows that the working point does not change.

We will demonstrate that the emittance is not changed either. The third order transformation matrix for a period is

$$M = \begin{pmatrix} m_{11} & qm_{12} & qm_{13} \\ m_{21}/q & m_{22} & m_{23} \\ 0 & 0 & 1 \end{pmatrix} \qquad (32)$$

where m_{11}, m_{12}, m_{13}, m_{21}, m_{22}, m_{23} are the matrix elements of the original storage ring (formula (32) can be proven as formula (28) in the Appendix). The dispersion function and its devaritive η, η' are

$$\eta = \frac{(1 - m_{22})qm_{13} + qm_{12}m_{23}}{2 - (m_{11} + m_{22})} = q\eta_0 \qquad (33)$$

and

$$\eta' = \frac{(1 - m_{11})m_{23} + (m_{21}/q)qm_{13}}{2 - (m_{11} + m_{22})} = \eta'_0 \qquad (34)$$

Therefore the H function is

$$\begin{aligned}
H &= \gamma\eta^2 + 2\alpha\eta\eta' + \beta\eta'^2 \\
&= (\gamma_0/q)(q\eta_0)^2 + 2\alpha_0 q\eta_0\eta_0' + q\beta_0\eta_0'^2 \\
&= qH_0
\end{aligned} \tag{35}$$

where H and H_0 are the H functions for the new and old storage ring, respectively. Then we get the emittance of the new storage ring as

$$\varepsilon_x = \frac{C_q\gamma_E^2 < H >_{dipole}}{J_x\rho} = \frac{C_q\gamma_E^2 q < H_0 >_{dipole}}{J_x q\rho} = \varepsilon_{x0} \tag{36}$$

$$\varepsilon_y = \frac{C_q\beta_{yn}}{2J_y\rho} = \frac{C_q(q\beta_{yn0})}{2J_y(q\rho)} = \varepsilon_{y0} \tag{37}$$

that is ,

$$\varepsilon = \varepsilon_0 \tag{38}$$

Hence the emittance is not changed. From

$$\Delta\nu = -\frac{1}{4\pi}\beta\delta K\Delta s = -\frac{1}{4\pi}(q\beta_0)\left(\frac{\delta K_0}{q^2}\right)(q\Delta s_0) = \Delta\nu_0 \tag{39}$$

the chromaticity of the new storage ring is also found to stay constant

$$\xi = \frac{\Delta\nu}{\Delta E/E} = \frac{\Delta\nu_0}{\Delta E/E} = \xi_0 \tag{40}$$

(3) Application Example for Scaling Law

Now we will give a practical example to show the scaling law. In the case of the HLS, the energy is 800MeV, TBA lattice, four periods, working point $\nu_{x0}=3.58$, $\nu_{y0}=2.58$, and the emittance is $\varepsilon_{x0} = 13.34\times10^{-8}m\cdot rad$.

Setting q=0.8, 1.2, 2, the strength of the dipoles are changed as $1/q = 1/0.8$, $1/1.2$, $1/2$, and the focusing strength of the quadrupoles as $1/q^2 = 1/0.8^2$, $1/1.2^2$, $1/2^2$. Then, the working point and the emittance will stay constant. The calculation results are shown in table 2.

Table 2. Parameters of the HLS for using scaling law

	Original Parameters	New Parameters		
	$q=1$	$q=0.8$	$q=1.2$	$q=2.0$
Circumference(m)	$C_0=66.1308$	$qC_0=52.90464$	79.35696	132.2616
Length of straight section (m)	$L_{D01}=1.6811$	$qL_{D01}=1.34488$	2.01732	3.3622
	$L_{D02}=0.32$	$qL_{D02}=0.256$	0.384	0.64
	$L_{D03}=1.0$	$qL_{D03}=0.8$	1.2	2.0
Dipole length(m)	$L_{B0}=1.1635$	$qL_{B0}=0.9308$	1.3962	2.327
Dipole field(T)	$B_0=1.2$	$B_0/q=1.5$	1.0	0.6
Quadrupole length(m)	$L_{Q0}=0.30$	$qL_{Q0}=0.24$	0.36	0.6
Focusing strength	$K_{01}=1.569180$	$K_{01}/q^2=2.451844$	1.089708	0.392295
	$K_{02}=-0.955667$	$K_{02}/q^2=-1.493230$	-0.663658	-0.238917
Of quadrupole	$K_{03}=-2.267100$	$K_{03}/q^2=-3.542344$	-1.574375	-0.566775
(m^{-2})	$K_{04}=3.070800$	$K_{04}/q^2=4.798125$	2.132500	0.767700
β function at mid-point of long straight section (m)	$\beta_{x0}=21.55$	$\beta_x=17.24$	25.86	43.10
Maximum dispersion fuction (m)	$\eta_{x0}=1.60$	$\eta_x=1.28$	1.92	3.20
Working point	$V_{x0}=3.58$	$V_x=3.58$	3.58	3.58
	$V_{y0}=2.58$	$V_y=2.58$	2.58	2.58
Beam emittance $(10^{-8} m \cdot rad)$	$\varepsilon_{x0}=13.34$	$\varepsilon_x=13.34$	13.34	13.34

The results in the table 2 are calculated by COMFORT code. From table 2, we can see that both, working point and emittance, remain unchanged when the lengths and strengths of elements are varied according to the scaling law. Note that the value of q can take any positive real number in principle. In practice, for values of q not too far from 1, reasonable results are obtained. However, if the value of q is too large or too small, then the parameters are no longer reasonable. From table 2, it can be seen that for q 0.8 or 1.2, the results are better while they are no longer good for $q = 2$. In addition, the scaling law can also be used for searching initial values of the parameters of new storage rings.

5. Synchrotron Radiation Integrals

Many of the important properties of the stored beam in an electron storage ring are determined by the synchrotron radiation integrals [8]. These integrals are conveniently used for calculating beam parameters.

In the usual linear approximation, the integrals are expressed in terms of the four functions of the azimuth coordinates: $\rho(s)$, n, $\beta(s)$, $\eta(s)$.

The synchrotron radiation integrals are defined by

$$I_1 = \oint \frac{\eta}{\rho} ds \tag{41}$$

$$I_2 = \oint \frac{1}{\rho^2} ds \tag{42}$$

$$I_3 = \oint \frac{1}{|\rho^3|} ds \tag{43}$$

$$I_4 = \oint \frac{(1-2n)\eta}{\rho^3} ds \tag{44}$$

$$I_5 = \oint \frac{H}{|\rho^3|} ds \tag{45}$$

where $\rho(s)$ is the radius of curvature of the design orbit, n the field index, $\beta(s)$ the radial betatron function, and $\eta(s)$ the dispersion function. The function $H(s)$ is defined as usually by

$$H(s) = \gamma\eta^2 + 2\alpha\eta\eta' + \beta\eta'^2$$

$$= \frac{1}{\beta}\left[\eta^2 + \left(\beta\eta' - \frac{1}{2}\beta'\eta\right)^2\right] \tag{46}$$

with $\beta' = d\beta/ds$, and $\eta' = d\eta/ds$. Note that the factor of $1/\rho$ appears in each integral, thus, straight sections or pure quadrupoles make no contribution. The integrals only depend on the guide fields.

The various performance parameters of storage rings can be expressed in terms of these integrals as follows:

(1) Momentum compaction factor α

$$\alpha = \frac{I_1}{L} \tag{47}$$

where L is the length of the design orbit.

(2) The energy loss U_0 from synchrotron radiation in one revolution

$$U_0 = \left[\frac{2}{3} r_e E^4 / (m_0 c^2)^3 \right] I_2 \tag{48}$$

where r_e is the electron classical radius, E the nominal energy, and $m_0 c^2$ the electron rest energy.

(3) The damping partition factors J_x and J_ε

$$J_x = 1 - \frac{I_4}{I_2} \tag{49}$$

$$J_\varepsilon = 2 + \frac{I_4}{I_2} \tag{50}$$

(4) The exponential damping coefficients α_x, α_z, α_ε

$$\alpha_x = \frac{r_e c \gamma^3}{3L} (I_2 - I_4) \tag{51}$$

$$\alpha_z = \frac{r_e c \gamma^3}{3L} I_2 \tag{52}$$

$$\alpha_\varepsilon = \frac{r_e c \gamma^3}{3L} (2I_2 + I_4) \tag{53}$$

(5) The root mean-square energy spread $\dfrac{\sigma_\varepsilon}{E}$

$$\left(\frac{\sigma_\varepsilon}{E}\right)^2 = C_q \gamma^2 \frac{I_3}{2I_2 + I_4} \tag{54}$$

(6) The beam emittance (horizontal) ε_x

$$\varepsilon_x = C_q \gamma^2 \frac{I_5}{I_2 - I_4} \tag{55}$$

where $\gamma = E/m_0 c^2$, and C_q is a constant defined as

$$C_q = \frac{55}{32\sqrt{3}} \cdot \frac{\hbar c}{m_e c^2} = 3.8319 \times 10^{-13} m$$

The synchrotron radiation integrals I_1, I_2, I_3, I_4, I_5 can be calculated by many codes for lattice design, for example COMFORT. Therefore using the synchrotron radiation integrals to calculate the various performance parameters of storage rings is very convenient.

For HLS, the synchrotron radiation integrals are (resultants were calculated using COMFORT code):

$$I_1 = 3.1720(m),$$
$$I_2 = 2.8276(m^{-1}),$$
$$I_3 = 1.2725(m^{-2}),$$
$$I_4 = -0.0664(m^{-1}),$$
$$I_5 = 0.4111(m^{-1}).$$

Substituting I_1, I_2, I_3, I_4, I_5 into formulas (47)—(55), we obtain

$$\alpha = 0.047966$$
$$U_0 = 16.3 keV$$
$$J_x = 1.0235, \; J_\varepsilon = 1.9765$$
$$\alpha_x = 47.29, \; \alpha_y = 46.20, \; \alpha_\varepsilon = 91.32$$
$$\frac{\sigma_\varepsilon}{E_0} = 0.4624 \times 10^{-3}$$
$$\varepsilon_x = 13.34 \times 10^{-8} \, m \cdot rad .$$

6. Dynamic Aperture

The dynamic aperture is defined as the maximum phase space amplitude within which particles do not get lost [9]. The dynamic aperture is the transverse margin of the particles' stable motions (Fig. 5). The most efficient way to determine beam stability characteristics for a particular lattice design is to

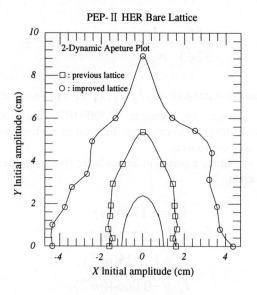

Figure 5 Dynamic aperture

perform numerical particle tracking studies. Widely used computer programs to perform such tracking studies include PATRICIA [10], MAD [11], and SIXTRACK [12].

It is well known that the negative chromaticity ξ will cause the head-tail instability in the electron storage ring for operation above transition energy. In general, one uses sextupoles to correct the chromaticities to slightly positive values in order to overcome the head-tail instability. As the sextupole is a non-linear element its non-linear field will distort the particle trajectory. The ensuing deformation and dispersion of the phase space volume filled by particles will lead to a reduction of the dynamic aperture. Therefore we have to choose an optimal correcting scheme by careful distribution of the sextupoles in the lattice design. In addition, the non-linear fields such as the octupole field used for Landau damping and the multipole fields due to magnet fabrication errors are also reducing the dynamic aperture. They must be considered synthetically in the lattice design. It is very important that there is a large enough dynamic aperture for storing high current and keeping long beam lifetime. When the lattice and its parameters have been determined, we should do our best to make the dynamic aperture as large as possible.

Appendix

The formula (28) is not difficult to demonstrate. For example, the original transformation matrix of a focussing quadrupole is

$$
M_0^{QF} = \begin{pmatrix} m_{11}^{QF} & m_{12}^{QF} \\ m_{21}^{QF} & m_{22}^{QF} \end{pmatrix} = \begin{pmatrix} \cos\sqrt{K_{0i}}\,L_{Q0i} & \sqrt{1/K_{0i}}\,\sin\sqrt{K_{0i}}\,L_{Q0i} \\ -\sqrt{K_{0i}}\,\sin\sqrt{K_{0i}}\,L_{Q0i} & \cos\sqrt{K_{0i}}\,L_{Q0i} \end{pmatrix}
$$

For the new focussing quadrupole, the transformation matrix is

$$
M^{QF} = \begin{pmatrix} \cos\sqrt{K_i}\,L_{Qi} & \sqrt{1/K_i}\,\sin\sqrt{K_i}\,L_{Qi} \\ -\sqrt{K_i}\,\sin\sqrt{K_i}\,L_{Qi} & \cos\sqrt{K_i}\,L_{Qi} \end{pmatrix}
$$

$$
= \begin{pmatrix} \cos\sqrt{K_{0i}/q^2}\,qL_{Q0i} & \sqrt{q^2/K_{0i}}\,\sin\sqrt{K_{0i}/q^2}\,qL_{Q0i} \\ -\sqrt{K_{0i}/q^2}\,\sin\sqrt{K_{0i}/q^2}\,qL_{Q0i} & \cos\sqrt{K_{0i}/q^2}\,qL_{Q0i} \end{pmatrix}
$$

$$
= \begin{pmatrix} \cos\sqrt{K_{0i}}\,L_{Q0i} & q\sqrt{1/K_{0i}}\,\sin\sqrt{K_{0i}}\,L_{Q0i} \\ -\sqrt{K_{0i}}\,\sin\sqrt{K_{0i}}\,L_{0i}/q & \cos\sqrt{K_{0i}}\,L_{Q0i} \end{pmatrix}
$$

$$= \begin{pmatrix} m_{11}^{QF} & qm_{12}^{QF} \\ m_{21}^{QF}/q & m_{22}^{QF} \end{pmatrix}$$

In the same way, we can get other transformation matrices, such as defocussing quadrupoles M^{QD}, bending magnets M^{B}, drift section M^{D}, etc. Then we will obtain the transformation matrix of a complete period (cell) by matrix multiplication, that is

$$M = M^{D} \cdot M^{QD} \cdot M^{B} \cdot M^{QF} \cdots$$

$$= \begin{pmatrix} m_{11}^{D} & qm_{12}^{D} \\ m_{21}^{D}/q & m_{22}^{D} \end{pmatrix} \cdot \begin{pmatrix} m_{11}^{QD} & qm_{12}^{QD} \\ m_{21}^{QD}/q & m_{22}^{QD} \end{pmatrix} \cdot \begin{pmatrix} m_{11}^{B} & qm_{12}^{B} \\ m_{21}^{B}/q & m_{22}^{B} \end{pmatrix} \cdot \begin{pmatrix} m_{11}^{QF} & qm_{12}^{QF} \\ m_{21}^{QF}/q & m_{22}^{QF} \end{pmatrix} \cdots$$

$$= \begin{pmatrix} m_{11} & qm_{12} \\ m_{21}/q & m_{22} \end{pmatrix}$$

This is the demonstration of formula (28).

The formula (32) can be proven in the same way as formula (28). For instance, the third order original transformation matrix of a focussing quadrupole is

$$M_0^{QF} = \begin{pmatrix} m_{11}^{QF} & m_{12}^{QF} & m_{13}^{QF} \\ m_{21}^{QF} & m_{22}^{QF} & m_{23}^{QF} \\ 0 & 0 & 1 \end{pmatrix}$$

$$= \begin{pmatrix} \cos\sqrt{K_{0i}}L_{Q0i} & \sqrt{1/K_{0i}}\sin\sqrt{K_{0i}}L_{Q0i} & \left(1-\cos\sqrt{K_{0i}}L_{Q0i}\right)/\rho_0 K_{0i} \\ -\sqrt{K_{0i}}\sin\sqrt{K_{0i}}L_{Q0i} & \cos\sqrt{K_{0i}}L_{Q0i} & \sin\sqrt{K_{0i}}L_{Q0i}/\rho_0\sqrt{K_{0i}} \\ 0 & 0 & 1 \end{pmatrix}$$

This formula can apply to any focussing elements, for pure quadrupole, $m_{13}^{F} = m_{23}^{F} = 0$. For the new focussing quadrupole, its third order transformation matrix is

$$
M^{QF} = \begin{pmatrix} \cos\sqrt{K_i}L_{Qi} & \sqrt{1/K_i}\sin\sqrt{K_i}L_{Qi} & \left(1-\cos\sqrt{K_i}L_{Qi}\right)\big/\rho K_i \\ -\sqrt{K_i}\sin\sqrt{K_i}L_{Qi} & \cos\sqrt{K_i}L_{Qi} & \sin\sqrt{K_i}L_{Qi}\big/\rho\sqrt{K_i} \\ 0 & 0 & 1 \end{pmatrix}
$$

$$
= \begin{pmatrix} \cos\sqrt{K_{0i}/q^2}\,qL_{Q0i} & \sqrt{q^2/K_{0i}}\sin\sqrt{K_{0i}/q^2}\,qL_{Q0i} & \left(1-\cos\sqrt{K_{0i}/q^2}\,qL_{Q0i}\right)\big/q\rho_0\,K_{0i}/q^2 \\ -\sqrt{K_{0i}/q^2}\sin\sqrt{K_{0i}/q^2}\,qL_{Q0i} & \cos\sqrt{K_{0i}/q^2}\,qL_{Q0i} & \sin\sqrt{K_{0i}/q^2}\,qL_{Q0i}\big/q\rho_0\sqrt{K_{0i}/q^2} \\ 0 & 0 & 1 \end{pmatrix}
$$

$$
= \begin{pmatrix} \cos\sqrt{K_{0i}}L_{Q0i} & q\sqrt{1/K_{0i}}\sin\sqrt{K_{0i}}L_{Q0i} & q\left(1-\cos\sqrt{K_{0i}}L_{Q0i}\right)\big/\rho_0 K_{0i} \\ -\sqrt{K_{0i}}\sin\sqrt{K_{0i}}L_{Q0i}\,/\,q & \cos\sqrt{K_{0i}}L_{Q0i} & \sin\sqrt{K_{0i}}L_{Q0i}\big/\rho_0\sqrt{K_{0i}} \\ 0 & 0 & 1 \end{pmatrix}
$$

$$
= \begin{pmatrix} m_{11}^{QF} & qm_{12}^{QF} & qm_{13}^{QF} \\ m_{21}^{QF}/q & m_{22}^{QF} & m_{23}^{QF} \\ 0 & 0 & 1 \end{pmatrix}
$$

In the same way, we can get the third order transformation matrices for every other element, such as defocussing quadrupoles M^{QD}, bending magnets M^{B}, drift section M^{D}, etc. Then we will obtain the transformation matrix for a complete period (cell) by matrix multiplication, that is

$$
M = M^{D} \cdot M^{QD} \cdot M^{B} \cdot M^{QF} \cdots
$$

$$
= \begin{pmatrix} m_{11}^{D} & qm_{12}^{D} & qm_{13}^{D} \\ m_{21}^{D}/q & m_{22}^{D} & m_{23}^{D} \\ 0 & 0 & 1 \end{pmatrix} \cdot \begin{pmatrix} m_{11}^{QD} & qm_{12}^{QD} & qm_{13}^{QD} \\ m_{21}^{QD}/q & m_{22}^{QD} & m_{23}^{QD} \\ 0 & 0 & 1 \end{pmatrix} \cdot \begin{pmatrix} m_{11}^{B} & qm_{12}^{B} & qm_{13}^{B} \\ m_{21}^{B}/q & m_{22}^{B} & m_{23}^{B} \\ 0 & 0 & 1 \end{pmatrix} \cdots
$$

$$
= \begin{pmatrix} m_{11} & qm_{12} & qm_{13} \\ m_{21}/q & m_{22} & m_{23} \\ 0 & 0 & 1 \end{pmatrix}
$$

This is the proof of formula (32).

References

[1] A. Jackson, The comparison of the Chasman-Green and triple bend achromat lattice, Particle Accelerator, Vol.22 (1987).

[2] H. Winick & G. Williams, Overview of synchrotron radiation sources worldwide, Synchrotron Radiation News, Vol.4, No.5 (1991) p23.

[3] L. C. Teng, Minimizing the emittance in designing the lattice of an electron storage ring, Fermilab TM-1269 (1984).

[4] L. C. Teng, Minimum emittance lattice for synchrotron radiation storage rings, ANL LS-17 (1985).

[5] Y. Jin, Scaling law in the lattice design for storage ring, High energy physics and nuclear physics, Vol.21, No.1 (1997) p75.

[6] Y. Jin, The Physics of Electron Storage Rings, Press of University of Science and Technology of China (2001).

[7] E. D. Courant and Snyder, Theory of the Alternating-Gradient Synchrotron, Annals of Physics, 3, (1958) p1.

[8] R. H. Helm, M. J. Lee, and P.L. Morton, Evaluation of synchrotron radiation integrals, IEEE Trans, Nucl. Sci., Ns20 (1973) p900.

[9] A. W. Chao, M. Tigner, Handbook of Accelerator Physics and Engineering, World Scientific Publishing Co., (1999).

[10] S. Kheifets, Studies in PEP and description of the computer code PATRICIA, SLAC-PUB-2922 (1982).

[11] C. Iselin, The MAD program, CERN-LEP, TH-85 (1986).

[12] F. Schmidt, SIXTRACK, CERN-SL 94-56(1997).

SPALLATION NEUTRON SOURCE AND OTHER HIGH INTENSITY PROTON SOURCES*

WEIREN CHOU

Fermi National Accelerator Laboratory
P.O. Box 500
Batavia, IL 60510, USA
E-mail: chou@fnal.gov

This lecture is an introduction to the design of a spallation neutron source and other high intensity proton sources. It discusses two different approaches: linac-based and synchrotron-based. The requirements and design concepts of each approach are presented. The advantages and disadvantages are compared. A brief review of existing machines and those under construction and proposed is also given. An R&D program is included in an appendix.

1. Introduction

1.1. *What is a Spallation Neutron Source?*

A spallation neutron source is an accelerator-based facility that produces pulsed neutron beams by bombarding a target with intense proton beams.

Intense neutrons can also be obtained from nuclear reactors. However, the international nuclear non-proliferation treaty prohibits civilian use of highly enriched uranium U^{235}. It is a showstopper of any high efficiency reactor-based new neutron sources, which would require the use of 93% U^{235}. (This explains why the original proposal of a reactor-based Advanced Neutron Source at the Oak Ridge National Laboratory in the U.S. was rejected. It was replaced by the accelerator-based Spallation Neutron Source, or SNS, project.)

A reactor-based neutron source produces steady higher flux neutron beams, whereas an accelerator-based one produces pulsed lower flux neutron beams. So the trade-off is high flux *vs.* time structure of the neutron beams. This course will teach accelerator-based neutron sources.

An accelerator-based neutron source consists of five parts:
1) Accelerators
2) Targets
3) Beam lines
4) Detectors

---

* This work is supported by the Universities Research Association, Inc., under contract No. DE-AC02-76CH03000 with the U.S. Department of Energy.

5) Civil construction

A project proposal includes 1) through 5), plus a cost estimate, a schedule and environment, safety and health (ES&H) considerations. This course will teach part 1) only, although part 2) is closely related to 1) and a critical item in the design of a spallation neutron source.

1.2. Parameter Choice of a Spallation Neutron Source

The requirements of neutron beams for neutron scattering experiments are as follows:

- Neutron energy: low, about a few milli electron volts.
- Neutron pulse: sharp, about 1 μs.
- Pulse repetition rate: 10-60 Hz.

When an intense proton beam strikes on a target (made of carbon or heavy metal), neutrons are produced via spallation. The production rate is roughly proportional to the power deposited on the target.

Proton energies between 1 and 5 GeV prove optimal for neutron production. At 1 GeV, each incident proton generates 20-30 neutrons.

The beam power P is the product of beam energy E, beam intensity N (number of protons per pulse) and repetition rate f:

$$P \text{ (MW)} = 1.6 \times 10^{-16} \times E \text{ (GeV)} \times N \times f \text{ (Hz)} \qquad (1)$$

Typical parameters of a modern high power spallation neutron source are:

- $P \sim 1$ MW
- $E \sim 1$ GeV
- $N \sim 1 \times 10^{14}$
- $f \sim 10\text{-}60$ Hz

1.3. Linac-based vs. Synchrotron-based Spallation Neutron Source

There are two approaches to an accelerator-based spallation neutron source: linac-based and synchrotron-based.

A linac-based spallation neutron source has a full-energy linac and an accumulator ring. It works as follows:

- A heavy-duty ion source generates high intensity H⁻ beams.
- A linac accelerates H⁻ pulses of ~1 ms length to ~1 GeV.
- These H⁻ particles are injected into an accumulator via a charge exchange process, in which the electrons are stripped by a foil and dumped, and the H^+ (proton) particles stay in the ring.
- This injection process takes many (several hundreds to a few thousands) turns.

- The accumulated protons are then extracted from the ring in a single turn onto a target. The pulse length is about 1 μs.
- This process repeats 10-60 times every second.

A synchrotron-based spallation neutron source has a lower energy linac and a rapid cycling synchrotron. It works differently.

- A heavy-duty ion source generates high intensity H⁻ beams.
- A linac accelerates H⁻ pulses of ~1 ms length to a fraction of a GeV.
- These H⁻ particles are injected into a synchrotron via the same charge exchange process.
- This injection process takes many (several hundreds) turns.
- The H^+ (proton) beam is accelerated in the synchrotron to 1 GeV or higher and then extracted in a single turn onto a target. The pulse length is about 1 μs.
- This process repeats 10-60 times every second.

Compared with a linac-based spallation neutron source, a synchrotron-based one has the following advantages:

- For the same beam power, it would cost less, because proton synchrotrons are usually less expensive than proton linacs.
- For the same beam power, it would have lower beam intensity, because the beam energy could be higher.
- Because the injected linac beam has lower power, the stripping foil is easier. Also, larger beam loss at injection could be tolerated.
- A major problem in a high intensity accumulator ring (a DC machine) is the e-p instability. However, this has never been observed in any synchrotron (an AC machine) during ramp.

The disadvantages of a synchrotron-based spallation neutron source include:

- AC machines (synchrotrons) are more difficult to build than DC machines (accumulators). In particular, the hardware is challenging, e.g., large aperture AC magnets, rapid cycling power supplies, field tracking during the cycle, eddy current effects in the coil and beam pipe, high power tunable RF system, etc.
- AC machines are also more difficult to operate than DC machines. Therefore, the reliability is lower.

One needs to consider all these factors when deciding which approach to take for a spallation neutron source.

1.4. Spallation Neutron Source vs. Other High Intensity Proton Sources

Spallation neutron sources are an important type of high intensity proton sources. However, a high intensity proton source may find many other applications. For example:

- To generate high intensity secondary particles for high-energy physics experiments, e.g., antiprotons (Tevatron p-pbar collider), muons (AGS, JHF), neutrinos (K2K, MiniBooNE, NuMI, JHF), kaons (CKM, KAMI, JHF), ions (ISOLDE), etc.
- To generate neutrino superbeams as the first stage to a neutrino factory and a muon collider. (Such a high intensity proton source is called a Proton Driver.)
- Nuclear waste transmutation (JHF, CONCERT).
- Energy amplifier (CERN).
- Proton radiography (AHF).

The design concept learned from this course can readily be applied to the design of other high intensity proton sources.

2. High Intensity Proton Sources: Existing, Under Construction, and Proposed

There are a number of high intensity proton sources operating at various laboratories over the world. There are presently two large construction projects: the SNS at the Oak Ridge National Laboratory in the U.S., and the JHF at the KEK/JAERI in Japan. Each has a construction budget of about 1.3 billion US dollars and is scheduled to start operation around 2005-2006. There are also numerous proposals for Proton Drivers and other high intensity proton sources. Table 1 is a summary based on a survey conducted during the Snowmass 2001 Workshop. [1]

Among the existing machines, the highest beam power from a synchrotron is 160 kW at the ISIS at Rutherford Appleton Laboratory in England. The highest beam power from an accumulator is 64 kW at the PSR at Los Alamos National Laboratory in the U.S.

The SNS is a linac-based spallation neutron source. The design beam energy is 1 GeV, beam power 1.4 MW. The JHF is a synchrotron-based facility. It has a 400 MeV linac, a 3 GeV rapid cycling synchrotron with a beam power of 1 MW, and a 50 GeV slow ramp synchrotron with a beam power of 0.75 MW.

Several proposals of proton drivers have been documented and can be found in Ref. [2]-[4].

Table 1. High intensity proton sources: existing, under construction, and proposed
(Snowmass 2001 survey)

Machine	Flux (10^{13}/pulse)	Rep Rate (Hz)	Flux[†] (10^{20}/year)	Energy (GeV)	Power (MW)
Existing:					
RAL ISIS	2.5	50	125	0.8	0.16
BNL AGS	7	0.5	3.5	24	0.13
LANL PSR	2.5	20	50	0.8	0.064
ANL IPNS	0.3	30	9	0.45	0.0065
Fermilab Booster (*)	0.5	7.5	3.8	8	0.05
Fermilab MI	3	0.54	1.6	120	0.3
CERN SPS	4.8	0.17	0.8	400	0.5
Under Construction:					
ORNL SNS	14	60	840	1	1.4
JHF 50 GeV	32	0.3	10	50	0.75
JHF 3 GeV	8	25	200	3	1
Proton Driver Proposals:					
Fermilab 8 GeV	2.5	15	38	8	0.5
Fermilab 16 GeV	10	15	150	16	4
Fermilab MI Upgrade	15	0.65	9.8	120	1.9
BNL Phase I	10	2.5	25	24	1
BNL Phase II	20	5	100	24	4
CERN SPL	23	50	1100	2.2	4
RAL 15 GeV (**)	6.6	25	165	15	4
RAL 5 GeV (**)	10	50	500	5	4
Other Proposals:					
Europe ESS (**)	46.8	50	2340	1.334	5
Europe CONCERT	234	50	12000	1.334	25
LANL AAA	-	CW	62500	1	100
LANL AHF	3	0.04	0.03	50	0.003
KOMAC	-	CW	12500	1	20
CSNS/Beijing	1.56	25	39	1.6	0.1

[†] 1 year = 1×10^7 seconds.
(*) Including planned improvements.
(**) Based on 2-ring design.

3. Design Concept of a Linac-based Spallation Neutron Source

A linac-based spallation neutron source has three major accelerator components:

- Linac front end
- Linac (full energy)
- Accumulator

We will discuss the design concept of each component in the following sections.

3.1. *Linac Front End*

The linac front end consists of an ion (H⁻) source, a pre-accelerator (Cockcroft-Walton or RFQ), a low energy beam transport (LEBT), and a chopper.

3.1.1. *H⁻ source*

H⁻ ions have almost been universally adopted for multi-turn injection from a linac to an accumulator ring. These ions are generated in an H⁻ source. There are several different types: surface-plasma source (magnetron), semi-planatron, surface-plasma source with Penning discharge (Dudnikov-type source), RF volume source, etc. This is a highly specialized field. There are regular conferences and workshops devoted to this topic. The main challenges are to provide ion beams with high brightness (i.e., high intensity and low emittance) and to operate at high duty factor with a reasonable lifetime.

3.1.2. *Cockcroft-Walton and RFQ*

The kinetic energy of the H⁻ particles from an ion source is about a few tens of keV. These particles are accelerated by a pre-accelerator, which can be either a Cockcroft-Walton or RFQ. The former has been in use for many years and has a maximum energy of about 750 keV. In a number of laboratories it has been replaced by the latter, which is a common choice of new accelerators. This is because an RFQ has higher energy (several MeV) and a much smaller physical size. Its beam has higher brightness and is bunched. (The Cockcroft-Walton needs a buncher.) The design issues of an RFQ include high beam current, high efficiency, small emittance dilution, and higher order mode (HOM) suppression.

3.1.3. *LEBT*

When an RFQ is used, one needs a low energy beam transport (LEBT) as a matching section between the ion source and the RFQ. It consists of lenses that focus the beam from the ion source, which is relatively large in radius and divergence. The LEBT also usually contains source diagnostics and provides the differential vacuum pumping between the source and the RFQ.

3.1.4. *Chopper*

The purpose of a chopper is to chop the beam so that it can properly fit into the RF bucket structure in an accumulator. This would greatly reduce the injection loss caused by RF capture. The requirements of a chopper are: short rise- and fall-time (10-20 ns), short physical length (to reduce space charge effects), and a flat top and a flat bottom in the field waveform (to reduce energy spread in the beam).

There are several different types of choppers:

- Transverse deflector: This is a traveling wave structure. It has short rise- and fall-time. The shortcoming is its physical size (about 1-meter long). It is used at the Los Alamos National Laboratory and the Brookhaven National Laboratory.

- Electric deflector: This is a split-electrode structure for deflecting the beam right after the LEBT. It was built at the Lawrence Berkeley National Laboratory and will be installed in the linac front end of the SNS project. [5]

- Beam transformer (energy chopper): This is a new type of chopper and is based on the fact that an RFQ has a rather small energy window. A pulsed beam transformer that provides 10% energy modulation to the beam in front of an RFQ can effectively chop the beam. It has short rise- and fall-time and a short physical length (about 10 cm). A prototype has been built by a KEK-Fermilab team and is installed at the HIMAC in Japan for beam testing. [6]

3.2. Linac

The linac is the main accelerator. Its function is to accelerate the H⁻ beam to full energy ($\sim$ 1 GeV) before injection into the accumulator. Because the particle velocity changes over a wide range during the acceleration ($\beta = 0.046$ at 1 MeV, $\beta = 0.875$ at 1 GeV), the linac is partitioned to several parts. Each part uses a different design to best match the corresponding β values.

3.2.1. Low energy part (below 100 MeV, $\beta < 0.4$)

Drift tube linac (DTL) is a common choice of this part. It is a matured technology and has been used in every proton linac over the world. A potential concern is that some vacuum tubes used to drive the RF cavities could have supply problem because the vendors may terminate their production.

There is also an effort to develop superconducting RF cavities (the so-called spoke cavity) for low β acceleration.

3.2.2. Medium energy part (100 MeV - 1 GeV, $0.4 < \beta < 0.9$)

This is the bulk part of the linac. There are two design choices. One is room temperature coupled-cell linac (CCL), another superconducting (SC) linac.

The CCL is a matured technology and has been used in all existing linacs (e.g., Fermilab, Los Alamos National Laboratory, Brookhaven National Laboratory, etc.). The highest energy using this technology reaches 800 MeV (LANSCE at Los Alamos). However, the new project SNS has decided to use an SC linac for good reason.

SC linacs have been proved reliable and efficient in electron machines (e.g., LEP and CEBAF). But still, it is a challenge for employment in proton machines when operating in short pulse mode and accelerating particles with different β values. In the past decade, SC linac technology has been making good and steady progress. [7] Compared with a room temperature linac, an SC linac has the following advantages:

- Higher accelerating gradient.
- Larger aperture (which is particularly important for high intensity beams).
- Lower operation cost.
- Lower capital cost if higher energy is required. (There is an energy threshold above which an SC linac becomes more economical.)

In addition to the SNS, CERN and KEK are also planning to use an SC linac in their future machines (SPL and JHF Stage 2 linac, respectively).

Among various challenges to an SC linac, a crucial one is the RF control. The allowable phase error ($< 0.5°$) and amplitude error ($< 0.5\%$) are demanding. One needs to investigate the choice of RF source (number of cavities per klystron), redundancy (off-normal operation with missing cavities), feedback and feedforward technique.

3.2.3. High energy part (above 1 GeV, $\beta > 0.9$)

In this range, particles travel at a velocity near that of the light and behave similar to electrons. An SC linac is an obvious and probably also the only choice from economical considerations. Several new high-energy proton linac proposals (2.2 GeV at CERN, 3 GeV at Los Alamos, and 8 GeV at Fermilab) have all picked this design.

3.3. Accumulator

As the name indicates, an accumulator is a ring that accumulates many turns of injected particles and ejects them in a single turn. The purpose is to convert long beam pulses (~ 1 ms) to short beam pulses (~ 1 μs) for experiments. It is a DC machine. Its hardware is more or less straightforward (a main advantage of the accumulator approach). But this by no means implies an *"easy"* machine. On the contrary, there are a number of challenges due to high beam power.

3.3.1. *Beam loss control*

This is the most challenging problem. For a 1 MW beam power, 1% beam loss would give 10 kW, which already exceeds the full beam power on the targets for most of the existing physics experiments. Therefore, allowable beam loss must be much lower than 1%.

There are two types of beam loss: controllable and uncontrollable. One uses specially designed collimators and dumps to collect the former so that the loss can be localized. The uncontrollable beam loss would spread over the entire machine and must be kept very low. The rule of thumb is that it must be below 1 W/m in order to make hands-on maintenance possible. For a 100-meter machine, 1 W/m gives the total uncontrollable beam loss of 100 W, which is 0.01% of the total beam power. This is a goal not impossible but very difficult.

In the PSR at Los Alamos, which is a 64 kW accumulator, the total beam loss is a fraction of a percent. Most of them are unstripped H^0 and H^- particles, which are collected by special beam dumps.

3.3.2. *Collimators and remote handling*

Collimators are a critical part of an accumulator. They are used to localize the beam loss and leave a majority part of the machine "clean."

Modern collimators use a 2-stage design. The primary collimator scatters the halo particles; the secondary collimator (which can be more than one) collects them. There is one set of collimators in each transverse plane. Longitudinal collimators are also used, which are placed in high dispersion regions. The design efficiency of collimators is 95% or higher.

The area near the collimators is very "hot" (highly radioactive). One must use remote handling for maintenance in this area. Robot arms and cranes are often employed. This should be an integral part in the machine design. Invaluable experiences can be learned from LANSCE (Los Alamos, U.S.) and PSI (Switzerland). These machines have been handling MW beams for years and have designed several remote-handling systems that work reliably. [8]

3.3.3. *H^- injection*

This is another difficult part of the design and has many technical issues involved.

- The stripping foil (usually made of carbon) must stand for high temperature and large shock waves. It must also have high stripping efficiency and a reasonable lifetime.

- The stripped electrons and unstripped H^0 and H^- particles must be collected.

- During the many-turn (hundreds to thousands) injection, the orbit bump needs to "paint" the particles in the phase space so that a uniform distribution can be obtained. This would reduce the space charge effect. The orbit bump also needs to minimize the average number of hits per particle on the foil.

- The beam emittance dilution due to Coulomb scattering from the foil should be kept under control.

- There are proposals from the KEK and Los Alamos for laser stripping. The R&D is being pursued.

3.3.4. Lattice

The main requirement is multiple long straight sections, which are used for, respectively, injection, extraction, RF and collimation.

3.3.5. e-p instability

This is a main beam dynamics problem in an accumulator. In the PSR at Los Alamos, e-p instability is the bottleneck limiting the beam power. When the beam intensity reaches a threshold, rapid beam oscillations (usually in one transverse plane) occur that leads to fast beam loss. This instability is believed to be caused by electrons trapped in the proton bunch gap. These electrons come mainly from secondary yield. When a primary electron hits the wall, secondary electrons are generated, which are accelerated by the proton beam and hit the opposite side of the wall, generating more electrons, so on and so forth, causing an avalanche.

This so-called electron cloud effect (ECE) has also been seen in electron storage rings, in particular in the two B-factories: PEP-II and KEK-B. These two machines effectively used solenoids to suppress this effect. However, solenoids or clearing electrodes appear to be less useful in the PSR. (This is a puzzle to be resolved.) Instead, the following measures have been found effective in raising the instability threshold in the PSR: beam scrubbing (which conditions the wall), inductive inserts (which make the proton bunch gap cleaner), and sextupoles (which couple the motion in the two transverse planes and stabilize the oscillation in one plane).

This is an active research field. [9]

3.3.6. *Hardware*

The challenge is the magnet, which must have large aperture in order to accommodate large beam size and beam halo. Good field quality is necessary to ensure large dynamic aperture.

Other technical systems, including power supplies, RF, vacuum and diagnostics, are relatively straightforward.

4. Design Concept of a Synchrotron-based Spallation Neutron Source

A synchrotron-based spallation neutron source also has three major accelerator components:

- Linac front end
- Linac (lower energy)
- Synchrotron

The designs of the linac front end and the linac are similar to that of a linac-based spallation neutron source. However, a synchrotron design is very different from an accumulator. Therefore, we will focus on the synchrotron in this section.

4.1. *Lattice*

There are two basic requirements on the design: a transition-free lattice, and several dispersion-free straight sections. For high intensity operation in proton synchrotrons, transition crossing is often a major cause of beam loss and emittance blowup. One should avoid it in the first place. Dispersion in the RF, which is placed in one or more straight sections, may lead to synchro-betatron coupling resonance and should also be avoided.

For a medium energy synchrotron (above ~ 6 GeV), regular FODO lattices (in which $\gamma_t \propto \sqrt{R}$, γ_t the lattice transition γ, R the machine radius) are ruled out because they would use too many cells to achieve a high γ_t. Otherwise a transition crossing is inevitable when the γ of the beam approaches γ_t during ramp. There are several lattices that can give either a high or an imaginary γ_t so that a transition crossing would not occur. For example, (a) a flexible momentum compaction (FMC) lattice, which has a singlet 3-cell modular structure with a missing or short dipole in the mid-cell; (b) a doublet 3-cell modular structure with a missing or short dipole in the mid-cell. Figure 1 is an example of (b), which is designed for a new 8 GeV synchrotron at Fermilab.

The choice of phase advance per module is of critical importance in this type of lattice. There are two reasons. (i) The chromaticity sextupoles are placed in the mid-cell, where the beta-function peaks and available space exists. In order to cancel the higher order effects of these sextupoles, they need to be paired

242

properly. (ii) The phase advance per arc in the horizontal plane must be multiple of 2π in order to get zero dispersion in the straights without using dispersion suppressors (which are space consuming). Other requirements in the lattice design include: ample space for correctors (steering magnets, trim quadrupoles, chromaticity and harmonic sextupoles, etc.), ample space for diagnostics, low beta and dispersion functions (to make the beam size small), large dynamic aperture (to accommodate beam halo), and large momentum acceptance (to allow for bunch compression when necessary).

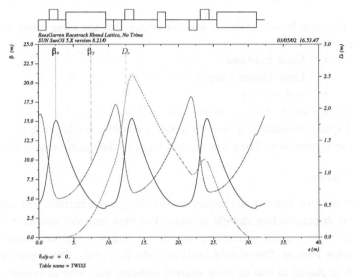

Figure 1. Lattice module of the Fermilab new 8 GeV synchrotron design. Each module has three doublet cells. The dipole in the mid-cell is short. The phase advance per module is 0.8 and 0.6 in the h- and v-plane, respectively. There are five modules in each arc.

4.2. Space Charge

Amongst various beam physics issues, the space charge is a major concern. It is often the bottleneck limiting the beam intensity in an intense proton source, in particular, in a synchrotron, because the injection energy is low.

A useful scaling factor is the Laslett tune shift

$$\Delta \nu = - (3r/2) \times (N/ \varepsilon_N) \times (1/\beta\gamma^2) \times B_f \qquad (2)$$

in which r is the classical proton radius (1.535×10^{-18} m), N the total number of protons, ε_N the normalized 95% transverse emittance, β and γ the relativistic factors, and B_f the bunching factor (ratio between peak and average beam current). It shows the space charge effect is most severe at injection because $\beta\gamma^2$ takes the minimum value. The situation becomes worse for high-intensity machines not only because the intensity is high but also because the injection time is long. Numerical simulation is the main tool to study this effect. A number of 1-D, 2-D and 3-D codes have been or are being written at many institutions. An example is shown in Figure 2. These codes are particularly useful to the design of the injection orbit bump current waveform for achieving uniform particle distribution in the beam, reducing emittance dilution and minimizing average number of hits per particle on the stripping foil during the phase space painting process. Several other measures, e.g., tune ramp, inductive inserts, quadrupole mode damper and electron beam compensation are under investigation for possible cures of the space charge effects.

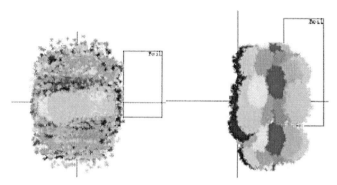

Figure 2. Space charge simulation using Track2D (by C. Prior). It shows the particle distribution after 45 turns injection in the Fermilab new 8 GeV synchrotron with (left) and without (right) the space charge effect.

4.3. Other Beam Dynamics Issues

In addition to the space charge, there are several other beam dynamics issues that need to be studied concerning an intense proton source.

- Electron cloud effect (ECE). This has been discussed in Section 3.3.5. It is interesting to note that, by far all reported ECE is either in DC machines (accumulators and storage rings) or AC machines in DC operation (i.e., on flat top or flat bottom). No ECE has been seen in AC machines during ramp. Does this imply that AC machines are immune to ECE? If true, this would be an important advantage of the synchrotron approach. However, this is solely an empirical observation.

Lack of a reliable theory for understanding and analyzing the ECE is a loophole that urgently needs to be filled.

- Microwave instability of bunched beam below transition. Because the machine will always operate below transition, the negative mass instability due to space charge would not occur. Would then this machine be immune to the microwave instability?

- Tune split. A split between the horizontal and vertical tunes is required in order to avoid the strong resonance $2v_x - 2v_y = 0$ that could be excited by the space charge. However, it is not clear how big the split needs to be. Does it have to be an integer? Or would a half-integer suffice?

4.4. Beam Loss, Collimation and Remote Handling

This part is similar to that for accumulators as discussed in Sections 3.3.1 and 3.3.2. There is, however, an important difference. Because the injected linac beam power is low (e.g., 6% of the full beam power if the acceleration range of the synchrotron is 16), higher beam loss at injection (which usually accounts for most of the total loss) can be tolerated. This is an advantage of the synchrotron approach.

4.5. Slow Extraction

In addition to one-turn extraction, synchrotrons are also used for experiments that require slow extraction (many turns). A critical issue is the efficiency. Although the efficiency of one-turn fast extraction can exceed 99%, it is much lower for multi-turn slow extractions. At high-intensity operation, the beam loss in existing machines during slow spill is usually around 4-5%. This is not acceptable for the next generation of high-intensity machines, in which the beam power will be 1 MW or higher and one percent loss would mean 10 kW or higher. This is a serious problem in the case of KAMI and CKM at the Fermilab Main Injector, and kaon and nuclear physics programs at the JHF. A recent ICFA mini-workshop was devoted to this topic. [10]

4.6. Hardware

4.6.1. Magnets

Magnets are one of the most expensive technical systems of a synchrotron. A critical parameter in the magnet design is the vertical aperture of the main bending magnets. The magnet cost is essentially proportional to the aperture. It should be large enough to accommodate a full size beam including its halo. The following criterion can be used in design:

$$A = \{3 \, \varepsilon_N \times \beta_{max} / \beta\gamma\}^{1/2} + D_{max} \times \Delta p/p + \text{c.o.d.} \qquad (3)$$

in which A is the half aperture, ε_N the normalized 95% beam emittance, β_{max} the maximum beta-function, D_{max} the maximum dispersion, $\Delta p/p$ the relative momentum spread, c.o.d. the closed orbit distortion. The parameter 3 is the estimated size of the beam halo relative to the beam size.

Because this is an AC machine, field tracking between the dipoles and quadrupoles at high field is an important issue. Trim quads or trim coils are needed. The peak dipole field should not exceed 1.5 Tesla. The peak quadrupole gradient is limited by the saturation at the pole root (not pole tip).

The choice of the coil turn number per pole is a tradeoff between the coil AC loss and voltage-to-ground. The former requires the use of many small size coils, whereas the latter requires the opposite, namely, small number of turns. There are two ways to compromise. One is to employ stranded conductor coils, as shown in Figure 3, which was adopted in the JHF 3 GeV ring design. Another is to connect several coils in parallel at the magnet ends, as done in the ISIS. The ratio of the AC *vs.* DC coil loss should be kept around 2-3. The voltage-to-ground should not exceed a few KV.

The aperture and good field region should include a rectangular area (instead of an elliptical area). This is because there will be a significant number of particles residing in the corners of the rectangle.

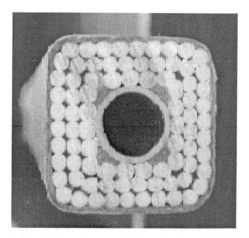

Figure 3. Stranded conductor coil for reducing coil AC loss.

4.6.2. *Power supplies*

This is another expensive technical system. There are several choices for the power supplies in a rapid cycling machine. (1) A single harmonic resonant system, e.g., the Fermilab Booster which resonates at 15 Hz. (2) A dual-harmonic resonant system, e.g., the Fermilab new 8 GeV synchrotron which uses a 15 Hz component plus a 12.5% 30 Hz component as shown below: [11]

$$I(t) = I_0 - I \cos(2\pi ft) + 0.125 \, I \sin(4\pi ft) \qquad (4)$$

in which f = 15 Hz, I_0 and I are two constants determined by the injection and peak current. The advantage of this system is that the peak value of dB/dt is decreased by 25%, which leads to a saving of the peak RF power by the same amount. (3) A programmable ramp system, e.g., the AGS Booster and AGS. Although this is a most versatile system (e.g., allowing for a front porch and a flat top), it is also most expensive.

4.6.3. *RF*

The RF system is demanding, because it must deliver a large amount of power to the beam in a short period. In addition, it must be tunable, because the particle revolution frequency increases during acceleration. Cavities with ferrite tuners have been in use for decades. Recently the development of the Finemet cavities at the KEK has aroused strong interest at many laboratories. Thanks to a US-Japan collaboration, Fermilab has built a 7.5 MHz, 15 kV Finemet cavity and installed it in the Main Injector for bunch coalescing. [12] The main advantages of the Finemet cores are high accelerating gradient and wide bandwidth. The former is especially important for high intensity small size rings, in which space is precious. The main concern, however, is its high power consumption. For example, the Fermilab Finemet cavity needs a 200 kW power amplifier to drive it. New types of magnetic alloys are under investigation for performance improvement.

4.6.4. *Vacuum*

Vacuum pipe for a rapid cycling machine is probably one of the most challenging items. Ceramic pipe with a metallic cage inside has been successfully employed at the ISIS. However, this is a costly solution, because it occupies a significant portion of the magnet aperture. Assuming the ceramic wall and the cage need a 1-in vertical space, a 4-in aperture magnet would have to increase its vertical gap by 25% to 5-in in order to accommodate this pipe. This would directly be translated to a 25% increase in the magnet and power supply costs, equivalent to tens of millions dollars.

Thin metallic pipe is an alternative. However, it must be very thin (several mils) in order to minimize the eddy current effects (pipe heating and induced magnetic field). Such a thin pipe is mechanically unstable under vacuum. Several designs have been tried to enhance its stability, including ceramic shields, metallic ribs and spiral lining. Prototyping of the first two designs did not work well. The third one looks promising and is currently under investigation. [13]

Another alternative is that the magnets employ external vacuum skins like those in the Fermilab Booster. Perforated metallic liners are used in the magnet gap to provide a low-impedance environment for the beam.

4.6.5. Diagnostics

In addition to the conventional diagnostics for measuring beam position, tune, profile, intensity and loss, intense proton sources have several specific requirements. A system that can diagnose beam parameters during multi-turn injection is highly desirable. The method for fast, accurate non-invasive tune measurement is being developed. A circulating beam profile monitor covering a large dynamic range with turn-by-turn speed will be crucial for studying beam halo. (A similar instrument has been developed for the linac beam halo experiment at Los Alamos. [7]) There was also an ICFA mini-workshop devoted to this topic. [14]

4.7. New Ideas

In the past several years, there are a number of new or revitalized ideas proposed to the high intensity proton source study. Here are a few examples:

4.7.1. Inductive inserts

They are made of ferrite rings and also can have bias current for impedance tuning. Their inductive impedance would fully or partially compensate the space charge impedance, which is capacitive. The first successful experiment was at the PSR. [15] Two ferrite modules made by Fermilab have been installed in the ring. They help increase the e-p instability threshold, which is a major bottleneck of that machine. Another experiment is going on at the Fermilab Booster.

4.7.2. Induction synchrotron

This is a longitudinally separated function machine. In other words, the longitudinal focusing and acceleration are carried out by two separate RF systems. The former uses barrier RF buckets, the latter a constant RF voltage curve. One useful feature of this type of machine is tunable bunch lengths. So the so-called superbunch acceleration could be possible. Because a superbunch is similar to a debunched beam, the peak beam current is low. Thus, the space charge effect can be reduced and beam intensity increased.

4.7.3. Barrier RF stacking

The application of Finemet and other magnetic alloys makes it possible to build broadband barrier RF cavities with high voltage (~10 kV or higher). They can be used to stack beams in the longitudinal phase space. This is particularly useful when the beam intensity of a synchrotron is limited by its injector (e.g., the intensity of the Fermilab Main Injector is limited by the Booster). Compared to the slip stacking, an advantage of barrier RF stacking is the greatly reduced beam loading effects due to a lower peak beam current. [16,17]

4.7.4. Fixed field alternating gradient (FFAG) accelerator

Although MURA first proposed this idea about 40 years ago, it was almost forgotten. Only the recent activities at the KEK brought it back to the world's attention. KEK has successfully built a 1 MeV Proof-of-Principle (PoP) proton FFAG and is building a 150 MeV one. [18] FFAG is an ideal machine for high intensity beams. Its repetition rate can be much higher than a rapid cycling synchrotron (in the range of kHz). One problem of the FFAG, however, is that it is difficult (if not impossible) to fit it into an existing accelerator complex, which usually consists of a linac and a cascade of synchrotrons.

4.7.5. Repetition rate increase in existing synchrotrons

This is a brute force approach but can be appealing because it is straightforward. For example, the Brookhaven National Laboratory has a proposal for increasing the AGS repetition rate from 0.5 Hz to 2.5 Hz. [19] The Fermilab Main Injector upgrade also includes a rep rate increase (from 0.53 Hz to 0.65 Hz). [20]

5. Design Concept of a Proton Driver

5.1. Differences between a Proton Driver and a Spallation Neutron Source

A proton driver is a high intensity proton source. It can be used as a spallation neutron source. But it can do more. It can generate neutrino superbeams and other high intensity secondary particles (muons, kaons, pions, antiprotons, etc.) for high-energy physics experiments. It can also be used as the first stage of a neutrino factory and a muon collider.

There are two main differences between a proton driver and a spallation neutron source.

- The beam energy of a proton driver is higher. A commonly used production target is carbon. For a carbon target, the π^- cross-section is much lower than π^+ when the proton beam energy is below 4 GeV. Therefore, for polarized muon experiments, a proton driver must be 4 GeV or higher. Furthermore, for neutrino oscillation experiments, a proton source with tunable energy in the range of several GeV up to about 100 GeV is preferred.
- The bunch length of a proton driver is shorter. The pion yield (i.e., number of pions per unit proton beam power) has a strong dependence on the proton bunch length. This is the only parameter that we have control to minimize the 6-D phase space volume of the pions. Moreover, to obtain highly polarized pion beams also requires short proton bunch length. The typical bunch length in a proton driver is a few ns (instead of μs as in a spallation neutron source).

5.2. How to Achieve Higher Energies

In a synchrotron-based design, this is not difficult. The energy covers a wide range: from as low as 3 GeV (JHF, 1 MW) to as high as 120 GeV (Fermilab Main Injector upgrade, 2 MW).

In a linac-based design, however, this is severely limited by the cost. The existing highest energy proton linac is the LANSCE (0.8 GeV) at Los Alamos. The SNS linac under construction at Oak Ridge is 1 GeV. There are proposals for 2.2 GeV (CERN), 3 GeV (Los Alamos) and 8 GeV (Fermilab) proton linacs. But none of these has become a construction project.

5.3. How to Obtain Short Bunch Lengths

In a synchrotron-based design, the bunch length is determined by the RF bucket length, i.e., by the RF frequency. A short bunch length implies the use of a high frequency RF system. However, sometimes there are good reasons to use low frequency RF (e.g., to limit the number of bunches). In this case, a bunch rotation technique can be used for compressing the bunch length.

It should be pointed out that there is a new beam dynamics problem associated with bunch rotation, namely, the path length dependence on momentum spread $\Delta p/p$ and space charge tune shift Δv. This is especially important for proton drivers, in which due to large momentum spread (a few percent) and large tune shift (a few tenth), the dependence of the path length ΔL on $\Delta p/p$ and Δv can no longer be ignored. In other words, the momentum compaction factor $\alpha = (\Delta L/L) / (\Delta p/p)$ cannot be treated as a constant during bunch rotation. It is dependent upon the momentum and amplitude of each particle. This will result in a longer bunch after rotation. Simulation study must take this effect into account.

In a linac-based design, a compressor ring (separate from an accumulator ring) will be needed in order to provide the required bunch length and bunch structure.

6. Summary

Two recent major spin-offs from high-energy accelerators are synchrotron light sources and high intensity proton sources. Both have found wide-range applications in the field of basic sciences (e.g., material science, molecular biology, chemistry, etc.) as well as in industrial research and development (e.g., chip technology, nano technology, medical and pharmaceutical research, etc.). Spallation neutron source is an important type of the latter.

There are two approaches to a spallation neutron source. One is linac-based, another synchrotron-based. Each approach has its pros and cons. The PSR at the Los Alamos National Laboratory and the SNS project at the Oak Ridge National Laboratory belong to the former, while the ISIS at the Rutherford Appleton Laboratory and the JHF project at the KEK/JAERI represent the latter. (Note that the JHF is a multi-purpose facility unlike the SNS, which serves solely as a neutron source.)

There are close connections between the design of a spallation neutron source and a proton driver. The latter is a strong contender for a near term construction project in the high-energy physics field in the U.S., Europe and Japan. The studies of the two types of machines benefit each other.

The work on high intensity proton sources has been a dynamic field in the accelerator world. There are numerous challenging problems as well as great expectations. Out of the world's three large accelerator projects currently under construction - LHC, SNS and JHF - two are high intensity proton sources. Several more have appeared on the horizon. We'd like to encourage young people to join this field and bring with them their energy, enthusiasm and fresh ideas.

Acknowledgements

The author would like to express his gratitude to the 3rd OCPA International Accelerator School for the invitation to give a lecture and for the hospitality during his stay in Singapore.

Appendix A

There have been numerous conferences and workshops on high intensity proton sources sponsored by the Beam Dynamics Panel of the International Committee for Future Accelerators (ICFA). For example, there are a series of ICFA mini-workshops on various specific topics, including transition crossing, particle losses, RF, beam loading, transverse and longitudinal emittance measurement and preservation, injection and extraction, beam halo and scraping, two-stream instability, diagnostics and space charge simulations. These mini-workshops can be found on the web http://www-bd.fnal.gov/icfa/workshops/workshops.html. Paper proceedings are also available from the workshop organizers.

There was an ICFA-HB2002 workshop in April 2002 at Fermilab, which covered almost all the aspects concerning high intensity proton sources. The web address is http://www-bd.fnal.gov/HB2002/.

There was an ECLOUD'02 workshop also in April 2002 at CERN for the study of electron cloud effect. The proceedings and presentations are posted at http://wwwslap.cern.ch/collective/ecloud02/.

An international workshop on induction accelerators took place in October 2002 at the KEK. The web address is http://conference.kek.jp/RPIA2002/.

Appendix B

In July 2001, about 1,200 physicists over the world gathered at Snowmass, Colorado, USA, for three weeks to discuss the future of high-energy physics. One specific topic was high intensity proton sources. A detailed 26-point R&D program was crafted. This program is directly related to the spallation neutron source work. The Executive Summary is attached, which can be used as guidance for planning future R&D. The full context can be found on the web: http://www-bd.fnal.gov/icfa/snowmass/.

Executive Summary of Snowmass2001 on High Intensity Proton Sources

The US high-energy physics program needs an intense proton source (a 1-4 MW Proton Driver) by the end of this decade. This machine will serve multiple purposes: (i) a stand-alone facility that will provide neutrino superbeams and other high intensity secondary beams such as kaons, muons, neutrons, and anti-protons (cf. E1 and E5 group reports); (ii) the first stage of a neutrino factory (cf. M1 group report); (iii) a high brightness source for a VLHC (cf. M4 group report).

Based on present accelerator technology and project construction experience, it is both feasible and cost-effective to construct a 1-4 MW Proton Driver. There are two PD design studies, one at FNAL and the other at the BNL. Both are designed for 1 MW proton beams at a cost of about US$200M (excluding contingency and overhead) and upgradeable to 4 MW. An international collaboration between FNAL, BNL and KEK on high intensity proton facilities addresses a number of key design issues. The sc cavity, cryogenics, and RF controls developed for the SNS can be directly adopted to save R&D efforts, cost, and schedule. PD studies are also actively pursued at Europe and Japan.

There are no showstoppers towards the construction of such a high intensity facility. Key research and development items are listed below ({} indicates present status). Category A indicates items that are not only needed for future machines but also useful for the improvement of existing machine performance; category B indicates items crucial for future machines and/or currently underway.

1) H⁻ source: Development goals - current 60–70 mA {35 mA}, duty cycle 6–12% {6%}, emittance 0.2 π mm-mrad rms normalized, lifetime > 2 months {20 days}. (A)

2) LEBT chopper: To achieve rise time < 10 ns {50 ns}. (B)

3) Study of 4-rod RFQ at 400 MHz, 100 mA, 99% efficiency, HOM suppressed. (B)

4) MEBT chopper: To achieve rise time < 2 ns {10 ns}. (B)

5) Chopped beam dump: To perform material study & engineering design for dumped beam power > 10 kW. (A)

6) Funneling: To perform (i) one-leg experiment at the RAL by 2006 with goal one-leg current 57 mA; (ii) deflector cavity design for CONCERT. (all B)

7) Linac RF control: To develop (i) high performance HV modulator for long pulsed (>1ms) and CW operation; (ii) high efficiency RF sources (IOT, multi-beam klystron). (all A)

8) Linac sc RF control: Goal - to achieve control of RF phase error < 0.5° and amplitude error <0.5% {presently 1°, 1% for warm linac}. (i) To investigate the choice of RF source (number of cavity per RF source, use of high-power

source); (A) (ii) to perform redundancy study for high reliability; (B) (iii) to develop high performance RF control (feedback and feedforward) during normal operation, tuning phases and off-normal operation (missing cavity), including piezo-electric fast feedforward. (A)

9) Space charge: (i) Comparison of simulation code ORBIT with machine data at FNAL Booster and BNL Booster; (ii) to perform 3D ring code bench marking including machine errors, impedance, and space charge (ORNL, BNL, SciDAC, PPPL). (all A)

10) Linac diagnostics: To develop (i) non-invasive (laser wire, ionization, fluorescent-based) beam profile measurement for H^-;(ii) on-line measurement of beam energy and energy spread using time-of-flight method; (iii) halo monitor especially in sc environment; (iv) longitudinal bunch shape monitor. (all A)

11) SC RF linac: (i) High gradients for intermediate beta (0.5 – 0.8) cavity; (A) (ii) Spoke cavity for low beta (0.17 – 0.34). (B)

12) Transport lines: To develop (i) high efficiency collimation systems; (A) (ii) profile monitor and halo measurement; (A) (iii) energy stabilization by HEBT RF cavity using feedforward to compensate phase-jitter. (B)

13) Halo: (i) To continue LEDA experiment on linac halo and comparison with simulation; (ii) to start halo measurement in rings and comparison with simulation. (all B)

14) Ring lattice: To study higher order dependence of transition energy on momentum spread and tune spread, including space charge effects. (B)

15) Injection and extraction: (i) Development of improved foil (lifetime, efficiency, support); (A) (ii) experiment on the dependence of H^0 excited states lifetime on magnetic field and beam energy; (B) (iii) efficiency of slow extraction systems. (A)

16) Electron cloud: (i) Measurements and simulations of the electron cloud generation (comparison of the measurements at CERN and SLAC on the interaction of few eV electrons with accelerator surfaces, investigation of angular dependence of SEY, machine and beam parameter dependence); (A) (ii) determination of electron density in the beam by measuring the tune shift along the bunch train; (A) (iii) theory for bunched beam instability that reliably predicts instability thresholds and growth rates; (A) (iv) investigation of surface treatment and conditioning; (A) (v) study of fast, wide-band, active damping system at the frequency range of 50–800 MHz. (B)

17) Ring beam loss, collimation, protection: (i) Code benchmarking & validation (STRUCT, K2, ORBIT); (A) (ii) engineering design of collimator and beam dump; (A) (iii) experimental study of the efficiency of beam-in-gap cleaning; (A) (iv) bent crystal collimator experiment in the RHIC; (B) (v) collimation with resonance extraction. (B)

18) Ring diagnostics: (i) Whole area of diagnosing beam parameters during multi-turn injection; (ii) circulating beam profile monitor over large dynamic range with turn-by-turn speed; (iii) fast, accurate non-invasive tune measurement. (all A)

19) Ring RF: To develop (i) low frequency (~5 MHz), high gradient (~1 MV/m) burst mode RF systems; (B) (ii) high gradient (50-100 kV/m), low frequency (several MHz) RF system with 50-60% duty cycle; (B) (iii) high-voltage (>100 kV) barrier bucket system; (B) (iv) transient beam loading compensation systems (e.g. for low-Q MA cavity). (A)

20) Ring magnets: (i) To develop stranded conductor coil; (ii) to study voltage-to-ground electrical insulation; (iii) to study dipole/quadrupole tracking error correction. (all B)

21) Ring power supplies: To develop (i) dual-harmonic resonant power supplies; (ii) cost effective programmable power supplies. (all B)

22) Kicker: (i) Development of stacked MOSFET modulator for DARHT and AHF to achieve rise/fall time <10-20 ns; (B) (ii) impedance reduction of lumped ferrite kicker for SNS. (A)

23) Instability & impedance: (i) To establish approaches for improved estimates of thresholds of fast instabilities, both transverse and longitudinal (including space charge and electron cloud effects); (ii) to place currently-used models such as the broadband resonator and distributed impedance on a firmer theoretical basis; (iii) impedance measurement based on coherent tune shifts *vs.* beam intensity, and instability growth rate *vs.* chromaticity, including that for flat vacuum chambers; (iv) to develop new technology in feedback implementation. (all B)

24) FFAG: (i) 3-D modeling of magnetic fields and optimization of magnet profiles; (ii) wide-band RF systems; (iii) transient phase shift in high frequency RF structures; (iv) application of sc magnets. (all B)

25) Inductive inserts: (i) Experiments at the FNAL Booster & JHF3; (A) (ii) programmable inductive inserts; (B) (iii) development of inductive inserts which have large inductive impedance and very small resistive impedance; (B) (iv) theoretical analysis. (B)

26) Induction synchrotron: (i) Study of beam stability; (ii) development of high impedance, low loss magnetic cores. (all B)

References

1. The Snowmass 2001 Workshop web site: http://www.snowmass2001.org/.
2. W. Chou, C. Ankenbrandt, and E. Malamud, editors, *"The Proton Driver Design Study,"* FERMILAB-TM-2136 (December 2000).
3. G.W. Foster, W. Chou, and E. Malamud, editors, *"Proton Driver Study II,"* FERMILAB-TM-2169 (May 2002).

4. M. Vretenar, editor, *"Conceptual Design of the SPL, a High-Power Superconducting H⁻ Linac at CERN,"* CERN 2000-012 (December 2000).
5. J. Staples et al., Proc. 1999 PAC (New York, USA), p. 1961.
6. W. Chou et al., *"Design and Measurements of a Pulsed Beam Transformer as a Chopper,"* KEK Report 98-10 (September 1998); W. Chou et al., Proc. 1999 PAC (New York, USA), p. 565; Y. Shirakabe et al., Proc. 2000 EPAC (Vienna, Austria), p. 2468.
7. T.P. Wangler, *"Linac Based Proton Drivers,"* AIP Conference Proc. 642, p. 43 (2002).
8. E. Wagner, *"Remote Handling and Shielding at PSI,"* AIP Conference Proc. 642, p. 265 (2002).
9. For references see http://wwwslap.cern.ch/collective/ecloud02/.
10. The 10th ICFA Mini-Workshop on Slow Extraction, October 15-17, 2002, Brookhaven National Laboratory, USA. The proceedings can be found on the web: http://www.agsrhichome.bnl.gov/ICFA2002/.
11. Chapter 5 of Ref. 3.
12. D. Wildman et al., Proc. 2001 PAC (Chicago, USA), p. 882.
13. Z. Tang, *"A New Kind of Vacuum Tube for Proton Driver,"* FERMILAB-TM-2188 (September 2002).
14. The 11th ICFA Mini-Workshop on Diagnostics, October 21-23, 2002, Oak Ridge National Laboratory, USA. The proceedings can be found on the web: http://www.sns.gov/icfa/.
15. M.A. Plum et al., *Phys. Rev.* **ST-AB 2**, 064210 (June 1999).
16. K. Koba and J. Steimel, *"Slip Stacking,"* AIP Conference Proc. 642, p. 223 (2002).
17. K.Y. Ng, *"Doubling Main Injector Beam Intensity using RF Barriers,"* AIP Conference Proc. 642, p. 226 (2002).
18. Y. Mori, *"Progress on FFAG Accelerators at KEK,"* ICFA Beam Dynamics Newsletter, No. 29, p. 20 (December 2002).
19. W.T. Weng and T. Roser, *"The AGS High Power Upgrade Plan,"* AIP Conference Proc. 642, p. 56 (2002).
20. Chapter 13 of Ref. 3.

RF ELECTRON LINAC AND MICROTRON

SHU-HONG WANG

Institute of High Energy Physics (IHEP)
Yuquan Road 19, Beijing 100039, China
E-mail: wangsh@sun.ihep.ac.cn

In this lecture note, the RF Electron Linac and the Microtron are introduced including the principles of acceleration, their basic features, main structures, design considerations and their various applications.

9. Introduction to the RF Electron Linac

9.1. *Properties of the RF Electron Linac*

Electrons can be resonantly accelerated, along an almost linear orbit, by an RF electric field. This accelerating facility is called the RF Electron Linac. The RF accelerating field is either a traveling wave in loaded waveguides, or a standing wave in loaded cavities.

The RF electron linac has the following features, compared with other types of accelerators:

♦ It has no difficulties with the beam injection (into the linac) and ejection (from the linac) compared with the circle / ring-type accelerators.

♦ It can accelerate electrons from low energy (a few tens of keV) to a very high energy (~ TeV). It is not like the dc high-voltage accelerator which has the dc voltage breakdown limitation, and also not like the electron ring-type accelerator which has a beam energy loss limitation caused by the synchrotron radiation.

♦ It can provide a high current (or high intensity) beam with transverse focusing and longitudinal bunching.

♦ It can work at a pulsed mode with any duty factor, and / or at a CW mode.

♦ It can be designed, installed and commissioned section by section.

♦ It is mostly equipped by the RF accelerating structures which are not easy to be operated and need to be maintained with high stability and reliability, and its construction and operation costs per unit beam power is expensive compared with circle / ring accelerators.

9.2. *Applications of the RF Electron Linac*

♦ To be used as an injector for synchrotrons, synchrotron radiation light sources and electron-positron colliders.

♦ Medical uses, such as radiotherapy and production of medical isotopes.

♦ Industrial irradiation for various materials and products.

♦ Linac-based Free Electron Laser (FEL).

♦ Electron-positron linear colliders.

2. Elementary Principles of the RF Electron Linac

2.1 *Acceleration with the RF Linac*

Assuming an RF electromagnetic (EM) field travels in a uniform cylindrical waveguide, its fundamental mode TM_{01} has the EM components of longitudinal electric field E and azimuthal magnetic field B, as shown in Figure. 1. Their distributions are analytically described with following expressions;

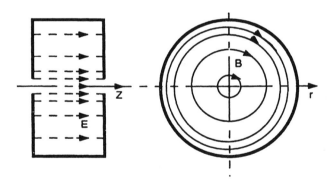

Figure 1.   EM field pattern of TM_{01} mode

$$E_z(r,z,t) = E_0 J_0(k_c r)e^{j\omega t - k' z} ,$$

$$E_r(r,z,t) = jE_0[1 - (\frac{\omega_{cr}}{\omega})^2]^{1/2} J_1(k_c r)e^{j\omega t - k' z} ,$$

$$E_\theta = 0 , \tag{1}$$

$$B_\theta(r,z,t) = j\mu_0 E_0 J_1(k_c r)e^{j\omega t - k' z} ,$$

$$B_r = B_z = 0 .$$

where J_0 and J_1 are zero-order and first-order Bessel functions, respectively, $k_c = \omega_{cr}/c$ is the wave number, its frequency is the waveguide's cutoff frequency ω_{cr}, and its phase velocity is the velocity of light. The cutoff frequency for a given radius R of the waveguide can be obtained from the boundary condition of $E_z(R) = 0$, that is, the first root of $J_0(k_c R) = 0$

$$k_c R = \frac{\omega_{cr}}{c} R = 2.405 \tag{2}$$

$k' = \alpha + jk_0$, where α is the field attenuation factor due to the RF loss on a resistive wall, $k_0 = \dfrac{\omega}{v_p}$ is a wave number with frequency ω and phase velocity v_p. Let us first consider the case of no power loss (ideal conductor, $\alpha = 0$), then its propagation property (dispersive relation) is as follows:

$$k_0{}^2 = (\frac{\omega}{c})^2 - k_c{}^2 = (\frac{\omega}{c})^2 - (\frac{\omega_{cr}}{c})^2. \tag{3}$$

It describes the relations among k_c, ω and k_0 in $J_0(k_c r)e^{j(\omega t - k_0 z)}$, as shown in Figure 2.

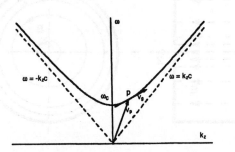

Figure 2 . Dispersion curve for a uniform waveguide

For TM_{01}-mode to exist in the waveguide, k_0 should be a real number, so that $\omega \geq \omega_{cr}$. This means that only the waves with $\omega \geq \omega_{cr}$ can be propagated in the waveguide. But their phase velocity is

$$v_p = \frac{\omega}{k_0} = \frac{c}{\sqrt{1-(\omega_{cr}/\omega)^2}} \geq c \tag{4}$$

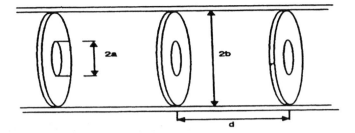

Figure 3. Disk-loaded traveling wave (TW) structure

Obviously, these waves can not resonantly accelerate electrons. To have an accelerating structure in which the propagated waves have $v_p \leq c$, we must modify the structure to slow down the v_p, for instance, by introducing a periodic disk-loaded structure, as shown in Figure 3. Then the wave amplitude is periodically modulated:

$$E_z(r,z,t) = E_{L_c}(r,z)e^{j(\omega t - k_0 z)} \tag{5}$$

where $E_{L_c}(r,z)$ is a periodic function with period d $= L_c$. This is Floquet's theorem: at the same locations in different periods, the amplitudes of the propagated field are the same but their phases differ by a factor of $e^{jk_0 L_c}$. We can express $E_{L_c}(r,z)$ as a Fourier series in z:

$$E_{Lc}(r,z) = \sum_{n=-\infty}^{\infty} E_n J_0(k_n r)e^{-j\frac{2\pi n}{L_c}z} \tag{6}$$

where the coefficients $E_n J_0(k_n r)$ are the solutions of the wave motion equation with cylindrical boundary conditions, so that

$$E_z(r,z,t) = \sum_{n=-\infty}^{\infty} E_n J_0(k_n r)e^{j(\omega t - k_n z)} \tag{7}$$

where $k_n = k_0 + 2\pi n / L_c$ is the wave number of the nth space harmonic, which has the phase velocity of

$$v_{np} = \frac{\omega}{k_n} = \frac{\omega}{k_0(1 + 2\pi n / k_0 L_c)} \leq c \tag{8}$$

With above expressions we find that

♦ A traveling wave (TW) consists of infinite space harmonic waves, as shown in Figure 4.

♦ Harmonic waves with $n > 0$ and propagating in the $+z$ direction are known as forward waves, and those with $n < 0$ and propagating in the $-z$ direction are called back waves.

♦ If each forward wave has the same amplitude and phase velocity as a back wave, then they form a standing wave. Therefore a method of analysis using space harmonic waves can be used to describe both the standing waves and the traveling waves.

♦ Because the various space harmonic waves have different phase velocities, only one of the harmonic waves can be used to resonantly accelerate particles. The fundamental mode ($n = 0$) generally has the largest amplitude and hence is used for acceleration.

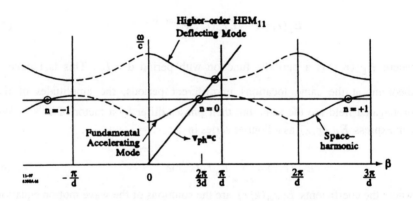

Figure 4. Brillouin diagram for a periodically loaded structure [5]

♦ When a TW is used to accelerate particles, a particle that " rides " on the wave at phase φ_0 and moves along the axis has an energy gain per period (L_c, cell length) of

$$\Delta W = e\, E_0 L_c cos\varphi_0 \tag{9}$$

where E_0 is the field on axis, averaged over a period

$$E_0 = \frac{1}{L_c} \int_0^{L_c} E_z(0, z)dz .$$ (10)

♦ Figure 4 also shows a second upper branch, which is one of an infinity of such high-order modes (HOMs), and intercepts the $v_p = c$ line. These modes are the so-called wake fields, which can be excited by the transversely offset beam.

2.2 Essential Parameters of a TW Accelerating Structure

2.2.1 Shunt-Impedance Z_s

The shunt-impedance per unit length of the structure is defined as

$$Z_s = \frac{E_a^{\ 2}}{-dP_w / dz} \quad (\text{M}\Omega/\text{m}).$$ (11)

It expresses that if we are given the RF power loss per unit length then we can know how high an electric field E_a can be established on the axis. Since $P_w \propto E_a^2$, therefore Z_s is independent of E_a and the power loss depends only on the structure itself which includes its configuration, dimension, material and operating mode.

2.2.2 Quality Factor Q

The unloaded quality factor of an accelerating structure is defined as

$$Q = \frac{\omega U}{-dP_w / dz}$$ (12)

where U is the stored energy per unit length of structure. The Q also describes the efficiency of the structure. With this definition one can see that given the stored energy, the higher the Q, the lesser is the RF loss; or given the RF loss and the higher the Q, the higher is the E_a (since $U \propto E_a^2$).

2.2.3 Z_s/Q

With the definitions of Z_s and Q, we have

$$Z_s/Q = E_a^2/\omega U .$$ (13)

This defines that for establishing a required electric field E_a the minimum stored energy required Z_s/Q is independent of power loss in the structure.

2.2.4 Group Velocity v_g

It is the velocity of the field energy traveling along the waveguide,

$$v_g = P_w/U$$ (14)

where P_w is the power flow defined by integrating the Poynting vector over a transverse plane within the inner diameter of the disk. For TM_{01}-mode, $P_w = \int_0^a E_r H_\theta 2\pi r dr$ here a is the iris radius, and for this mode, $E_r \propto r$ and $H_\theta \propto r$, so that $v_g \propto a^4$.

2.2.4 Attenuation Constant

We define

$$\tau_0 = \int_0^{L_s} \alpha(z)dz$$ (15)

as the attenuation constant, where $\alpha(z)$ is the attenuation per unit length of the structure, as mentioned in section 2.1. This is one of the most important parameters for TW structures, since it defines the ratio of output power to input power for an accelerating section (of length L_s), and determines the power loss per unit length

$$P_{out} = P_{in}e^{-2\tau_0} \quad \text{and} \quad \frac{dP_w}{dz} = \frac{P_{in}}{L_s}(1 - e^{-2\tau_0})$$ (16)

It is clear that the larger the τ_0, the smaller is the output power and hence the higher the rate of power use. On the other hand, a smaller τ_0 gives a larger group velocity of the structure and thus a larger inner radius of the disk ($v_g \propto a^4$) and a larger transverse acceptance. Finally, τ_0 should be chosen by a compromise between these two effects. The residual output power is absorbed by a load installed at the end of the section, as shown in Figure. 5.

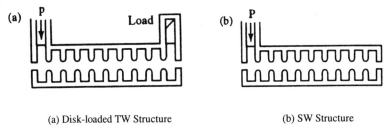

(a) Disk-loaded TW Structure   (b) SW Structure

Figure 5. Power absorber at the end of a TW section

2.2.6 Working Frequency

The working frequency is one of the basic parameters of the structure, since it affects most of the other parameters according to the following scaling laws:

♦ Shunt-impedance $\quad Z_s \propto f_0^{1/2}$,

♦ Quality Factor $\quad Q \propto f_0^{-1/2}$,

♦ Total RF peak power $\quad P_{tot} \propto f_0^{-1/2}$,

♦ Minimum energy stored $\quad Z_s / Q \propto f_0$,

♦ RF energy stored $\quad U \propto f_0^{-2}$,

♦ Power filling time $\quad t_F \propto f_0^{-3/2}$,

♦ Transverse dimension of structure a and $b \propto f_0^{-1}$.

The final choice of f_0 is usually made by adjusting all of the above factors and by considering the available RF source as well. Most electron linacs work at a frequency of about 3000 MHz (S-band), e.g., 2856 MHz ($\lambda \approx 10.5$ cm) for the SLAC linac and many others.

2.2.7 Operation Mode

Here we define the operation mode, which is specified by the RF phase difference between two adjacent accelerating cells. For instance 0-mode, $\pi/2$-mode, $2\pi/3$ -mode and π-mode are the operation modes that have the phase differences of 0, $\pi /2$, $2\pi/3$ and π, respectively, between two adjacent cells as shown in Figure. 6.

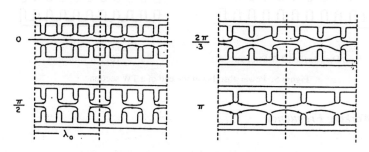

Figure 6.   Operation modes

For a disk-loaded TW structure the optimum operation mode is the $2\pi/3$-mode as it has the highest shunt-impedance as indicated in Figure. 7.

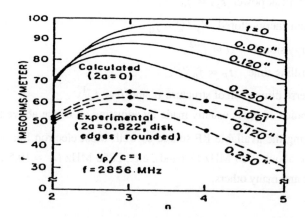

Figure 7 Shunt-impedance vs. operation modes in TW structure [2]

3. Traveling Wave Accelerating Structure

There is no firm rule with which to decide whether a traveling wave or a standing wave structure is to be chosen. However, traveling wave structure is usually used when dealing with short beam pulses and when particle velocities approach the velocity of light, as is the case with electrons.

3.1 Constant Impedance Structure

With the definitions of structure parameters, the RF power distribution along the linac section is

$$\frac{dP_w}{dz} = -\frac{\omega P_w}{Qv_g} = -2\alpha_0 P_w \tag{17}$$

where α_0 is the attenuation per unit length of structure, $\alpha_0 = \dfrac{\omega}{2Qv_g}$. If the structure is uniform along the z axis, from the above equations, we have

$$E_a{}^2 = \frac{\omega Z_s}{Qv_g} P_w \quad \text{and} \quad \frac{dE_a}{dz} = -\frac{\omega E_a}{2Qv_g} = -\alpha_0 E_a \tag{18}$$

For a uniform structure, α_0=constant, $E_a(z) = E_0 e^{-\alpha_0 z}$ and $P_w(z) = P_0 e^{-2\alpha_0 z}$. Thus in a constant-impedance structure $E_a(z)$ and $P_w(z)$ decrease along the z axis in a section. At the end of a section with length L_s

$$E_a(L_s) = E_0 e^{-\tau_0} \quad \text{and} \quad P_w(L_s) = P_0 e^{-2\tau_0} \tag{19}$$

where $\tau_0 = \alpha_0 L_s = \dfrac{\omega L_s}{2Qv_g}$ is the section attenuation. The energy gain of an electron that "rides" on the crest of the accelerating wave and moves to the end of section is

$$\Delta W = e \int_0^{L_s} E_a(z)dz = eE_0 L_s \frac{1 - e^{-\tau_0}}{\tau_0}$$

Using $E_0^2 = 2Z_s \alpha_0 P_{in}$ (P_{in} =input power), then

$$\Delta W = e\sqrt{2Z_s P_{in} L_s} \cdot (\frac{1-e^{-\tau_0}}{\sqrt{\tau_0}}) \tag{20}$$

For an optimized design of a constant impedance structure, we should maximize ΔW. Given P_{in} and L_s we make

$$Z_s \Rightarrow \text{maximum} \quad \text{and} \quad (\frac{1-e^{-\tau_0}}{\sqrt{\tau_0}})_{\max} \Rightarrow \tau_0 = 1.26 \tag{21}$$

Given L_s and Q, we can obtain the optimized group velocity v_g. Obviously the smaller v_g, the bigger is τ_0, and the bigger v_g, the lower E_0. An effective way to control v_g is to adjust the inner radius a of the disk along the section. On the other hand, the power filling time of a waveguide is $t_F = L_s / v_g = 2\pi\tau_0 / \omega$. To decrease t_F, then τ_0 should be < 1.26.

3.2 Constant Gradient Structure

To keep $E_a = E_0 =$ constant along the structure, the structure is not made uniform, so that $\alpha_0 = \alpha_0(z)$. The question is how to determine $\alpha_0(z)$. Let us change the radii of the structure, a and b, to vary v_g and to keep frequency constant along the section, also to keep the variations of Q and Z_s along z so small that they can be neglected, then we have

$$dP_W / dz = -2\alpha_0(z)P_w \quad \text{and} \quad P_{L_s} = P_0 e^{-2\tau_0} \tag{22}$$

where $\tau_0 = \int_0^{L_s} \alpha_0(z)dz$ is a section attenuation. Since $E_a^2 = -Z_s \dfrac{dP_w}{dz}$, to keep $E_a =$ constant, we need $dP_w / dz =$ const, so that

$$P_w(z) = P_0 + \frac{P_{L_s} - P_0}{L_s} z = P_0 \left[1 - \frac{1-e^{-2\tau_0}}{L_s} z \right] \tag{23}$$

Thus, in a constant gradient structure, P_w decreases linearly along the structure. With $dP_W / dz = -2\alpha_0(z)P_W$ and $v_g(z) = \omega / 2Q\alpha_0(z)$, we have

$$\alpha_0(z) = \frac{1}{2L_s} \cdot \frac{1 - e^{-2\tau_0}}{1 - \frac{z}{L_s}(1 - e^{-2\tau_0})} \tag{24}$$

and

$$v_g(z) = \frac{\omega L_s}{Q} \cdot \frac{1 - \frac{z}{L_s}(1 - e^{-2\tau_0})}{1 - e^{-2\tau_0}} \tag{25}$$

Thus, in a constant gradient structure, the $v_g(z)$ also decreases along the structure in the same way as $P_W(z)$. The energy gain for an on-crest particle is

$$\Delta W = e \int_0^{L_s} E_a(z)dz = eE_0 L_s$$

With

$$E_0^2 = -Z_s \frac{dP_{L_s}}{dz} = \frac{Z_s P_0}{L_s}(1 - e^{-2\tau_0})$$

we obtain

$$\Delta W = e\sqrt{Z_s P_0 L_s (1 - e^{-2\tau_0})} \tag{26}$$

To have ΔW_{max}, we should have $Z_s \Rightarrow$ maximum and $\tau_0 \Rightarrow$ maximum, all power should be lost in the structure. On the other hand, we should also consider the filling time, $t_F = 2Q\pi/\omega$, and τ_0 should be chosen by a compromise between several effects. An example of a SLAC constant gradient structure is shown in Figure 8. Each section is designed to be a tapered structure: $2b \approx 8.4$ to 8.2 cm, $2a \approx 2.6$ to 1.9 cm, $v_g/c \approx 0.021$ to 0.007, $L_s = 3.05$ m, and $Z_s \approx 57$ MΩ/m.

The advantages of the constant gradient structure are its uniform power loss and lower average peak surface field, thus most TW electron linacs are designed as constant gradient structures.

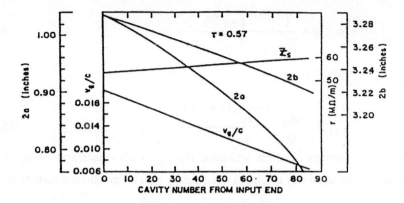

Figure 8. Parameters of a SLAC constant gradient structure [2]

4. Standing Wave Accelerating Structure

4.1 Standing wave for acceleration

A direct and a reflected sinusoidally varying wave, traveling with the same velocity, but in opposite directions, combine to create a standing wave (SW) pattern. If the amplitudes of the direct and reflected wave are A and B, respectively, the SW pattern has maximum A+B and minimum A-B, distant from each other by $d = \pi / 2k_0$, with $k_0 = \omega / v_p$. The average amplitude of the SW pattern is A, hence the same as the direct traveling wave (TW). Such a SW pattern is not useful since the reflected wave only dissipates power traveling backwards and does not contribute to the acceleration of particles.

However, SW accelerators use both the direct and reflected waves to accelerate particles. It can be understood from Figure 4 that at the points where the direct and reflected space harmonics join, they have the same phase velocity, and if this velocity is synchronous with the particle, both harmonics contribute to the acceleration.

From Figure 4 again, one finds that SW accelerators operate either at the lowest or at the highest frequency of the pass band, where $k_n L = N\pi$, and $N = 0, \pm 1$. That means the operation modes in SW accelerators is either 0 or π.

4.2 Stabilized SW accelerating structure

In addition to the high accelerating efficiency, the structure should have a high stability as well in the operation.

4.2.1 Properties of a structure with single-periodic chain

A linac structure usually consists of a series of periods to provide a space harmonic wave that has a phase velocity equal to the particle velocity for accelerating particles resonantly. In this case the structure has only one kind of period, or the so-called single-periodic chain. It is also known that in a cavity many "separated modes" can be excited, such as TM_{010}, TM_{011},... TM_{01n}, and so on. They have different frequencies and amplitudes. On the other hand, in a single-periodic chain, many "operating modes" can be excited, that are defined by their phase shift between adjacent periods, such as 0, $\pi/2$, π-modes. These two kinds of modes are related to and different from each other. To simply study the properties of a single-periodic structure, an equivalent circuit is usually used. In this analytical way, each period of the structure is described by an equivalent circuit with lumped parameters (L, C, R), and all periods in the structure are coupled with each other along with those lumped parameters.

Assuming that a structure consists of N periods (cells) in the chain, and is terminated by a half cell at both ends. Each mode's frequency, amplitude, and phase can be obtained by solving N+1 coupled equations. With the periodic property of the chain, the nth solution of these equations can be

$$X_n = A_n \cos n\varphi_n , \qquad n = 0, 1, 2,...N \qquad (27)$$

where A_n is a constant and φ_n is the phase shift between adjacent cells. For a SW structure, we have $X_0 = X_N$ at the two ends of the chain so that

$$N\varphi = q\pi, \qquad\qquad q = 0, 1, 2,....N. \qquad\qquad (28)$$

Then, the coupled equations have their solutions of Eigen function (field amplitude)

$$X_n{}^q = A_0 \cos(\frac{qn\pi}{N}) \qquad\qquad (29)$$

and Eigen value (frequency)

$$\omega_q = \frac{\omega_a}{\sqrt{1 + k_c \cos(q\pi / N)}} \qquad\qquad (30)$$

where ω_a is the resonant frequency in each cell. These equations describe the dispersion relation, thus we see

♦ If a chain consists of N+1 cells, then only N+1 modes can be excited in the chain.

♦ Each ω_q corresponds to a mode which has a phase shift of $\varphi = q\pi/N$ between adjacent cells.

♦ The number q for the modes of 0, $\pi/2$ and π-modes are 0, N/2 and N, respectively, and only when N is even, there is a $\pi/2$-mode in the chain.

♦ The band width of the chain is $\omega_b = \omega_\pi - \omega_0$, as shown in Figure 9. Usually the coupling constant $k_c << 1$ (e.g 5%), then $\omega_b \approx k_c\omega_a$, is proportional to k_c.

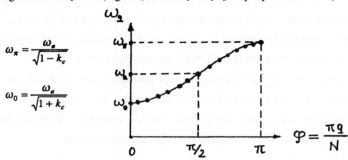

Figure 9. Band width and modes

♦ The mode separation between two adjacent modes is varied with mode's number. From (30), it is easy to have the following expressions of the mode separations at 0-, $\pi/2$- and π - modes:

$$(\frac{\Delta\omega}{\omega_a})_{0,\pi} = \frac{k_c\pi^2}{4N^2}, \quad \text{and} \quad (\frac{\Delta\omega}{\omega_a})_{\pi/2} = \frac{k_c\pi}{2N} \tag{31}$$

Thus the biggest mode separation is at $\pi/2$-mode and it is proportional to k_c/N^2, while the smallest one is at 0- and π-modes.

♦ The group velocity of each mode can be easily obtained as well, for instance $v_g = \frac{d\omega}{dk} \approx 0$ for 0 and π-modes, and $v_g = \frac{d\omega}{dk} = \omega_a k_c L_c / 2$ for $\pi/2$-mode. Thus, the $\pi/2$ mode has the biggest group velocity and hence the fastest energy propagation in the structure. This is very helpful to overcome the beam loading effect, particularly for high current beam acceleration.

♦ π-mode is located at the edge of the pass band. Both of the forward wave of n = 0 and its backward wave's harmonic wave of n = 1 make the contributions to accelerate particles, see Figure 4. Hence, it has the highest shunt impedance compared with all other modes. On the contrary, $\pi/2$-mode has the lowest shunt impedance, since all of its backward waves have the phase velocities in the opposite direction to particle's motion, so they do not make any contribution to the acceleration and are just lost on the cavity wall. A brief summary of mode's comparison is listed in Table 1.

Table 1. A brief summary of mode's comparison

	π-modes	$\pi/2$-mode
Effective shunt impedance	maximum	minimum
Mode's separation	minimum	maximum
Group velocity	minimum	maximum
Field distortion by perturbation	maximum	minimum

From Table 1, one can see that, for a SW structure, 0-mode or π-mode has the highest shunt-impedance, but the lowest group velocity. Thus, it has a high accelerating efficiency, but may not be stable in operation. On the other hand, the π /2-mode has the lowest shunt-impedance, but the largest group velocity, and thus a lower accelerating efficiency, but high operation stability. To have a structure with both, high efficiency and high stability, the solution is to use a so-called biperiodic structure which combines the advantages of π-mode and π/2-mode.

4.2.2 Properties of a biperiodic chain

The purpose of introducing a biperiodic chain is to have an operation mode which can combine all advantages of π/2 and π-modes, and hence to form a very effective and very stable accelerating structure. Let us introduce a coupling periodic chain to the accelerating periodic chain and form a biperiodic chain. By adjusting the coupling chain to make its pass band to be coupled resonantly with the accelerating pass band, there is also a faster group velocity and a bigger mode separation for the 0-mode or π-mode, as shown in Figure 10. An example of this kind of structure of an electron linac is the so-called coupled-cavity linac (CCL).

4.3 Coupled-Cavity Linac (CCL)

The CCL is operated at the π/2-mode of the biperiodic chain. In this structure, the coupling element is also of cavity type. There are two types of CCL for the electron linac. One is the on-axis-coupled structure, where each coupling cavity is located on axis and between two accelerating cavities, as indicated in Figure 11 (b). Another one is the so-called side-coupled structure, as shown in Figure 11 (c).

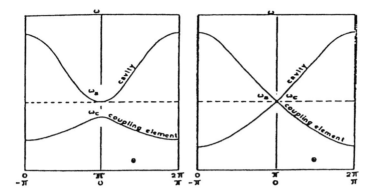

Figure 10. Resonant coupling of accelerating passband with coupling passband

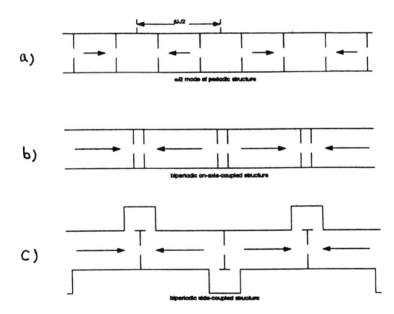

Figure 11 $\pi / 2$ -mode operation of a cavity resonator chain [3]

and in Figure 12, where each coupling cavity is alternately located at the side of the accelerating cavity. The coupling factors for both on-axis-coupled and side-coupled structures are adjusted by modifying the slot size between accelerating cavity and coupling cavity. In the side-coupled structure, the traveling ways for beam and for the RF field are separated, so that both the accelerating cavity and the coupling cavity can be optimized independently from each other.

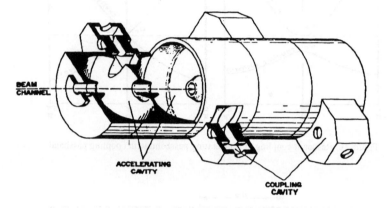

Figure 12   Side-Coupled Cavity structure

5.   Electron Pre-injector Linac

As we have mentioned at the beginning of this lecture, electron linacs are widely used as the injectors of synchrotrons, SR light sources and ring-type electron-positron colliders, linac based FEL, radiotherapy machines, and electron-positron linear colliders. All of these electron linacs must have a pre-injector at the beginning. Although these pre-injectors may be different from each other for the various uses they consist of the most basic and common components of the electron linac as explained in the following.

5.1 Electron Gun and Beam Bunching System

Figure 13 shows a schematic layout of an injector linac. Two types of electron pre-injector are commonly used: a dc high voltage gun with a bunching (velocity modulation) system (as shown in Figure 13), and an RF gun followed by a short accelerating structure.

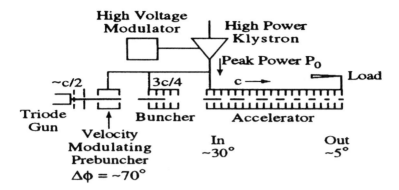

Figure. 13   Schematic layout of an electron injector linac [5]

The dc electron gun has a cathode (thermionic or photo-cathode) and an anode. It produces electrons with pulse lengths of 1 μs to several μs and a beam energy of 50 keV to 200 keV. If the gun uses a thermionic cathode, then a wire-mesh control grid is needed to form a beam pulse, which normally works at a voltage of about minus 50V with respect to the cathode. The cathode consists of some oxide and is heated to reduce the work function of electron emission. Beam dynamics in the gun is usually simulated with the EGUN code.

Since the electrons from the gun have the velocities of $v < c$ (e.g. $\sim$ 0.5 c), the electron bunch can be shortened by using a bunching system that modulates the electron velocity with an RF field in the cavity or in the waveguide bunchers, followed by a drift space. The RF is phased with respect to the electron beam such that the front electrons experience a decrease in energy and the back electrons experience an increase in energy. Beams can be bunched to about 10° of the fundamental frequency. The bunched beam is accelerated to about 20~50 MeV before the space charge effects can be neglected in these system.

The first stage of bunching the beam from the dc high voltage gun is commonly accomplished by using SW single-cavity bunchers, followed by some TW bunchers for further bunching and acceleration. Usually the first few cells of the TW buncher have $v_p < c$ in order to synchronize with the beam.

If the gun uses a photo-cathode, then the electrons are produced by the photo-electric effect, using a laser pulse incident on the cathode, and no grid in the gun.

The RF gun is followed by an accelerating structure, since the electrons from the cathode are soon bunched by the RF field. The RF gun consists of one or more SW cavities with the cathode installed in the upstream wall of the first cavity. Compared with the dc gun, the RF gun has the advantage of quickly accelerating electrons to relativistic velocity (about 5 to 10 MeV), which avoids collective effects such as the space-charge effect and provides a shorter bunch length and lower beam emittance at the cathode. However, the RF gun has some time-dependent effects due to its time-dependent RF field, which may dilute the performance of electron bunches.

5.2 RF Power Source

Another essential component of the electron linac is the high power RF source which in most machines are klystrons with their associated high voltage modulators. Magnetrons are used in low energy electron linacs with single sections. Because of the high peak power required (say 10 - 80 MW), these klystrons work at low duty cycles, for example, with a pulse repetition rate of about 100 Hz and pulse lengths of a few μs.

To increase the peak power and hence to increase the accelerating gradient, some linacs use an RF pulse compression system, such as the so-called SLED (SLAC Energy Development). The RF pulse compressor is used to compress a longer pulse at a lower power into a shorter pulse with higher power.

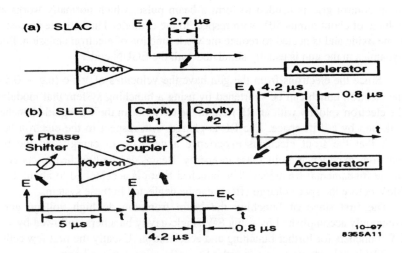

Figure 14. SLAC RF system (a) without SLED, (b) with SLED [5]

The principle of SLED can be seen in Figure 14, where part (a) shows non-SLED operation, with the RF power from the klystron directly transmitted to the linac. Part (b) shows the SLED system, which has two major components: a π-phase shifter on the drive side of the klystron, and two high-Q (Q_0 = 100,000) cavities on the output side of the klystron, operated at TE_{015}-mode and connected to a 3-db coupler. During the first part of the pulse, the phase of the RF drive signal is reversed and the RF cavities fill up with energy at that phase. The fields emitted by the cavities (E_e) add to the field reflected by the cavity coupling irises (RE_{in})

$$E_{out} = E_e + RE_{in} \tag{32}$$

and the power flows toward the accelerator sections.

5.3 Other Subsystems

Similar to most other accelerators, the RF electron linac has many other subsystems, such as beam transverse focusing system (usually use of quadrupole magnets), RF drive system, RF phase control system, water cooling system, alignment system, beam monitor and instrumentation system, vacuum system and machine operation control system.

6. Introduction to the Microtron

The microtron is a cyclic electron accelerator for the energy range from some tens of MeV (circular / classical microtron) to a few hundred MeV (racetrack microtron). The electrons are accelerated by a high frequency (S-band or L-band) electric field in a resonant cavity or by a linac, and are recirculated by a homogeneous and constant magnetic field.

The microtrons are only used to accelerate the electron or positrons, since this type of accelerator is restricted to the particles whose kinetic energy can be increased by an amount comparable to their rest energy during one single pass through the accelerating cavity. The microtrons have the advantages of delivering a high beam current with small beam emittance and small energy spread. These basic features of the microtron will be described below.

The circular microtron was first proposed by V. I. Veksler in 1944 [6] and was first constructed in Ottawa, Canada, in 1948 [7]. Due to inefficient injection methods resulting in very low beam currents, the microtron disappeared for about a decade. After the 2nd world war, in the USSR, however, the microtron was studied theoretically, and by the 1960's it was highly developed and extensively used. An improved injection method was developed by O.Wernholm in 1964 [8], leading to the beam current increase by an order of magnitude. During this time period research and development programs were also carried out in Sweden, Italy, UK, Canada and USA.

The first practical machine which was sector-focused and single-cavity, called the racetrack microtron (RTM) was built at the University of Western Ontario in 1961 [9]. The first multi-cavity standard RTM was built at the University of Illinois around 1975 [10]. Then the Scanditronix commercially available RTM 50 and RTM 100 were developed from the accelerator concepts in Lund and Stockholm. The largest racetrack microtron now in the world is the MAMI

278

project at the University of Mainz [12,13]. It consists of three cascaded microtrons, with the final energy of 855 MeV.

7. Circular Microtrons [11]

7.1 Operation Principle

The layout of the circular microtron is as shown in Figure 15. In the circular microtron (some time called classical microtron or electron cyclotron) the condition for synchronous acceleration is obviously:

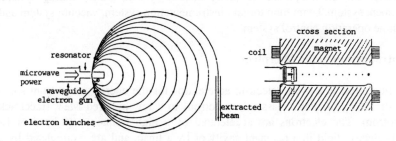

Figure 15 The layout of the circular microtron

$$T_n = k_n T_{rf} \tag{33}$$

where T_n is electron's n^{th} revolution time period, T_{rf} is the RF period of the accelerating cavity and k_n is an integer. This condition is actually based on a relation among the magnetic field (B), the frequency (f_{rf}) and the amplitude (V_{rf}) of the accelerating voltage. For the first revolution (it also depends on the injection energy), the revolution time in the first orbit is

$$T_1 = \frac{2\pi}{ec^2 B} E_1 = \mu T_{rf} \tag{34}$$

where μ is an integer, and E_1 is the total energy in the first orbit

$$E_1 = E_0 + E_i + \Delta E \tag{35}$$

Here, E_0 is the electron's rest energy (0.511 MeV), E_i is the injection energy, and ΔE is the energy gain in the cavity. For the other orbits, the revolution time must increase with an integer number of RF periods relative to the previous orbit

$$\Delta T = \frac{2\pi}{ec^2 B} \Delta E = \nu T_{rf} \qquad (36)$$

where Δ T is the difference in the revolution time between the adjacent two orbits, and ν is an integer. Combining (34), (35), and (36), we find the energy gain for the resonant acceleration

$$\Delta E = (E_0 + E_i)\frac{\nu}{\mu - \nu} = \frac{ec^2 B}{2\pi f}\nu \qquad (37)$$

where obviously $\mu = 2,3,4.....$; $\nu = 1,2,3,.....$, and $\mu > \nu$. The total energy in the nth orbit is then

$$E_n = \Delta E(\frac{\mu}{\nu} + n - 1) \qquad (38)$$

and magnetic field

$$B = \frac{2\pi f}{ec^2}\frac{\Delta E}{\nu} = \frac{2\pi f}{ec^2}\frac{E_0 + E_i}{\mu - \nu} \qquad (39)$$

For the fundamental operation mode $\nu = 1$ and $\mu = 2$, $\Delta E = E_0 + E_i$ (kinetic energy gain per revolution $\approx$ rest energy), and B = B_{max}. So for a given final energy, this leads to the most compact microtron. However the B field is still rather small. For f = 3 GHz, B = 0.107 T and E_f = 20 MeV, the diameter of the last orbit is about 1.3 m.

In a resonant circular accelerator, the particle's bunch number in the orbit could be a harmonic number of the fundamental frequency of the accelerating cavity. If $\nu = 1$ and $\mu = 2$, then in the first orbit there are two bunches, and in each of the other orbits the bunch number will be increased by one relative to the previous orbit, also as shown in Figure 15.

7.2 Beam Injection

In the first microtron, the injection system (as shown in Figure 16, OTTAWA 1948) was based on field emission from the lips of the accelerating cavity, resulting in a very low current (average currents of the order of 10 nA). O.Wernholm developed an injector consisting of a compact coaxial electron gun,

280

located outside the accelerating cavity. With this system an average current of 50 µA at 5 MeV was reached.

More efficient injection methods were developed by P. Kapitza and V. N. Melekhin in 1959 [21], as shown in Figure 16 (Moscow 1959). These methods make it possible to vary the electron energy by modifying the first orbit energy

$$E_1 = E_0 + m\Delta E \tag{40}$$

where 1< m <2.

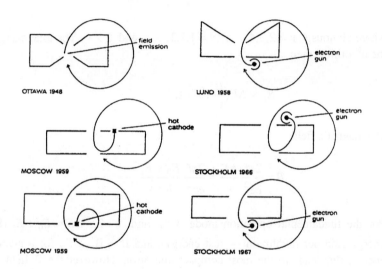

Figure 16 Injection systems

Introducing

$$B_0 = \frac{2\pi f E_0}{ec^2} \tag{41}$$

we obtain the cyclotron field for the electron at RF frequency f (from (39))

$$\frac{B}{B_0} = \frac{\Delta E}{E_0}\frac{1}{v} \tag{42}$$

Using (34) and (40), we find

$$\frac{\Delta E}{E_0} = \frac{v}{\mu - mv} \quad \text{and} \quad \frac{B}{B_0} = \frac{1}{\mu - mv} \tag{43}$$

S.P. Kapitza and V. N. Melekhin have demonstrated that the electrons can be accelerated over a fairly wide range of m, thus giving a variable final energy. In the fundamental mode (v =1 and μ =2), m has to be in between 1 and 2. For example, if m = 3/2, then $\dfrac{\Delta E}{E_0} = \dfrac{B}{B_0} = 2$, reducing the orbit radius by a factor of 2 for a given energy.

7.3 Beam Extraction

The electron beam extraction from a microtron is rather simple, since the orbit separation is large (about 3.2 cm) in the fundamental mode. The usual extraction method is by insertion of a field-free channel (usually a steel tube) tangential to the orbit to be extracted, as shown in Figure 17.

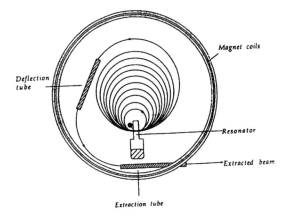

Figure 17 Beam extraction from a circular microtron

7.4 Phase Stability

As the revolution period for the electrons increases with energy, in the microtron, the phase stable region is located on the negative slope of the accelerating field. Since the phase shift per orbit is large (2 π in the fundamental mode), the phase oscillations can be well described by difference equations. The relations between the phase shift φ_{n+1} at the cavity from orbit n to n+1 and the energy error $\delta E_{n+1} / \Delta E$ for a particle are

$$\varphi_{n+1} = \varphi_n + 2\pi v \frac{\delta E_n}{\Delta E} \tag{44}$$

$$\frac{\delta E_{n+1}}{\Delta E} = \frac{\delta E_n}{\Delta E} + \frac{\cos \varphi_{n+1} - \cos \varphi_s}{\cos \varphi_s} \tag{45}$$

where the index s denotes the synchronous particle. These relations can be used to find the size and shape of the phase stable region by inserting different initial conditions φ_1 and δE_1 into numerical calculations. The stable phase can be found by studying small deviations from the synchronous phase, i.e., with linearized equations (44) and (45). The result will be a matrix equation

$$\begin{bmatrix} \delta \varphi_{n+1} \\ \delta E_{n+1} \end{bmatrix} = \begin{bmatrix} 1 & 2\pi v / \Delta E \\ -\Delta E \tan \varphi_s & 1 - 2\pi v \tan \varphi_s \end{bmatrix} \begin{bmatrix} \delta \varphi_n \\ \delta E_n \end{bmatrix} \tag{46}$$

for which the stable condition can be found with $[traceM] < 2$. So for the stable phases we require:

$$0 < \pi v \tan \varphi_s < 2 \tag{47}$$

which leads to the following stable phases:

$$\begin{aligned}
v = 1: & \quad 0 < \varphi_s < 32.5^0 \\
v = 2: & \quad 0 < \varphi_s < 17.7^0 \\
v = 3: & \quad 0 < \varphi_s < 12.0^0
\end{aligned} \tag{48}$$

With φ_s at the center of the phase-stable region, we get the largest phase-stable area for $v = 1$ and $\varphi_s = 16.25^0$, within which stable phase oscillations occur. In general these oscillations maintain a constant phase and energy width of the electron bunches, making the energy spread inversely proportional to the number of orbits, as shown in Figure 18. For example, if a microtron works in the fundamental mode ($v = 1$) with $\varphi_{ma} = 32.5^0$ and its injection energy is 50 keV, then one can estimate the energy spread

$$\delta E = \frac{32.5^0}{360^0}(0.511MeV + 0.05MeV) = 51keV ,$$

resulting in a relative energy spread of 0.25% for a final energy of 20 MeV, which is a typical figure for a circular microtron.

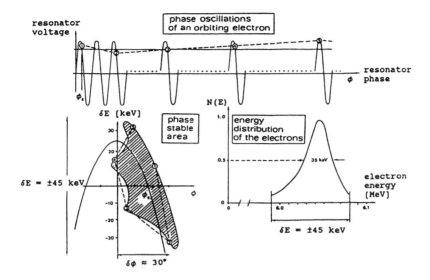

Figure 18   Phase motion in the circular microtron

7.5 Transverse Stability

With a homogeneous magnetic field there are no problems with the horizontal stability. To make a net focusing in the vertical direction, one can have the electric quadrupoles formed by the electric field gradient in the apertures of the accelerating cavity. This gradient is such that it is focusing at the entrance, but defocusing at the exit from the cavity. Since the focusing strength of an electric quadrupole is inversely proportional to electron's momentum, and also because the synchronous phase is on the falling part of the RF-period, hence these lead to a net focusing action on electrons, which is sufficient for vertical stability.

7.6 The RF System

The usable range of frequencies for the microtron RF system is limited by the requirement of a large magnetic field for compactness, leading to a high RF frequency (see Equations. (41)), and the mechanical difficulties of building small accelerating cavities. Most of the microtrons' RF systems operate in S-band, with a wavelength of about 10 cm. Due to the limitation of the cavity cooling, microtrons are usually operated in a pulsed regime. The most common power source is the pulsed magnetron or klystron. The typical parameters are power levels at several MW, with pulse lengths up to 5 μs and duty factors of the order of 0.1%.

284

8. Racetrack Microtron

There are some limitations in the circular microtron:
1) For $E_f \geqslant 20 \sim 40$ MeV, it becomes heavy and large due to the low cyclotron field (for $f = 3$ GHz, $B_0 = 0.11$ T);
2) The many and long orbits impose severe restrictions on the homogeneity of the magnetic field of a large single magnet. An efficient way to overcome these limitations is offered by the racetrack microtron (RTM) [11].

In the RTMs, the single magnet is replaced by two "end magnets", and the cavity is replaced by a linac, as shown in Figure 19.

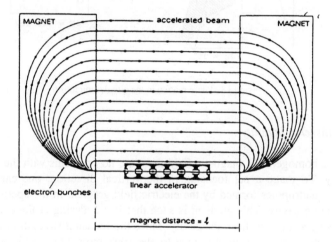

Figure 19 The racetrack microtron

8.1 Operation Principle

In the RTMs, the revolution time in the first orbit is

$$T_1 = \frac{2\pi}{ec^2 B}(E_0 + E_i + \Delta E) + \frac{2l}{c} = \mu\tau = \frac{\mu}{f} \qquad (49)$$

where l is the length of straight section between the magnets. Compared to Equation (36), the increase in revolution time between orbits n and n+1 must satisfy

$$\Delta T = \frac{2\pi}{ec^2 B} \Delta E = v\tau \tag{50}$$

Since the wavelength $\lambda = c\tau = c/f$, we have the resonant energy gain

$$\Delta E = \frac{(E_0 + E_i)v}{\mu - v - 2l/\lambda} \tag{51}$$

with $\mu \gg 2$ and $v = 1,2,3\ldots$ The magnetic field is given by Equation (39), or in terms of the cyclotron field Equation (41) as

$$\frac{B_0}{B} \frac{\Delta E}{E_0} = v \tag{52}$$

The orbit radius ρ_n in the nth orbit is

$$2\pi \frac{\rho_n}{\lambda} = \frac{B_0}{B} \frac{E_n}{E_0} = (n+i)v \tag{53}$$

where $i = (E_0 + E_i)/\Delta E$. From Equation (51), the energy gain in the RTM can be much larger than in the circular microtron, since the straight section l can be chosen "freely", except that l has to be such that μ becomes an integer. With this big energy gain, the magnetic field becomes several times the cyclotron field, making the accelerator compact for a high final energy.

As an example, if the designed parameters are an orbit number = 20, E_f =100 MeV, f = 3 GHz, λ = 10 cm, B_0=0.107 T, v =1 (for maximum phase stable area), ΔE = 5 MeV and E_I = 50 keV, then we have B = 1.05 T, ρ_1 = 1.8 cm , ρ_{20} = 32 cm. With l = 1 m, then μ = 21, which is much larger than for the circular microtron.

8.2 Beam Injection

The electrons are either injected directly from the gun (non-relativistic electrons), or from an injector linac (high energy, say a few MeV, relativistic electrons). The second approach becomes a necessity when the energy gain in the linac becomes high.

There are two problems related to the first orbit in an RTM: 1) The orbit radius may not be large enough for the first orbit to clear the linac structure. 2) The electrons in the first orbit may not be sufficiently relativistic, so that some correction of the first orbit path length is required. To cure these problems, R.

286

Alvinsson and M. Eriksson suggested to reflect the first orbit geometry [9], so that the electrons after the first pass through the linac are displaced by a magnet system and then directly reflected back into the linac by the main dipole, as shown in Figure 20. The acceleration phase during the first pass can be independently adjusted by sliding the linac along its axis, without affecting the phase for subsequent orbits.

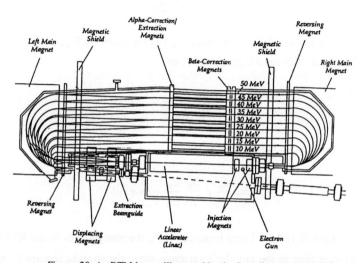

Figure 20 An RTM layout illustrated by the Scanditronix RTM50

8.3 Beam Extraction

The beam extraction from any orbit of RTM is uncomplicated, since the orbits are well separated in the RTM. The extraction is usually carried out with a small magnet deflecting the orbit into an extraction channel.

8.4 Phase Stability

The relations for the phase motion in the circular microtron, Equations (44) and (45), are valid also for the RTM, with the energy spread inversely proportional to the orbit number. However, Equations (44) and (45) are derived under the assumption of extreme relativistic particles and without focusing of the particles. Actually the focusing modifies the phase motion and the conditions for extreme relativistic particles may not be fulfilled for the first few orbits. Performing numerical calculations with these factors taken into account is therefore the best approach to study the phase stability in the RTMs.

8.5 Transverse Stability

For having the transverse stability, a solution of a separated-function magnet is usually employed due to the main dipole geometry in the RTM. There are two different approaches to arrange the beam optics in an RTM: 1) to place focusing elements in the drift section of each orbit, and in this way the betatron phase advance can be kept constant for all orbits; 2) to put the focusing elements on the linac axis. Since the focusing strength $K \propto \dfrac{1}{B\rho}\dfrac{\partial B}{\partial r}$, in order to avoid over-focusing in the first orbit, the focusing in the last orbit becomes weak. The beam optics therefore will be working close to integer resonances, with high sensitivity to dipole errors. However, these errors can easily be corrected by small dipoles in each orbit in the RTM. A problem in all RTMs is the vertical defocusing effect caused by fringing fields of the dipoles, with a focusing length of the same order of magnitude as the bending radius in the magnet. This can be solved by adding the reversing magnets suggested by H. Babic and M. Sedlacek in 1967 [15], and now used in most RTMs.

8.6 RF System

The typical parameters of the RF system for the RTMs are f = 1.3 - 3 GHz, RF power source with klystron, $\tau_p = 1 - 5\mu s$, duty factor = 0.1% - a few %, standing wave injector linac, dE/dz = 2 -15 MV/m, ΔE = 1-5 MeV, I_b = 1-100µA. RTMs can also be operated in cw mode.

9.   Applications for Microtrons

9.1   Microtrons for the studies on nuclear physics.

For example, the Mainz Microtrons (MAMI) are the cw electron accelerators for the studies on nuclear physics. It consists of three cascaded RTMs with a 3.5 MeV injector linac. The output energies from these three RTMs are 14 MeV, 180 MeV and 855 MeV, respectively. Table 2 shows the main parameters of these RTMs [13].

Table 2 The main parameters for MAMI accelerators

Stage No.	RTM - 1	RTM- 2	RTM - 3	Units
Input energy	3.455	14.35	179.5	MeV
Output energy	14.35	179.5	854.6	MeV
No. of orbits	18	51	90	
Magnet distance	1.668	5.597	12.86	m
Magnetic field	0.1026	0.555	1.2842	T
Max. orbit diameter	0.964	2.166	4.432	m
Weight per magnet	1.3	43	450	t
Gap width	6	7	10	cm
Linac length	0.80	3.55	8.87	m
No. of klystrons	1	2	5	
RF beam power	1.2	17	68	kW
Energy gain	0.599	3.24	7.50	MeV

The linacs for these three RTMs are on-axis-coupled SW structure with beam aperture 14 mm diameter. Beams can be extracted from each even numbered return path of RTM3, i.e. $E(out) = 180 + 2n \times 7.5 MeV$, with n=1, 2, 3,..... 45. The final beam performance: 850 MeV, 100 μA, energy spread 30 keV(FWHM), energy drifts 100 keV (FWHM), horizontal emittance 13 π-mm-mrad (1 σ), and vertical emittance 1.7 π-mm-mrad (1 σ).

9.2 Injectors to storage rings and synchrotrons.

The excellent energy spread and small emittance from microtrons simplify the multi-turn injection [14]. Some synchrotron radiation facilities have adopted microtrons as the injectors on account of its cost, better beam quality and smaller machine size compared with other conventional accelerators such as the linacs and the synchrotrons. Here are some examples:

- A 20 MeV, 20 mA (pulsed, 1-2 μs) circular microtron (22 orbits) as the injector of 450 MeV/700 MeV synchrotrons at INDUS-1, CAT, Indore[16].
- A 100 MeV racetrack microtron as the injector of Helios 1(operated at IBM East Fishkill in 1989-98 and recently has been moved into Jefferson Lab.) and as the injector of Helios 2 and of the Microtron Undulator Radiation Facility (MURF) at the SSLS[17].
- A 150 MeV racetrack microtron as the injector of SR facility HiSOR at Hiroshima University[18].

9.3 Industrial & Medical uses and Free Electron Laser Applications

Industrial and medical applications include radiography and electron radiation treatment, and radiation therapy [19], where the high energy and good beam quality is employed for scanned beam techniques [20]. Also, due to its fine beam qualities, the RTM microtron is one of the candidates to produce the free electron laser.

References

1. Lapostolle and A. Septier, Eds., Linear Accelerators, *North Holland and Wiley*, 1970.

2. G. A. Loew and R. Talman, Elementary principles of linear accelerators, *AIP Conf. Proc.* 105, 1983.

3. T. P. Wangler, Principles of RF Linear Accelerators, *John Wiley & Sons, Inc.* 1998.

4. D.H. Whittum, Introduction to Electrodynamics for Microwave Linear Accelerators, *SLAC-PBU-7802*, April 1998.

5. A. W. Chao and M. Tigner, Eds., Handbook of Accelerator Physics and Engineering, *World Scientific Publishing Co., Inc.*, 1998.
   The relevant Sections in this Handbook are as follows:

   1.6.10   G.A. Loew, Linear accelerator for electrons , p. 26.

   2.4   H.G. Kirk, R. Miller and D. Yeremian, Electron gun and pre-injector, p. 99.

   7.3.5   G.A. Loew, Normal Conducting υ_p = c Linac Structures , p. 516.

   6.7   Z.D. Farkas, RF Pulse Compression, p. 374.

   7.1.1   A.D. Yeremian , R.H. Miller, Electron gun and preinjector, p. 419.

   2.5.1   K. Thompson , K. Yokoya, Collective effect in high energy electron linacs, p. 103.

6. V. I.Veksler, *Proc.USSR Acad. Sci.* **43** (1944) 346 and *J. Phys. USSR* **9**(1945)153.

7. W.J.Henderson, et al, *Nature 162* (1948) 699.

8. O. Wernholm, *Arkiv. Fys. 26* (1964) 527.

9. E. Brannen,et al, *J. Appl. Phys. 32* (1961) 1179.

10. A. O. Hanson,et al, *SLAC Conf.* (1974) 151.

11. Per Lidbjoerk, CERN Accelerator School, *CERN 94-01*, 971.

12. H. Herminghaus, et al, *IEEE Trans. NS-30,* **4** (1983) 3274.

13. http://www.kph.uni-mainz.de/B1/parameters.html.

14. R. Alvinsson and M. Eriksson, Royal Inst. of Tech., Stockholm, Report *TRITA-EPP*-76-07 (1976).

15. H.Babic and Sedlacek, *Nucl. Instrum. Methods* **56** (1967) 170.
16. G.K. Sahoo, et al, *Proc. APAC'98* (1998), 274.
17. H.O.Moser, et al, *Proc. APAC'01* (2001), 31.
18. K. Yoshida, et al, *Proc. APAC'98* (1998), 653.
19. A.Brahme and H. Svensson, *Acta Rad. Onc.* **18** (1979) 244.
20. M. Karlsson, et al, *Med. Phys.* **19** (1992) 307.
21. P. Kapitza and V. N. Melekhin, The Microtron. *Harwood Academic Publishers*, London, (1978).
22. R.E.Rand, Recirculating electron accelerators. *Harwood Academic Publishers*, (1984).

COLLECTIVE BEAM EFFECTS IN STORAGE RINGS [*]

GUO ZHIYUAN

Institute of High Energy Physics
Chinese Academy of Sciences, Beijing, 100039, China
E-mail: guozy@mail.ihep.ac.cn

Collective beam effects are very important phenomena which may cause beam instabilities and limit the beam performance in storage rings at higher beam currents. These effects have been studied analytically, experimentally, and by computer simulation. In this lecture, the wake field and impedance, the transverse and longitudinal beam instabilities, the single bunch and multi bunch effects, as well as the tune shift and bunch lengthening are treated. Moreover, the macro particle model, the perturbation formalism, and Landau damping are introduced. The discussion is focused on storage rings.

1. Introduction

Charged particle accelerators are devices which can offer beams for high energy physics or other research fields. The beam in an accelerator should be stable for these studies. The collective beam instability is a very important topic for the operation of accelerators.

Particle motion in an accelerator is affected by the principal electro-magnetic guide fields, synchrotron radiation, and electro-magnetic fields of the particle beam. The principal electro-magnetic guide fields include the magnetic fields of dipoles, quadrupoles, sextupoles, octupoles, and solenoids as well as the radio frequency (RF) electric field. The electro-magnetic fields of the particle beam include space charge effects and wake fields.

Particle losses in an accelerator during operation may be caused by single particle effects as well as collective effects of the beam. Single particle effects include initial conditions of injection, stability of linear lattice, tolerance effects of the electro-magnetic guide fields, nonlinear effects and time-dependant parameters, etc.

The single beam collective effects can be divided into the coherent and incoherent collective effects. The coherent collective effects are wake fields and impedance, tune shift (transverse and longitudinal), beam instabilities (transverse and longitudinal) and two stream beam interactions, etc. The incoherent

---
[*] This work is supported by National Natural Science Foundation of China (19975056)

collective effects are random interactions between charged particles, beam-gas scattering (elastic and inelastic interaction), particles trapping effect, etc.

Collective effects can also be produced in the beam-beam (colliding beams) interaction. Two beams cross each other during colliding. The main effect of this interaction is the tune shift which arises from the disturbed particle motion caused by the space charge forces. This effect probably leads to the instability of particle motion, therefore, we see the beam-beam limit on the colliders.

The particle losses finally result into a finite beam lifetime. The beam lifetime is usually defined as the time during which the beam intensity decays to a certain fraction (say 1/e) of its initial value

$$I = I_0 e^{-t/\tau} \qquad \text{or} \qquad \tau = -\frac{I}{dI/dt}. \qquad (1)$$

In modern electron storage rings, the total beam lifetime can be longer than hundreds hours, corresponding to about 10^{11} turns of the particles around the ring or even more.

2. Wake field and Impedance

A moving charged particle will generate electro-magnetic fields by the interaction of the particle with the structural surroundings of the beam tube which are not perfectly conducting smooth pipes or other cavity-like vacuum components. In the relativistic limit, there is electro-magnetic field which decays with time only behind the particle. We call this field the *wake field*. Stronger beam intensity can induce stronger wake fields which may lead to beam instability.

We consider the case in which a beam with charge q moves along the axis of an axially symmetric vacuum chamber, and a test charge e follows the beam at a fixed relative distance. The beam and the test charge are assumed to move with the speed of light $v \cong c$. The Lorentz force experienced by the test charge is

$$F = e[E + \frac{v}{c} \times B]$$

or written in components with $|v/c \approx 1|$

$$F_s = eE_s$$
$$F_\theta = e(E_\theta + B_r) \qquad (2)$$
$$F_r = e(E_r - B_\theta)$$

where E and B denote the electric and the magnetic field component, respectively, affecting the following charge e. An example of a wake field sketch is shown in Figure 1.

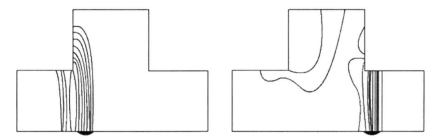

Figure 1. The sketch of the wake field.

2.1. Free Space

In the case of a charge q moving with a speed close to the speed of light, the field will be perpendicular to the direction of the particle motion. From Gauss' and Ampere's law we have

$$B_\theta = \frac{2q}{r}\delta(s - ct)$$

$$E_r = \frac{2q}{r}\delta(s - ct)$$

(3)

Electric force and magnetic force cancel each other exactly, in fact, the charge experiences no Lorentz force. In high energy electron storage rings, v≅c, the space charge force can be omitted in the single beam case.

2.2. Perfect conducting pipe

In the case of a charge q surrounded by a perfect conducting pipe, moving along the axis of the round pipe with the speed of light, the force experienced by the charge is $F=0$. For instance, if a charge moves along the axis of the pipe with an offset a in the $\theta=0$ direction, the charge density and current density can be decomposed as

$$\rho = \sum_{m=0}^{\infty} \frac{I_m}{\pi a^{m=1}(1 + \delta_{m0})}\delta(s - ct)\delta(r - a)\cos m\theta \qquad (4)$$

where $\delta_{m0}=1$ and $I_m=q$,   if $m=0$,

         $\delta_{m0}=0$ and $I_m=a^m q$,   if $m\neq 0$.

The above formulas show that the point charge is distributed as a thin ring with radius a and with a cos $m\theta$ angular distribution, as shown in the Figure 2[1].

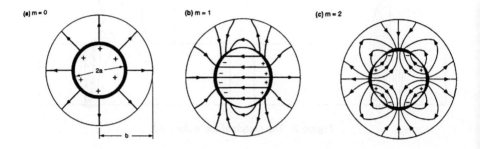

Figure 2. The distribution of the charge and current density.

By solving Maxwell equations with proper boundary conditions, the electro-magnetic fields can be obtained

$$E_r = \frac{2I_m}{1+\delta_{m0}}\delta(s-ct)\cos m\theta \begin{cases} (\dfrac{1}{b^{2m}}-\dfrac{1}{a^{2m}})r^{m-1}, & r<a, \\[2mm] \dfrac{1}{r^{m+1}}+\dfrac{r^{m-1}}{b^{2m}}, & a<r<b, \end{cases}$$

$$E_\theta = \frac{2I_m}{1+\delta_{m0}}\delta(s-ct)\sin m\theta \begin{cases} -(\dfrac{1}{b^{2m}}-\dfrac{1}{a^{2m}})r^{m-1}, & r<a, \\[2mm] \dfrac{1}{r^{m+1}}-\dfrac{r^{m-1}}{b^{2m}}, & a<r<b, \end{cases} \qquad (5)$$

$$B_r = -E_\theta,$$

$$B_\theta = E_r.$$

We can see from these equations, the fields have an angular distribution of cos$m\theta$ and sin$m\theta$, and $\delta(s-ct)$ in their longitudinal distribution. The electric force and the magnetic force eliminate each other exactly, so the following particle experiences no electro-magnetic force.

In analyzing the cases of a charge in free space, and in a perfectly conducting pipe, we have an important conclusion that the collective beam instability can only occur under two conditions, namely, the beam must not be ultra-relativistic, or its surrounding must not be a perfect conducting smooth pipe.

2.3. *Resistive wall*

For simplicity, we assume that the pipe wall has infinite thickness, the beam is moving with the speed of the light and the charges are distributed as given above. In cylindrical coordinates, we can write down Maxwell's equations as

$$\frac{1}{r}\frac{\partial(rE_r)}{\partial r} + \frac{1}{r}\frac{\partial E_\theta}{\partial \theta} + \frac{\partial E_s}{\partial s} = 4\pi\rho,$$

$$\frac{1}{r}\frac{\partial B_s}{\partial \theta} - \frac{\partial B_\theta}{\partial s} - \frac{1}{c}\frac{\partial E_r}{\partial t} = \frac{4\pi}{c}j_r,$$

$$\frac{\partial B_r}{\partial s} - \frac{\partial B_\theta}{\partial r} - \frac{1}{c}\frac{\partial E_\theta}{\partial t} = \frac{4\pi}{c}j_\theta,$$

$$\frac{1}{r}\frac{\partial(rB_\theta)}{\partial r} - \frac{1}{r}\frac{\partial B_r}{\partial \theta} - \frac{1}{c}\frac{\partial E_s}{\partial t} = \frac{4\pi}{c}j_s,$$

$$\frac{1}{r}\frac{\partial(rB_r)}{\partial r} - \frac{1}{r}\frac{\partial B_\theta}{\partial \theta} + \frac{\partial B_s}{\partial s} = 0,$$

$$\frac{1}{r}\frac{\partial E_s}{\partial \theta} - \frac{\partial E_\theta}{\partial s} + \frac{1}{c}\frac{\partial B_r}{\partial t} = 0,$$

$$\frac{\partial E_r}{\partial s} - \frac{\partial E_s}{\partial r} + \frac{1}{c}\frac{\partial B_\theta}{\partial t} = 0,$$

$$\frac{1}{r}\frac{\partial(rE_\theta)}{\partial r} - \frac{1}{r}\frac{\partial E_r}{\partial \theta} + \frac{1}{c}\frac{\partial B_s}{\partial t} = 0.$$

(6)

We assigned that the combined variable $z = s\text{-}ct$ is the relative longitudinal displacement from the beam, $z > 0$ is ahead of the beam, $z < 0$ is behind the beam. Due to the causality, no wake field is produced ahead of the beam, i.e., in the region $z > 0$. Using the boundary conditions and the charge distribution as before, we make inverse Fourier transforms to obtain the fields as

$$E_s = \frac{I_m}{\pi b^{2m+1}} \sqrt{\frac{c}{\sigma}} r^m \cos m\theta \frac{1}{|z|^{3/2}},$$

$$E_r = -\frac{3I_m}{4\pi b^{2m+1}} \sqrt{\frac{c}{\sigma}} \frac{1}{m+1} r^{m-1} \cos m\theta (r^2 + b^2) \frac{1}{|z|^{5/2}},$$

$$E_\theta = -\frac{3I_m}{4\pi b^{2m+1}} \sqrt{\frac{c}{\sigma}} \frac{1}{m+1} r^{m-1} \sin m\theta (r^2 - b^2) \frac{1}{|z|^{5/2}}, \qquad (7)$$

$$B_s = -\frac{I_m}{\pi b^{2m+1}} \sqrt{\frac{c}{\sigma}} r^m \sin m\theta \frac{1}{|z|^{3/2}},$$

$$B_r = -E_\theta - \frac{2I_m}{\pi b^{2m+1}} \sqrt{\frac{c}{\sigma}} mr^{m-1} \sin m\theta \frac{1}{|z|^{1/2}},$$

$$B_\theta = E_r - \frac{2I_m}{\pi b^{2m+1}} \sqrt{\frac{c}{\sigma}} mr^{m-1} \cos m\theta \frac{1}{|z|^{1/2}}.$$

These expressions are valid for the region behind the beam. We can see here that the beam dimension a does not appear explicitly in the field expression, this indicates that the wake field is independent of the detailed shape of the beam distribution for a given mth moment of the beam.

The charged particle generates the electro-magnetic fields on the structural wall, the following charged particle experiences the Lorentz force

$$\mathbf{F} = e[\mathbf{E} + \frac{v}{c} \times \mathbf{B}]. \qquad (8)$$

In general, in the high energy storage ring, we are only interested in the average effect of the wake force, but not the details. For example, in Beijing Electron Positron Collider, BEPC, the time during which particles travel over a structure of the beam pipe is about 10^{-9} second (a few ns), the radiation damping time of electron is about 10^{-2} second (a few ten ms).

2.4. Wake Function

Considering the case of resistive wall, a test charge e moves, with a speed of light, in an axially symmetric vacuum pipe at a fixed relative distance z to the beam. The test charge experiences a Lorentz force due to the wake field behind the beam. At high energies, as the beam and the test charge travel a distance over a wall structure, the net effect on the test charge can be obtained by integrating the force through the distance along the wall structure, that is the average effect

$$\overline{F} \equiv \int_{-L/2}^{L/2} ds F .$$ (9)

For an axially symmetric environment, $F_\perp$ and F_r are proportional to $\cos m\theta$, B_s and F_θ are proportional to $\sin m\theta$, and the integral of the Lorentz force can be written as

$$\int_{-L/2}^{L/2} ds\vec{F}_\perp = -eI_m W_m(z) mr^{m-1}(\hat{r}\cos m\theta - \hat{\theta}\sin m\theta),$$

$$\int_{-L/2}^{L/2} ds F_{II} = -eI_m W'_m(z)r^m \cos m\theta,$$ (10)

$$\int_{-L/2}^{L/2} dse B_\theta = eI_m W'_m(z)r^m \sin m\theta,$$

where $W_m(z)$ is a function of z and $W'_m(z)$ is the derivative respect to z. The function $W_m(z)$ is called the **wake function**. It describes the shock response of the vacuum chamber environment to a δ function beam which carries an mth moment.

It may be more convenient to call $W_m(z)$ the transverse wake function, and $W_m'(z)$ the longitudinal wake function. The dimensionality of $W_m(z)$ is L^{-2m} in cgs units. Similar with the concept of the electric potential, the integrals on the left hand side of the equations are called the **wake potentials**.

From above expressions, we find that the transverse gradient of the longitudinal wake potential is equal to the longitudinal gradient of the transverse wake potential

$$\nabla_\perp \int_{-L/2}^{L/2} ds F_{II} = \frac{\partial}{\partial z} \int_{-L/2}^{L/2} ds\vec{F}_\perp .$$ (11)

This expression is referred to as the *Panofsky-Wenzel (P-W) theorem.*

Since typically $b >> a$ in a real machine, i.e., the beam size is much smaller than the size of the beam pipe transversely, the lower modes usually dominate. The wake W_1 is often loosely referred to as the transverse wake function, similarly, W_0' is referred to as the longitudinal wake function.

The most important characters of the longitudinal wake function are $W_m'(z) = 0$, if $z > 0$ and $W_m'(0^-) \geq |W_m'(z)|$, for all z.

2.5. Coupling impedance

We can use a concept of coupling impedance to describe the interaction between the moving charged particles and the beam wake field. The wake field has been described as a response to the δ-function beam in time domain. By performing a Fourier transformation on the wake field, we can transfer it to the frequency domain. The expression of wake function in frequency domain is defined as *coupling impedance.*

In general, the beam current can be described as a multi-pole moment

$$j_m(s,t) = j_m e^{i(ks-\omega t)}.$$ (12)

Similar to the relation between resistance, voltage, and current, we can define the coupling impedance in accelerator physics, i.e., the Fourier transformation on the wake function

$$V = \overline{E}_s = -j_m Z_m^{II} r^m \cos m\theta,$$

$$Z_m^{II}(\omega) = \int_{-\infty}^{\infty} \frac{dz}{c} e^{-i\omega z/c} W_m'(z),$$ (13)

$$Z_m^{\perp}(\omega) = i \int_{-\infty}^{\infty} \frac{dz}{c} e^{-i\omega z/t} W_m(z).$$

In above equations, the minus sign means that the longitudinal wake field is opposite to the direction of the particle motion, i means the transverse wake field is perpendicular to the direction of the particle motion. The dimensionality of transverse impedance is ΩL^{-2m+1}, and the dimensionality of longitudinal impedance is ΩL^{-2m}.

Similarly to the wake function, in high energy accelerators, the lower modes of impedance usually dominate. The impedance $Z_1^\perp$ is sometimes referred to as the transverse impedance, the dimensionality is Ω/m; and Z_0^{II} is referred to the longitudinal impedance, its dimensionality is Ω.

By inverse Fourier transformation, we can also construct the wake function from the impedance

$$W_m'(z) = \frac{1}{2\pi} \int_{-\infty}^{\infty} d\omega e^{i\omega z/c} Z_m^{\text{II}}(\omega) , \tag{14}$$

$$W_m(z) = \frac{-i}{2\pi} \int_{-\infty}^{\infty} d\omega e^{i\omega z/c} Z_m^\perp(\omega) . \tag{15}$$

We have a relationship between the longitudinal impedance and the transverse impedance from the Panofsky-Wenzel theorem as

$$Z_m^{\text{II}}(\omega) = \frac{\omega}{c} Z_m^\perp(\omega) . \tag{16}$$

Normally, the longitudinal impedance can be modeled by an equivalent parallel RLC resonator circuit as

$$W(t) = \frac{R_s \omega_R}{Q} e^{-\alpha} \cos \omega_R t, \qquad\qquad \alpha = \frac{\omega_R}{2Q}, \tag{17}$$

$$Z_m^{\text{II}}(\omega) = \frac{R_s}{1 + iQ(\dfrac{\omega_R}{\omega} - \dfrac{\omega}{\omega_R})} . \tag{18}$$

We have the corresponding transverse impedance referred to this resonator circuit from the P-W theorem

$$Z_m^\perp = \frac{c}{\omega} \frac{R_s}{1 + iQ(\dfrac{\omega_R}{\omega} - \dfrac{\omega}{\omega_R})} . \tag{19}$$

We may then find a few general characteristics of impedance as

(a). $Z_m^{\text{II}*}(\omega) = Z_m^{\text{II}}(-\omega)$ and $Z_m^{\perp*}(\omega) = -Z_m^\perp(-\omega)$.

$\operatorname{Re} Z_m^{\text{II}}$ and $\operatorname{Im} Z_m^\perp$ are even functions of ω,

$\operatorname{Im} Z_m^{\text{II}}$ and $\operatorname{Re} Z_m^\perp$ are odd functions of ω.

(b). Since the beam can not gain energy from the pipe structure, only the real part of the impedance causes an energy loss of the beam. We then conclude that

$$\operatorname{Re} Z_m^{\text{II}}(\omega) \geq 0 \qquad \text{for all } \omega$$

$$\operatorname{Re} Z_m^{\perp}(\omega) \begin{cases} \geq 0 & \omega > 0, \\ \leq 0 & \omega < 0, \end{cases} \tag{20}$$

(c). Considering the importance of lower mode impedance, normally we have the following relationship between the longitudinal and transverse impedance

$$Z_1^{\perp} \sim \frac{2c}{b^2 \omega} Z_0^{\text{II}} \qquad \text{or} \qquad Z_1^{\perp} \sim \frac{2R}{b^2} \frac{Z_0^{\text{II}}}{n}. \tag{21}$$

2.5.1. Broad band impedance and narrow band impedance

Related to the *RLC* resonator circuit, the model for the impedance is given by equation (18). For a cavity of radial size ~b, in the region of the frequency $\omega \leq \dfrac{c}{b}$ and in the case of $Q \approx 1$, the impedance corresponds to the **broad band impedance**. In this case, as the filling time and decay time, in which the wake field is induced, is short, for example, $Q \approx 1$, $\tau \approx 1.5\,ns$, the effect on the beam represents the short range effect, i.e., single bunch effect.

Similarly, for the frequency region of $\omega \leq \dfrac{c}{b}$, and in the case of $Q \gg 1$, the impedance corresponds *a* **narrow band impedance**. The filling time and decay time, in which the wake field is induced, is long, for example $\tau = \dfrac{1}{\alpha} = \dfrac{2Q}{\omega_R} \approx 100\,\mu s$, the long range effect, i.e., multi-bunch effect. It may be induced by the cavity structure.

2.5.2. Diffraction model impedance

The impedance, in the high frequency region of $\omega \gg \dfrac{c}{b}$, resembles the **diffraction model**. The lowest mode of the longitudinal impedance can be written as

$$Z_0^{//}(\omega) = \left(1 + \mathrm{sgn}(\omega)i\right)\frac{Z_0}{2\pi^{3/2}}\frac{1}{b}\sqrt{\frac{cg}{|\omega|}}, \tag{22}$$

where g is the gap of the cavity. For any mode, it may be written as

$$Z_m^{II}(\omega) = \frac{\omega}{c}Z_m^{\perp}(\omega) = \left(1 + \mathrm{sgn}(\omega)i\right)\frac{Z_0}{\pi^{3/2}}\frac{1}{b^{2m+1}}\sqrt{\frac{cg}{|\omega|}}. \tag{23}$$

We can then see in the region of $\omega \gg \dfrac{c}{b}$,

the longitudinal impedance $\quad Z_m^{II}(\omega) \propto \omega^{-1/2}$, $\tag{24}$

and the transverse impedance $\quad Z_m^{\perp}(\omega) \propto \omega^{-3/2}$.

In the region $\omega \le \dfrac{c}{b}$, the diffraction model can be matched smoothly with the broad band resonator model.

2.5.3. Resistive Wall

This is in the low frequency region and is not important in the contribution to the impedance of high energy storage ring except for the very large scale of the ring.

2.6. Loss factor

As a beam passes through a beam pipe, it loses a certain amount of energy to the impedance. This energy loss is referred to as the *parasitic loss* of the beam. In circular accelerators, if the impedance is modeled by an *RLC* resonator, and in the case of a Gaussian bunch distribution, the parasitic loss power can be estimated as

$$P_{parasitic} = -\frac{\Delta E}{T_0} = \begin{cases} -\dfrac{q^2 c^3}{8\pi^{3/2}Q\omega_R\sigma_z^3}\left|\dfrac{Z_0^{//}}{n}\right|, & \text{for long bunch} \\[2mm] -\dfrac{q^2\omega_R^2}{4\pi}\left|\dfrac{Z_0^{//}}{n}\right|, & \text{for short bunch} \end{cases} \tag{25}$$

The long bunch means the bunch length is longer than the resonant wavelength of the impedance, and the short bunch is shorter than it. ΔE is the parasitic energy loss, T_0 is the revolution period in the accelerator.

We define the loss factor as $k = \Delta E / q^2$, the dimensionality is V/pC. Conversely, we can use above expression to estimate the impedance by measuring the parasitic loss of a stored beam. The loss factor can be obtained in the experiment by measuring the relation between the RF phase and the beam intensity.

Most of the energy stored in the wake field ends up as heat on the vacuum chamber walls. But under unfavorable conditions, the wake field energy can be transferred systematically back to the beam motion, which is the reason of the beam instability.

2.7. Acquirement of the impedance

After we understand the wake field and the impedance, the demanding task is how we can get detailed information on the wake function and the impedance. This is very important for practical accelerator design and operation. The information on impedance can help us to control beam instability and to improve beam performance in a storage ring.

2.7.1. Analytic method to calculate impedance

This requires the analytical solution of the boundary problem of the electromagnetic fields. The analytic expression derived according to the impedance definition is helpful for better understanding, but only simple geometric components could be solved in this way.

The analytic method is often used in the case of regular geometries of the vacuum components, such as BPM, taper, regular cavities, etc. Many accelerator physicists did a lot of work in this area. A few examples of solvable electromagnetic fields are listed below.

The wake function can be expanded for even symmetry structures into $e^{im\theta}$ with $s > 0, m \geq 1$

$$W_m^{\mathrm{II}}(s) = (\frac{r_1}{b})^m (\frac{r}{b})^m \cos m\theta W_m'(s),$$ (26)

$$W_m^{\perp}(s) = (\frac{r_1}{b})^m (\frac{r}{b})^{m-1} (\hat{r}\cos m\theta - \hat{\theta}\sin m\theta) W_m'(s),$$ (27)

where r_1 is the off-centered distance of the excitation charge, r_1 and r are much smaller than the radius b of the beam pipe. The lower modes are dominant

$$W_{\mathrm{II}}(s) = W_0'(s),$$ (28)

$$W_\perp(s) = \frac{r_1}{b} W_1(s). \tag{29}$$

The results show that the longitudinal wake function is independent on the particle transverse position. The lowest mode of the transverse wake function is proportional to the particle transverse displacement.

We may simplify some of the real structures such as to apply the analytical method in an approximate way to estimate the impedance. For the resistive wall, we assume that the wall thickness is greater than the skin depth, i.e., $t >> d$, then

$$Z_0^\parallel(\omega) = \frac{[1 - \mathrm{sgn}(\omega)i]R}{\sigma_c \delta b} = \frac{[1 - \mathrm{sgn}(\omega)i]R\sqrt{Z_0|\omega|}}{b\sqrt{2c\sigma_c}},$$

$$Z_1^\perp(\omega) = \frac{2c}{b^2\omega} Z_0^\parallel(\omega). \tag{30}$$

In simplification, we have

$$\frac{Z_0^\parallel}{n} = [\mathrm{sgn}(\omega) - i]\frac{R}{b}\sqrt{\frac{Z_0\omega_0}{2c\sigma_c}}|n|^{-\frac{1}{2}},$$

$$Z_1^\perp = [\mathrm{sgn}(\omega) - i]\frac{R}{b^3}\sqrt{\frac{2cZ_0}{\sigma_c\omega_0}}|n + v_\beta|^{-\frac{1}{2}}.$$

$$\tag{31}$$

For button type BPM, the longitudinal impedance can be written as

$$\mathrm{Im}\left(\frac{Z_0^\parallel}{n}\right) \approx -\frac{\alpha_m\omega_0}{\pi c^2 b^2}, \tag{32}$$

where

$$\alpha_m = \frac{2}{3}\left(\frac{\pi}{4}\right)^2 \frac{a^3}{\ln\dfrac{2\pi a}{w} + \dfrac{\pi a}{2w} - \dfrac{7}{3}}. \tag{33}$$

Putting together all possible contributions of the vacuum components, the complete impedance model of an accelerator can be described as

$$Z^\parallel(\omega) = i\omega L + R_s + B[1 - i\,\mathrm{sgn}(\omega)]\sqrt{|\omega|} + A\frac{1 + i\,\mathrm{sgn}(\omega)}{\sqrt{|\omega|}}. \tag{34}$$

The first term in the above equation is the low frequency inductive impedance, the second term is the resistance, the third term is the resistive wall, and the fourth term depends on high frequency like the impedance of a single cavity.

For the BEPC parameters, the impedances are

$$Z^{\parallel}(\omega) = i421.38 \times 10^{-9} \omega + 1237.8 + 1.025 \times 10^{-4} [1 - i\,\text{sgn}(\omega)]\sqrt{|\omega|}$$
$$+ 2.071 \times 10^{7} [1 + i\,\text{sgn}(\omega)] \frac{N_{rf}}{\sqrt{|\omega|}} \tag{35}$$

$$Z^{\perp}(\omega) = \frac{2c}{b^2} \frac{Z^{\parallel}(\omega)}{\omega}. \tag{36}$$

If the bunch length is 4.2 cm, then the impedance which the bunch experiences is

$$\frac{Z^{\parallel}}{n} = 4.37\,i + 2.45\,(\Omega), \qquad Z^{\perp} = 0.134\,i + 0.075\,(M\Omega/m). \tag{37}$$

The longitudinal low-frequency impedance of BEPC is

$$\left.\left|\frac{Z_{\parallel}}{n}\right|\right|_{0} \approx 4.0\Omega \tag{38}$$

2.7.2. Numerical method to calculate impedance

The numerical way to calculate the electro-magnetic field using a computer code by the network method can be applied to the axial symmetric structures which are most frequently used for the beam pipe.

Many computer codes have been developed and used to study impedance as described in the following. The code TBCI can be used to calculate the wake field in time domain. URMEL can be used to calculate the impedance in frequency domain. The code ABCI can be used to calculate the wake field and impedance of two-dimensional structures. MAFIA is much more powerful as it can be used to calculate the wake field and impedance of two-dimensional and three-dimensional structures.

2.7.3. Measurement method to estimate the impedance

Measurement methods to obtain the impedance are available for single components and for the integral beam pipe. A single component can be measured by a pulse current. The integral measurement needs to observe the beam characteristics straight forward.

The measurement of a single component relies on simulating a beam. A pulse current passes through a thin wire which is surrounded by the structure. From the measured changes of the pulse current before and after it passes through the structure the impedance can be estimated. This method is not restricted by the structure complexity, but it is not easy to reach higher accuracy. The impedance can be estimated in this way as

$$Z(\omega) = -2iR_0 \frac{\Delta \tilde{i}(\omega)}{\tilde{i}_0(\omega)}. \tag{39}$$

The total impedance of a storage ring can be estimated by the measurement of the beam parameters as the follows.

(1). Measurement of bunch length versus beam current which is the bunch lengthening effect. The scaling law can then be determined by the measurement. For example, above threshold of microwave instability, the scaling law of the BEPC bunch lengthening is

$$\sigma_l = a \left(\frac{\alpha_p I}{E v_s^2} \right)^{1/b}, \qquad a = 0.651, \qquad b = 3.49. \tag{40}$$

We obtained the relevant impedance parameters of the BEPC machine as $\left| \frac{Z_{//}}{n} \right|_0 \cong 4.0\Omega$. We may also estimate the impedance from the following analytic equation

$$\sigma_l = R \left(\frac{\sqrt{2\pi} I \alpha_p}{F k_b v_s^2 E} \left(\frac{Z_\parallel}{n} \right)_0 \right)^{1/3}. \tag{41}$$

(2). Measurement of the transverse coherent oscillation frequency shift versus beam current. Generally, the vertical tune can be measured more precisely. The relation measured at BEPC is $\Delta v_y = 0.00024\ I$. Then, the impedance can be estimated from

$$\Delta v = \frac{ecR^2 I}{2\sqrt{2} k_b v E \sigma_l b^2 \omega_0} \left| \frac{Z_\parallel}{n} \right|_0. \tag{42}$$

Comparing the above methods to estimate the impedance we note that, in the computational method, some real factors have to be ignored, in the analytical approach, some simplifications have to be made, and, with a measurement, experimental tolerances are involved and a high accuracy is not easy achievable. Thus, in an accelerator design or for an existing machine, we may combine the results from these different methods and so obtain a relatively accurate value of the impedance.

3. Collective Beam Instabilities

The phenomenon of single beam instability was first observed at SLAC in 1962. Dr. Sessler and other experts started to explain the instability as the resistive wall effect in 1965. In 1974, Dr. F. Sacherer of CERN developed the instability theory, which analyzed the single beam instability in accelerator based on the Vlasov equation.

Single beam instability, or single beam collective effect, is the collective behavior of the multi-particle in one beam, which can be divided to two groups as coherent motion and incoherent motion. The main difference between the coherent motion and incoherent motion is the frequency relativity of the particles motion. It is easy to observe and measure the behavior of the coherent motion, but it is very difficult to see the particle incoherent motion due to the random action in the frequency.

3.1. *Particle incoherent collective effect*

The incoherent collective effects we discuss here are the behavior of the interaction between individual particles in a bunch. The following phenomena are included.

3.1.1 *Elastic scattering by nucleus*

--- between beam particles and gas atom nucleus

In elastic scattering, the moving particles experience an impact action, the amplitude of the particles oscillation increases, the effective cross section is

$$\sigma_1 = \frac{4r_e^2 Z^2}{\gamma^2} \frac{\pi}{2} \left(\frac{\langle \beta \rangle}{a} \right)^2. \tag{43}$$

We can see that $\sigma_1 \propto \gamma^{-2}$, this means the effect reduces quickly when the energy increases.

3.1.2 *Inelastic scattering by nucleus*

--- between beam particles and gas atom nucleus

Also called bremsstrahlung, it causes the particles to lose energy and the energy spread to grow. The effective cross section is

$$\sigma_2 = \frac{4r_e^2 Z^2}{137} \frac{4}{3} \ln\left(\frac{183}{Z^{1/3}}\right) \ln\left(\frac{1}{\varepsilon_{RF}} - \frac{5}{8}\right). \tag{44}$$

Obviously, the cross section does not depend on the energy, so it may affect particles in high energy accelerators as well.

3.1.3 *Elastic scattering by electrons*

--- between beam particles and electrons

As the particles transfer energy to the electrons in the residual gas, the energy spread of the particles grows, eventually beyond the RF stable region in the phase space resulting in a particle loss. The effective cross section is

$$\sigma_3 = \frac{2\pi \cdot r_e^2 Z}{\gamma} \frac{1}{\varepsilon_{RF}}.$$

(45)

Here, the cross section is inversely proportional to the energy, so it should not be strong in high energy accelerator.

3.1.4 *Inelastic scattering by electrons*

--- between beam particles and electrons

In the inelastic scattering between the particles and the electrons, the electrons emit photons and lose part of their energy. The cross section is

$$\sigma_4 = \frac{4r_e^2 Z}{137} \frac{4}{3} (\ln\frac{2.5\gamma}{\varepsilon_{RF}} - 1.4)(\ln\frac{1}{\varepsilon_{RF}} - \frac{5}{8}) \tag{46}$$

Note that σ_4, similar to σ_2, does not depend much on the energy, so it could also affect high energy accelerators effectively.

The interaction between the particles and the molecules of the residual gas causes a reduction of the beam life time. The beam life time as caused by gas scattering can be written as

$$\frac{1}{\tau} = -\frac{1}{N}\frac{dN}{dt} = 3.47 \times 10^{34}\sigma \cdot p \;\; \left(m^2, Torr\right) hour^{-1} \qquad (47)$$

where $\sigma = \sigma_1 + \sigma_2 + \sigma_3 + \sigma_4$. For the BEPC, at 2 GeV, the beam life time is estimated to about 25 hours resulting from the interaction between the particles and the residual gas molecules.

3.1.5 Ion trapping

During the interaction between the particles and the residual gas, the molecules of the gas can be ionized. The positive ions trapped by the beam interact with the particles, resulting in particle loss and shortening of beam life time. The ion trapping effect is related to the kind of the gas molecules and depends on the gas pressure.

3.1.6 Intra-beam scattering

The Coulomb scattering occurs due to the different speed of the particle betatron oscillation. It leads to energy transferred from the transverse to the longitudinal oscillation, increasing its amplitude eventually beyond the longitudinal stable region. The life time limited by this effect is named Touschek life time

$$\frac{1}{\tau_T} = -\frac{Nr_0^2 c}{8\gamma^2 \pi \sigma_z \sigma_x \sigma_s}\left(\frac{\gamma mc}{\Delta p}\right)^3 D(\xi), \qquad (48)$$

where $D(\xi)$ is a function of $\gamma, \Delta p, \sigma_x,$ and β_x. For example, in the BEPC at 2.8 GeV, and at 66 mA, the Touschek life time is about 55 hours.

3.2. Particle coherent collective effect

Frequency correlation, the main character of the multi-particle coherent motion, is used to distinguish the coherent from the incoherent motion. Because of the ensemble motion, the coherent motion is easier to be observed and measured in the accelerators by the instrumentation system. The single beam instability, as discussed, is the multi-particle coherent collective effect. In our lecture, we just focus on the bunched beam instability in high energy circular accelerators.

4. Macroparticle model

As the beam is a large multi-body system consisting of commonly more than 10^{10} particles, we do not need to know the details of every particle, instead, our concern is the macroscopic behavior of the beam. The physical model which we establish should not only describe the physical nature of the objects correctly, but can also provide some simplified methods. For this purpose, the macroparticle model is introduced to deal with the beam instability.

In the macroparticle model, a quantity of macroparticles is used to describe the beam system that consists of a large quantity of particles. Every macroparticle represents a number of real particles. The scalar physical quantity of one macroparticle, such as mass and charge, is the scalar sum of all the particles contained in it. The vector physical quantity, such as displacement and velocity, is the vector sum of the particles barycenter, etc.

The number and distribution of macroparticles in the model depend on the system's real situation. They not only describe the whole characters of the system, for instance, mass, charge, center of mass, etc., but also represent the distribution inside the system. The interactions among the particles and the beam-environment interaction follow their respective.

When we study the beam instability analytically it is convenient to represent one bunch as one macroparticle or two, i.e., one-particle model or two-particle model. The advantage of the simplified one-particle and two-particle models is that it can provide an intuitive picture of the instability mechanisms.

When we use numerical simulation to explore the instability, one bunch can be regarded as an ensemble of hundreds or thousands of macroparticles, the number of macroparticles is limited by the computer capability normally.

When we studied the wake fields generated by a beam in vacuum chamber of an accelerator, we assumed that the particle distribution within the beam is rigid. For a one-particle model, the bunch distribution is rigid without any detail structure. A one-particle model can simply describe the beam-environment interaction. A two-particle model offers the opportunity of looking into the instability mechanisms associated with the internal degrees of freedom in the beam distribution.

For the case of a linac, the interaction between the beam and wake field is often illustrated using a model in which the beam is represented as two macroparticles. Each of the two macroparticles is considered to be a rigid point charge of $Ne/2$. The obvious convenience in this way is that the longitudinal motion of particles in a bunch can be neglected.

As for the linac, two macroparticles are considered fixed in their relative longitudinal position in one bunch. But for the circular accelerators, as the revolution circle time of the beam is longer than the period of the longitudinal synchrotron oscillations, the longitudinal motion of particles must be involved. Therefore, the instability of a bunched beam in the circular accelerators can be divided into transverse instability and longitudinal instability.

We studied the various instability mechanisms using highly simplified macroparticles models in which the particle beam was only modeled either as a single point charge, or as two point charges interacting with each other through wake fields, but we do not consider any real internal distribution structure in the bunch.

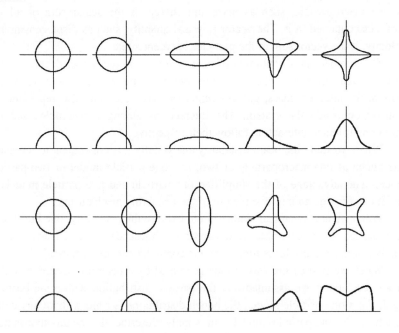

Figure 3. Schematic of beam instability oscillation modes.

However, there are some limitations using the simplified macroparticle models to study the beam instabilities. One limitation is that their quantitative predictions cannot be accurate, in fact they are rather crude. Another limitation is that the instabilities are treated each individually, even it may be described as a more general treatment that allows different kinds of beam instabilities to be incorporated into one framework. Still another limitation which is perhaps more serious is the fact that some instabilities observed in the circular accelerators involve higher oscillation modes in the longitudinal structure of the beam that cannot be properly treated by the simplified models. The shapes of a few collective modes of the longitudinal beam motion are represented schematically in Figure 3. We can see the dipole mode, the quadrupole mode, the sextupole mode and the octupole mode oscillation in the phase space and in the real space.

4.1. Longitudinal Robinson instability

Now, we discuss the longitudinal Robinson instability in circular accelerators, as an example to illustrate the one-particle model, and then we will illustrate the two-particle model by the strong head-tail instability.

When a bunch passes through an RF cavity the beam particles interact with the electro-magnetic field in the cavity. Different particle positions correspond to different phases of the field, and different energy exchanges. This instability mechanism was first analyzed by Dr. Robinson in LBL, so it was called the Robinson instability.

The physical mechanism of the Robinson instability is described below. The impedance of the RF cavity is modeled by the resonator model

$$Z^{\parallel}(\omega) = \frac{R_s}{1 + iQ(\dfrac{\omega_r}{\omega} - \dfrac{\omega}{\omega_r})} \tag{49}$$

The relation of the energy and the frequency is

$$\eta = -\frac{\Delta\omega/\omega}{\Delta p/p} = \frac{1}{\gamma_{tr}^2} - \frac{1}{\gamma^2} = \alpha_p - \frac{1}{\gamma^2}, \tag{50}$$

where γ_{tr} the transition energy of the particle in accelerators. We discuss this instability for particle energies above and below transition energy, respectively.

Above transition energy, $\gamma > \gamma_{tr}$, and for $\omega_r < h\omega_0$ (right-hand part of Fig. 4) a particle with $\Delta E > 0$ and, accordingly, $\Delta\omega < 0$, sees an increased impedance of the cavity as $R(h\omega_0 - \Delta\omega) > R(h\omega_0)$. Thus, it transfers some more energy to the cavity. Conversely, for a particle with $\Delta E < 0$ and, correspondingly, $\Delta\omega > 0$. At the corresponding frequency, the impedance of the cavity appears smaller as given by $R(h\omega_0) > R(h\omega_0 + \Delta\omega)$. This particle transfers some less energy to the cavity. We conclude that above transition energy, $\gamma > \gamma_{tr}$, and if

312

$\omega_r < h\omega_0$, the beam is stable under the influence of the Robinson effect. Therefore, the resonance frequency of the fundamental cavity mode needs to be detuned a bit lower than the integral multiple of the revolution frequency. In contrast, if $\omega_r > h\omega_0$, still above transition energy $\gamma > \gamma_{tr}$, the Robinson effect will cause the beam to be unstable.

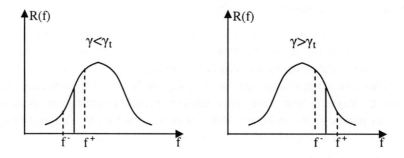

Figure 4. The Robinson instability.

Similarly, below transition energy $\gamma < \gamma_{tr}$ (left-hand part of Fig. 4) the beam is stable if $\omega_r > h\omega_0$, and unstable if $\omega_r < h\omega_0$. Therefore, the resonant frequency of the fundamental cavity mode needs to be tuned a bit higher than the integral multiple of the revolution frequency.

For beams of different energy, the above analysis indicates the correct frequency region of the fundamental cavity mode which should be chosen in the operation of an accelerator to make the beam stable against the Robinson effect.

Physically, the real reason of the Robinson effect is that the revolution frequency of particles whose energy is $p = p_0 + \Delta p$ is not ω_0, but $\omega_0(1 - \eta \Delta p/p)$. So by detuning the resonant frequency to the right position, the beams will be damped by the Robinson effect and, thus, be stable.

The equations of the particle longitudinal motion, in which the impedance is included, can be written as

$$\ddot{\phi} + \frac{I_n^2 \omega_s \left(Z^+ - Z^-\right)}{4I_0 V \cos\phi_s} \dot{\phi} + \omega_s^2 \phi = 0, \tag{51}$$

$$\phi(t) = \phi_0 e^{-\alpha} \cos \omega_s t , \qquad \alpha = \frac{I_n^2 \omega_s (Z^+ - Z^-)}{8 I_0 V \cos \phi_s} , \tag{52}$$

where $I_n \approx 2I_0 , Z^+ = Z(h\omega_0 + \omega_s) , Z^- = Z(h\omega_0 - \omega_s)$.

We can then use these equations to calculate the growth time or the damping rate of the Robinson instability,

$$\tau = \frac{1}{\alpha} \tag{53}$$

When the beam interacts with the higher modes of the cavity, Robinson's instability can also occur correspondingly. We may write the higher modes impedance by the resonator model as

$$Z_H^{//}(\omega) = \frac{R_{sH}}{1 + i Q_H \left(\dfrac{\omega_{rH}}{\omega} - \dfrac{\omega}{\omega_{rH}} \right)}. \tag{54}$$

The same analysis as for the fundamental cavity mode applies to the higher RF modes, so we can obtain the stable condition of the Robinson instability (H is an integer) as follows. Above transition $\gamma > \gamma_{tr}$, the beam is stable if $\omega_{rH} < H\omega_0$, and unstable if $\omega_{rH} > H\omega_0$. Below transition $\gamma < \gamma_{tr}$, the beam is stable if $\omega_{rH} > H\omega_0$, and unstable if $\omega_{rH} < H\omega_0$.

Usually, the frequency width of higher RF modes is narrow (Q_H is larger), the resonant frequency ω_{rH} is higher, and it is sensitive to the temperature of the cavity. Therefore, for good accelerator performance, the temperature of the RF cavity should be controlled strictly if the higher modes in the cavity are not suppressed very well, such as to prevent higher RF modes causing Robinson instability.

4.2. Strong head-tail instability

The strong head-tail instability in circular accelerators is a transverse instability. We take the strong head-tail instability here as an example to understand how the two-particle model is applied to the beam instability analysis. To illustrate the mechanism of the strong head-tail instability using the macroparticle model, a bunch is considered as consisting of two macroparticles, each with a charge of $Ne/2$, N being the total particle number in the bunch. The strong head-tail instability is sometimes also called the transverse mode coupling instability or transverse turbulent instability.

The position of the two macroparticles is at the head and the tail of the bunch, respectively, and each of them executes synchrotron oscillations. We assume their synchrotron oscillations have equal amplitude, but opposite phase. During a half period of the oscillation, $0<s/c<T_s/2$, where T_s is the synchrotron oscillation period, particle 1 leads particle 2 in the direction of the motion, the equations of motion for these two macroparticles are

$$y_1'' + \left(\frac{\omega_\beta}{c}\right)^2 y_1 = 0. \tag{55}$$

$$y_2'' + \left(\frac{\omega_\beta}{c}\right)^2 y_2 = \frac{Nr_0W_0}{2\gamma C} y_1.$$

Similarly, during another half period of the oscillation, $T_s/2<s/c<T_s$, we have the same equation with indices 1 and 2 exchanged. We can continue to apply above equations to the following motion with exchanging the indices every half period. We assume here the initial amplitude y_1, y_2 is small, as the amplitude will increase exponentially if the beam instability occurs.

We have also assumed, for simplicity, that the wake function is a constant in the above equations, and vanishes before the beam completes one revolution or during the time when another bunch is coming, that is

$$W_1(z) = \begin{cases} -W_0 & \text{if } 0 > z > -(\text{bunch length}) \\ 0 & \text{otherwise} \end{cases} \tag{56}$$

The property of wake function requires that $W_0>0$. This short range wake function corresponds to the broad band impedance. This is different with the Robinson instabilities, in which the long range wake function corresponds to the narrow band impedance. In this sense, the longitudinal Robinson instabilities are multi-turn phenomena, and the strong head-tail instability is essentially a single-turn effect or single bunch effect.

We now analyze the stability conditions of the two-particle beam system. The solution of the equation of motion is simply a harmonic oscillation, that is

$$\tilde{y}_1(s) = \tilde{y}_1(0)e^{-i\omega_\beta s/c} \tag{57}$$

where $\tilde{y}_1 = y_1 + i\dfrac{c}{\omega_\beta}y_1'$.

Substituting y_1 into the second of the equations (55) yields the solution for y_2 as equation (58)

$$\tilde{y}_2(s) = \tilde{y}_2(0)e^{-i\omega_\beta s/c} + i\frac{Nr_0W_0c}{4\gamma C\omega_\beta}\left[\frac{c}{\omega_\beta}\tilde{y}_1^*(0)\sin\frac{\omega_\beta s}{c} + \tilde{y}_1(0)se^{-i\omega_\beta s/c}\right].$$

The first two terms describe the betatron oscillation, the third term, proportional to s, is the resonantly driven response.

For $\omega_\beta \gg \omega_s$, the second term on the right hand side of the above expression can be neglected because it is much smaller than the third term. We can then write the solution for the equations of motion during a half oscillation period $0<s/c<T_s/2$ in a matrix form as

$$\begin{bmatrix}\tilde{y}_1 \\ \tilde{y}_2\end{bmatrix}_{s=cT_s/2} = e^{-i\omega_\beta T_s/2}\begin{bmatrix}1 & 0 \\ i\Lambda & 1\end{bmatrix}\begin{bmatrix}\tilde{y}_1 \\ \tilde{y}_2\end{bmatrix}_{s=0}, \tag{59}$$

where

$$\Lambda = \frac{\pi Nr_0W_0c^2}{4\gamma C\omega_\beta\omega_s}. \tag{60}$$

The time progress during another half period of the oscillation $T_s/2<s/c<T_s$ can be similarly obtained by exchanging indices 1 and 2 in the above analysis. Therefore the total transformation for one full synchrotron period can be written as

$$\begin{bmatrix}\tilde{y}_1 \\ \tilde{y}_2\end{bmatrix}_{cT_s} = e^{-i\omega_\beta T_s}\begin{bmatrix}1-\Lambda^2 & i\Lambda \\ i\Lambda & 1\end{bmatrix}\begin{bmatrix}\tilde{y}_1 \\ \tilde{y}_2\end{bmatrix}_{s=0}. \tag{61}$$

In the following time of the particle motion, the solution vector is repeatedly transformed by the matrix.

Stability of the system is thus determined by the eigenvalues of this matrix. The two eigenvalues for the two modes (a + mode and a − mode) are

$$\lambda_\pm = e^{\pm i\phi}, \qquad \sin\frac{\phi}{2} = \frac{\Lambda}{2}. \tag{62}$$

Stability requires ϕ is real, which is fulfilled if $\left|\sin\dfrac{\phi}{2}\right| \le 1$ or $\Lambda \le 2$. For the weaker beams, $\Lambda \ll 1$, that is $\phi \approx \Lambda$. The beam would become unstable when ϕ approaches π as Λ approaches 2.

The above analysis indicates that if $\Lambda > 2$, during half a synchrotron period, the motion of the trailing particle may grow by an amount exceeding twice the amplitude of the oscillating leading particle. If $\Lambda \le 2$, the growth amplitudes acquired during half a synchrotron oscillation period when particle is trailing do not accumulate, and the beam is stable. This threshold behavior is very different from the linac case, in which the beam, at least its head, is always stable, as the particles change their positions in the longitudinal phase space only little.

From the Λ expression, we know that the larger the current, the more unstable becomes the beam, and the higher the energy, the higher is the threshold, i.e., beam instability is inversely proportional to the energy. It also shows up in the fact that Λ is inversely proportional to ω_s, this means synchrotron oscillation is thus an effective stabilizing mechanism in circular accelerators. Another point to note is that Λ is also inversely proportional to ω_β, that is, proportional to the β-function. Therefore, it will be helpful for increasing the instability threshold to reduce the impedance located at β_{max}.

From the analysis of the strong head-tail instability modeled by two-particle model, we obtain the oscillation modes as

positive mode $\qquad \omega_\beta + l\omega_s - \dfrac{\phi}{2\pi}\omega_s \qquad\qquad$ l even

negative mode $\qquad \omega_\beta + l\omega_s + \dfrac{\phi}{2\pi}\omega_s \qquad\qquad$ l odd

As the beam intensity increases, the mode frequencies shift, when the frequencies of two modes merge into each other, the beam becomes unstable. The frequency shift caused by the instability could be readily observed in larger scale accelerators. This transverse head-tail oscillation can be detected using a streak camera.

From the above analysis, the initial slope of this frequency with respect to the beam intensity is

$$\left(\frac{d\omega_\beta}{dN}\right)_{N=0} = -\frac{r_0 W_0 c^2}{8\gamma C \omega_\beta}. \tag{63}$$

By measuring the instability threshold or by measuring the initial slope of the betatron frequency, information on the wake field or the impedance can be obtained. For example, at BEPC, we measured $\Delta\omega_\beta / I \approx -0.00024 / mA$, as mentioned before, so the impedance of BEPC was estimated to be about 4 Ohm.

Another interesting example for the application of the macroparticle model is the luminosity variation of colliding beams influenced by the longitudinal coherent oscillation of the particles. The phenomena can be observed by the wall current monitor, which may include the dipole mode and higher modes. When two beams are colliding, the longitudinal coherent oscillation may reduce the luminosity. During BEPC operation, we have observed this dependence, and the analysis is explained.

5. Perturbation formalism based on the Vlasov equation

One could, of course, increase the number of macroparticles in the simplified model, but when there are more than two macroparticles in the system, the analysis along this line becomes cumbersome. A computer simulation can be used to extend the model to anywhere from three to several thousand macroparticles, but then dealing with 10^{12} particles in this way seems hopeless, and it will be limited by the computer capability.

In the macroparticle model, the particle representation usually describes the instability motion in time domain, but the modes can represent the frequency distribution of the instability. The same results can be found from a different approach. In practice, the mode representation offers a formalism that can be used systematically to treat the instability problem in many cases, it can be used to obtain analytic results for arbitrarily high mode numbers. The basic mathematical tool in this method is the ***Vlasov equation***.

In principle, the mode representation and the particle representation of the beam motion are identical. The mode frequencies and patterns can be obtained from the Vlasov equation by using the perturbation formalism.

The stability of the beam requires that all modes are stable. If any one of the modes shows the potential of growing exponentially, the beam will be unstable. An analysis of the modes, therefore, leads to the stability criterion for the beam. The early works on the mode analysis of the beam instability were developed by F. Sacherer of CERN.

The Vlasov equation can give the particle distribution in which the nonlinear effects are involved. To simplify, we need to linearize the equation in order to search for the beam oscillation modes. In fact, when the instability occurs, particles will not immediately be lost from the beam, but the size of the beam will increase, such as bunch lengthening, energy spread widening, emittance growth, and so on.

The Vlasov equation describes the collective behavior of a multi-particle system under the influence of electro-magnetic forces. To construct the Vlasov equation, one starts with the single-particle equations of motion

$$\dot{q} = f(q, p, t)$$

$$\dot{p} = g(q, p, t)$$

The behavior of a particle at a time is represented by a point in the phase space. The motion of a particle can be described by the motion of its representative point in the phase space, e.g., for a particle executing a simple harmonic motion, the representative point in phase space traces out an ellipse.

In a conservative deterministic system, the particle trajectory in phase space is completely determined by the initial conditions (q, p) at time $t=t_0$. Two particles having the same initial conditions must have exactly the same trajectory in phase space. In other words, in a conservative system, trajectories of particles can never intersect, i.e., no particles run in or run out the boundary of the phase space.

If the system is conservative, i.e., if the system is not influenced by any damping or diffusion effects from any external sources, or the sum of internal damping and antidamping effects vanishes in the system, then we have

$$f = \frac{\partial H}{\partial p} \qquad \text{and} \qquad g = -\frac{\partial H}{\partial q}, \tag{64}$$

where H is the Hamiltonian. Thus,

$$\frac{\partial f}{\partial q} + \frac{\partial g}{\partial p} = 0 \tag{65}$$

The area conservation property in the phase space can be derived from these conditions, its shape may be distorted, but its area remains constant. We assume a rectangular $\Delta q \, \Delta p$ box in phase space, and the box is small enough so that the same numbers of particles is contained in adjacent boxes of the same size. The box is also large enough to contain a significant amount of particles. Let the number of particles enclosed by the box be $\psi(q, p, t)\Delta q\Delta p$, where ψ is the phase space density distribution depending on q, p, and t that is normalized by

$$\int_{-\infty}^{\infty} dq \int_{-\infty}^{\infty} dp\, \psi(q,p,t) = N \qquad (66)$$

Based on the condition that particles cannot flow in or flow out of a certain region in phase space, at a later time $t+\Delta t$, the shape of the area may be changed, but the size of the area remains constant, so we obtain

$$\frac{\partial \psi}{\partial t} + f \frac{\partial \psi}{\partial q} + g \frac{\partial \psi}{\partial p} = 0 \qquad (67)$$

The above expression is the Vlasov equation. It can also be written in the form

$$\frac{d\psi}{dt} = 0, \qquad \text{or} \qquad \psi = \text{constant in time.} \qquad (68)$$

Note that, in the equation, f and g are given by external forces. We assume that there are no collisions between the particles in the system. However, if a particle interacts more strongly with the collective fields of the other particles than with its individual nearest neighbors, the Vlasov equation still can be applied, if one treats the collective fields as the external fields. This in fact is the basis to study the collective instabilities using the Vlasov equation.

One special case in which the Vlasov equation can be solved exactly, is when the system can be described by a Hamiltonian which does not have any explicit time dependence. In the derivation of the Vlasov equation, we have assumed that there is no significant diffusion or external damping effects. This is usually a good approximation for proton beams. But for electron beams, synchrotron radiation contributes to both damping and diffusion. Strictly speaking, our results obtained using the Vlasov equation, cannot be applied to study the electron beam instabilities in storage ring. However, if the instability occurs in a time much shorter than the damping time or diffusion time, the Vlasov treatment can apply also to electron beam instability approximately.

In most cases, we can apply the perturbation form to the beam instability problem by linearizing the Vlasov equation to obtain the approximate analytic solution.

At first, we assume the wake field has vanished, and let the beam have an initial distribution in phase space ψ_0. Being an equilibrium distribution, ψ_0 is only a function of r, that is,

$$\psi_0 = \psi_0(r) \qquad (69)$$

where we introduce polar coordinates, for example, in the longitudinal case, we use

$$z = r \cos \phi \tag{70}$$

$$\frac{\eta c}{\omega_s} \delta = r \sin \phi \tag{71}$$

here r is related to the unperturbed Hamiltonian

$$H = \frac{\omega_s r^2}{2} \tag{72}$$

The coordinate system is illustrated in Figure 5.

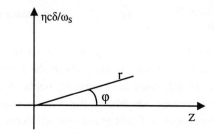

Figure 5. The coordinate system of equations (70), (71).

Now, we suppose that there is a disturbance due to the wake fields in the initial stable distribution. Then, we write

$$\psi(r, \phi, s) = \psi_0(r) + \psi_1(r, \phi) e^{-i\Omega s/c} \tag{73}$$

assuming here that the disturbance has a single frequency Ω, i.e., it contains only a single mode of oscillation. We will consider the disturbance as a small effect, that is, $\psi_1 \ll \psi_0$.

In fact, the mode frequency Ω and the mode distribution ψ_1 are not arbitrary. The disturbance ψ_1 first generates a wake field. Being an oscillation mode, the additional disturbance in the beam distribution caused by this wake field must have the same pattern as the original disturbance ψ_1. The beam-wake system, therefore, has to be solved *self-consistently*. As a result, the frequency distribution can only be associated with each value of Ω, and then there is a well-defined distribution ψ_1.

When ψ_1 is projected onto the z-axis, we then obtain the longitudinal distribution

$$\rho_1(z)e^{-i\Omega s/c} = \int_{-\infty}^{\infty} d\delta\psi_1(r,\phi)e^{-i\Omega s/c} \tag{74}$$

This $\rho_1(z)$ is the distribution signal which can be observed at a fixed location in an accelerator. After one revolution, the beam observed at the same location has a distribution as

$$\rho_1(z)e^{[-i\Omega(s/c - T_0)]}. \tag{75}$$

The wake field excited by ρ_1 produces a retarding voltage. The voltage seen by a particle at z can be written as

$$V(z,s) = e \int_{-\infty}^{\infty} dz' \sum_{k=-\infty}^{\infty} \rho_1(z')e^{-i\Omega[(s/c)-kT_0]}W_0'(z - z' - kcT_0) \tag{76}$$

In writing down this expression, we have included the multi-turn wake fields and have used the causality property

$$z > 0 \quad \text{then} \quad W_0'(z) = 0.$$

We now take Fourier transforms

$$\tilde{\rho}(\omega) = \int_{-\infty}^{\infty} dz e^{-i\omega z/c} \rho(z) \tag{77}$$

$$\rho(z) = \frac{1}{2\pi c} \int_{-\infty}^{\infty} d\omega e^{i\omega z/c} \tilde{\rho}(\omega) \tag{78}$$

Then we deal with the wake function similarly

$$\tag{79}$$

The voltage becomes $W_0'(z) = \dfrac{1}{2\pi} \displaystyle\int_{-\infty}^{\infty} d\omega e^{i\omega z/c} Z_0''(\omega)$

$$V(z,s) = \frac{e}{T_0} e^{-i\Omega s/c} \sum_{p=-\infty}^{\infty} \tilde{\rho}_1(p\omega_0 + \Omega) e^{i(p\omega_0+\Omega)z/c} Z_0''(p\omega_0 + \Omega) \qquad (80)$$

where $\omega_0 = \dfrac{2\pi}{T_0}$.

Then, the Vlasov equation reads

$$\frac{\partial \psi}{\partial s} - \eta \delta \frac{\partial \psi}{\partial z} + \frac{\omega_s^2}{\eta c^2} z \frac{\partial \psi}{\partial \delta} - \frac{e}{T_0 Ec} V(z,s) \frac{\partial \psi}{\partial \delta} = 0 \qquad (81)$$

where E is the energy of the beam particles. Rewriting the Vlasov equation in the polar coordinate system as equations (70) and (71), the two middle terms in the equation can be simplified, yielding

$$\frac{\partial \psi}{\partial s} + \frac{\omega_s}{c} \frac{\partial \psi}{\partial \phi} - \frac{e}{T_0 Ec} V(z,s) \frac{\partial \psi}{\partial \delta} = 0 \qquad (82)$$

We now substitute the distribution and the retarding voltage into the above equation, and linearize it by keeping only the first order terms to obtain the linearized Vlasov equation as

$$-i\Omega \psi_1 + \omega_s \frac{\partial \psi_1}{\partial \phi} - \frac{\eta r_0 c}{\gamma T_0^2 \omega_s} \sin \phi \psi_0'(r)$$

$$\times \sum_{p=-\infty}^{\infty} \tilde{\rho}_1(p\omega_0 + \Omega) e^{i(p\omega_0+\Omega)z/c} Z_0''(p\omega_0 + \Omega) = 0 \qquad (83)$$

Having this linearized Vlasov equation, we are now ready to discuss the collective modes of beam motion under the influence of beam-impedance interaction.

We expand the disturbance ψ_1 on the modes first. This is possible because ψ_1 must be periodic in ϕ with period 2π, that is

$$\psi_1(r,\phi) = \sum_{l=-\infty}^{\infty} \alpha_l R_l(r) e^{il\phi} \tag{84}$$

Substituting the above equation into the linearized Vlasov equation and integrating over ϕ from 0 to 2π, with the relationship between ρ and ψ_1, we obtain the integral equation

$$(\Omega - l\omega_s)\alpha_l R_l(r) = -i \frac{2\pi r_0 c}{\gamma T_0^2} l \frac{\psi_0'(r)}{r} \sum_{l'=-\infty}^{\infty} \int_0^\infty r' dr' \alpha_{l'} R_{l'}(r') i^{l-l'}$$

$$\times \sum_{p=-\infty}^{\infty} \frac{Z_0''(\omega')}{\omega'} J_l\left(\frac{\omega' r}{c}\right) J_{l'}\left(\frac{\omega' r'}{c}\right) \tag{85}$$

where J_l is a Bessel function. We can see in the equation that the impedance and the undisturbed distribution ψ_0 are also included explicitly.

The aim here is to find the disturbed distribution ψ_1, i.e., R_l, α_l and Ω. This is obviously not easy to do in general. We may introduce some special distribution for getting the analytic solution, for example, using a "water-bag" model

$$\psi_0(r) = \begin{cases} 0 & \text{for} \quad r > \hat{z}, \\ \dfrac{N\eta c}{\pi \hat{z}^2 \omega_s} & \text{for} \quad r < \hat{z}, \end{cases} \tag{86}$$

where N is the total number of the particles in the distribution.

We assume that any perturbation on a water-bag beam will occur around the edge of the distribution, this is $R_l(r) \propto \delta(r - \hat{z})$. The integral equation can then be reduced to a set of mode equations.

To ease the discussion, we suppose that if $N=0$, i.e., in the zero beam intensity limit, the solution for the lth mode is

$$\alpha_{l'}^{(l)} = \delta_{ll'} \qquad l' = 0, \pm 1, \pm 2, \ldots\ldots$$

$$\Omega^{(l)} = l\omega_s$$

Then, in the case of the beam intensity being nonzero but still weak, the lth mode frequency can be found as

$$\Omega^{(l)} - l\omega_s = i\frac{2Nr_0\eta c^2}{\gamma T_0^2 \omega_s \hat{z}^2} l \sum_{p=-\infty}^{\infty} \frac{Z_0''\left(p\omega_0 + l\omega_s\right)}{\omega'} J_l^2\left(\frac{\left(p\omega_0 + l\omega_s\right)\hat{z}}{c}\right) \quad (87)$$

Now, we finally obtain the self-consistent complex mode frequencies depending on the interaction between the beam with weak intensities and the impedance. In particular for the model of water-bag distribution, the real part of $\Omega^{(l)}$ gives the mode frequency shift $\Delta\Omega^{(l)}$, the imaginary part gives the instability growth rate $\dfrac{1}{\tau^{(l)}}$.

The above model analysis is an example of the use of the linearized Vlasov equation to study beam instabilities. Even though there are some assumed conditions in the analysis, the solution still has the general characteristics of the problem.

As an example, for $l=1$, let us assume the beam bunch is short enough, so that the wake field lasts much longer than the bunch length, and we want to recover the growth rate of the Robinson instability, the same result with that of the macroparticle model

$$\frac{1}{\tau^{(l)}} = \frac{1}{(l!)^2}\left(\frac{h\omega_0 \hat{z}}{2c}\right)^{2l-2} \frac{Nr_0\eta h\omega_0}{2\gamma T_0^2 \omega_s}$$
$$\times \left[\operatorname{Re} Z_0''\left(h\omega_0 + l\omega_s\right) - \operatorname{Re} Z_0''\left(h\omega_0 - l\omega_s\right)\right] \quad (88)$$

For $l = 1$ mode, we rewrite the impedance in resonator model form, then the frequency shift and growth rate of Robinson instabilities can be written as

$$\tau^{-1} \approx \frac{4Nr_0\eta R_s Q^2 \Delta\omega}{\pi\gamma T_0 h} \quad (89)$$

$$\Delta\Omega \approx -\frac{12Nr_0\eta R_s Q^3 v_s \Delta\omega}{\pi\gamma T_0 h^2} \quad (90)$$

Note that this is much more general than the one-particle model analysis results. It can be applied to higher order modes and arbitrary bunch length. So, it is more general to analyze the modes of beam instability by Vlasov's equation.

It should be mentioned that the water-bag model is a particularly simplified model that allows us to solve the radial modes $R_l(r)$. The radial mode structures may be degenerated, and some information may be lost.

We discuss another example to analyze the bunched beam longitudinal instabilities using Vlasov's equation

$$\frac{d\psi}{dt} = \frac{\partial\psi}{\partial t} + \dot{\phi}\frac{\partial\psi}{\partial\phi} + \ddot{\phi}\frac{\partial\psi}{\partial\dot{\phi}} = 0 \tag{92}$$

Linearizing the equation of particle longitudinal motion

we decompose the distribution into the stationary part and the disturbed part where ψ_0 is the stationary distribution and ψ_1 is the disturbed distribution and

$$\psi = \psi_0 + \psi_1 e^{j\omega t} \tag{93}$$

$\psi_1 \ll \psi_0$, ω is the complex coherent oscillation frequencies. Changing the coordinates to polar coordinates, the linearized Vlasov equation can be written as

$$j\omega\psi_1 - \omega_s\frac{\partial\psi_1}{\partial\theta} + \frac{F}{\omega_s}\sin\theta\frac{d\psi_0}{dr} = 0 \tag{94}$$

The disturbing force F in the equation comes from the wake fields which are generated by the beam interacting with the impedance, it can be written as

$$F = \frac{I\alpha e\omega_s^2}{\beta^2 v_s^2 E}\sum_p Z_p\lambda_p e^{jp\phi} \tag{95}$$

where $Z_p = \dfrac{Z(\omega_p)}{\omega_p}$, $\omega_p = p\omega_0 + \omega_s$, and $\lambda(\omega)$ is the Fourier transform of the disturbed distribution.

In circular accelerators, the particle motion and particle distribution are periodic functions, so the disturbed distribution function can be written as

$$\psi_1(r,\theta) = \sum_m R_m(r)e^{jm\theta} \tag{96}$$

Integrating the equation of motion over θ we find

$$j(\omega - m\omega_s)R_m(r) = j^m \frac{m\omega_s}{r} \frac{d\psi_0}{dr} \frac{I\alpha e}{\beta^2 v_s^2 E} \sum_p Z_p \lambda_p J_m(pr). \tag{97}$$

The coherent frequency shift can be obtained as

$$\Delta\omega_m = \frac{m\omega_s}{1+m} \frac{I_0}{3B^3 hV_T \cos\phi_s} \frac{\sum_p \dfrac{Z_p}{p} h_m(p)}{\sum_p h_m(p)} \tag{98}$$

where B is the bunching factor, $B = \dfrac{\sigma_l}{2\pi R}$, $V_T = V_{RF} - V_0$, V_0 is the retarding voltage, $\lambda_p(m)$ is the spectral density function of the mth mode and pth harmonic wave.

The beam stability criteria are: beam stable if Im $\Delta\omega > 0$, and beam unstable if Im $\Delta\omega < 0$. The growth time is $\tau \propto (\text{Im } \Delta\omega)^{-1}$.

Note that there is a limit in this method which uses Vlasov's equation to solve the beam instabilities, that is, the results are only applicable at the **onset** of the instability. Once the instability starts to grow, or its amplitude is increasing, it is no longer applicable in this way.

6. Beam instability studies in a storage ring

6.1. *Bunch lengthening*

The natural bunch length is the rms Gaussian distribution of the particle in the zero beam current limit,

$$\sigma_{l0} = \frac{\alpha_p R}{V_s} \sigma_e \tag{99}$$

There are two mechanisms that cause the bunch lengthening as follows.

6.1.1 *Potential well distortion*

When the beam current is below the threshold of the microwave instability, the inductive part of the impedance is the main reason that leads to the bunch lengthening. This effect can be estimated by the formula as

$$\left(\frac{\sigma_l}{\sigma_{l0}}\right)^3 - \left(\frac{\sigma_l}{\sigma_{l0}}\right) + I_b \frac{e\alpha_p \operatorname{Im}\left[\left(Z^{\parallel}/n\right)_{eff}\right]}{\sqrt{2\pi}v_{s0}^2 E}\left(\frac{R}{\sigma_{l0}}\right)^3 = 0 \qquad (100)$$

6.1.2 *Microwave instability*

When the beam current is above the threshold of the microwave instability, it is dominated even though the potential well distortion is still affected. The bunch lengthening can be estimated by means of the following formula

$$\sigma_l = \left(\frac{e\alpha_p I_b R^3}{\sqrt{2\pi}N_b E v_{s0}^2}\left[\left|\frac{Z^{\parallel}}{n}\right|_{crit} + \left(\frac{Z^{\parallel}}{n}\right)_{pot}\right]\right)^{1/3} \qquad (101)$$

Generally, *Boussard's* formula is also used to indicate the bunch length approximately as

$$\sigma_l = \left(\frac{e\alpha_p I_b R^3}{\sqrt{2\pi}E v_{s0}^2}\left|\frac{Z^{\parallel}}{n}\right|_0\right)^{1/3} \qquad (102)$$

In practice, there is a scaling law in each accelerator of the bunch lengthening. By measuring the bunch length and its lengthening in the BEPC, we obtain the corresponding scaling law of bunch lengthening above the threshold of microwave instability as

$$\sigma_l(cm) = 0.651 \times \left(\frac{I_b(mA)\alpha_p}{E(GeV)v_s^2}\right)^{1/3.49} \qquad (103)$$

This scaling law of bunch lengthening in BEPC is show in Figure 6.

328

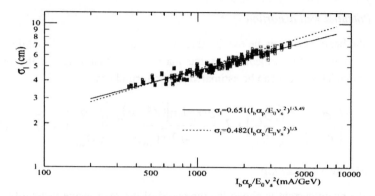

Figure 6. The scaling law of bunch lengthening in BEPC.

6.2. Head-tail instability

In our analysis of the strong head-tail instability, we assumed that the betatron and the synchrotron motions are decoupled from each other. If we include their coupling, i.e., a chromaticity effect, the betatron oscillation frequency of a particle which depends on the energy error $\delta = \Delta E / E$, can be written as $\omega_\beta(\delta) = \omega_\beta(1 + \xi\delta)$, where ξ is the chromaticity which is controlled by sextupoles in circular accelerators. The head-tail phase caused by the nonzero chromaticity is

$$\chi = \frac{\xi\omega_\beta\hat{z}}{c\eta} \qquad (104)$$

In the strong head-tail instability, once the beam current exceeds the threshold, the instability increases very quickly. But the head-tail instability can occur at much lower beam currents, in fact, there is no threshold.

The growth rate of the l th mode of the head-tail instability can be written as

$$\frac{1}{\tau^{(l)}} \approx -\frac{Ne^2c}{2ET_0^2\omega_\beta} \sum_{p=-\infty}^{\infty} \text{Re}\, Z_1^\perp(\omega') J_l^2\left(\frac{\omega'\hat{z}}{c} - \chi\right) \qquad (105)$$

where $\omega' = p\omega_0 + \omega_\beta + l\omega_s$.

For $\chi \ll 1$, the above expression is simplified to

$$\frac{1}{\tau^{(l)}} = -\frac{Ne^2 cW_1}{ET_0^2 \omega_\beta} \chi \frac{2}{\pi^2 (4l^2 - 1)} \qquad (106)$$

where $W_l < 0$ is a constant wake function. Then, the stability conditions of the head-tail effect are

$$\frac{\xi}{\eta} > 0 \qquad for \ l = 0$$

$$\frac{\xi}{\eta} < 0 \qquad for \ l \geq 1 \qquad (107)$$

It is obvious that the beam is stable only as $\xi = 0$. But in fact, the $l=0$ mode of the head-tail instability is much stronger than the other modes due to various damping effects. In the general case, all of the other higher modes of the head-tail instability effect to the beam are much weaker in a circular accelerator. In high energy electron storage rings, as $\eta > 0$, in normal operation, $\xi > 0$ is often set to control the lowest mode head-tail instability to keep the beam stable.

The tune shift and growth rate of the $l=0$ mode of the head-tail instability can then be written as

$$\Delta \nu = \frac{eR^2 I |Z_\perp|_0}{4\sqrt{2} k_b \nu E \sigma_s} \qquad and \qquad \frac{1}{\tau} = -\frac{ec^2 I |Z_\perp|_0 \xi}{4\sqrt{2} k_b \alpha E \omega_r \sigma_s}. \qquad (108)$$

In BEPC, if $\xi < 0$, the head-tail instability could occur obviously even at $I < 0.1$ mA.

6.3. Multi-bunch instability

In the linearization of Vlasov's equation, we can derive the distribution function $\psi(r,\theta)$ analytically corresponding to the relationship between the electro-magnetic fields and the impedances. We can then obtain the frequency shift expression by integrating the equation over (r,θ)

$$\Delta \omega_m = \frac{j}{1+m} \frac{\beta e}{\gamma m_0} \frac{I}{2Q\omega_0 L} \frac{\sum Z_\perp(\omega) h_m(\omega - \omega_\xi)}{h_m(\omega - \omega_\xi)} \qquad (109)$$

where the transverse impedance is

$$Z_\perp(\omega) = \frac{2R}{b^2} \frac{Z''(\omega)}{n} \quad (\Omega/m) \tag{110}$$

In the uniform multiple bunch case, i.e., the same space is kept between each bunch and there is the same number of particles in every bunch, we can evaluate the frequency shift also by the above formula, but the sum of all frequency shifts is required to compare with the criterion.

The phase relation among the bunches is $\Delta\phi = \dfrac{2\pi}{M} n$, where M is the bunch number, n is the number of coupling oscillation between the multiple bunches, and

$$\omega_n = pM\omega_0 + n\omega_0 + \begin{cases} m\omega_s \\ m\omega_s + \omega_\beta \end{cases} \tag{111}$$

6.4. *Computer code to estimate beam instabilities*

There are many computer codes such as BBI and ZAP that are used to calculate the bunched beam instabilities in high energy accelerators according to the analysis. The impedance and the beam current limitation due to the instability are estimated in these programs. The following contents are involved in the program: single bunch head-tail instability, transverse and longitudinal instabilities (single bunch and multi bunches), bunch lengthening effect and energy widening, mode coupling, i.e., strong head-tail instability, tune shift, Landau damping, and radiation damping.

The coherent oscillation modes can be determined according to the computation result from which we may find those dominant impedance parts which are the strongest causes of the beam instability. It is helpful for us to find practical ways to improve the beam stability and to increase the beam current.

7. Landau damping

We did not consider any damping effect in the analysis of the beam instability, neither in the macroparticle analysis nor in the model analysis from Vlasov 's equation. But in the practical operation of an accelerator, the particle motion is always influenced by damping effects that can be used to stabilise the beam. The damping effects may include radiation damping, non-linear damping, Landau damping, and the feed back effect etc. Except these damping effects estimated for real machine, we discuss the process of Landau damping in the following.

As we have seen, there are many different kinds of beam instabilities, but in most cases of an accelerator operation, the beam is quite stable. One of the important reasons is the Landau damping effect. The key feature of Landau damping is the frequency spread in the beam which may arise from the dependence of the betatron oscillation frequency on the energy, i.e., chromaticity, from any non-linear effect such as the frequency dependence on the oscillation amplitude in the longitudinal oscillation, and from some other sources.

We now discuss the physical process of Landau damping. Consider a simple harmonic oscillator which has a natural frequency ω, let this oscillator be driven by a sinusoidal force of frequency Ω. The equation of motion is $\ddot{x} + \omega^2 x = A \cos \Omega t$ with the initial conditions $x(0) = \dot{x}(0) = 0$. The solution of the equation is

$$x(t > 0) = -\frac{A}{\Omega^2 - \omega^2}\left(\cos \Omega t - \cos \omega t\right) \tag{112}$$

In the above equation, the $\cos\Omega t$ corresponds to the driving force, and the $\cos\omega t$ term is matched to the initial conditions.

The solution can also be written as the following form, if the initial conditions are not explicit involved

$$x(t) = -\frac{A}{\Omega^2 - \omega^2}\cos \Omega t \quad or \quad x(t) = -\frac{A}{\Omega^2 - \omega^2}e^{-i\Omega t}. \tag{113}$$

The solution contains a singularity at $\omega = \Omega$. We now consider an ensemble of oscillators, each oscillator represents a single particle in the beam, and the ensemble represents the beam. We assume the oscillators do not interact with each other and have a spectrum of the natural frequency ω with a distribution $\rho(\omega)$ satisfying

$$\int_{-\infty}^{\infty} d\omega \rho(\omega) = 1 \tag{114}$$

Starting at time $t = 0$, let this ensemble of particles be affected by the driving force $A\cos\Omega t$ with all particles having initial conditions $x(0) = \dot{x}(0) = 0$. We are interested in the ensemble average of the response which is given by

$$\langle x \rangle (t > 0) = -\int_{-\infty}^{\infty} d\omega \rho(\omega) \frac{A}{\Omega^2 - \omega^2} (\cos \Omega t - \cos \omega t) \qquad (115)$$

For simplicity, let us consider a narrow beam spectrum around a frequency ω_x, which close to the driving frequency Ω, i.e., $\Omega \approx \omega_x$. The response is then written as

$$\langle x \rangle (t) = -\frac{A}{2\omega_x} \int d\omega \rho(\omega) \frac{1}{\Omega - \omega} (\cos \Omega t - \cos \omega t) \qquad (116)$$

Changing variables from ω to $u = \omega - \Omega$, leads to

$$\langle x \rangle (t) = \frac{A}{2\omega_x} \left[\cos \Omega t \int_{-\infty}^{\infty} du \rho(u + \Omega) \frac{1 - \cos ut}{u} \right.$$
$$\left. + \sin \Omega t \int_{-\infty}^{\infty} du \rho(u + \Omega) \frac{\sin ut}{u} \right]. \qquad (117)$$

As the coefficients of the $\cos\Omega t$ and $\sin\Omega t$ terms are time dependent, we may show that they approach well-behaved limits, namely

$$\lim_{t \to \infty} \frac{\sin ut}{u} = \pi \delta(u) \qquad (118)$$

$$\lim_{t \to \infty} \frac{1 - \cos ut}{u} = P.V.(\frac{1}{u}) \qquad (119)$$

As time increases, we note that the peak value increases and the width decreases. If we are not interested in the transient effects immediately following the onset of the driving force, we can obtain the average behavior of the ensemble of oscillators as

$$\langle x \rangle(t) = \frac{A}{2\omega_x}\left[\cos\Omega t \; P.V. \int d\omega \frac{\rho(\omega)}{\omega - \Omega} + \pi\rho(\Omega)\sin\Omega t\right] \quad (120)$$

However, the amplitude of the ensemble motion does not increase with time. The oscillator system absorbs energy from the driving force, but the absorbed energy is not distributed more or less uniformly to each oscillator. Their frequencies are incoherent, i.e., the particles interact with the driving force individually on their own frequency. Figure 7 shows the single particle responses for different particle frequencies [1].

From the figure, we see that a particle with ω=Ω, being resonantly driven, increases in amplitude as t increases continuously. Any other particle with ω far away from Ω gets out of resonance after a time of $\pi/(\omega - \Omega)$ approximately. At time $2\pi/(\omega - \Omega)$, this particle returns to the initial position, giving back all its energy to the driving force in a beating process.

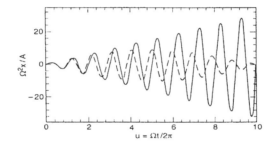

Figure 7. Response of the resonant and a non-resonant particle to the driving force.

The reason that the total amplitudes of the ensemble of oscillators are not time dependent is that the particles with frequencies near the driving frequency Ω continuously absorb energy from the driving force.

The region of time applicable for Landau damping is

$$\frac{1}{\delta\omega} \gg t \gg \frac{1}{\Delta\omega} \quad (121)$$

where $\Delta\omega$ is the frequency spread in the oscillators or the frequency spread of the beam spectrum, $\delta\omega$ is the frequency spacing between two nearest particles. One may have $\delta\omega \cong \Delta\omega/N$ with N particles in the beam. For example, taking $N=10^{11}$ and $\Delta\omega=10^3$ s^{-1}, the time is limited to the range between 1 ms and 10^8 s.

Applying above analysis to the collective beam instabilities in circular accelerators, the Landau damping effect on the instability can be issued.

Consider a one-particle model in which the bunched beam is a single macroparticle of charge Ne, but the N individual particles have different frequencies, i.e., they have a spread $\Delta\omega$ in their natural frequencies. The driving force on the individual particles comes from the wake field caused by center-of-charge displacement of the beam, and then the equation of motion can be written as

$$y''(s) + \left(\frac{\omega}{c}\right)^2 y(s) = -\frac{Nr_0}{\gamma C} \sum_{k=1}^{\infty} \langle y \rangle (s - kC) W_1 (-kC) \qquad (122)$$

where $\langle y \rangle (s) = B e^{-\Omega s/c}$.

Applying the result of the Landau damping analysis and the impedance, we have

$$\langle y \rangle = i \frac{BNr_0c}{2\omega_\beta \gamma T_0^2} \left[\sum_{p=-\infty}^{\infty} Z_1^\perp (p\omega_0 + \omega_\beta) \right] e^{-i\Omega s/c}$$

$$\times \left[\text{P.V.} \int d\omega \frac{\rho(\omega)}{\omega - \Omega} + i\pi\rho(\Omega) \right]. \qquad (123)$$

The mode frequency Ω is not arbitrary in view of the driving force in this solution. For some special spectrum example, we can write

$$\Omega = \omega_\beta + \xi_1 - i\Delta\omega \qquad (124)$$

where ξ_1 is related to the impedance. The stability condition for the beam (or for a set of oscillators) therefore becomes $\text{Im}\,\xi_1 < \Delta\omega$.

For the practical accelerator operation, we may only know the approximate information on the frequency spectrum, and our aim may be just a rough estimate whether the collective instability is Landau damped. For those purposes, without details of the spectrum, we introduce a simplified stability criterion as

$$\Delta\omega_{1/2} > \frac{\sqrt{3}Nr_0c}{2\omega_\beta\gamma T_0^2}\left|\sum_{p=-\infty}^{\infty}Z_1^\perp\left(p\omega_0 + \omega_\beta\right)\right| \tag{125}$$

where $\Delta\omega_{1/2}$ is the half width at half spread of the frequency spectrum.

The larger the frequency spread, the stronger the Landau damping is. The damping time due to the Landau damping effect can be estimated as

$$\tau^{-1} = \left|\Delta\omega_{1/2}\right|. \tag{126}$$

Acknowledgments

Special thanks to Dr. Alex Chao for his continuous help to study this subject in the study group that was organized in IHEP and involved many students from Tsinghua University and the University of Science and Technology of China, and for his encouragement to give a lecture in this school.

The contents of this lecture mainly refer to the textbook "Physics of Collective Beam Instabilities in High Energy Accelerators" by Alex Chao and are combined with observation studies in BEPC for years. Many more references can also be found in the textbook. The author wants to thank the colleagues of the BEPC team for their cooperation in the studies.

References

1. A. Chao, *Physics of Collective Beam Instabilities in High Energy Accelerators, John Wiley & Sons, Inc.* (1993).
2. Edited by M. H. Blewett, *Theoretical Aspects of the Behaviour of Beams in Accelerators and Storage Rings, CERN Report* 77-13 (1977).
3. Z. Y. Guo et al, *Beam Instability Studies in the BEPC, EPAC2000 Proceedings*, Vienna (2000).

DESIGNING SUPERCONDUCTING CAVITIES FOR ACCELERATORS

HASAN PADAMSEE
Cornell University
Ithaca, NY 14853
E-mail: hsp3@cornell.edu

Rapid advances in the performance of superconducting cavities have made RF superconductivity a key technology for accelerators that fulfill a variety of physics needs: high energy particle physics, nuclear physics, neutron spallation sources, and free electron lasers. New applications are forthcoming for frontier high energy physics accelerators, radioactive beams for nuclear astrophysics, next generation light sources, intense proton accelerators for neutron, neutrino and muon sources.

1. Introduction

The goal of this paper is to discuss design choices for superconducting cavities for various accelerator applications. Two classes of considerations govern structure design. The particular accelerator application forms one class, and superconducting surface properties the other. Designing a superconducting cavity is a strong interplay between these two classes. Typical accelerator driving aspects are the desired voltage, the duty factor of accelerator operation, beam current or beam power. Other properties of the beam, such as the bunch length, also play a role in cavity design. Typical superconducting properties are the microwave surface resistance and the tolerable surface electric and magnetic fields. These properties, which are also discussed in references [1,2,3], set the operating field levels and the power requirements, both RF power as well as AC operating power, together with the operating temperature.

Accordingly, the plan of this article is to briefly describe a number of distinct accelerator applications and the structures which emerged. To understand the evolution we first discuss the key electromagnetic properties of accelerating structures leading to an analysis of the power requirements of superconducting accelerators. The behavior of superconducting surfaces exposed to high surface electric and magnetic fields provide a guide to tailoring the cavity shape to achieve desirable values of key properties for accelerator performance. Both accelerator and surface issues govern the choice of the cavity shape, beam aperture, number of cell per structure as well as the choice of the RF frequency. Mechanical properties also play a role in the design aspects. Finally, input and output power coupling issues interact with cavity design, but these are covered in reference [4].

The discussion here is an overall summary and review of design aspects. We refer the reader to the reference text [5] and review article [6] for a more thorough discussion of many of the design topics and their intimate relationships to the physics of RF superconductivity.

Figure 1 shows a variety of superconducting accelerating cavities, ranging in frequency from 200 MHz to 3000 MHz and ranging in number of cells from one to nine. Most are cavities fabricated from pure sheet niobium and some, especially at frequencies below 500 MHz, are made of copper sputtered with a micron thin layer of niobium. Cavity fabrication issues are discussed in reference [3]. All the cavities of Figure 1 are intended for accelerating particles moving at nearly the velocity of light, i.e. $v/c = \beta \approx 1$. Accordingly, the period of a long structure (or the accelerating gap) is $\lambda/2$, where λ is the RF wavelength. Particles moving at $v \approx c$ will cross the gap in exactly half an RF period to receive maximum acceleration.

Figure 1. A spectrum of superconducting cavities.

2. Accelerator Requirements and Example Systems

Superconducting cavities have found successful application in a variety of accelerators spanning a wide range of accelerator requirements. High current storage rings for synchrotron light sources or for high luminosity, high energy physics with energies of a few GeV call for acceleration voltages of less than 10 MV, and carry high CW beam currents up to one amp. Figure 2 [7] shows the accelerating structure based on a 500 MHz, single cell cavity that evolved for the Cornell storage ring CESR/CHESS. The cavity was fabricated from pure sheet

338

niobium. Four such systems provide the needed voltage of 7 MV and beam power of more than one MW. Similar systems are under construction to upgrade the beam current of the existing Taiwan Light Source (SRRC), and for the new Canadian Light Source (CLS). The accelerating gradient choice for all these cases is 7 MV/m or less.

Figure 2. 3D-CAD drawing of the CESR superconducting cavity cryomodule (top). 500 MHz Nb cavity (bottom).

Near the energy frontier, LEP-II at CERN called for an accelerating voltage for nearly 3 GV to upgrade the beam energy from 50 to 100 GeV per beam, with a beam current of a few mA. With a frequency choice of 350 MHz, dominated by higher order mode (HOM) power loss and beam stability considerations, a 4-cell structure emerged [8]. To build 300 such units there was considerable savings in material cost by fabricating the cavity out of copper and coating it with niobium by sputtering. The LEP-II cavities (Figure 3) operated successfully at an average gradient of 6 MV/m.

A one GeV CW linac forms the basis of CEBAF, a 5-pass recirculating accelerator providing 5-6 GeV CW beam for nuclear physics [9]. The total circulating beam current is a few mA. Developed at Cornell, the 5-cell, 1500 MHz cavities (Figure 4) are also fabricated from solid sheet niobium. CEBAF cavities operate at an average accelerating field of 6 MV/m.

Figure 3. A 4-cell, 350 MHz Nb-Cu cavity for LEP-II

Figure 4. A pair of 5-cell Nb cavities developed at Cornell for CEBAF.

All the above accelerators run CW at 100% duty factor. The first pulsed superconducting linac will be for the Spallation Neutron Source (SNS) at Oak Ridge. 6-cell niobium cavities at 804 MHz will accelerate a high intensity ($\approx$10 mA) proton beam from 200 MeV to 1000 MeV. Figure 5 shows the medium β =0.64 cavity that resembles a $\beta = 1$ cavity that is squashed [10]. The duty factor for SNS is 6% and the RF pulse length is one ms. With recent improvement in cavity gradients the anticipated gradient is near 15 MV/m. Besides spallation neutron sources SNS technology could become suitable for high intensity proton linacs for various applications, such as transmutation of nuclear waste or generation of intense muon beams.

The dream machine for the future will be a 500 GeV energy frontier linac colliding electrons and positrons, upgradable to one TeV. As we will see, refrigerator power considerations drive the duty factor of operation to one percent. The average beam current is about 10 μA. A 9-cell niobium cavity design (Figure 6) has emerged from the TESLA collaboration [11]. With gradients improving steadily over the last decade, the choice of 25 MV/m will lead to 20 km of cavities for the 500 GeV machine. TESLA technology is likely to become the basis for the free electron lasers providing high brightness beams with wavelengths from the infra-red to ultraviolet and ultimately X-rays.

For the far future, acceleration of muons will also benefit from superconducting cavities [12]. A neutrino factory providing an intense neutrino beam from decaying muons may be the first step towards a muon collider that will penetrate the multi-TeV energy scale. At low energies (< a few GeV), where the muons have a large energy spread, the RF frequency has to be very low, e.g 200 MHz, leading to gigantic structures. Once again, economics will favor thin film Nb-Cu cavities over sheet Nb cavities. For comparison, a single cell Nb-Cu cavity at 200 MHz (Figure 1) dominates the size of superconducting cavities for the variety of accelerator applications discussed.

Figure 5. β = 0.6, 6-cell cavity for SNS, frequency 804 MHz

Figure 6. 1300 MHz 9-cell cavity for TESLA

3. Basics of Accelerating Structures

3.1. Accelerating field

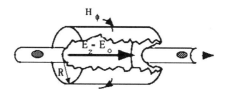

Figure 7. Pill-box resonator

Only simple structures can be calculated analytically, such as a cylinder with no beam holes (Figure 7), referred to as the "pill-box cavity." For our purposes, the analytic calculations of a simple cylindrical cavity are convenient to define the important performance parameters of superconducting cavities. For a cylinder of length d and radius R using cylindrical co-ordinates (ρ, ϕ, z), the electric (E_z) and magnetic (H_ϕ) fields for the TM_{010} mode are given by

$$E_z = E_0\, J_0\left(\frac{2.405\rho}{R}\right) e^{-i\omega t}, \quad H_\phi = -i\sqrt{\frac{\varepsilon_0}{\mu_0}}\, E_0\, J_1\left(\frac{2.405\rho}{R}\right) e^{-i\omega t} \tag{3.1}$$

where all other field components are zero. J_0 and J_1 are Bessel functions. The angular resonant frequency is given by

$$\omega_{010} = \frac{2.405\, c}{R}$$

which is independent of the cavity length.

First we determine the accelerating field, E_{acc}. Assume an electron travelling nearly at the speed of light (c). It enters the cavity at time t = 0 and leaves at a time t = d/c. To receive the maximum kick from the cavity, the time it takes the particle to traverse the cavity is to equal 1/2 an RF period (T_{RF}), i.e.

$$t = \frac{d}{c} = \frac{1}{2}\, T_{RF} = \frac{\pi}{\omega}$$

In this case, the electron always sees a field pointing in the same direction. The accelerating voltage (V_{acc}) for a cavity is

$$V_{acc} = \left| \int_{z=0}^{z=d} E_{el}\, dz \right|$$

For an electron accelerator with energy >10 MeV, it is sufficiently accurate to use v = c, so that t(z) = z/c. Thus

$$V_{acc} = \left| \int_{z=0}^{z=d} E_z \left(\rho = 0, z \right) e^{i\omega\, z/c} dz \right|$$

$$V_{acc} = E_0 \left| \int_{z=0}^{z=d} e^{i\omega z/c} dz \right| = d\, E_0 \, \frac{\sin\left(\dfrac{\omega d}{2c}\right)}{\dfrac{\omega d}{2c}} = d\, E_0\, T$$

At 1.5 GHz RF frequency, $d = \lambda/2 = 10$ cm, V_{acc} simplifies to

$$V_{acc} = (2/\pi)\, d\, E_0$$

The average accelerating electric field (E_{acc}) that the electron sees during transit is given by

$$E_{acc} = \frac{V_{acc}}{d} = \frac{2E_0}{\pi}$$

3.2. Peak Fields

To maximize the accelerating field, it is important to minimize the ratios of the peak fields to the accelerating field by selecting a suitable cavity geometry. For the TM_{010} accelerating mode in a pill-box cavity

$$E_{pk} = E_0, \quad H_{pk} = \sqrt{\frac{\varepsilon_0}{\mu_0}}\, J_1(1.841)E_0 = \frac{E_0}{647\,\Omega}$$

Thus we obtain the following ratios:

$$\frac{E_{pk}}{E_{acc}} = \frac{\pi}{2} = 1.6, \quad \frac{H_{pk}}{E_{acc}} = 2430 \frac{A/m}{MV/m} = 30.5 \frac{oersted}{MV/m}$$

Figure 8 shows the electric and magnetic field profiles for a real cavity shape. The peak field ratios for a realistic structure are much larger than for the pill-box. For example, for the TESLA cavity, $E_{pk}/E_{acc} = 2.0$ and $H_{pk}/E_{acc} = 42$ Oe per MV/m.

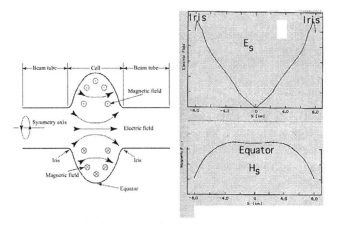

Figure 8. Electric and magnetic field profiles for a single cell cavity.

3.3. Power Losses and Q_0

In order to support the fields in the cavity, currents flow within a thin surface layer of the cavity walls. If the surface resistance is R_s, the power dissipated per unit area (P_a) due to Joule heating is

$$P_a = \frac{1}{2} R_s H^2$$

The two most salient characteristics of an accelerating cavity are its average accelerating field, E_{acc}, and the intrinsic quality factor Q_0. We just discussed E_{acc}. The quality (Q_0) is related to the power dissipation by the definition

$$Q_0 = \omega \frac{Energy\, stored}{power\, dissipated} = \frac{\omega U}{P_c}$$

where U is the stored energy and P_c is the dissipated power. Since the energy stored in the electric field is equal to that stored in the magnetic field, the total energy in the cavity is given by the power dissipated as,

$$U = \frac{1}{2} \mu_0 \int_v |H|^2 dv \qquad P_c = \frac{1}{2} R_s \oint_s |H|^2 ds$$

where the first integral is taken over the volume of the cavity, and the second over the surface. Thus

$$Q = \frac{\omega \mu_0 \int_v |H|^2 \, dv}{R_s \oint_s |H|^2 \, ds} \qquad Q = \frac{G}{R_s} \qquad G = \frac{\omega \mu_0 \int_v |H|^2 \, dV}{\oint_s |H|^2 \, ds}$$

Here G is called the "geometry factor" for the cavity shape. For a pill box in the TM_{010} mode, G = 257 Ω. Scaling arguments show that the ratio of the integrals $\int_v |\mathbf{H}|^2 \, dV / \int_s |\mathbf{H}|^2 \, ds$ must scale linearly with a, or alternatively, inversely with the mode frequency. Therefore the geometry constant, G, only depends on the cavity shape and not its size. A typical observed surface resistance for a well prepared superconducting Nb cavity is R_s = 20 nΩ. Thus we have a Q_0 value of

$$Q = \frac{G}{R_s} = 1.3 \times 10^{10}$$

For a typical cavity length of d = 10 cm and RF frequency of 1.5 GHz, the cavity radius is R = 7.65 cm. For an accelerating voltage of 1 MV, the following values result for the important features of a superconducting cavity

$$E_{acc} = \frac{V_{acc}}{d} = 10 \text{ MV/m}$$

$$E_{pk} = E_0 = \frac{\pi}{2} E_{acc} = 15.7 \text{ MV/m}$$

$$H_{pk} = 2430 \frac{A/m}{MV/m} E_{acc} = 24.3 \text{ kA/ m} = 305 \text{ Oersted}$$

$$U = \frac{\pi \, \varepsilon_0 \, E_0^2}{2} J_1^2 (2.405) \, d \, R^2 = 0.54 \text{ J}$$

$$P_c = \frac{\omega \, U}{Q} = 0.4 \text{ W}$$

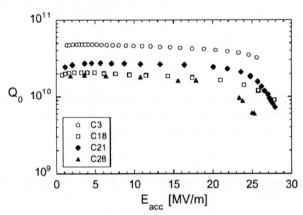

Figure 9. Q_0 vs E curves for 9-cell, 1.3 GHz cavities. When X-rays are also present, the Q drop at high fields is attributed to field emission [11].

The performance of a superconducting cavity is evaluated by measuring the Q_0 as a function of the cavity field level. These curves bear tell-tale signs of the activities inside the cavity. Figure 9 shows Q_0 vs E curves for some high performance TESLA multi-cell structures.

3.4. Shunt Impedance

Two important quantities for cavity design are the shunt impedance R_a and the geometric shunt impedance R_a/Q. R_a is defined in analogy to Ohm's law from

$$R_a = \frac{V_{acc}^2}{P_c}$$

the losses in a cavity at a given accelerating voltage.

Ideally, we want the shunt impedance to be large for the accelerating mode so that the dissipated power is small. This is particularly important for copper cavities, where the wall power dissipation is a major issue and we wish to have as large an accelerating field as possible. For the TM$_{010}$ mode we have

$$R_a = \frac{4\,\mu_0\,d^2}{\pi^3\,R_s\,\varepsilon_0\,J_1^2\,(2.405)\,R[R+d]} = 2.5 \times 10^{12}\,\Omega$$

From the definition of Q the ratio R_a/Q turns out to be

$$\frac{R_a}{Q} = \frac{V_{acc}^2}{\omega\,U}$$

The geometric shunt impedance is independent of the surface resistance. The geometric shunt impedance is crucial for determining the beam-cavity interaction in the fundamental and HOMs. Since the ratio V_{acc}^2/U scales inversely with the cavity's linear dimensions, R_a/Q is independent of cavity frequency, and only depends on the cavity geometry

$$\frac{R_a}{Q} = 150\frac{d}{R} = 196\Omega$$

Beam holes reduce the shunt impedance and enhance the peak surface fields relative to the pill-box case so that for a realistic cavity shape, R_a/Q drops by a factor of 2. A typical number for the R/Q of an sc cavity cell is 100 Ω.

Finally we define the shunt impedance per unit length, r/Q, which gives the power dissipated per unit length at a given accelerating field E_{acc}. r/Q increases linearly with frequency.

$$r/Q = \frac{R/Q}{L} \qquad\qquad \frac{P}{L} = \frac{E_{acc}^2}{(r/Q)\cdot Q_0}$$

For real structures with contoured shapes, beam apertures and beam pipes, it is necessary to use field computation codes, such as MAFIA and Microwave

Studio. Figure 10 shows the electric and magnetic fields computed by Microwave Studio for the accelerating mode of a pillbox cavity with a beam hole, and for a round wall cavity. Such codes are also necessary for computing the fields in the higher order modes of a cavity that can have an adverse effect on beam quality or cause instabilities. Figure 11 shows the electric and magnetic fields of the first monopole HOM. Beam induced voltages are also proportional to the R/Q of HOMs.

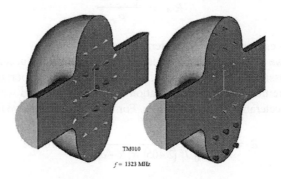

Figure 10. Electric (left) and magnetic (right) fields for a round cavity with beam holes.

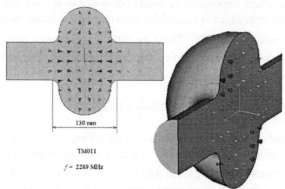

Figure 11. Electric (left) and magnetic (right) fields for the first monopole HOM.

3.5. Multicell Cavities

A multicell cavity is a structure with several cells coupled together. As with any set of coupled oscillators there are multiple modes of excitation for the full structure for every given mode of a single cell. The frequencies (f) of these modes can be cast via Equation 1 in terms of the single cell resonant freqency

(f_0) and a cell-to cell coupling strength (k) via a coupled LC oscillator model (Figure 12). A measurement of the highest and lowest frequencies directly yields the value of the cell-to-cell coupling (Equation 2). Out of the N modes in the TM_{010} pass band of an N-cell structure, the accelerating mode is the one where the fields in the neighboring cells are π radians out of phase with each other so that each cell provides the same acceleration kick to a velocity-of-light particle that crosses each gap in one-half RF period (Figure 12). Having equal fields in

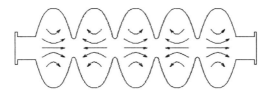

: Sketch of the electric field lines of the π-mode of a 5-cell :

$$\omega_0^2 = 1/LC, \; k = C/C_k, \; \gamma = C/C_b$$

Figure 12. Circuit model for a 5-cell cavity

each cell maximizes the overall accelerating voltage and minimizes the peak fields in each cell. A flat field profile is only achieved when the cells are properly tuned relative to each other. As Equations (3.5.1) and (3.5.2) show, a large number (N) of cells or a small cell-to-cell coupling (k) reduces the spacing between the accelerating mode and its nearest neighbor, making tuning more difficult, and making the field profile more sensitive to any cell-to-cell frequency differences that arise from manufacturing tolerances. Therefore, in an application that demands high total voltage (e.g TESLA, $\approx$ TeV), a high number of cells is desirable, and it is important to increase the aperture and decrease the cell-to-cell distance to increase the cell-to-cell coupling coefficient.

$$\left(\frac{f_m}{f_0}\right)^2 = 1 + 2k\left[1 - \cos\left(\frac{m\pi}{N}\right)\right] \tag{3.5.1}$$

If we measure $f^{(N)}$ and $f^{(1)}$, this becomes

$$k = \frac{\frac{1}{2}\left[\left(f^{(N)}\right)^2 - \left(f^{(1)}\right)^2\right]}{2\left(f^{(1)}\right)^2 - \left(f^{(N)}\right)^2\left[1 - \cos\left(\pi/N\right)\right]}. \tag{3.5.2}$$

3.6. *Traveling Wave versus Standing Wave*

Superconducting structures operate in the standing wave mode for which the power required to establish the fields is commensurate with wall losses. Hence, the RF power that must be supplied is usually comparable to the beam power. Since the peak fields to accelerating field ratios are high in the standing wave, it is worth examining how much gain in peak fields can be expected by operating in the traveling wave mode. Figure 13 shows the reduction in peak surface magnetic field for traveling wave mode operation in the $2\pi/3$, $\pi/4$ and $\pi/3$ modes. Although the 25% magnetic field reduction seems attractive at first glance, it comes with the price of smaller gap cells and thus more cells per meter. Consequently, the cost of such structures would increase substantially over the standing wave structure, resulting in little overall gain in the total capital cost of a traveling wave superconducting accelerator. Moreover, the traveling wave power required to establish the fields would need to be dumped at the end of the structure or recirculated. The first option is wasteful, defeating the main advantage of superconducting cavities, while the second option would increase structure complexity and cost.

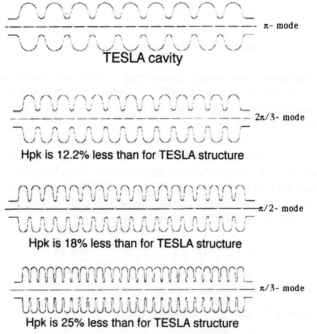

Figure 13. Structure geometry for travelling wave and accompanying decrease of H_{pk}.

4. RF Superconductivity Basics: Surface Resistance

As explained in the reference text [5], the remarkable properties of superconductivity are attributed to the condensation of charge carriers into Cooper pairs, which then move without friction; hence the zero resistance. At T = 0 K, all charge carriers condense. At higher temperatures, pairs break up. The fraction of unpaired carriers increases exponentially with temperature, as $e^{-\Delta/kT}$, until none of the carriers are paired above T_c – the normal conducting state. Here 2Δ is the energy gap of the superconductor, i.e., the energy needed to break up the pairs. In this simplified picture, known as the London two-fluid model, when a DC field is turned on, the pairs carry all the current, shielding the applied field from the normal electrons. Electrical resistance vanishes since Cooper pairs move without friction.

In the case of RF currents, however, dissipation does occur for all T > 0 K, albeit very small compared to the normal conducting state. While the Cooper pairs move without friction, they do have inertial mass. For high frequency currents to flow, forces must be applied to bring about alternating directions of flow. Hence an AC electric field will be present in the skin layer, and it will continually accelerate and decelerate the normal carriers, leading to dissipation, proportional to the square of the RF frequency. A simplified form of the temperature dependence of Nb for $T_c/T > 2$, and for frequencies much smaller than $2\Delta/h \approx 10^{12}$ Hz is

$$R_s = A(1/T)\ f^2 exp(-\Delta(T)/kT) + R_0$$

Here, A is a constant that depends on material parameters, as discussed in reference [1]. Based on the very successful BCS theory, expressions for the superconducting surface impedance have been worked out in terms of material parameters. The operating temperature of a superconducting cavity is usually chosen so that the temperature dependent part of the surface resistance is reduced to an economically tolerable value. R_0, referred to as the residual resistance, is influenced by several factors such as the ambient DC magnetic field environment of the cavity or the overall hydrogen gas content of the Nb material.

5. Example 1, Power Considerations for Storage rings

For a few GeV electron storage ring as in a light source or a B-factory, consider a 500 MHz single cell cavity of the geometry in Figure 14 (left), with half wavelength gap about 0.3m, R/Q = 89 Ohm, and G = 270 Ohm. Start with a modest CW voltage of one MV from the cavity, at a gradient about 3 MV/m. For a copper cavity, the Q_0 is 45,000 from the surface resistance of copper ($R_n = \sqrt{\pi f \mu_0 \rho} = 6m\Omega$). The shunt impedance R_a = 4 MΩ, giving a dissipated power of 250 kW. This would result in overheating of the copper cell. Water-cooled copper cavities at this frequency can safely dissipate about 40 kW. To bring dissipation to tolerable level it is essential to either lower the gradient or to raise the Ra/Q. Figure 14 (right) shows a normal conducting cell shape with $R_a/Q = 265$ Ω, for which the dissipation drops to 80 kW/cell. For a typical klystron efficiency of 0.5, the AC wall power is 160 kW per cell. But with the

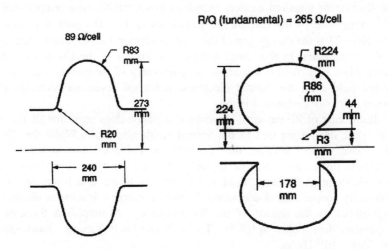

Figure 14   A comparison of typical shapes used for superconducting (left) and normal conducting (right) cavities.

small beam hole, HOMs cannot propagate down the beam pipe and many protrusions have to be added to the cell to remove HOMs which make harmful interactions with the beam (Figure 15). While the small beam hole helps to raise the R/Q of the fundamental mode as desired, it also raises the R/Q of the HOMs, increasing the danger of beam instabilities. Figure 16 compares the R/Q of HOMs for the large and small beam hole cases.

For a superconducting Nb cavity of the shape of Figure 14, the BCS surface resistance at 4.2 K leads to a Q of 2×10^9, R = 0.5 TΩ, and dissipated power in the wall of about 2 W. Taking into account refrigerator efficiency (at 4.2 K) of 1/350, the AC wall plug power is 0.7 kW. In this case the static heat leak and cryogen transfer lines will dominate the dynamic heat load. Even if these contributions total 45 W, the AC wall power due to the refrigerator will be remain a factor of 10 smaller than for the copper cavity. The power economy of superconductivity opens the possibility of raising the gradient to say 10 MV/m, and reducing the number of cells by a factor of 3. Together with the small HOM impedance of the large beam hole the reduced number of cells improves beam quality, avoids beam instabilities and allows higher currents.

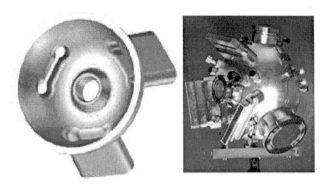

Figure 15. Inside view of a copper cell showing reentrant iris and apertures for HOM couplers (left). Full view of a copper cell as for PEP-II (SLAC) with HOM couplers [13] (right).

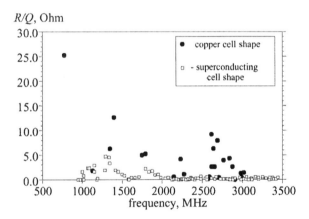

Figure 16. A comparison of the shunt impedances of a copper cell shape with a supercon-ducting cavity cell shape [7].

To complete the picture, electrons in a 5 GeV storage ring have a synchrotron radiation energy loss of about one MV per turn. For one amp beam current, the beam power that must be supplied through the cavities is one MW. The corresponding AC wall-plug power is 2 MW due to klystron efficiency. In CESR, the over-voltage factor to overcome quantum fluctuations from synchrotron radiation is about 7, so that the total voltage requirement is not 1 MV, but 7 MV. This can be met by four single cell units (Figure 1), each providing about 2 MV at a gradient of 7 MV/m. Each unit has a dynamic heat load of 50 W and a static heat plus transfer line heat load of 50 W. The total cryogenic heat load is 400 W, corresponding to an AC power demand of 140 kW. The copper cavity solution (as for PEP-II at SLAC) would call for 9 units (as in Figure 15), each dissipating 80 kW. Taking into account the klystron efficiency (0.5), the structure associated AC power is 1.44 MW, the same factor of 10 higher than the superconducting case, as discussed. But on adding the AC power associated with the beam power (1 MW) to both solutions, one obtains AC power equal to 3.44 MW for the normal-conducting case versus 2.14 MW for the superconducting case. The large beam power reduces the superconducting advantage from a factor of 10 to just 60%. The impedance reduction remains the dominant advantage.

6.   RF Superconductivity Basics: high field behavior

As mentioned, the accelerating field, E_{acc}, is proportional to the peak electric (E_{pk}) as well as magnetic field (H_{pk}) on the surface of the cavity. Therefore important fundamental aspects of superconducting cavities are the maximum surface fields that can be tolerated without increasing the microwave surface resistance substantially, or without causing a catastrophic breakdown of superconductivity. The ultimate limit to the accelerating field is the RF critical magnetic field above which the superconducting phase can no longer exist. The RF critical field is related to the thermodynamic critical field. In the process of a phase transition to the normal conducting state, a phase boundary must be nucleated. Because of the rapidly changing RF fields (ns time scale), it is possible for the Meissner state to persist above the thermodynamic critical field (H_c) for Type I superconductors, and above the lower critical field (H_{c1}) for Type II superconductors. Such a metastable situation can be expected up to a superheating critical field, $H_{sh} > H_c$ (Type I) $> H_{c1}$ (for type II). It is important to note that the RF critical field does *not* depend on H_{c2}. Therefore, high field magnet materials, such as Nb-Ti, do not offer correspondingly higher operating fields for superconducting cavities. Indeed for RF superconductivity, it is essential to always operate in the Meissner state. For the most popular superconductor, niobium, H_{sh} is about 0.23 T. These surface fields translate to a maximum accelerating field of 55 MV/m for a typical niobium. The exact values depend on the detailed structure geometry.

Typical cavity performance is significantly below the theoretically expected surface field limit. One important phenomenon that limits the achievable RF magnetic field is ``thermal breakdown" of superconductivity, originating at sub millimeter-size regions of high RF loss, called defects. When the temperature outside the defect exceeds the superconducting transition temperature, T_c, the losses increase substantially, as large regions become normal conducting (see Figure 17). Measures to overcome thermal breakdown are to improve the thermal conductivity of niobium by purification or to use thin films of niobium on a copper substrate cavity.

The Q_0 vs E_{acc} curve (Figure 9) only gives information on the average behavior of the RF surface. To resolve the local distribution of RF losses and identify various mechanisms, temperature mapping proves to be a powerful diagnostic technique. A chain of rotating carbon thermometers, or an array of fixed thermometers, samples the temperature of the outer wall of the cavity. The temperature map of Figure 17 shows a hot spot that leads to thermal breakdown, and the SEM micrograph reveals a 50 μm culprit copper particle.

In contrast to the magnetic field limit, we know of no theoretical limit to the tolerable surface electric field. Fields up to 220MV/m have been imposed on a superconducting niobium cavity without any catastrophic effects [14]. However, at high electric fields, an important limitation to the performance of superconducting cavities arises from the emission of electrons from local spots in the high electric field regions of the cavity. This is a problem endemic to all high voltage devices. Power is absorbed by the electrons and deposited as heat upon impact with the cavity walls. Copious amounts of X-rays are emitted due to bremsstrahlung. At high fields, the exponential drop in Q_0 with field suggests that field emission is the dominant limiting mechanism, provided X-rays are also observed (Figure 9).

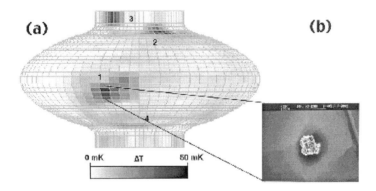

Figure 17. (a) Temperature map at 400 Oe of a 1.5 GHz, single cell cavity showing heating at a defect site, labelled #1 and field emission sites labelled #2, 3, and 4. (b) SEM micrograph of the RF surface taken at site #1 showing a copper particle [5].

Figure 18 shows electron trajectories in one RF period and the heating due to their impact on the cavity wall. A typical emitter is a microparticle contaminant. When emission grows intense at high electric fields it can even initiate thermal breakdown. In many cases, intense field emission eventually leads to momentary voltage breakdown of the vacuum in the cavity. This has mostly a beneficial effect for superconducting cavities, known as conditioning. After a voltage breakdown event, it is usually possible to raise the electric field until field emission grows intense once again at another spot on the cavity surface. We have learned much about the nature of field emission sites and made progress in techniques to avoid them as well as to destroy them by conditioning with high voltage breakdown [5].

In the early stages of the development of superconducting cavities, a major performance limitation was "multipacting". This is a resonant process in which an electron avalanche builds up within a small region of the cavity surface due to a confluence of several circumstances. With the invention of the round wall cavity shape, multipacting is no longer a significant problem for velocity-of-light structures. The essential idea to avoid multipacting is to gradually curve the outer wall of the cavity – hence the round wall profile.

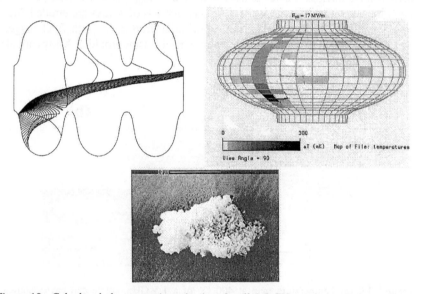

Figure 18. Calculated electron trajectories in a 3-cell 1.5 GHz cavity operating at E_{pk} = 50 MV/m. Here the emitter is located in an end cell, just below the iris, where the surface electric field is 44 MV/m. Note that a significant number of electrons emitted during the early part of the RF cycle bend back and strike the wall near the emitter (top left). Temperature map of the heating by impacting electrons (top right). A micron size foreign particle found at an emission site (bottom).

If the shape is not rounded, one-surface multipacting will severely limit the cavity performance (Figure 19). An electron emitted from one region of the surface (usually the outer cylindrical wall) travels in a cyclotron orbit in the RF magnetic field, and returns to near its point of origin. Upon impact it generates a secondary electron which mimics the trajectory of the primary. An exponential build up occurs if the round-trip travel time of each electron is an integer multiple of an RF period, i.e., the electron returns in the same phase of the RF period when it is generated. For the build up to persist, the secondary emission coefficient must be greater than one. This is true for a niobium surface when the electron energy is between 50 eV and 1000 eV [5]. During their excursion into the RF fields, the electrons must gain enough energy from the electric field to generate secondaries on impact. When these conditions are met, an electron avalanche occurs to absorb the RF power, making it impossible to raise the fields by increasing the incident RF power. The electrons impact the cavity walls, which leads to a large temperature rise, thermal breakdown, and in some cases a momentary gas discharge. When the cell shape is rounded (Figure 19) the electrons are forced to the equator region where the electric field is too low for the electrons to gain sufficient energy to regenerate. The avalanche is arrested.

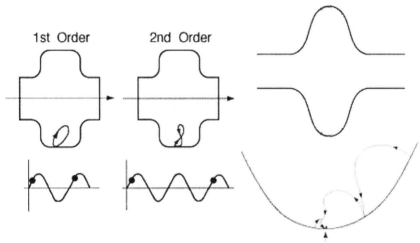

Figure 19. In a cavity with a nearly pill-box like shape, electrons can multiply in the region shown. The electron returns in an integer multiple of RF periods. The number of RF periods is called the order of multipacting (left). The start and end phases of the electron that returns in one and two RF periods. A travel time of one RF period results in first order multipacting (center). With a rounded cavity shape, the electrons drift to the zero field region at the equator. Here the electric field is so low that the secondary cannot gain enough energy to regenerate. Multipacting stops (right).

7. Choice of Cavity Shape

There are many factors which influence the cavity shape. Multipacting is a key factor that governs the overall rounded contour of the cavity profile. Beam dynamics considerations control the size of the aperture. To lower the peak electric field it is necessary to round the iris region with circular or elliptical arcs. Peak magnetic field considerations influence the shape of the cavity in the large diameter (equator) region where the magnetic field is strongest. Elliptical arc segments increase the strength of the cavity against atmospheric load (see mechanical considerations below) and also provide a slope for efficient rinsing of liquids during surface etching and cleaning.

Mechanical considerations also influence the cavity shape. To prevent the cavity from deforming excessively or even collapsing under atmospheric load it is important to avoid flat regions in the cavity profile (Figure 20). Elliptical

Figure 20. Flat (left) versus elliptical (center) cavity profile. Stiffening rings for the TESLA cavity (right).

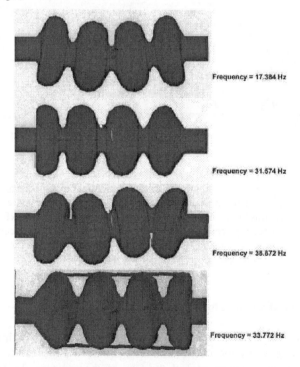

Figure 21. Mechanical resonant modes of a 4-cell, 200 MHz cavity with 8 mm wall thickness. The low resonant frequencies spell trouble in the form of microphonics. Reducing the number of cells or stiffening is essential.

segments tend to be strongest. For pulsed operation at high gradients, the Lorentz force may be strong enough to cause cell deformation demanding larger thickness walls or stiffening rings near the iris (see Figure 20). Typical detuning coefficients are in the range of a few $Hz/(MV/m)^2$. Additional stiffening of multicell structures may be necessary to raise the frequency of mechanical vibration modes. As an extreme example, Figure 21 shows exaggerated structure deformations for the lowest frequency vibration modes for a 200 MHz 4-cell cavity.

8. Superconductor Choice

For a material to be useful for accelerators, the primary requirements are a high transition temperature, T_c, and a high RF critical magnetic field, H_{sh}. Among the elemental superconductors, niobium has the highest T_c and the highest RF critical field. Accordingly, it is a most attractive choice for accelerator cavities. Successful cavities have been made from sheet Nb, or by sputtering Nb onto a copper cavity. The realm of superconducting compounds has been much less explored because of technical complexities that govern compound formation. In looking at compound candidates, it is important to select a material for which the desired compound phase is stable over a broad composition range so that formation of the compound is more tolerant to variations in experimental conditions, making it possible to achieve the desired single phase over a large surface area. Nb_3Sn is a promising material. The T_c is 18 K and the RF critical field is 0.4T, twice as high as for Nb. On fundamental grounds, the higher field opens up the possibility of accelerating gradients higher than allowed for niobium cavities. However, the performance for Nb_3Sn cavities to date is far lower than for niobium cavities. The new HTS are even further from the microwave performance level desired for application to accelerators.

A strong motivation for using thin films of Nb on copper cavities is to provide increased stability against thermal breakdown of superconductivity. The thermal conductivity of copper at 4.2 K is between 300 to 2000 W/m-K, depending on the purity and annealing conditions, as compared to the thermal conductivity of 300 RRR niobium, which is 75 W/m-K at 4.2 K. The cost saving of niobium material is another potential advantage, significant for large-size, e.g. 350 MHz cavities, as for LEP-II, or for future projects, which aspire to make 200 MHz cavities, where each cell is more than one meter in diameter.

9. Example Linear Colliders, Gradient and Power Issues

Consider a one TeV center-of-mass energy linear collider, the ambition of many international accelerator collaborations. Based on the latest progress in TESLA gradients (Figure 22), we can confidently select E_{acc} = 30 MV/m [15], resulting in a 33 km active length linac. TESLA cryomodules have achieved a filling factor of 0.75, so that the real estate length of superconducting linac would be 45 km. Because of a higher gradient choice (50 MV/m, loaded gradient), a normal conducting linear collider would be shorter, about 25 km. At a Q_0 of 10^{10} we determine a dynamic heat load of 90 W/m to yield a preposterous, total dynamic heat load of 3 MW at 2 K. The capital cost for such a titanic refrigerator would exceed 10 billion dollars, and the AC power to run it would exceed 3 GW, comparable to a nuclear power plant. Hence a superconducting TeV energy linac must run in the pulsed mode with a duty factor of about one per cent, cutting the dynamic load to just 1 W/m, and the total dynamic cryogenic load to 33 kW.

Or we must look toward improvements that could lead to Q values of 10^{11}. Such Q values have in fact been reached [16] in single cells at 1.6 K (see Figure 23), but remain to be demonstrated in full scale structures inside accelerator cryomodules.

Continuing with the one per cent duty factor scenario, other important heat loads are the static heat and a fraction of the HOM power deposited at low temperature. Assuming another one watt/m for these contributions, and taking into account the entire accelerator length of 44 km, the grand total cryogenic heat load is 33kW + 44 kW = 77 kW. At 2 K the refrigerator efficiency is 1/750, leading to an AC wall plug power load of 58 MW.

As in the case of the storage ring example above, we now include the beam power. For a superconducting linear collider, the typical value for the total beam power is 30 MW [11]. At a klystron efficiency of 70%, the RF installation then calls for an AC power of 43 MW. The grand total AC power becomes 100 MW. An important figure of merit, the efficiency of AC power to beam power conversion, is an attractive 30%. Typical efficiencies of normal-conducting linac options range around 10% due to the large RF power needed to fill copper structures to high gradients [17]. Beam powers are also kept low to manage HOMs. Hence a superconducting linac opens the route to high luminosity via high beam powers, rather than squeezing the spot size at the collision point to nano-meter dimensions as for normal conducting linacs.

The peak beam power for the superconducting linac is 30 MW divided by the duty factor, which comes out to 3 GW, or roughly 100 kW/m. Here is another advantage of the superconducting option. Because of the long filling time allowed by the low cavity wall losses, cavities can be filled slowly (ms) reducing the peak RF power requirement compared to a normal conducting linear collider. Here filling times must be of the order of ns, and peak power in the order of hundred MW per meter to achieve high gradients (70 MV/m). The peak total RF power falls in the multi-Terawatt regime.

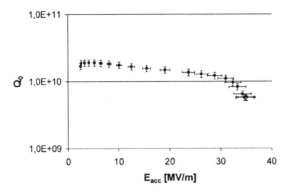

Figure 22. A TESLA cavity which reached $E_{acc} = 35$ MV/m after electropolishing [15].

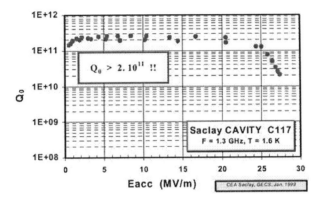

Figure 23. A single cell cavity which reached $E_{acc} = 25$ MV/m at a Q_0 greater than 10^{11} [16].

10. Conclusions

When designing a superconducting cavity, many trade-offs must be made between accelerator requirements and cavity performance issues. Reviewing the choice of RF frequency, a high frequency is better because structures are smaller, easier to handle and the per meter structure cost is lower. Small surface area means fewer defects that can cause thermal breakdown and fewer emitters that can cause field emission. Higher R/Q (ohm per meter) means reduced dynamic heat load to liquid helium for the same operating gradient and the same length of accelerator. Smaller volume, high frequency structures contain less stored energy at a given field and lead to reduced AC wall plug power. Finally, a topic we have not discussed here, higher frequency structures have less capture of field emitted electrons in the beam pipe, reducing the possibility of polluting the beam. Power economy considerations make it important to keep superconducting cavities clean enough that dark currents are very low.

A low frequency choice is better because the superconducting state (BCS) surface resistance (α f^2) is lower resulting in higher Q at fixed operating temperature. The R/Q (per meter) of HOMs are lower which is better for beam stability and HOM power loss. Since each cell is longer there are fewer cells per meter. For the same number of cells per structure the end associated costs are less.

Reviewing considerations that govern aperture choice, a large aperture is better because of larger beam clearance and greater cell-to-cell energy coupling which reduces field non-uniformity caused by errors in cell shape, allowing a larger number of cells per cavity unit. A large aperture also improves coupling of power from input coupler to cells in the presence of beam loading. For higher mode effects, a large aperture reduces wakefields, both long- and short-range, resulting in better beam quality and allowing higher beam current. On the other hand, a small aperture is better because it reduces E_{pk}/E_{acc} and H_{pk}/E_{ac}. The resulting higher R/Q in accelerating mode also means lower RF heat load into liquid helium.

Turning to number of cells, a small number of cells per structure unit makes it easier to obtain a flat field profile and leads to less power for each window/coupler to deliver. Again, a smaller cavity area means fewer defects that can lead to thermal breakdown and fewer emitters that cause field emission. There is less chance of trapped HOMs and it is easier to remove HOM from couplers on the beam pipe which intercept fields in the end cells only. On the other hand, a large number of cells is better because it minimizes system costs by reducing wasted space between cells and minimizes the effects of fringing field (which lower R/Q).

To recap the salient results of the storage ring and linear collider examples, superconducting cavities excel in applications requiring continuous wave (CW) or long-pulse accelerating fields above a few million volts per meter (MV/m). Since the ohmic power loss in the walls of a cavity increases as the square of the accelerating voltage, copper cavities become uneconomical when the demand for high CW voltage grows with particle energy. A similar situation prevails in applications that demand long RF pulse length, or high RF duty factor. Here superconductivity brings immense benefits. The surface resistance of a superconducting cavity is many orders of magnitude less than that of copper. After accounting for the refrigerator power needed to provide the liquid helium operating temperature, a large net gain factor remains to provide many advantages.

Copper cavities are limited to gradients well below a few MV/m in CW and long-pulse operation because the capital cost of the RF power and the AC-power operating cost become prohibitive. For example, several MW/m of RF power are required to operate a copper cavity at 5 MV/m. There are also practical limits to dissipating high power in the walls of a copper cavity. The surface temperature becomes excessive causing vacuum degradation, stresses and metal fatigue due to thermal expansion. On the other hand, copper cavities offer higher accelerating fields ($\approx$ 50 MV/m) for short pulse (μs) and low duty factor (< 0.1%) applications. For such applications, it is important to provide abundant peak RF power, e.g., 100 MW/m, in order to reach the high fields.

There is another important advantage that SRF cavities bring to accelerators. The presence of accelerating structures has a disruptive effect on the beam, limiting the quality of the beam in aspects such as energy spread, beam halo, or even the maximum current. Because of their capability to provide higher voltage, SRF systems can be shorter, and thereby impose less disruption. Due to their high ohmic losses, the geometry of copper cavities must be optimized to provide a high electric field on axis for a given wall dissipation. This requirement tends to push the beam aperture to small values, which disrupts beam quality. By virtue of low wall losses, it is affordable to design an SRF cavity to have a large beam hole, reduce beam disruption and provide high quality beams for physics research.

There has been much progress in understanding the gradient and Q_0 limitations in superconducting cavities. Through better understanding, new techniques have been developed to overcome the limitations. Producing high gradients and high Q_0 with Nb cavities demands excellent control of material properties and surface cleanliness. As a result of the improved understanding and the invention of new treatments, there has been much progress in reducing the spread in gradients that arises from the random occurrence of defects and emitters. Prescreening the starting material by eddy current scanning reduces the number of defects that can cause thermal breakdown. High RRR, high thermal conductivity Nb reduces the impact of any remaining defects. It will be important to aim for higher RRR in large area cavities, where there is a high chance of defects and contamination. High pressure rinsing greatly reduces the number of field emitters. High pulsed power processing destroys accidental field emitter contaminants. This technique will continue to be necessary in order to realize - in accelerators - the high intrinsic gradient potential of SRF cavities. There is now excellent prognosis for reaching 25 – 35 MV/m for future colliders. The road to 40 MV/m is opening up. Although most successful cavities are based on Nb, some exploratory work has been carried out on other materials.

Acknowledgments

The author is deeply indebted to Valery Shemelin for providing results presented here from Microwave Studio calculations and for peak fields in traveling wave structures.

References

1. P. Schmueser, *Proceedings of the CERN Accelerator School on Superconductivity and Cryogenics for Accelerators and Detectors* (2002), Erice, Italy, Ed. S.Russenschuck and G.Vandoni.

2. H. Safa, *Proceedings of the CERN Accelerator School on Superconductivity and Cryogenics for Accelerators and Detectors* (2002), Erice, Italy, Ed. S.Russenschuck and G.Vandoni.

3. D. Proch, *Proceedings of the CERN Accelerator School on Superconductivity and Cryogenics for Accelerators and Detectors* (2002), Erice, Italy, Ed. S.Russenschuck and G.Vandoni.

4. R. Parodi, *Proceedings of the CERN Accelerator School on Superconductivity and Cryogenics for Accelerators and Detectors*, (2002), Erice, Italy, Ed. S.Russenschuck and G.Vandoni.

5. Padamsee H., Knobloch J. and Hays T., *RF Superconductivity for Accelerators* (John Wiley and Sons, New York, 1998).

6. Proch D., Rep. Prog. Phys. 61, IOP Pub. (1998) p. 1.

7. Belomestnykh S., *Proceeding of the 1999 Particle Accelerator Conference* ed. by A. Luccio et al (1999) p. 272.

8. Brown P. et al., *Proceedings of the 9th Workshop on RF Superconductivity* (1999) ed. By B. Rusnak, paper MOA001.

9. Reece C. et al., *Proceedings of the 9th Workshop on RF Superconductivity* (1999) ed. By B. Rusnak, paper MOA004.

10. Ciovati et al, *Proceedings of the 10th Workshop on RF Superconductivity* (2001) ed. By S. Noguchi, paper PT016.

11. Trines D., *Proceedings of the 9th Workshop on RF Superconductivity* (1999) ed. By B. Rusnak, paper FRA 003.

12. Padamsee H, *Proceedings of the 10th Workshop on RF Superconductivity* (2001) ed. By S. Noguchi, paper FRA 003.

13. Rimmer R et al, *Proceeding of the 1997 Particle Accelerator Conference* ed. by M. Comyn et al (1997) p. 3004.

14. Delayen J. Shepard K. W.. Appl. Phys. Lett., **57**, p. 514 (1990).Tajima T., *Proceeding of the 1999 Particle Accelerator Conference* ed. A. Luccio et al (1999) p. 440.

15. Lijle L, DESY, private communication.

16. Safa, H, *Proceedings of the 10th Workshop on RF Superconductivity* (2001) ed. S. Noguchi, paper TL003.

17. Loew, International Linear Collider Technical Report (1995).

ACCELERATOR MAGNETS: DIPOLE, QUADRUPOLE AND SEXTUPOLE

C. S. HWANG

National Synchrotron Radiation Research Center (NSRRC),
101 Hsin-Ann Road, Hsinchu Scienced-Based Industry Park,
Hsinchu 30077, Taiwan
E-mail: cshwang@srrc.gov.tw

The main characteristics of the magnet design and performance of lattice magnets including dipole, quadrupole, and sextupole magnets, will be presented. The design and construction constraints of the conventional, steel iron yoke, and the direct-current magnets are also discussed. The equations of the ideal pole shapes of the dipole, quadrupole, and sextupole, are used to design the magnets. Several two- or three-dimensional calculation codes are introduced to simulate the magnets. Moreover, a brief description of the injection magnets is also presented. Finally, field measurement and mapping methods are introduced and the results from field measurements by means of a Hall probe are discussed.

1. Introduction

Synchrotron light sources have been built to include various types of magnets to store accelerated particles (electrons or positrons). The demands on magnet design technology have increased because of the need for high-quality bending and focusing elements to provide increasingly precise beams for third generation light sources. Thus, the skills required to design and fabricate conventional magnets are important standard techniques. Many important text books [1, 2, 3, 4] have discussed the design and construction of magnets. Accelerator magnets include superconducting magnets, pure permanent magnets, hybrid structure magnets, fast-pulsed magnets with ferromagnets, and specialized magnets. However, the scope of this article is limited to conventional accelerator magnet design and the basic accelerator type of electromagnet in the synchrotron radiation source. Their properties and roles are discussed. The main accelerator magnets are dipole magnets for electron bending, quadrupole magnets for focusing electrons, and sextupole magnets for controlling the electron beam's chromaticity. These magnets fall into the category of iron dominated dipoles, quadrupoles and sextupoles, which, in general, operate with fixed fields. The design of magnets must address both the physics of the components and the practical engineering constraints. Magnetic field calculation code is used to predict flux density distributions and iterative techniques are used to design a pole face and coils. The required parameters of an accelerator magnet are also very important. This article will also address gradient magnets and magnets for injection and extraction. Magnet measurements are also discussed in this article. Several standard text books [2, 4, 5, 6] consider the various methods of magnetic field measurement. Only the line integral field quality of multipole magnets (dipoles, quadurpoles and sextupoles) is characterized by using rotating coils and the point field distribution is mapped by using Hall probe measurement methods.

2. Magnetic features of various accelerator magnets

2.1 Lorentz force

$$F = eE + \frac{e}{c}(v \times B) \tag{2.1.1}$$

For relativistic particles with velocity $v \approx c$ the magnetic field causes the dominant force term since a magnetic field of 1 T provides the same bending force as an electric field of 300 million V/m. The features of magnets are:

* Magnets in storage ring
  Dipole   -   bending and radiation
  Quadrupole   -   focusing or defocusing
  Sextupole   -   chromaticity correction
  Corrector   -   small angle correction

2.2 Magnetic field characteristics of lattice magnets

In a common definition of the magnetic flux as

$$\Phi_M(x, y, z) = \sum_{m=0}^{\infty} \frac{(B_m(z) + i A_m(z))(x+iy)^{m+1}}{(m+1)} \tag{2.2.1}$$

two sets of orthogonal multipoles have amplitude constants B_m and A_m which represent the normal and skew multipole components. The field strength can be derived as,

$$\vec{B}(x, y, z) = -\vec{\nabla}\Phi(x, y, z) \tag{2.2.2}$$

Accordingly, the general magnetic field equations including only the most commonly used upright multipole components in Cartesian coordinates is given by

$$B_x = A_1 \cdot y + A_2 \cdot xy + \frac{A_3}{6} \cdot (3x^2 y - y^3) + \dots \tag{2.2.3}$$

$$B_y = B_0 + B_1 \cdot x + \frac{B_2}{2} \cdot (x^2 - y^2) + \frac{B_3}{6} \cdot (x^3 - 3xy^2) + \dots \tag{2.2.4}$$

where B_0, B_1 (A_1), B_2 (A_2), and B_3 (A_3) denote the harmonic field strength components and are called the normal (skew) dipole, quadrupole, sextupole and so on.

Furthermore, magnetic spherical harmonics are derived from Maxwell's equations [1]. Maxwell's equations for magnetostatics are:

$$\vec{\nabla} \cdot \vec{B} = 0 \, ; \qquad (2.2.5)$$

$$\vec{\nabla} \times \vec{H} = \vec{j} \, ; \qquad (2.2.6)$$

Given no current excitation of $j = 0$, the magnetic flux density B is found as

$$\vec{B} = -\vec{\nabla}\Phi_M \qquad (2.2.7)$$

Therefore, Laplace's equation can be obtained as,

$$\nabla^2\Phi_M = 0 \qquad (2.2.8)$$

where Φ_M is the magnetic scalar potential.

Consider the two-dimensional case (constant in the z direction), and solve in cylindrical coordinates (r,θ). In practical magnetic applications, the scalar potential becomes,

$$\Phi_M = \sum_n \left(a_n r^n \cos n\theta + b_n r^n \sin n\theta \right), \qquad (2.2.9)$$

with n integral and a_n, b_n functions of geometry, yielding components of magnetic flux density,

$$B_r = \sum_n \left(n a_n r^{n-1} \cos n\theta + n b_n r^{n-1} \sin n\theta \right) \qquad (2.2.10)$$

$$B_\theta = \sum_n \left(- n a_n r^{n-1} \sin n\theta + n b_n r^{n-1} \cos n\theta \right) \qquad (2.2.11)$$

2.3 Contour equation and magnetic flux density of the ideal pole shape of a dipole magnet

Figures 1 and 2 show the equipotential lines and the magnetic flux of a dipole magnet. The scalar potential and magnetic flux density in cylindrical and Cartesian coordinates are as follows [1,7,8].

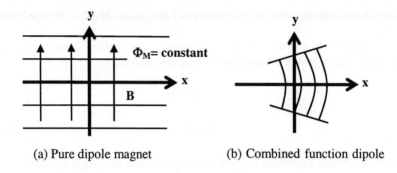

(a) Pure dipole magnet   (b) Combined function dipole

Figure1 Equipotential lines of a dipole magnet. (a) pure dipole magnet (b) combined function dipole magnet.

Cylindrical Coordinates	Cartesian Coordinates
$B_r = a_1 \cos\theta + b_1 \sin\theta$;	$B_x = a_1$ or $B_x = a_1 y$
$B_\theta = -a_1 \sin\theta + b_1 \cos\theta$;	$B_y = b_1$ or $B_y = b_1 x$
$\Phi_M = a_1 r \cos\theta + b_1 r \sin\theta$.	$\Phi_M = a_1 x + b_1 y$ or
	$\Phi_M = (a_1 + b_1) xy$

Therefore, $a_1 = 0$ and $b_1 \neq 0$ define a vertical dipole field, $b_1 = 0$ and $a_1 \neq 0$ a horizontal dipole field (which is rotated by $\dfrac{\pi}{2 \times 1}$).

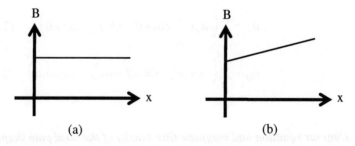

(a)   (b)

Figure 2 Magnetic flux density versus transverse coordinate x. (a) Pure (b) Combined function dipole magnets.

2.4 Contour equation and magnetic flux density of the ideal pole shape of a quadrupole magnet

Figures 3 and 4 show the equipotential lines and the magnetic flux of a quadrupole magnet. The scalar potential and magnetic flux density, described in cylindrical and Cartesian coordinates, are as follows [1,9,10].

Cylindrical Coordinates

$$B_r = 2a_2 r \cos 2\theta + 2b_2 r \sin 2\theta \ ;$$

$$B_\theta = -2a_2 r \sin 2\theta + 2b_2 r \cos 2\theta \ ;$$

$$\Phi_M = a_2 r^2 \cos 2\theta + b_2 r^2 \sin 2\theta \ ;$$

Cartesian Coordinates

$$B_x = 2(a_2 x + b_2 y)$$

$$B_y = 2(-a_2 y + b_2 x)$$

$$\Phi_M = a_2(x^2 - y^2) + 2b_2 xy$$

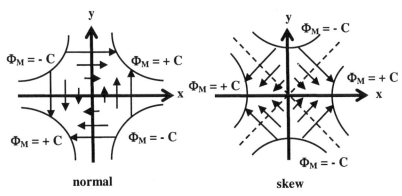

normal   skew

Figure 3 Equipotential lines of a quadrupole magnet (normal and skew).

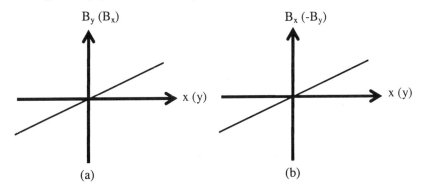

(a)   (b)

Fig.4 Magnetic flux density distribution versus transverse coordinate. (a) normal (b) skew quadrupole magnets.

These are quadrupole distributions, where $a_2 = 0$ gives a normal quadrupole

368

field. Then, $b_2 = 0$ yields a skew quadrupole field (pole rotated by $\dfrac{\pi}{2 \times 2}$). Figure 5 and the table present the multipole field error analysis.

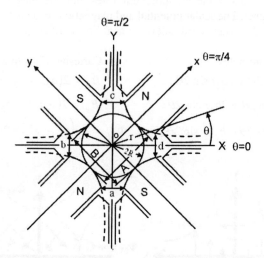

Figure 5 Practical quadrupole. The shape of the pole surfaces is not a true hyperbola and the poles are truncated laterally to leave space for coils.

(1) Sextupole (n=2)	1. a=b=c≠ d,	-----Forbidden term
	mechanical asymmetric	
	2. a≠b≠c≠ d	
	3. one bad coil	
(2) Octupole (n=3)	A=B, a=d, b=c, but a≠ b	-----Forbidden term
(3) Decapole (n=4)	Tilt one pole piece	-----Forbidden term
(4) Dodecapole (n=5)	1. lateral truncation	-----Allow term
	2. a=b=c=d, A=B≠ 2R	-----Forbidden term
(5) 20-pole (n=10)	1. lateral truncation	-----Allow term
	2. a=b=c=d, A=B≠ 2R	-----Forbidden term

n= 2(2m+1) m=0,1,2,3 -----, n is the allow term for the quadrupole magnets.

2.5 Contour equation and magnetic flux density of the ideal pole shape of a sextupole magnet

Figures 6 and 7 present the equipotential lines and the magnetic flux of a sextupole magnet. The scalar potential and magnetic flux density in the cylindrical and Cartesian coordinates are as follows [1,11,12].

Cylindrical Coordinates

$B_r = 3a_3 r^2 \cos 3\theta + 3b_3 r^2 \sin 3\theta$;

$B_\theta = -3a_3 r^2 \sin 3\theta + 3b_3 r^2 \cos 3\theta$;

$\Phi_M = 3a_3 r^3 \cos 3\theta + 3b_3 r^3 \sin 3\theta$;

Cartesian Coordinates

$B_x = 3a_3 \left(x^2 - y^2\right) + 6b_3 xy$

$B_y = -6a_3 xy + 3b_3 \left(x^2 - y^2\right)$

$\Phi_M = a_3 \left(x^3 - 3y^2 x\right) + b_3 \left(3yx^2 - y^3\right)$

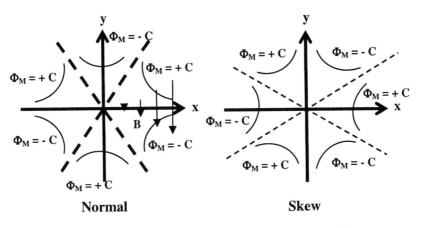

Normal   **Skew**

Figure 6 Equipotential lines of a sextupole magnet (normal and skew).

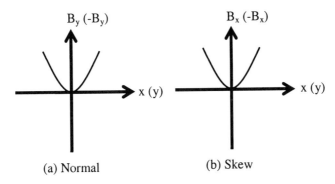

(a) Normal   (b) Skew

Fig.7 Magnetic flux density distribution versus transverse coordinate. (a) normal (b) skew sextupole magnets.

For $a_3 = 0$, $b_3 \neq 0$, $B_y \propto x^2$, $B_y = 3b_3 \left(x^2 - y^2\right)$ give a normal sextupole. For $a_3 \neq 0$, $b_3 = 0$, $B_y = -6a_3 xy$ yield a skew sextupole (pole rotated by $\dfrac{\pi}{2 \times 3}$).

Figures 8 and 9 and the table show a multipole field error analysis. N= 3(2m+1) and m=0,1,2,3 -----, where n is the allowed term components of the sextupole magnet. If the dipole field (n=1) changes, the field for the allowed term components of sextupole magnet (n=2m+1) would also be changed.

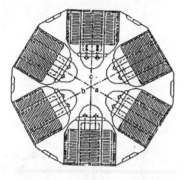

Figure 8 Dipole perturbation in a sextupole field generated by geometrical asymmetry

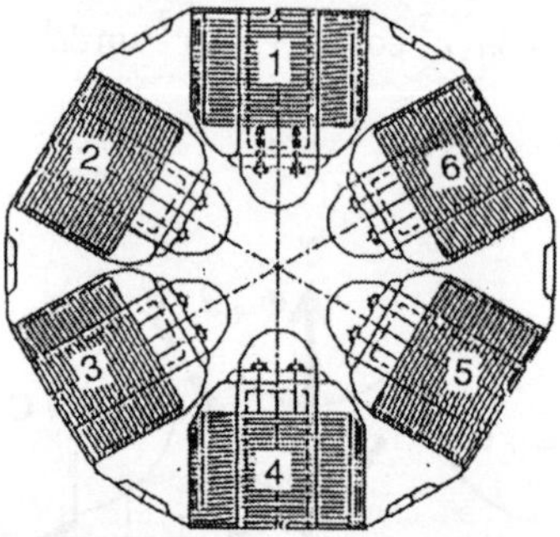

Figure 9 Dipole and skew octupole perturbation in a sextupole field generated by a bad coil (1 or 4 is bad coil).

(1) Octupole (n=4)   : 1. a=b≠ c,   -----Forbidden term
                                       mechanical asymmetric
                                     2. a≠b≠ c
                                     3. bad coil
(2) Decapole (n=5)   : a=b=c≠ 2R   -----Forbidden term
(3) 18-pole (n=9)   : 1. lateral truncation   -----Allow term
                                     2. a=b=c≠ 2R   -----Forbidden term
(4) 30-pole (n=15)   : 1. lateral truncation   -----Allow term
                                     2. a=b=c≠ 2R   -----Forbidden term

3. Magnet field calculation

3.1 *Ideal pole shapes are equal magnetic lines* [1]

At the steel boundary, with no currents in the steel,

$$\bar{\nabla} \times \bar{H} = 0 \qquad (3.1.1)$$

Stoke's theorem is applied to a closed loop that enclose the boundary (shown in Figure 9)

$$\iint \left(\bar{\nabla} \times \bar{H} \right) \cdot d\bar{S} = \oint \bar{H} \cdot d\bar{\ell} \qquad (3.1.2)$$

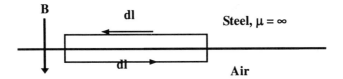

Figure 9 Ideal pole shapes boundary between steel and air.

Hence around the loop,

$$\oint \vec{H} \cdot d\vec{\ell} = 0 \qquad (3.1.3)$$

But for infinite permeability in the steel, H = 0, while outside the steel, H = 0 parallel to the boundary. Therefore B in the air adjacent to the steel is normal to the steel surface and is equal at all points on the surface. Therefore, from $\vec{B} = -\vec{\nabla}\Phi_M$, the steel surface is an iso-scalar-potential line.

3.2 Magneto-motive force in a magnetic circuit [1]

Stoke's theorem for vector A is, (see figure below)

$$\oint A \cdot d\ell = \iint \left(\vec{\nabla} \times \vec{A} \right) \cdot d\vec{S} \qquad (3.2.1)$$

which is applied to

$$\vec{\nabla} \times \vec{H} = \vec{j} \qquad (3.2.2)$$

Then, for any magnetic circuit, Eq. (3.2.2) is put into Eq. (3.2.1) to obtain

$$\oint \vec{H} \cdot d\vec{\ell} = NI \qquad (3.2.3)$$

where NI is the total Amp-turns through the loop dS.

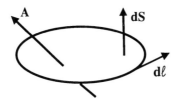

3.3 Determination of ampere-turns in a dipole [1]

The magnetic flux density B is approximately constant around loop l and magnet gap g (see figure below)

$$H_{iron} = H_{air} / \mu \qquad (3.3.1)$$

$$B_{air} = \mu_0 \, NI / (g + l/\mu) \qquad (3.3.2)$$

where g and l/μ are the reluctance of the gap and the iron. Ignoring iron reluctance,

$$NI = B \, g / \mu_0 \qquad (3.3.3)$$

Thus, we would obtain the result of Eq. (3.3.4).

$$B = NI\mu_0 / g \qquad (3.3.4)$$

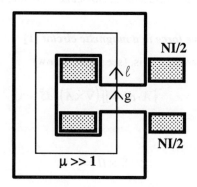

3.4 Determination of ampere-turns in a quadrupole magnet [1]

The magnetic circuit is shown below and the quadruple pole profile equation is,

$$xy = R^2 / 2. \qquad (3.4.1)$$

On the transverse x-axis, the magnetic flux density can be expressed as,

$$B_y = gx \qquad (3.4.2)$$

where g is the gradient field strength (T/m). At large x, where the field B is vertical,

$$NI = B \cdot \ell = \left(gx\right)\left(R^2 / 2x\right)/ \mu_0 \qquad (3.4.3)$$

$$NI = g\left(R^2 / 2\mu_0\right) \qquad (3.4.4)$$

Therefore, the gradient field strength can be expressed as,

$$g = 2\mu_0 NI / R^2 \qquad (3.4.5)$$

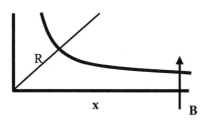

3.5 *Determination of ampere-turns in a sextupole magnet*

The magnetic circuit is similar to that the quadrupole magnet shown above and the pole equation of sextupole magnet is,

$$3x^2 y - y^3 = \pm R^3. \qquad (3.5.1)$$

On the transverse x-axis, the magnetic flux density can be expressed as,

$$B_y = g_s x^2 \qquad (3.5.2)$$

where g_s is the sextupole strength (T/m^2). At large x, where the vertical field B is strong,

$$NI = B \cdot \ell = g_s x^2 \cdot \left(\frac{R^3}{3x^2 \mu_0}\right) = \frac{g_s R^3}{3\mu_0} \qquad (3.5.3)$$

Therefore, the sextupole strength can be expressed as

$$g_s = \frac{3\mu_0 NI}{R^3} \qquad (3.5.4)$$

3.6 *Comparison of four commonly used magnetic computation codes*

A number of standard codes is available for designing the poles and the coils. "MAGNET" and "POISSON" are simple, two dimensional codes while "TOSCA" and "RADIA" are complex three-dimensional codes. "MAGNET" and "POISSON" are similarly able to design magnets. TOSCA is a state-of-the-art, three-dimensional package that is maintained and updated as a commercial package. It can be used for DC and AC field calculations. The software is available from the company, and training courses are offered to familiarize both beginners and more experienced designers with the wide range of features of the program. However, the code is very expensive. The "RADIA" [13] code is associated with the mathematics utility to build a model easily and uses the quick-time utility to view and rotate a 3D structure. It is a 3D code with fast calculation. However, it can not be used in AC field calculation. The advantage and disadvantage of the four standard codes are compared below [1].

Advantages	Disadvantages
MAGNET:	
Quick and easy to learn and simple to use;	Only 2D predictions;
Small CPU requirement;	Batch processing only-slows down problem turn-round time;
Fast execution time.	Inflexible rectangular lattice;
	Inflexible data input;
	Geometry errors possible from interaction of input data with lattice;
	No pre- or post-processing;
	Poor performance in high saturation.
POISSON:	
Similar computation as MAGNET;	Harder to learn for users;
Interactive input/output is possible;	Just only two-dimensional predictions.
More input options with greater flexibility;	
Flexible lattice eliminates geometry errors;	
Better handing of local saturation;	
Some post processing available.	
TOSCA:	
Three dimensional prediction package and working in PC;	Training course needed for familiarization;
Accurate prediction of distribution and strength in 3D;	Expensive to purchase;
Extensive pre- and post-processing;	Large computer needed;
Multipole function calculation	Large use of memory;
For static & DC & AC field calculation	CPU time is huge for non-linear 3D problem.
Easy to view 3D structure and field distribution.	
RADIA:	
Full three dimensional package;	Larger computer needed;
Accurate prediction of distribution and strength in 3D;	Large use of memory;
With quick-time to view and rotate 3D structure;	DC field calculation;
Easy to build model with mathematic;	Can down load from ESRF website.
Fast calculation in PC.	

4.   Magnet geometries and pole shim

The field distribution of the three dimensional fringe fields at the ends of the magnet is treated qualitatively in the following discussions [1,7]. Fringe field length is typically longer in the center for an unchamfered pole end. The field will be saturated at the edge easily. Therefore, a chamfer will be performed to reduce the higher harmonic strength. To increase the good field region, the shim on yoke shown in the figures is necessary.

4.1   *Magnet shims*

4.1.1 *Dipole magnet*

'C' Type dipole magnet:

Advantages:
  Easy access for installation;
  Easy for field measurement;
  Easy to install vacuum chamber.

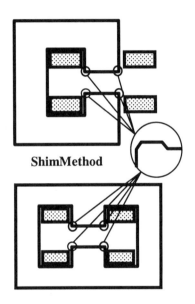

Disadvantaged:
  Pole shims are needed;
  Asymmetric field distribution (small);
  Less rigid.

ShimMethod

'H' Type dipole magnet:

Advantages:
  Symmetry of field features;
  More rigid;
Disadvantages:
  Also needs shims;
  Access problems for installation.
  Not easy for field measurement.

4.1.2 *Quadrupole and Sextupole Type Magnet [1,8,9]*

The typical shims for quadrupole and sextupole magnets are shown below. The quadrupole and sextupole shims are generated from a tangent to the hyperbolic pole and project from the point on the pole face and then to terminate at the extended pole side. The pole shim is to avoid field saturation in the region at high flux density. The shim dimension is determined by the field quality of the good field region.

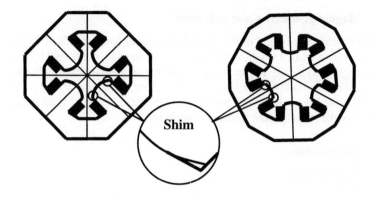

4.2 *Magnet chamfer for avoiding field saturation at magnet edge*

The chamfering geometry is shown in Figure 10. The characteristics of the chamfer are as below:

(a) Chamfering at both sides of dipole magnet to compensate for the sextupole components and the allowed harmonic terms.

(b) Chamfering at both sides of the magnet edge to compensate for the 12-pole (18-pole) on quadrupole (sextupole) magnets and their allowed harmonic terms.

(c) Chamfering means to cut the end pole along 45° on the longitudinal axis (z-axis) to avoid the field saturation at magnet edge.

(d) Define magnetic length more precisely.

(e) Prevent saturation.

(f) Control transverse distribution.

(g) Prevent flux entering iron normal to lamination (vital for ac magnets)

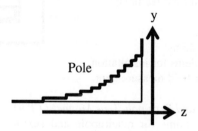

Figure 10. Chamfer geometry on longitudinal axis.

5. Magnets for injection and extraction system

The injection and extraction system include the A.C. field and pulse magnets like septum and kicker [1,3]. In general, the injection and extraction magnet systems of accelerators need two types of magnets – kicker and septum magnets. The kicker magnets are used for the fast temporary displacement of the closed orbit. The septum magnets provide the necessary deflection to the incoming or outgoing beam to match the angle of the beam transfer line to the accelerator orbit. The magnet design criteria that are to minimize eddy losses in low frequency magnets are first considered. Therefore, the material choice of steel in the magnet yoke is very important. The eddy losses in conductors are estimated as:

(1) Rectangular conductor without cooling hole, resistivity ρ, width a, cross section A. In an a.c. field B sin ωt, the power loss per m is $P = \omega^2 B^2 A a^2 / (12\rho)$

(2) Circular conductor, diameter d, power loss $P = \omega^2 B^2 A d^2 / (16\rho)$

For example, a 5 mm square copper conductor in a 1 T peak field strength, 10 Hz a.c. field has a total power loss of $P = 8.5$ kW / m. The cooling formula is $P = (\phi \Delta T)/14.33$ where P is the dissipated power (kW), ϕ the water flow (l/min), and ΔT the temperature increase (centigrade).

The criterion related to the eddy current in septum magnets is the skin depth in a conductor. If the resistivity is ρ, the permeability μ, and the frequency ω, the skin depth is given by $\delta = \sqrt{(2 \rho / \omega\mu\mu_0)}$.

5.1 Kicker magnet design

The magnet can be placed inside or outside the accelerator vacuum system. In general, it is placed inside the vacuum system. If the magnet is outside the vacuum chamber, a high resistivity ceramic chamber as well as a ferrite core are necessary to avoid the skin effect.

Ferrite Core

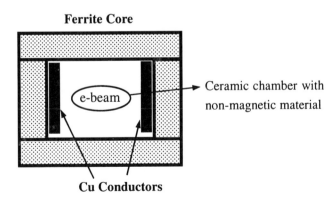

Ceramic chamber with non-magnetic material

Cu Conductors

5.2 Alternative septum magnet designs

The magnet is usually located inside the vacuum system and deflects the incoming or outgoing electron beam. The yoke is made from low carbon steel. The conductor is placed inside the magnet and should be as thin as possible.

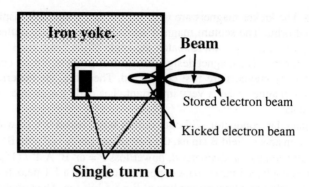

Single turn Cu

6. Classification of field measurement and analysis methods

The gradient field inhibits the Nuclear Magnet Resonance (NMR) measurement method. A stable and precise Hall probe [14] or other ultra stable point probe which can be used to measure the large gradient magnetic field would be one choice. Therefore, the Hall probe point mapping method [14] can be used for combined function bending magnets as well as for any other multipole magnets. Moreover, we believe that the mapped field data contain much more information on the magnetic field than integrated field values obtained by other methods. It serves as a general purpose method [5,6,15] for magnetic field measurement on different kinds of magnets. But there is also a need of other measurement systems for the magnetic field measurement of different kinds of magnet, for example, the moving coil or search coil. Setting up a fully automatic measurement [16,17] and analysis system to reduce time consumption and human error is a necessity. The choices of measurement system are dependent on the features of magnet. The purposes of the magnetic field measurement are

- to make sure the magnet field strength and quality fulfills the specification,
- to calibrate the relationship between the electron energy E(B), magnetic field strength B(I) and the excitation current I,
- to show the statistic distribution of the total magnets for the mass production and to sort the magnets in order to arrange them in the appropriate location,
- to obtain a reduced but sufficient number of magnetic parameters allowing simulation of the behavior, and
- to define fiducial references on the magnets for precise alignment in the machine (if necessary).

The accuracy and precision of the measurement system depend on the required tolerance of the experiment and precision of each component of the system. Some different measurement systems are necessary to match the different kinds of magnet. Be careful to choose the environment of the measurement room such as to keep a constant ambient temperature and to avoid floor vibrations as well as EMI.

6.1 Measurement methods

The measurement methods are summarized below and the relative accuracy range of each method is shown in Figure 12 [5].

1. The fluxmeter method (induction method)
   Based on the induction law this method is the oldest of the currently used methods for magnetic measurement, yet it can be very precise.
   a. Search coil
      Suitable for point field measurement
   b. Rotating harmonic coil
      Suitable for the analysis of the harmonic field components
   c. Moving or fixed stretched wire
      To measure the static or varying magnet flux
   d. Helmholtz coil
      To measure the homogeneous field or the three-dimension magnetic moment of permanent magnet.

2. The Hall effect method
   Advantage: High resolution of point field and inhomogeneous field measurement
   Disadvantage: moderate accuracy and susceptible to environmental influences.

3. Nuclear Magnetic Resonance (NMR) method
   Advantage: not only for calibration purposes, but also for high precision field mapping
   Disadvantage: can not measure inhomogeneous field.

4. Fluxgate magnetometer (FM)
   Advantage: offering a linear measurement and well suited for static operation.
   Disadvantage: restricted to low field.

380

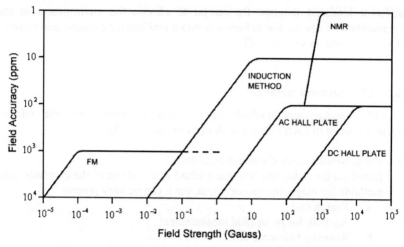

Figure 12. Measurement methods in accuracy and ranges.

6.2 Operating principle of the Hall probe

The Hall generator is a solid-state sensor which provides an output voltage V_h proportional to the magnetic flux density B_y. As suggested by its name, this device relies on the Hall effect. The Hall effect is the generation of a voltage across a sheet of conductor when a current I_c is flowing and the conductor is placed in a magnetic field. R_h is the planar Hall coefficient. Figure 13 states the relationship among the current, induced voltage, and the magnetic field and the relevant equation is shown below. If the field is inhomogeneous, the induced field error is given by Eq. (6.2.3). This method is useful mainly for measuring the large curvature dipole magnet.

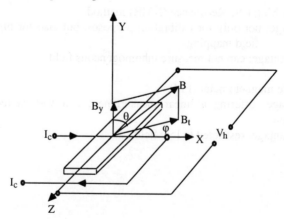

Figure 13. Operating principle of Hall probe devices.

$$V_h = V_0 + R_h B_y I_c - P B_t^2 I_c \sin(2\varphi) \tag{6.2.1}$$

where $B_t = B_o \sin\theta$, $B_y = B_o \cos\theta$, $P = K_2(\alpha_t - \alpha_l)$ \hfill (6.2.2)

$$\Delta B = (1/24)L^2 \frac{d^2 B(x)}{dx^2} (1/24)L^2 \frac{d^2 B(x)}{dx^2}\bigg|_{x=0} + (1/1920L^4 \frac{d^4 B(x)}{dx^4}\bigg|_{x=0} + \cdots \tag{6.2.3}$$

where p is the planar Hall coefficient. The planar Hall coefficient is changed due to the anisotropy between the transversal α_t and the longitudinal α_l changes in the magnetic resistance change when the magnetic field is changed in the sensor's plane.

6.3 Analyzing data from the Hall probe system

The mapping method [14,18] measures either the vertical field $B_y(x,y)$ or the horizontal field $B_y(x,y)$. To perform the harmonic field analysis, the position (x,y,s) and the field $B_y(x,y)$ are recorded, and then the field is integrated along the s orbits. The features of a normal magnet depend on the magnet pole face and its magnetic field equation which can be solved by using Laplace's equation in the two dimensional (x,y) plane. The magnetic field equation can be expressed as

$$B_y(x,y) = \sum_{n=1}^{n}\sum_{m=0}^{m} N_{n,m} H_{n,N} x^{n-2m} y^{2m} + \sum_{n=1}^{n}\sum_{m=0}^{m} S_{n,m} H_{n,S} x^{n-2m-1} y^{2m-1} \tag{6.3.1}$$

$$N_{n,m} = \frac{(-1)^m n!}{(n-2m)!(2m)!} \tag{6.3.2}$$

$$S_{n,m} = \frac{(-1)^m n!}{(n-2m-1)!(2m+1)!} \tag{6.3.3}$$

where $N_{n,m}$ and $S_{n,m}$ are the weighting factors, $n-2m \geq 0$ and $n-2m-1 \geq 0$. The field strength $H_{n,N}$ is the normal multipole strength and $H_{n,S}$ is the skew multipole strength and n represents the harmonic number. Also eq. (6.3.1) can be transformed into cylindrical coordinates as follows. The vertical field $B_y(x,y)$ can be represented as

$$B_y(r,\theta) = \sum_{m,n} H_{nm} r^n \left[Sin(n\theta - \alpha_{nm}) \right] \tag{6.3.4}$$

$$H_{nm} = \left(H_{n,N}^2 + H_{n,S}^2 \right)^{1/2} \tag{6.3.5}$$

$$\alpha_{nm} = Sin^{-1}\left(H_{n,N} / H_{nm} \right) \tag{6.3.6}$$

where H_{nm} is the combined harmonic field strength, α_{nm} the orientation angle and θ the phase angle. From the magnetic field equation (6.3.1), the skew dipole field is an unknown term, unless the horizontal field $B_x(x,y)$ is measured. The mapping data of the magnetic field strength $B_x(x,y,s)$ or $B_y(x,y,s)$ are substituted into Eq. (6.3.1). The harmonic field strength $H_{n,N}$ and $H_{n,S}$ can be obtained by performing a non-linear least square fit. The results can be substituted into Eqs. (6.3.5) and (6.3.6) to obtain the combined harmonic field strength H_{nm} with an orientation angle α_{nm}.

6.4 Measurement philosophy and data obtained by the rotating coil method

The basic device for determining the field quality is a flat radial coil, rotating about the axis of the magnet, as shown in Figure 14. This method [15] is used for multipole magnets. The data can be obtained by Fast Fourier Transform (FFT) to obtain the multipole components using Eqs. (6.4.1) and (6.4.2).

$$B_r(r,\theta) = \sum_{m=1}^{\infty} B_{ref} \left(\frac{r_1}{r_{ref}} \right)^{m-1} \left[Sin(m\theta - \alpha_m) \right] \tag{6.4.1}$$

$$\sum_{n=1}^{\infty} a_n \cos(n\theta) + b_n \sin(n\theta) = \sum_{n=1}^{\infty} \frac{2r_1^n N}{n} \left[A_n \cos(n\theta) + B_n \sin(n\theta) \right] \sin\left(\frac{n\Delta\theta}{2} \right)$$

$$\tag{6.4.2}$$

where $\Delta\theta$ is the angle difference between each sampling point, N is the number of wire turns, a_n is the normal component and b_n is the skew component.

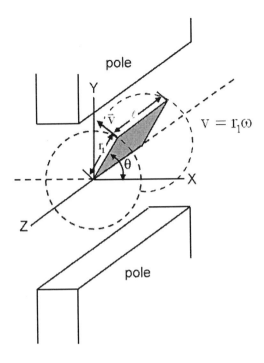

Figure 14. Single loop asymmetric coil in multipole field measurement.

6.5 *Field quality control for mass production*

The sequence of main tasks for making field measurements for field quality control and magnet mass production quality control is given below.

1) Precise alignment of each magnet (need some monument for position reproducibility)
2) Excitation current cycle to overcome hysteresis
3) Measure
   - (a) the central field $B_y(x,y)$
   - (b) integrated field $\int B_y(x,y)ds$
   - (c) field magnetic length BL/B_0
   - (d) the tilt $\Delta\theta = \int B_x(x,y)ds / \int B_y(x,y)ds$
   - (e) the normal term and skew term of the higher central and integrated multipoles.
4) Determine the field quality, including
   - (a) the distribution of $\Delta(\int B_y(x,y)ds)/\int B_y(x,y)ds$
   - (b) the good field region
   - (c) the comparison of the measured results with the specified tolerances
   - (d) the statistical distribution among mass produced magnets

5) Determine
   (a) the influence of surrounding materials
   (b) the effect of μ_r and H_C
   (c) the effect of eddy currents
   (d) the effect of initial magnetization

The characteristics of field quality control on the central pole region and the whole magnet are shown below.

(a) In the central pole region

Dipole: calculate and plot $\dfrac{B_y(x) - B_y(0)}{B_y(0)}$ for a pure dipole

or $\dfrac{B_y(x) - \left(B_y(0) + g\ x\right)}{B_y(0)}$ for a combined function dipole magnet; g is

the gradient field in a combined dipole magnet

Quadrupole: calculate and plot $\dfrac{dB_y(x)}{dx} = g(x)$ and $\dfrac{g(x) - g(0)}{g(0)}$; g is the

quadrupole strength of the quadrupole magnet.

Sextupole: calculate and plot $\dfrac{d^2 B_y(x)}{dx^2} = S(x)$ and $\dfrac{S(x) - S(0)}{S(0)}$; S is the

sextupole strength of the sextupole magnet.

(b) For a whole magnet with integral field

Dipole: $\dfrac{\int B_y(x)ds - \int B_y(0)ds}{\int B_y(0)ds}$ or $\dfrac{\int B_y(x)ds - \left[\int B_y(0)ds + \int gxds\right]}{\int B_y(0)ds}$

Quadrupole: $\dfrac{\int g(x)ds - \int g(0)ds}{\int g(0)ds}$

Sextupole: $\dfrac{\int S(x)ds - \int S(0)ds}{\int S(0)ds}$

7. Field measurement and analysis

Several examples explain the results of magnetic measurement and analysis. The dipole, quadrupole, and sextupole magnets are all measured using the Hall probe method. The field measurements are compared with and without the shim mechanism.

7.1 Combined function dipole magnet [7, 8, 18]

Figures 15 and 17 compare the magnetic field distribution over the central pole region between design and construction. The parameters of a dipole magnet are shown in reference [7]. The combined function dipole magnet has (Figure 15) a quadrupole field strength (Figure 17). The measurement shows that the result is quite consistent with that of the field calculation. The Hall probe mapping method also gives the half effective length distribution measurement (Figure 16), revealing that the center of the field in the longitudinal direction is –0.23 mm [7]. The pole face is bent but symmetric. Figure 18 shows the good field region of the integral field with and without shim. After 2-2 shims were added on the magnet edges, the good field region is much wider than that obtained without shim (Figure 18) and the strong sextupole strength at both ends is lower (Figure 22) [7]. Figures 19, 20 and 21 reveal two different measurement methods using the Hall probe, to obtain the magnetic field features [18]. A strong negative focusing field is present at the two edges in the radial mapping trajectory, corresponding to the thin lens effect from the pole face rotation. Figure 21 shows that the sextupole strength is very strong at the magnet edges of the magnet, but rather weak in the center region. The edge effect can be treated separately as a thin lens located on both sides of the dipole magnet [18].

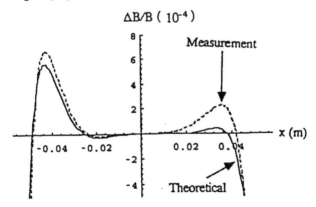

Figure 15. Field deviation as a function of x, according to the two dimensional theoretical calculation, and the measurements for the combined function bending magnet. The field deviation is defined as $\Delta B / B = \lfloor B(x) - (a_1 + a_2 x) \rfloor / (a_1 + a_2 x)$. (From reference [14])

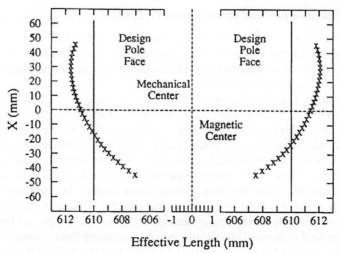

Figure 16. Half the effective length distribution on the two sides of the combined function dipole magnet. (From reference [12])

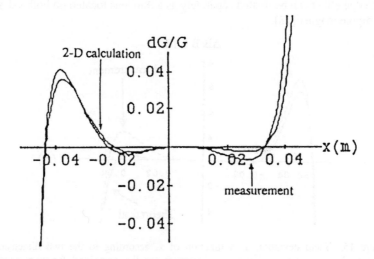

Figure 17. Gradient field deviation according to the 2-D calculation and measurements at the magnetic center. (From reference [12])

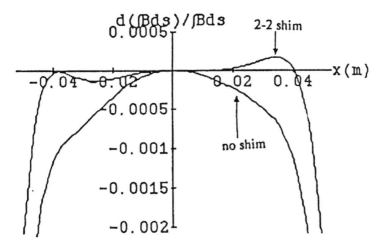

Figure 18. Measurement of the integrated field deviation after and before 2-2 shims. (From reference [12])

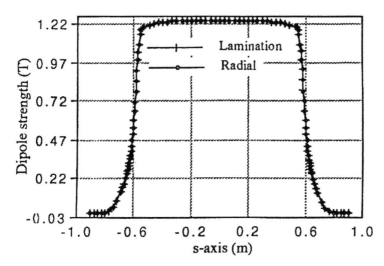

Figure 19. Dipole field distribution of the radial and lamination mapping in the beam direction of the beam. Here, radial means the azimuthal direction that is orthogonal to the electron beam direction (From reference [18])

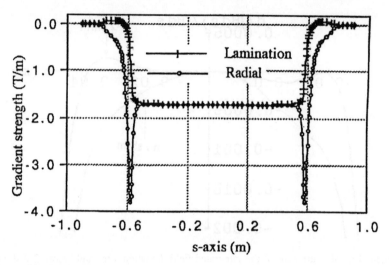

Figure. 20. Gradient field distribution of radial and lamination mapping along the beam direction. (From reference [18])

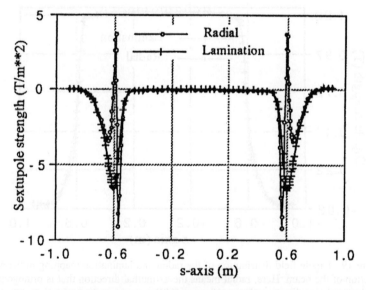

Figure 21. Sextupole distribution of the radial and lamination mapping along the beam direction. (From reference [18])

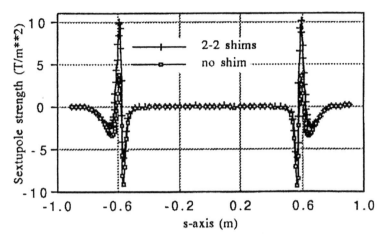

Figure 22. Measured sextupole field distribution along the beam direction of the after and before 2-2 shims. (From reference [18])

7.2 *Quadrupole magnet* [9, 10, 19, 20]

Figure 23 reveals the calculated and measured results for the quadrupole magnet whose parameters are in reference [9]. The normal field gradient of 12 T/m is operated at an excitation current of 184 A. The field quality of the quadrupole was calculated by varying the shim's width and thickness. The final shape of the pole has a relative gradient field error of $<5 \times 10^{-4}$ over a width of 30 mm in the midplane [9]. The magnet appears quite symmetric in Figure 23, however, the good field region is too short because of the strong negative integrated strength of the dodecapole in the edge region. Hence, the magnetic edge must be cut at $45°$, as a chamfer, to compensate for the negative dodecapole strength [19].

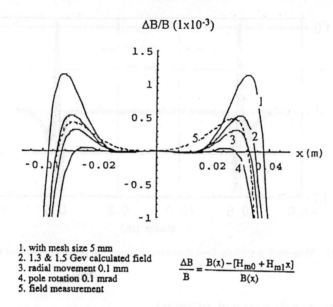

1. with mesh size 5 mm
2. 1.3 & 1.5 Gev calculated field
3. radial movement 0.1 mm
4. pole rotation 0.1 mrad
5. field measurement

$$\frac{\Delta B}{B} = \frac{B(x) - [H_{m0} + H_{m1}x]}{B(x)}$$

Figure 23. Field deviation as a function of transverse axis, according to the 2-D calculation and field measurement using a Hall probe on the midplane. (From reference [9])

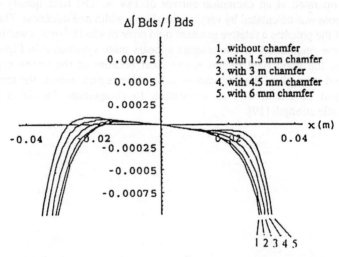

Figure 24. Deviation of integrated field with thickness of chamfering. (From reference [9]).

7.3   *Sextupole magnet* [11, 12, 19]

Reference [11] lists the parameters of the sextupole magnet. The magnets were measured using the Hall probe mapping system. Figure 25 shows the field deviation in the magnetic center, according to the 2-D design calculation and field measurements. The sextupole field strength is 380 T/m² at an excitation current of 183 A. The results show that the designed and constructed sextupole magnets are sufficiently similar [11]. Figure 26 presents the integral field profile in the longitudinal direction. The good field region is demonstrated in this figure as meeting the specification that it should be within 2x10⁻³ over ±28 mm. However, if a chamfer is included at the edge, the good field region is wider and the integral eighteen-pole field strength will be reduced.

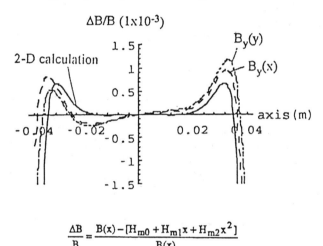

$$\frac{\Delta B}{B} = \frac{B(x) - [H_{m0} + H_{m1}x + H_{m2}x^2]}{B(x)}$$

Figure 25. Field deviation in the center of magnet, according to calculation and the measurements. (From reference [11])

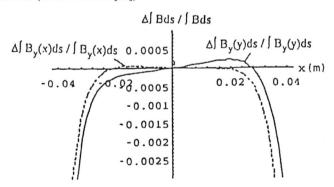

Figure 26. Integral field deviation of the field strength distribution in the longitudinal direction. (From reference [11])

Acknowledgement

The author would like to thank the SRRC people for supporting this article. Especially, it is a pleasure to thank the magnet group people in SRRC, C. H. Chang, T. C. Fan, F. Y. Lin, H. H. Chen, and M. H. Huang, to work together for various helpful discussions.

Reference:

[1] Neil Marks, CERN Accelerator School, fifth general accelerator physics course, Editor by S. Turner, CERN 94-01, 26 January,1994, Vol. II.

[2] Handbook of accelerator physics and engineering, World Scientific publishing Co. Pte. Ltd., edited by Alexander Wu Chao and Maury Tigner (1998).

[3] Jack Tanabe, Lectures of conventional magnet design, USPAS, Tucson, Arizona, January, 2000.

[4] Synchrotron Radiation Sources- A Primer, World Scientific publishing Co. Pte.

[5] CERN Accelerator School, Magnetic measurement and alignment, Editor by S. Turner, CERN 92-05, 15September,1992.

[6] CERN Accelerator School, Measurement and alignment of accelerator and detector magnets, Editor by S. Turner, CERN 98-05, 4 August 1998.

[7] C. S. Hwang, C. H. Chang, G. J. Hwang, T. M. Uen, and P. K. Tseng, 1996, July, "*Features of The Magnetic Field of a Rectangular Combined Function Bending Magnet*", IEEE Trans. on magn., 32, No. 4, (1996) 3032-3035.

[8] C. H. Chang, C. S. Hwang, M. H. Huang, F. Y. Lin, P. K. Tseng, and G. J. Hwang, "*Construction and Measurement of the Bending Magnets for the SRRC Transfer Line*", The Forth European Particle Accelerator Conference EPAC"94, June 27 - July 1 (1994).

[9] C. H. Chang, H. H. Chen, C. S. Hwang, G. J. Hwang, and P. K. Tseng, 1994, July, "*Design, Construction and Performance of the SRRC Quadrupole Magnets*", IEEE Trans. on magn., 30, No. 4, (1994) 2241-2244.

[10] C. H. Chang, C. S. Hwang, J. Y. Hsu, G. J. Hwang, and P. K. Tseng, "*Skew Quadrupole for SRRC Storage Ring*", The 9th conference on magnetism and magnetic technologies, June, Taiwan (1994).

[11] C. H. Chang, C. S. Hwang, C. P. Hwang, G. J. Hwang, and P. K. Tseng, 1994, July, "*Design, Construction and Measurement of the SRRC Sextupole Magnets*", IEEE Trans. on magn., 30, No. 4, (1994) 2245-2248.

[12] C. S. Hwang, C. H. Chang, T. C. Fan, T. M. Uen, and P. K. Tseng,1996, October, "*Magnetic Field Features of Various Magnetic Circuits on the Sextupole Magnet*", Chinese Journal of Physics, 34, No.5, (1996) 1227-1236.

[13] P. Elleaume, O. Chubar, J. Chavanne, "Computing 3D magnetic field

from insertion devices", Particle Accelerator conference (PAC97), Aancouver, p.3509, May 1997.

[14] C. S. Hwang, F. Y. Lin, G. J. Jan, and P. K. Tseng,1994, August, *"High-Precision Harmonic Magnetic-Field Measurement and Analysis Using a Fixed Angle Hall Probe"*, Rev. Sci. Instrum., 65(8), (1994) 2548-2555.

[15] T. C. Fan, C. S. Hwang, and P. K. Tseng, "*A Simple Magnet Multipole Measurement System*", The Fourth European Particle Accelerator Conference EPAC'94, June 27 - July 1 (1994).

[16] C.S. Hwang, F.Y. Lin, P.K. Tseng, 1999, August, "*A PC based real-time Hall probe automatic measurement system for magnetic fields*", *IEEE Trans. on Instrum. and Meas, Vol. 48, No. 4, (1999)*.

[17] Poh- Kun Tseng, and Ching-Shiang Hwang, "*The Analysis Methods of Hall Probe Mapping System for the Combined Function Dipole Magnet*", XVth International Conference On High Energy Accelerators 1, 598 - 601 (1992).

[18] C. S. Hwang, C. H. Chang, P. K. Tseng, T. M. Uen, and Joel Le Duff, December, "*Different Kinds of Analysis Methods on a Rectangular Combined Function Bending Magnet*", Nucl. Instrum. Meth. in Physics A383, (1996) 301-308.

[19] C. S. Hwang, C. H. Chang, T. C. Fan, P. K. Tseng, and G. J. Hwang, "*Results Analysis of Measurement of the Multipole Magnet with Hall Probe*", The Forth European Particle Accelerator Conference EPAC'94, June 27 - July 1 (1994).

[20] Ching-Shiang Hwang, Fu-Yuen Lin, Tai-Ching Fan, and Poh- Kun Tseng, "*Quadrupole Magnetic Measurement Results of The SRRC Storage Ring by the Hall Probe Mapping System*", XVth International Conference On High Energy Accelerators, 1, 602 - 604 (1992).

EMITTANCE AND COOLING[*]

LEE C. TENG

Argonne National Laboratory
Advanced Photon Source
9700 South Cass Avenue
Argonne, IL 60439, USA
E-mail: teng@aps.anl.gov

We start with the motion of a system of particles for which:
- The particles are noninteracting
- Each particle forms a closed system (e.g., no synchrotron radiation, no rf acceleration, etc.)
- In most cases, the motion is Hamiltonian.

The most important and useful collective parameter for an ensemble of particles is the "emittance." Generally, beams with low emittances are more desirable for use in physics experiments. For a closed system as defined above the emittance is a constant of motion. Interactions with external systems caused by errors in design, construction, or operation of the accelerator system tend to result in undesirable growths in beam emittances. Proper external interactions can be introduced to reduce the emittances. But such interactions are tricky and must be cleverly designed, elaborately constructed, and precisely operated. The procedure to reduce the emittance is called "cooling."

1. Formulation of Emittance

Emittance is basically a one-degree-of-freedom concept. It is a single parameter specification of the distribution of phase points of beam particles in a two-dimensional (2-D) phase plane (x, x'). The two-component position vector of the phase point of particle i is written as

$$X_i = \begin{pmatrix} x_i \\ x_i' \end{pmatrix}. \tag{1}$$

The symplectic conjugate of X is defined as

$$X^+ \equiv \widetilde{X}\widetilde{S} = (x' - x), \tag{2}$$

---

[*] This work is supported by the U.S. Department of Energy, Office of Basic Energy Sciences, under Contract No. W-31-109-ENG-38.

where ~ means "transpose" and where

$$S \equiv \begin{pmatrix} 0 & 1 \\ -1 & 0 \end{pmatrix} = \text{symplectic unit matrix.} \qquad (3)$$

The symplectic conjugate of a row vector such as X^+ is defined as

$$\left(X^+ \right)^+ \equiv S\widetilde{X^+} = S\widetilde{X}\widetilde{S} = SSX = -X . \qquad (4)$$

A distribution of phase points is generally described by moments:

$$\begin{cases} \text{1st moments (centroid)} & \overline{x_i}, \quad \overline{x_i'} \\ \text{2nd moments (size)} & \overline{x_i^2}, \quad \overline{x_i x_i'}, \quad \overline{x_i'^2} \\ \text{3rd moments (skewness)} & \overline{x_i^3}, \quad \overline{x_i^2 x_i'}, \quad \overline{x_i x_i'^2}, \quad \overline{x_i'^3} \\ \text{etc.,} \end{cases} \qquad (5)$$

where a bar denotes averaging over all particles. Under the "averaging bar" the subscript i may be omitted. It is most convenient and useful to define a second moment matrix

$$E \equiv -\overline{XX^+} = -\overline{\begin{pmatrix} x \\ x' \end{pmatrix}\left(x' - x \right)}$$

$$= \begin{pmatrix} -\overline{xx'} & \overline{x^2} \\ -\overline{x'^2} & \overline{xx'} \end{pmatrix} \qquad \text{(Traceless).} \qquad (6)$$

For a single-parameter specification of the "size" of the distribution one takes

$$\varepsilon^2 \equiv |E| = \overline{x^2}\,\overline{x'^2} - \overline{xx'}^2 , \qquad (7)$$

where ε is called the rms emittance or just the emittance. (Later we will show that $|E|$ is positive definite.) We can now write for E

$$E = \varepsilon \begin{pmatrix} \alpha & \beta \\ -\gamma & -\alpha \end{pmatrix} \equiv \varepsilon J. \tag{8}$$

With this definition we get the conventional relations:

$$\begin{cases} \overline{x^2} = \varepsilon\beta, & -\overline{xx'} = \varepsilon\alpha, & \overline{x'^2} = \varepsilon\gamma, & \text{and} \\ |J| = \beta\gamma - \alpha^2 = 1, & & J^2 = -I. \end{cases} \tag{9}$$

The symplectic conjugate of E is defined as

$$E^+ \equiv S\tilde{E}\tilde{S} = \begin{pmatrix} \overline{xx'} & -\overline{x^2} \\ \overline{x'^2} & -\overline{xx'} \end{pmatrix} = -E. \tag{10}$$

Thus we have

$$\begin{cases} E^+E = EE^+ = |E|I = \varepsilon^2 I, \\ E + E^+ = (\mathrm{Tr}E)I = 0. \end{cases} \tag{11}$$

Using the first of Eq. (11) we can now show that $|E| \geq 0$

$$\varepsilon^2 = \frac{1}{2}\mathrm{Tr}\left(E^+E\right) = -\frac{1}{2}\mathrm{Tr}\left(\overline{XX^+}\ \overline{XX^+}\right)$$

$$= -\frac{1}{2N^2}\sum_{ij}\mathrm{Tr}\left[\begin{pmatrix} x \\ x' \end{pmatrix}_i (x'-x)_i \begin{pmatrix} x \\ x' \end{pmatrix}_j (x'-x)_j\right]$$

$$= -\frac{1}{2N^2}\sum_{ij}\mathrm{Tr}\left[(x'-x)_i \begin{pmatrix} x \\ x' \end{pmatrix}_j (x'-x)_j \begin{pmatrix} x \\ x' \end{pmatrix}_i\right]$$

$$= \frac{1}{2N^2}\sum_{ij}\left(x'_i x_j - x_i x'_j\right)^2$$

$$= \overline{x^2} \, \overline{x'^2} - \overline{xx'}^2 \qquad \text{(N} \equiv \text{no. of particles)}, \qquad (12)$$

where we have used the fact that the Trace is invariant to cyclic permutations of the product elements. The next to the last expression shows that $|E| \geq 0$.

We note also that a general matrix M is called symplectic if

$$M^+ M = M M^+ = I \quad \text{or} \quad |M| = 1 \qquad (13)$$

The linear motion of a particle corresponds to a transformation of its phase vector by a transfer matrix M. For a Hamiltonian motion M is symplectic. Under the transformation (subscript 1 denotes quantities after transformation)

$$X_1 \equiv MX \qquad X_1^+ = X^+ M^+ \qquad (14)$$

and

$$E_1 \equiv -\overline{X_1 X_1^+} = M\left(-\overline{XX^+}\right)M^+ = MEM^+, \qquad (15)$$

so that

$$\varepsilon_1^2 = |E_1| = \left|MEM^+\right| = |M|^2 |E| = \varepsilon^2. \qquad (16)$$

This shows that the emittance is an invariant for a Hamiltonian motion.

The second moment matrix E contains all three second moments, hence can be represented by a second moment ellipse that also has three parameters: area, ellipticity, and orientation. The equation of the ellipse is

$$-X^+ E^{-1} X = 1, \qquad E^{-1} \equiv \text{inverse of E}. \qquad (17)$$

Here X is the running variable. Since $E^{-1} = \dfrac{E^+}{|E|} = \dfrac{-E}{|E|}$ we can also write the equation as

$$\begin{cases} X^+EX = |E| = \varepsilon^2 \quad \text{or} \quad X^+JX = \varepsilon \quad \text{or} \\[2ex] \gamma x^2 + 2\alpha xx' + \beta x'^2 = \varepsilon. \end{cases} \tag{18}$$

To get the area of the ellipse we go to the principal coordinates (ξ,ξ') by an orthonormal transformation R such that $X = R\Xi = R\begin{pmatrix} \xi \\ \xi' \end{pmatrix}$. This gives

$$X^+EX = \Xi^+R^+ER\Xi = \tilde{\Xi}(\tilde{R}\tilde{S}ER)\Xi. \tag{19}$$

The transformation R diagonalizes $\tilde{S}E$ to give

$$\tilde{R}\tilde{S}ER = \begin{pmatrix} b^2 & 0 \\ 0 & a^2 \end{pmatrix}. \tag{20}$$

The equation of the ellipse is, then

$$b^2\xi^2 + a^2\xi'^2 = |E| = |\tilde{R}\tilde{S}ER| = a^2b^2 \tag{21}$$

or

$$\frac{\xi^2}{a^2} + \frac{\xi'^2}{b^2} = 1. \tag{22}$$

Thus, the area of the ellipse is $\pi\,ab$ and the emittance is given by

$$\varepsilon = |E|^{\frac{1}{2}} = ab = \frac{1}{\pi}\,(\text{area}). \tag{23}$$

This second-moment ellipse is also called the rms emittance ellipse or just the emittance ellipse.

For a Hamiltonian motion we have a bilinear invariant for the motion of every particle i

$$W_i \equiv X_i^+JX_i = \gamma x_i^2 + 2\alpha x_i x_i' + \beta x_i'^2. \tag{24}$$

Under transformation M we have (subscript 1 denotes transformed quantities).

$$W_{il} = X_{il}^+ J_1 X_{il} = X_i^+ M^+ MJM^+ MX_i = X_i^+ JX_i = W_i, \qquad (25)$$

where we used the fact that for a Hamiltonian motion M is symplectic and $M^+M = I$. Equation (25) shows that W_i is an invariant of motion. Averaging W_i over all particles we get the useful relationship

$$\overline{W_i} = \gamma \overline{x_i^2} + 2\alpha \overline{x_i x_i'} + \beta \overline{x_x'}^2 = 2(\beta\gamma - \alpha^2)\varepsilon = 2\varepsilon. \qquad (26)$$

Many different multipliers are used to define the "emittance" by different laboratories and for different applications. These are listed here, expressed in terms of the area A of the second-moment ellipse studied above.

Emittance	% of particles contained	Application
$\varepsilon = \dfrac{1}{\pi} A$ (defined here)	14.7	Electron
$\pi\varepsilon = A$	39.3	Area of ellipse
$2\pi\varepsilon = 2A$	63.2	
$4\pi\varepsilon = 4A$	86.5	CERN
$6\pi\varepsilon = 6A$	95.0	Fermilab
$(\ln 10)\, 2\pi\varepsilon = 4.6\, A$	90.0	Linac

2. Emittance Growth

2.1. *Discrete Process – Mismatch*

A mismatch typically occurs at transfer of the beam from one accelerator to another. After transfer the mismatched beam distribution (i.e., the emittance ellipse) will initially tumble around. Nonlinearities will cause the tumbling distribution to smear (filamentation) and eventually settle into a larger steady-state distribution, namely a larger emittance. We will examine here a few examples of mismatches.

2.1.1 Orbit mismatch (aiming error)

An incident beam with matched second moments, hence the matched emittance

$$\varepsilon^2 = \overline{x_i^2}\,\overline{x_i'^2} - \overline{x_i x_i'}^2 \tag{27}$$

is injected on a trajectory missing the design orbit by δx and $\delta x'$. After reaching steady-state the new emittance is

$$(\varepsilon + \delta\varepsilon)^2 = \overline{(x_i + \delta x)^2}\,\overline{(x_i' + \delta x')^2} - \overline{(x_i + \delta x)(x_i' + \delta x')}^2. \tag{28}$$

All terms linear in x_i or x_i' will average to zero and we get

$$2\varepsilon\delta\varepsilon = \overline{x_i'^2}\delta x^2 - 2\overline{x_i x_i'}\delta x\delta x' + \overline{x_i^2}\delta x'^2. \tag{29}$$

But $\overline{x_i^2} = \beta\varepsilon$, $-\overline{x_i x_i'} = \alpha\varepsilon$, $\overline{x_i'^2} = \gamma\varepsilon$, so we have finally

$$\delta\varepsilon = \frac{1}{2}\left(\gamma x^2 + 2\alpha\delta x\delta x' + \beta\delta x'^2\right). \tag{30}$$

This is independent of ε and shows that the emittance growth, $\delta\varepsilon$, is just $\dfrac{1}{2\pi}$ of the area of the emittance ellipse passing through (δx, $\delta x'$). This is shown in Figure 1.

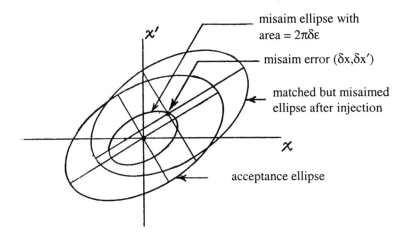

Figure 1. Emittance growth due to misaiming at injection.

2.1.2 Optics mismatch (distribution or second-moment error)

We assume that a beam with betatron parameters α and β is injected into an accelerator with "matched" lattice betatron parameters $\alpha+\delta\alpha$ and $\beta+\delta\beta$. For particle i before injection the betatron motion is given by the invariant

$$W_i = \gamma x_i^2 + 2\alpha x_i x_i' + \beta x_i'^2. \tag{31}$$

After injection the change in its motion is given by

$$\delta W_i = \delta\gamma x_i^2 + 2\delta\alpha x_i x_i' + \delta\beta x_i'^2, \tag{32}$$

where $\delta\gamma$ is to be expressed in terms of $\delta\alpha$ and $\delta\beta$. This is

$$\delta\gamma = \frac{1+(\alpha+\delta\alpha)^2}{\beta+\delta\beta} - \frac{1+\alpha^2}{\beta}$$

$$= \frac{1}{\beta}(2\alpha\delta\alpha - \gamma\delta\beta) + \frac{1}{\beta^2}(\gamma\delta\beta^2 - 2\alpha\delta\beta\delta\alpha + \beta\delta\alpha^2), \tag{33}$$

where in the last expression we have expanded to second-order terms in $\delta\alpha$ and $\delta\beta$. We average δW_i over all particles and substitute from Eq. (9) and Eq. (26). This gives

$$2\delta\varepsilon = \left(\beta\delta\gamma - 2\alpha\delta\alpha + \gamma\delta\beta\right)\varepsilon. \tag{34}$$

Substituting $\delta\gamma$ from Eq. (33) we get the emittance increment

$$\frac{\delta\varepsilon}{\varepsilon} = \frac{1}{2\beta}\left(\gamma\delta\beta^2 - 2\alpha\delta\beta\delta\alpha + \beta\delta\alpha^2\right). \tag{35}$$

2.2. Continuous process – Diffusion

During acceleration and storage (in a storage ring) there are always small non-Hamiltonian disturbances causing emittance growth. Although minute and random, these disturbances occur densely in time so that their effects can be approximated as being continuous and that their time-average effects can be considered constant over a reasonable length of time.

We describe the distribution of phase points in the phase plane by the distribution function

$$\psi = \psi(x,x';t) = \psi(w,\phi;t), \tag{36}$$

where

$$\begin{cases} w = \gamma x^2 + 2\alpha xx' + \beta x'^2 = \dfrac{x^2 + (\alpha x + \beta x')^2}{\beta} \\ \tan\phi = \dfrac{\alpha x + \beta x'}{x} \end{cases} \tag{37}$$

are the Floquet "polar" coordinates. The time evolution of the distribution function is given by the diffusion equation

$$\frac{\partial\psi}{\partial t} = D\frac{\partial}{\partial w}\left(w\frac{\partial\psi}{\partial w}\right), \tag{38}$$

where

$$D \equiv \frac{d\overline{W_i}}{dt} = \text{diffusion constant.}$$

In this expression $\overline{W_i}$ = average over all particles of their "invariants" of motion (which are, now, slowly and continuously growing), and D is sensibly constant over the time interval concerned.

The diffusion equation (38) gives the detailed time variation in the phase-point distribution. In most cases, however, one is interested only in the emittance growth. This is given simply by

$$\delta\varepsilon = \frac{1}{2}\delta\overline{W_i} = \frac{1}{2}D\delta t . \tag{39}$$

In fact, one generally derives $\delta\overline{W_i}$ first (hence $\delta\varepsilon$) in order to get the diffusion constant D. We illustrate this by the following examples.

2.2.1 Quantum excitation from synchrotron radiation

This was studied in section 2.2 of the course on "Synchrotron Radiation." There it was shown that the quantum nature of the emission of synchrotron radiation leads to the diffusion constants

$$D_a \equiv \frac{d\overline{w_a}}{dt} = \frac{d}{dt}\left\langle \overline{A_a^2} \right\rangle = 4\frac{Q_a}{\tau} , \tag{40}$$

where subscript a stands for x, y, or E, and we have omitted the subscript i for individual particles. The quantities Q_a and τ are as defined in section 2.2 of "Synchrotron Radiation."

2.2.2 Residual gas scattering

A stored beam will suffer multiple scattering from the residual gas molecules in the vacuum chamber. We denote the plane projected rms scattered angle of a beam particle by $\theta_{rms}(\ell)$, where ℓ is the distance traversed in the scattering medium. From a fitting to Moliére scattering theory one gets

$$\theta_{rms}(\ell) = \frac{13.6\,\text{MeV}}{pv}\, z\sqrt{\frac{\ell}{X_o}}\underbrace{(1+0.038\ln\frac{\ell}{X_o})}_{\text{negligible}} , \tag{41}$$

where p, v, and z are the momentum, velocity, and charge number, respectively, of the beam particle, and X_o is the radiation length of the scattering medium (here the residual gas). From

$$W_i = \gamma x_i^2 + 2\alpha x_i x_i' + \beta x_i'^2 \tag{42}$$

we get for an increment $\delta x_i'$

$$\delta W_i = 2(\alpha x_i + \beta x_i')\delta x_i' + \beta \delta x_i'^2, \tag{43}$$

and (again omitting subscript i)

$$\delta \overline{W} = \beta \overline{\delta x'^2} = \frac{R}{v}\theta_{rms}^2, \tag{44}$$

where R = radius and v = tune of the storage ring. The emittance growth is then

$$\delta \varepsilon = \frac{1}{2}\delta\overline{W} = \frac{Rz^2}{2v}\left(\frac{13.6\,\mathrm{MeV}}{pv}\right)^2 \frac{\ell}{X_o}. \tag{45}$$

Generally, this is all one wants. But one can also get the diffusion constant

$$D = \frac{d\overline{W}}{dt} = v\frac{d\overline{W}}{d\ell} = \frac{Rz^2}{v}\left(\frac{13.6\,\mathrm{MeV}}{pv}\right)^2 \frac{v}{X_o} \tag{46}$$

and solve the diffusion equation.

3. Emittance Reduction (Cooling)

3.1. *Temperature*

We use the word "cooling" to describe the reduction of the emittance or the average kinetic energy of the random motion of the particle. In a reference frame (the beam rest-frame) moving at the average velocity of the beam particles the motion of the particles is totally random (zero average), and the temperature T as defined in statistical mechanics is, in one degree of freedom (dof), say, x

$$\frac{1}{2}kT_x = \frac{1}{2}\overline{mv_x^{*2}} = \text{average K.E.} \quad \text{(nonrelativistic)}, \quad (47)$$

where * denotes quantities in the beam rest-frame. Normally T_x is measured in units of Kelvin, K, the Boltzmann constant k has the value $k = 8.6 \times 10^{-5}$ eV/K. It is customary in this application to set $k = 1$ and measure T in energy units namely $T_x = m\overline{v_x^{*2}}$.

In three dof we then have

$$\frac{3}{2}kT = \frac{1}{2}\overline{mv^{*2}} = \frac{1}{2}m\left(\overline{v_x^{*2}} + \overline{v_y^{*2}} + \overline{v_z^{*2}}\right)$$

$$= \frac{1}{2}k\left(T_x + T_y + T_z\right)$$

or

$$T = \frac{1}{3}\left(T_x + T_y + T_z\right) = \text{average of one dof Ts.} \quad (48)$$

In terms of lab-quantities we have

$$\begin{cases} v_x^* = \gamma v_x = \gamma vx', \\ v_y^* = \gamma v_y = \gamma vy', \\ v_z^* = \gamma^2 \delta v = v\dfrac{\delta p}{p}. \end{cases} \quad (49)$$

And the beam emittances are

$$\varepsilon_x = \frac{\overline{x'^2}}{\gamma_x}, \qquad \varepsilon_y = \frac{\overline{y'^2}}{\gamma_y}, \tag{50}$$

where γ_x and γ_y are the Courant-Snyder parameters. Altogether this gives

$$\begin{cases} T_x = m\overline{v_x^{*2}} = m\gamma^2 v^2 \overline{x'^2} = \left(m\gamma^2 v^2\right)\gamma_x \varepsilon_x, \\[2mm] T_y = m\overline{v_y^{*2}} = m\gamma^2 v^2 \overline{y'^2} = \left(m\gamma^2 v^2\right)\gamma_y \varepsilon_y, \\[2mm] T_z = m\overline{v_z^{*2}} = \left(mv^2\right)\overline{\left(\frac{\delta p}{p}\right)^2}. \end{cases} \tag{51}$$

It is frequently true that $T_x \cong T_y$. Hence we define a transverse temperature $T_\perp \equiv \frac{1}{2}(T_x + T_y)$, which is generally quite different from the longitudinal temperature $T_\parallel \equiv T_z$.

3.2. Synchrotron radiation damping (cooling)

This subject was studied in the course on "Synchrotron Radiation." There the emittance-damping rates (cooling rates) were shown to be

$$\frac{1}{\tau_a} = -\frac{1}{\overline{A_a^2}} \frac{d\overline{A_a^2}}{dt} = \frac{2}{\tau} J_a, \tag{52}$$

where a = x, y, or E and where the standard damping rate $\frac{1}{\tau}$ and the partition factors J_x, J_y, and J_E are as defined in section 2.1 of "Synchrotron Radiation."

We note that there are essentially two separate effects leading to cooling – the acceleration effect and the orbit-dispersion effect. The dispersion effect is peculiar to orbits in a ring accelerator. The acceleration effect is quite general. This is the effect that the transverse momentum of a beam particle is continually reduced by a succession of processes during which the particles lose momentum (and energy) either isotropically or in its instantaneous direction of motion but has the energy restored by an accelerating system that imparts momentum only in the forward direction of the beam axis. This general feature is employed in several other cooling schemes as will be discussed later.

3.3. Stochastic cooling (electronic feedback cooling)

We discuss here only the transverse (x) cooling. To correct the position and angle of the centroid (first moments) motion of a beam using a feedback system is straightforward and frequently applied. In 1980 Simon Van der Meer of CERN, Geneva, showed that the beam width (second moments or emittance) can also be reduced by a fast feedback system. This is because particles at larger excursions contribute more to the beam–centroid signal. With a gain of g after one feedback correction the position of the kth particle is

$$x_{k1} = x_k - g\bar{x} \,, \tag{53}$$

where $\bar{x}$ is the centroid position (subscript k omitted), and subscript 1 denotes quantities after the correction. Squaring Eq. (53) gives

$$x_{k1}^2 = (x_k - g\bar{x})^2 = x_k^2 - 2g\bar{x}x_k + g^2\bar{x}^2 \,. \tag{54}$$

We can write $\bar{x}$ and $\bar{x}^2$ as

$$\begin{cases} \bar{x} = \dfrac{1}{N_s}\sum_i x_i = \dfrac{1}{N_s}x_k + \dfrac{1}{N_s}\sum_{i \ne k} x_i \,, \\[4mm] \bar{x}^2 = \dfrac{1}{N_s^2}\left(\sum_i\sum_j x_i x_j\right) = \dfrac{1}{N_s^2}\left(\sum_i x_i^2\right) + \underbrace{\dfrac{1}{N_s^2}\left(\sum_{i \ne j}\sum x_i x_j\right)}_{\parallel \atop 0} \\[4mm] \quad\;\; = \dfrac{1}{N_s}\overline{x^2} \,, \end{cases} \tag{55}$$

where N_s is the number of particles in the sample. Equation (54) then becomes

$$x_{k1}^2 = x_k^2 - \frac{2g}{N_s}x_k^2 - \frac{2g}{N_s}x_k\left(\sum_{i \ne k} x_i\right) + \frac{g^2}{N_s}\overline{x^2} \,. \tag{56}$$

Taking average of Eq. (56) over all particles k we get

$$x_1^2 = \overline{x^2} - \frac{2g}{N_s}\overline{x^2} - \frac{2g}{N_S}\underbrace{\left(\sum_{i\neq k}\sum x_i x_k\right)}_{0} + \frac{g^2}{N_s}\overline{x^2}. \qquad (57)$$

The emittance damping rate is then

$$\frac{1}{\tau} = -\frac{1}{\varepsilon}\frac{d\varepsilon}{dt} = -\frac{1}{x^2}\frac{\overline{dx^2}}{dt} = \frac{1}{TN_s}\left(2g - g^2\right), \qquad (58)$$

where T is the revolution period (assuming one correction per turn). The first term in the parenthesis, 2g, is the coherent self-feedback and the second term, g^2, is the incoherent noise contribution from all other particles. In this ideal situation the maximum damping rate is $\dfrac{1}{TN_s}$ at g = 1. To Eq. (58) we have to make the following modifications for the practical case.

    a.   Noise in the feedback electronics is defined as

$$U = \frac{\text{noise power}}{\text{signal power}}. \qquad (59)$$

    This should be included in the coefficient of g^2.

    b.   The sample size N_s or the sample time T_s is determined by the bandwidth W of the electronics to be $T_s = \dfrac{1}{2W}$. The sample size is then $N_s = N\dfrac{T_s}{T}$, where N is the total number of particles in the ring.

    c.   We like N_S small to get high damping rate, hence we need broadband electronics. But for $N_S < N$ we need "mixing" in order to reach all the particles in the ring. If $\delta T \equiv$ spread in T due to momentum spread in beam, we can define a mixing parameter

$$M \equiv \frac{T_s}{\delta T}. \qquad (60)$$

If $\delta T = T_s$, M = 1, the sample will be totally mixed every turn. Generally $\delta T < T_s$ and M > 1. This should also be included in the coefficient of the noise term to replace unity.

With all these modifications included we get

$$\frac{1}{\tau} = \frac{2W}{N}\left[2g - g^2(M+U)\right].$$ (61)

The ideal case is when $U = 0$ and $M = 1$. In general, the damping rate is maximum when

$$g = g_{max} = \frac{1}{M+U} \ll 1$$ (62)

showing that the optimum gain is much less than 1. The maximum damping rate is then

$$\left(\frac{1}{\tau}\right)_{max} = \frac{2W}{N}\frac{1}{M+U}.$$ (63)

For a numerical example we take $W = 5 \times 10^9$ s^{-1} (5 GHz electronics), $N = 10^{10}$ (antiprotons, say) and $M+U \approx 10$. This gives

$$\left(\frac{1}{\tau}\right)_{max} \approx 0.1\,s^{-1}.$$ (64)

Many fancy electronics innovations have been developed and used to make stochastic cooling systems serviceable and efficient. It is gratifying that with currently available fast electronics the damping rate is adequate even for rather large practical samples.

3.4. Electron cooling

This was first proposed in 1966 by G. I. Budker of BINP, Novosibirsk. In this scheme a cold (low emittance, low random kinetic energy) electron beam is made to travel and overlap with the proton beam. Traveling together means same average β and γ for both beams. Thus the electron beam energy should be 1/1836 times that of the proton beam. In the rest frame of both beams the cold electrons form a nearly stationary electron medium through which the warm protons traverse randomly. The protons are then slowed down by Coulomb interactions with the electrons until they come to energy equilibrium with the much cooler electron gas (equipartition of energy). The emittance of the proton beam is thus reduced at the expense of increasing the emittance of the electron beam.

Since we will be using the beam rest-frame a great deal, to simplify notation we shall drop the star * for indicating rest-frame quantities. Instead we will specify in parenthesis after each equation whether it is in the beam rest-frame (R) or the lab-frame (L).

When all electrons are at rest in the beam rest-frame (T = 0) the slowing down of the proton by Coulomb interactions with the electrons is given by the Bethe-Bloch (Bohr) formula for the "stopping power" (mean rate of energy loss)

$$\left(\frac{dE}{ds}\right)_p = -\frac{4\pi(\ln\Lambda)}{m}\frac{e^4}{v_p^2}$$

$$= \text{friction force on proton,} \qquad \text{(R) (65)}$$

where

$\mathcal{E}$ = kinetic energy,   s = distance traveled,

$\ln\Lambda$ = Coulomb logarithm $\cong$ 12,

m, e = electron mass and charge,

n = electron density,

and we used subscript p to indicate proton quantities (no subscript = electron)

If the electrons have a distribution $f(\vec{v})$ in velocity (generally Maxwellian), the stopping power is modified by a velocity potential, which can be obtained by an analogy to the electrostatic potential of a distribution in position, $\rho(\vec{r})$, of electrons. For easy comparison we recapitulate here the electrostatic potential formulation.

When all N electrons are located at the origin (r=0), the force on a proton at position $\vec{r}_p$ is given by

$$\vec{F}(\vec{r}_p) = -Ne^2\frac{\vec{r}_p}{r_p^3} = -e\vec{\nabla}_{rp}\phi, \qquad (66)$$

where the potential is defined as $\phi(\vec{r}_p) = -\frac{Ne}{r_p}$ (electronic charge = -e). When the N electrons are distributed in position as $\rho(\vec{r})$ (normalized $\int\rho(\vec{r})d^3\vec{r} = 1$), the potential becomes

$$\phi\left(\vec{r}_p\right) = -Ne \int \frac{\rho(\vec{r})}{\left|\vec{r}_p - \vec{r}\right|} d^3\vec{r} . \qquad (67)$$

This gives the force

$$\vec{F}\left(\vec{r}_p\right) = -e\vec{\nabla}_{rp}\phi = -Ne^2 \int \frac{\vec{r}_p - \vec{r}_p}{\left|\vec{r}_p - \vec{r}\right|^3} \rho(\vec{r}) d^3\vec{r} \qquad (68)$$

and the force gradient

$$\vec{\nabla}_{rp} \cdot \vec{F} = -e\nabla_{rp}^2\phi = -4\pi Ne^2 \rho\left(\vec{r}_p\right). \qquad (69)$$

Similarly, when all n electrons are at v = 0 (stationary), the force on the proton is

$$\vec{F}\left(\vec{v}_p\right) = -Ke^2 n \frac{\vec{v}_p}{v_p^3} \quad \text{where} \quad K \equiv 4\pi r_e c^2 \, (\ln\Lambda). \qquad \text{(R) (70)}$$

Thus, when the electrons are distributed in velocity as $f(\vec{v})$ $\left(\text{normalized} \int f(\vec{v}) d^3\vec{v} = 1\right)$, we can write a velocity potential (Rosenbluth potential)

$$V\left(\vec{v}_p\right) = -ne \int \frac{f(\vec{v})}{\left|\vec{v}_p - \vec{v}\right|} d^3\vec{v} , \qquad \text{(R) (71)}$$

which gives the friction force on the proton

$$\vec{F}\left(\vec{v}_p\right) = -Ke\vec{\nabla}_{vp} V = -Ke^2 n \int \frac{\vec{v}_p - \vec{v}}{\left|\vec{v}_p - \vec{v}\right|^3} f(\vec{v}) d^3\vec{v} . \qquad \text{(R) (72)}$$

Thus, the velocity damping rate is

$$\frac{1}{\tau_{vp}} = \frac{1}{m_p} \vec{\nabla}_{vp} \cdot \vec{F} = -\frac{Ke}{m_p} \nabla_{vp}^2 V = -4\pi \frac{Ke^2 n}{m_p} f\left(\vec{v}_p\right). \qquad \text{(R) (73)}$$

One can assume the electron velocity distribution to be Maxwellian, except here the longitudinal and the transverse distributions are generally different. So we will write

$$f(\vec{v}) = \frac{e^{-\frac{mv_\parallel^2}{2T_\parallel}} e^{-\frac{mv_\perp^2}{T_\perp}}}{(2\pi)^{3/2}(T_\parallel/m)^{1/2}(T_\perp/m)}.$$

(R) (74)

Hence

$$\frac{1}{\tau_{vp}} = \frac{4\sqrt{2\pi}r_e r_p cn(\ln\Lambda)}{\left(\frac{T_\parallel}{mc^2}\right)^{\frac{1}{2}}\left(\frac{T_\perp}{mc^2}\right)} e^{-\frac{mv_{p\parallel}^2}{2T_\parallel}} e^{-\frac{mv_{p\perp}^2}{T_\perp}}.$$

(R) (75)

The proton velocity v_p is generally smaller than the electron velocity v. (At equipartition $\dfrac{v_p^2}{v^2} = \dfrac{m}{m_p} = \dfrac{1}{1836}$.) So we can put the exponential factor equal to unity. We need also to transform back to the lab-frame with (restore superscript star, *, for beam rest-frame quantities)

$$t = \gamma t^*, \qquad n = \gamma n^* = \frac{j}{e\beta c},$$

(76)

and the emittance damping rate is related to the velocity damping rate by

$$\frac{1}{\tau_{\varepsilon p}} = \frac{1}{2}\frac{1}{\tau_{vp}} = \frac{1}{2\gamma}\frac{1}{\tau_{vp}^*}.$$

(77)

So we get finally

$$\frac{1}{\tau_{\varepsilon p}} = \frac{2\sqrt{2\pi}r_e r_p (j/e)(\ln\Lambda)}{\beta\gamma^2\left(\frac{T_\parallel}{mc^2}\right)^{\frac{1}{2}}\left(\frac{T_\perp}{mc^2}\right)}\eta,$$

(L) (78)

where η is the fraction of the ring circumference covered by electron cooling beam.

The $\dfrac{1}{\beta\gamma^2}$ dependence of Coulomb interaction makes the electron cooling rate impractically low at proton energies much above 1 GeV. Take for example the electron cooling ring at Fermi National Accelerator Laboratory, which has the following parameters:

$$\beta = 0.566 \qquad \left(\begin{array}{l} E_p = 200 \text{ MeV} \\ E_e = 109 \text{ keV} \end{array} \right)$$
$$\gamma = 1.213$$
$$T_\parallel = 0.2 \text{ eV} \qquad j_e = 1.5 \text{ A/cm}^2$$
$$T_\perp = 0.3 \text{ eV} \qquad \eta = 0.05,$$

which gives

$$\frac{1}{\tau_{\varepsilon p}} = 4.0 \text{ sec}^{-1}. \qquad \text{(L)} \quad (79)$$

We mention here a few additional features of electron cooling without giving quantitative analyses.

    a.   The protron velocity $\vec{v}_p$ also has a distribution. Averaging over $\vec{v}_p$ is, however, straightforward. In most cases, it suffices to just evaluate the cooling rate for the average $\vec{v}_p$.

    b.   Magnetized fast cooling – The beams overlapping region is frequently enclosed in a solenoid so that the electron beam is confined by the solenoid field $\vec{B}_\parallel$. Without $\vec{B}_\parallel$, after a Coulomb interaction with a proton the large scattering angle of the electron will send it away from the protons. With a strong $\vec{B}_\parallel$, the gyroradius of the electron after a scattering is smaller than the Debye shielding length so that the same electron can interact with the protons over and over again leading to a greatly enhanced cooling rate.

    c.   For reasonably achievable electron density n, proton beams of kinetic energy < 1 GeV (electron beam energy of $\lesssim$ 500 keV) can be cooled with a good rate. Energy recovery from the electron beam is necessary. Different electron systems are proposed and used for different energy ranges:

1. Low energy: $\mathscr{E}_p < 1$ GeV, $\mathscr{E}_e < 500$ keV. In this range one can use a Cockcraft-Walton electron accelerator with energy recovery.
2. Medium energy: $\mathscr{E}_p < 40$ GeV, $\mathscr{E}_e < 20$ MeV. One can use a Van de Graaff electron accelerator with energy recovery.
3. High energy: $\mathscr{E}_p < 200$ GeV, $\mathscr{E}_e < 100$ MeV. One can contemplate using an electron storage ring.

So far, however, all electron cooling systems are used in the low-energy range.

3.5. Ionization cooling

In ionization cooling one tries to simulate the process comprising the synchrotron radiation cooling. The emission of synchrotron radiation reduces the energy of the electron and takes away the corresponding momentum along the direction of motion. When the energy loss is replenished by the rf-accelerating system that imparts momentum only in the forward direction, a small amount of the transverse momentum component is removed, leading thereby to cooling.

When a heavy particle goes through an energy-attenuating target it loses energy by ionizing the target atoms and, in so doing, loses momentum along the direction of motion in exactly the same manner as the process of emitting synchrotron radiation. The only different is, in this case, the particle also suffers random scattering by the target atoms. Thus it is now a competition between the damping (cooling) due to the ionization energy-momentum loss (plus rf acceleration) process and the excitation (heating) due to the scattering process. One finds that in some special circumstances and arrangements (e.g., the muon collider) the cooling effect indeed surpasses the heating effect, at least during the initial application of the process.

RF SYSTEMS FOR LIGHT SOURCE STORAGE RINGS

Z. T. ZHAO

Shanghai Synchrotron Radiation Facility
2019 Jialuo Road ,Shanghai 201800,P.R.China
E-mail: zhaozt@ssrc.ac.cn

The RF system is one of the sophisticated components in light source storage rings. Its extensive contents cover a broad range of physical and technical topics. This lecture does not intend to give an exhaustive introduction to all aspects of the RF system, but focuses on its basics, including RF acceleration, cavity fundamentals, steady beam loading effects, key components of RF systems, and harmonic RF systems. The transient beam loading, cavity higher order modes issues, and the corresponding coupled bunch instabilities are not discussed in this lecture.

1. Introduction

The RF system is one of the main accelerator sub-systems. It transfers the mains power to beam power and accelerates the charged particle to higher energy or keeps a constant energy of the charged particle in storage ring. In light source storage rings, the RF system not only provides the energy to electron bunches, but also determines some beam parameters such as bunch length and Touschek life time. Moreover, as one of the major RF system components, the RF cavity interacts with the circulating beam via beam coupling impedance and may cause beam instabilities. This lecture intends to introduce briefly the principle of RF acceleration and the fundamentals of RF systems for light source storage rings. It cannot be exhaustive, but restricts itself to basic concepts of the RF physics and technology that exist in every RF system. In section 2, the RF acceleration in storage rings is reviewed with an introduction to phase stability and RF dependent machine parameters. Section 3 and 4 present RF cavity fundamentals and steady beam loading effect of RF systems. Section 5 describes the light source storage ring RF system and its main components via a few existing examples and aims at giving an overview of this main accelerator sub-system. Finally, section 6 introduces the main principle and examples of the high harmonic RF acceleration for controlling bunch length and providing Landau damping. There are some other rich and important topics on RF system not discussed in this lecture, e.g. transient beam loading issue, cavity higher order modes (HOM) impedances and HOM induced coupled bunch instabilities. These issues are particularly important for the high current storage rings. The interested reader may refer to the literature given [1~2].

2. RF Acceleration in Storage Rings

2.1. *RF Acceleration and Phase Stability in Storage Rings*

When traveling in an electromagnetic field with speed of v, an electron experiences the Lorentz force which can be written as

$$\vec{F} = -e\vec{E} - e(\vec{v} \times \vec{B}) \tag{2.1}$$

The electromagnetic field does work on the electron, and its rate is given by

$$\vec{F} \cdot \vec{v} = -e\vec{E} \cdot \vec{v} - e(\vec{v} \times \vec{B}) \cdot \vec{v} . \tag{2.2}$$

Since $e(\vec{v} \times \vec{B}) \cdot \vec{v}$ is always zero, only the electric field can exchange energy with the electron and accelerate it. There are three types of electric field used in electron acceleration, i.e. direct-voltage (static or pulsed) electric field, inductive electric field and RF electric field [3]. In a static electric field,

$$\oint \vec{E} \cdot dl = 0, \tag{2.3}$$

while in an RF electric field,

$$\oint \vec{E} \cdot dl \neq 0 . \tag{2.4}$$

This indicates that RF electric fields can be used in circular particle acceleration. RF electromagnetic fields are established in traveling wave or standing wave accelerating structures. In light source storage rings, the standing wave resonant cavities are used exclusively to accelerate electrons. The standing wave electric field in an acceleration gap can be expressed as

$$E_z(r, z, t) = E_z(r, z)\sin(\omega t + \phi) , \tag{2.5}$$

thus the energy gain on axis over the acceleration gap distance g is

$$\Delta W = e\int_{-g/2}^{g/2} E_z(0, z)\sin(\omega t + \phi)dz . \tag{2.6}$$

With the synchronous condition $z=ct$ and $E_z(0,z)$ being an even function with respect to the geometric centre of the acceleration gap, this can be written in the form

$$\Delta W = eV_c \sin\phi , \tag{2.7}$$

$$V_c = \int_{-g/2}^{g/2} E_z(0, z)\cos\omega t dz \tag{2.8}$$

where V_c is the maximum accelerating voltage of the RF structure.

In a light source storage ring, electrons circulate on the orbit defined by magnetic fields. They lose their energy due to synchrotron radiation and gain the energy back while passing through accelerating cavities. As shown in Figure 2.1, as long as the RF voltage is larger than a required minimum, there exists an RF focusing process which provides bunching of the beam and captures electrons within RF buckets. Furthermore, there is an ideal electron in each electron bunch which has the nominal energy E_0, always travels along the closed orbit, and passes through the RF cavity at a constant RF phase ϕ_s called the synchronous phase. This ideal electron is called the synchronous particle. It is characterized by an orbital revolution time T_0, and an orbital revolution frequency ω_0. The integer h, defined by ω_{RF}/ω_0, is called the harmonic number of the storage ring.

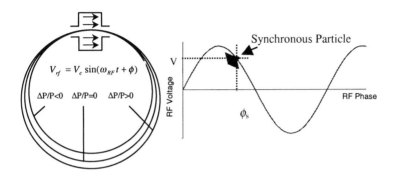

Figure 2.1 Synchrotron phase and phase focusing.

The electron energy loss on each turn U_0, the RF voltage and the synchronous phase follow the relation

$$U_0 = eV_c \sin \phi_s \,. \tag{2.9}$$

The stable synchronous phase for electron storage rings is in the range of $\pi/2 < \phi_s < \pi$. Electrons in the bunch travel along the ring and execute longitudinal oscillations around the synchronous particle due to energy or phase deviations. For example, an electron with the same initial RF phase ϕ_s, but a somewhat higher energy than the synchronous particle, will take more time to make one revolution around the ring. Thus, it will arrive at the RF cavity at an RF phase ϕ $> \phi_s$ and gain less energy through the cavity. In contrast, a lower energy electron will need less time for one revolution, and will arrive at the RF cavity at an RF phase $\phi < \phi_s$, thus gaining more energy. This principle inherent to RF acceleration is the so-called phase stability of an accelerator. The longitudinal beam dynamics can analytically describe this principle via electron synchrotron motion in detail. Analysis shows that the energy oscillation and the phase oscillation can be expressed as [4]

$$\frac{d\Delta\phi}{dt} = \alpha_c \omega_{RF} \frac{\Delta E}{E_0} \,, \tag{2.10}$$

$$\frac{d\Delta E}{dt} = \frac{eV_c \sin \phi - U_0}{T_0} - 2\alpha_E \Delta E \,, \tag{2.11}$$

where $\Delta E = E - E_0$ is the energy deviation• $\Delta\phi = \phi - \phi_s$ the phase deviation, α_c the momentum compaction factor and α_E the energy (or phase) oscillation damping rate from synchrotron radiation. In small amplitude oscillation, the phase equation can be written as

$$\frac{d^2\Delta\phi}{dt^2} + 2\alpha_E \frac{d\Delta\phi}{dt} + \omega_s^2\Delta\phi = 0 \,, \tag{2.12}$$

where ω_s is the synchrotron angular frequency and can be written as

$$\omega_s = \omega_0 \sqrt{\frac{e\alpha_c V_c h \cos\phi_s}{2\pi E_0}} \,. \tag{2.13}$$

In large amplitude oscillation case, there exists a phase space region within which electrons perform the stable longitudinal motion. This phase space determines an energy acceptance which can be expressed as [5]

$$\frac{\Delta E_{max}}{E_0} = \sqrt{\frac{U_0 F(q_V)}{\pi\alpha_c h E_0}} \,, \tag{2.14}$$

$$F(q_V) = 2\left[\sqrt{q_V^2 - 1} - \cos^{-1}\left(\frac{1}{q_V}\right)\right], \tag{2.15}$$

where $q_V = eV_c/U_0$ is the over voltage factor.

The energy acceptance can be enlarged by choosing lower RF frequency or increasing RF voltage of storage rings. However, higher RF voltage means more expenses of RF power dissipated on the RF cavities.

2.2. RF Dependent Parameters

Some important beam parameters depend strongly on the RF voltage and the resonant frequency of the cavities in light source storage rings, especially the bunch length, the quantum life time and Touschek life time. These issues impose the design requirement on the RF system and make it not only energize the electron beam, but also control the beam performance, such as providing longitudinal focusing, bunch length and energy acceptance as well as beam life time controlling, and moreover, providing damping to the longitudinal beam oscillations.

When circulating electrons emit synchrotron radiation in storage rings, the excitation and damping effects on the energy oscillation of these electrons exist simultaneously and lead to an energy spread of the bunched electrons. This energy spread σ_E together with synchrotron angular frequency in turn determines the bunch length of these electrons, which can be written as [5]

$$\sigma_l = \frac{\alpha_c c}{\omega_s} \frac{\sigma_E}{E_0}. \tag{2.16}$$

From (2.13), the bunch length becomes

$$\sigma_l = \frac{c\sigma_E}{\omega_{RF}} \sqrt{\frac{2\pi\alpha_c h}{E_0 eV_c \cos\phi_s}} \tag{2.17}$$

The quantum excitation of synchrotron radiation has an impact on beam life time if the energy loss of an electron due to the emission of photons exceeds the energy acceptance. This electron will be lost as its energy oscillation extends beyond the stable region. This loss rate is characterized by the longitudinal quantum life time [5]

$$\tau_q = \frac{E_0 T_0}{U_0} \frac{e^\xi}{2\xi}, \quad \xi = \frac{1}{2}\left(\frac{\Delta E_{\max}}{\sigma_E}\right)^2 \tag{2.18}$$

In light source storage rings, however, it is the Touschek life time that has the determining impact on the machine parameter optimizations. Touschek effect, i.e. large-angle intrabeam scattering, dominates the beam life time in low and medium energy electron storage rings or high energy electron storage rings operated at few bunches mode. The Touschek life time is strongly dependent on bunch volume and energy acceptance, and can be expressed as [6]

$$\tau_T = \frac{8\pi <\sigma_x\sigma_y> \sigma_l\gamma^2}{r_e^2 cN_b D(\varepsilon)} \left(\frac{\Delta E_{\max}}{E_0}\right)^3, \tag{2.19}$$

where $<\sigma_x\sigma_y>$ is the transverse beam cross-section, σ_l the bunch length, r_e the classical electron radius, N_b the number of electrons per bunch, $D(\varepsilon)$ the Touschek effect function expressed by

$$D(\varepsilon) = \sqrt{\varepsilon}\left[-\frac{3}{2}e^{-\varepsilon} + \frac{\varepsilon}{2}\int_\varepsilon^\infty \frac{\ln\mu}{\mu}e^{-\mu}d\mu + \frac{1}{2}(3\varepsilon - \varepsilon\ln\varepsilon + 2)\int_\varepsilon^\infty \frac{e^{-\mu}}{\mu}d\mu\right], \quad (2.20)$$

$$\varepsilon = \frac{1}{2\gamma^2}\frac{<\beta_x>}{\varepsilon_x}\left(\frac{\Delta E_{\max}}{E_0}\right)^2, \tag{2.21}$$

where $<\beta_x>$ is the average horizontal beta function, ε_x the horizontal emittance and γ the relativistic energy factor.

The Touschek life time depends obviously on bunch charge, bunch volume and energy acceptance, and thus, in turn, implicitly on RF voltage and RF frequency through the bunch number, bunch length, and energy acceptance.

2.3. *Beam Power and Required RF Voltage*

In storage rings, the RF power delivered from the high power RF source to the beam is determined by beam current and electron energy loss per turn due to synchrotron radiation and wake fields. It can be simply written as

$$P_b = I_b U_0 / e = I_b V_c \sin \phi_s . \qquad (2.22)$$

The required RF voltage depends apparently on the electron energy loss per turn, the electron bunch length and RF frequency. It can be derived from equations (2.17) and (2.22), and is given by

$$V_c = \sqrt{\left(\frac{P_b}{I_b}\right)^2 + \left(\frac{2\pi\alpha_c hc^2\sigma_E^2}{eE_0\omega_{RF}^2\sigma_l^2}\right)^2} . \qquad (2.23)$$

In light source storage rings, energy acceptance is the crucial parameter which dominates the required RF voltage, and becomes a pre-condition to design the RF system for getting longer Touschek life time. When the RF over voltage factor is much larger than 1, $F(q_v) \approx 2q_v\text{-}\pi$, simpler relations among RF voltage, bunch length and energy acceptance can be derived from equations (2.14) and (2.17), which are

$$V_c = \frac{\pi\omega_{RF}\alpha_c E}{2e\omega_0}\left(\frac{\Delta E_{max}}{E}\right)^2 + \frac{\pi U_0}{2e} , \qquad (2.24)$$

$$\sigma_l = \frac{2c\sigma_E}{\omega_{RF}}\left[\left(\frac{\Delta E_{max}}{E}\right)^2 + \frac{\omega_0 U_0}{\omega_{RF}\alpha_c E}\right]^{-\frac{1}{2}} . \qquad (2.25)$$

3.   Fundamentals of RF Cavities

3.1. *Resonant Modes and Electromagnetic Fields*

An accelerating field can be efficiently established in an electromagnetic resonator, namely, an RF cavity. Cylindrical resonators and coaxial resonators are commonly used RF cavities for accelerating electrons in light source storage rings. The coaxial cavities are often used at low frequencies below 100MHz, and the cylindrical cavities are the dominant ones at frequencies above 100MHz, covering the majority of cavity scenarios in light source storage rings. Therefore, this lecture will mainly discuss the details of cylindrical cavities.

As sketched in Figure 3.1, the practical RF cavity used in SRS [5] is a cylindrical metal box with two beam ports on the cylindrical axis. In addition, there are a few more ports on the cavity body for installing a tuner, an input power coupler, and HOM couplers. The accelerating field is excited through this input coupler which transfers the RF power from the high power RF amplifier to the cavity for making up the cavity dissipated power and the beam power extracted by electron bunches passing through the RF cavity.

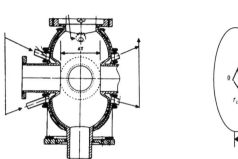 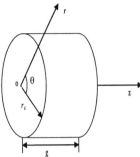

Figure 3.1 Daresbury SRS cavity [7]　　Figure 3.2 A pill box cavity

Most of the practical cavities have evolved geometrically and electrically from an ideal pill box cavity which reveals the main characteristics of RF cavities. Figure 3.2 shows a pill box cavity with radius r_c and length g where the standing wave electromagnetic field satisfying its boundary conditions can be established by proper external excitation via coupling devices. From electrodynamics we know that this pill box cavity has an infinite number of resonant modes that can be classified into transverse magnetic (TM) and transverse electric (TE) modes. These eigenmodes are orthogonal modes, and there is no energy exchange among these modes in the cavity made of highly conductive materials. The field components of TM_{mnp} modes in this pill box cavity can be written as [8]

$$E_z = E_0 J_m(k_{mn}r)\cos m\theta \cos(k_p z),$$

$$E_r = -\frac{k_p}{k_{mn}} E_0 J'_m(k_{mn}r)\cos m\theta \sin(k_p z),$$

$$E_\theta = \frac{mk_p}{k_{mn}^2 r} E_0 J_m(k_{mn}r)\sin m\theta \sin(k_p z),$$

$$B_z = 0,$$

$$B_r = -j\frac{m\omega_{mnp}}{k_{mn}^2 c^2}\frac{E_0}{r} J_m(k_{mn}r)\sin m\theta \cos(k_p z),$$

$$B_\theta = -j\frac{\omega_{mnp}}{k_{mn} c^2} E_0 J'_m(k_{mn}r)\cos m\theta \cos(k_p z),$$

(3.1)

where $k_p = p\pi/g$, $k_{mn} = \chi_{mn}/r_c$, and χ_{mn} is the n-th root of the Bessel function $J_m(x)$, and ω_{mnp} the resonant frequencies of the TM_{mnp} modes.

The field components of TE_{mnp} modes are [8]

$$E_z = 0,$$

$$E_r = j\frac{m\omega_{mnp}}{k'^2_{mn}}\frac{B_0}{r}J_m(k'_{mn}\, r)\sin m\theta \sin(k_p z),$$

$$E_\theta = j\frac{\omega_{mnp}}{k'_{mn}\, c}B_0 J'_m(k'_{mn}\, r)\cos m\theta \sin(k_p z),$$

$$B_z = B_0 J_m(k'_{mn}\, r)\cos m\theta \cos(k_p z),$$

$$B_r = \frac{k_p}{k'_{mn}}B_0 J'_m(k'_{mn}\, r)\cos m\theta \cos(k_p z),$$

$$B_\theta = -\frac{mk_p}{k'^2_{mn}}\frac{B_0}{r}J_m(k'_{mn}\, r)\sin m\theta \cos(k_p z)$$

(3.2)

where $k'_{mn} = \chi'_{mn}/r_c$, χ'_{mn} are the zeros of the derivatives of the corresponding Bessel functions and ω_{mnp} the resonant frequencies of the TE_{mnp} modes.

The TM_{010} mode with the lowest resonant frequency in the pill box cavity is termed as fundamental mode, its electric field along the cylindrical axis is well suitable for particle acceleration and exclusively used as accelerating mode. The remaining resonant modes are called higher order modes (HOM). In accelerator terminology, TM_{onp} modes are termed as longitudinal modes and the other resonant modes as transverse. The HOM can be excited by the accelerated particle bunches in accelerators, and the beam induced HOM fields may cause longitudinal and transverse beam instabilities.

Practical RF cavities are designed and constructed with various structures and techniques which include single cell and multi cell cavities, normal conducting, and superconducting cavities. The cell's geometric shapes are optimized from specific considerations, such as high power efficiency and high accelerating gradient. In practical cavities, pill box, re-entrant, bell or spherical shapes are the commonly chosen geometries. Moreover, the ports on the cavity body for accommodating beam pipe, input coupler, vacuum pump, tuners and HOM dampers, distort the electromagnetic field distributions. Therefore, the fundamental mode and the HOM field components cannot be simply expressed by analytic formulae, but can be calculated with computer codes like MAFIA, HFSS, SUPERFISH, etc.

3.2. Cavity Figures of Merit

Several important cavity parameters are chosen and defined to characterize the cavity performance and its resonant modes. Among them some reflect intrinsic properties of the cavity itself, and others reveal the properties of the cavity interacting with electron beam.

3.2.1. Resonant frequency and ratios of the peak surface fields to the accelerating field

The resonant frequency and the ratios of maximum surface electromagnetic fields to accelerating field, E_{ps}/E_{acc} and H_{ps}/E_{acc}, are intrinsic parameters of a cavity which are determined by its geometrical dimensions. For getting higher accelerating gradient, these field ratios are optimized to be as low as possible. For the TM_{010} mode of a pill box cavity, the resonant frequency is independent on cavity length and can be analytically expressed as

$$\omega_{010} = \frac{2.405c}{r_c}. \tag{3.3}$$

When the cavity length equals one half of the RF wavelength, the field ratios are

$$E_{ps}/E_{acc} = \frac{\pi}{2}, \qquad H_{ps}/E_{acc} = \frac{\pi}{2}\sqrt{\frac{\varepsilon_0}{\mu_0}}J_1(1.841). \tag{3.4}$$

3.2.2. Quality factor and decaying time

The unloaded quality factor is a universal figure of merit which measures the energy storage efficiency of each resonant mode of a cavity. It is defined as the energy stored in the cavity over the energy dissipated on the cavity's inner surface in one RF period and equals the number of oscillations a resonator will go through before consuming its stored energy. The analytic definition is

$$Q_0 = \frac{\omega U}{P_c}. \tag{3.5}$$

where U is the stored energy, P_c the cavity wall loss, and ω the resonant angular frequency. For the TM_{010} mode of the pill box cavity,

$$Q_0 = \frac{2.045Z_0 g}{2R_s(r_c + g)}, \tag{3.6}$$

where $Z_0 = \sqrt{\mu_0/\varepsilon_0}$ is the impedance of free space, and R_s the cavity surface resistance. When a cavity couples with an external system and loads, another quality factor, external Q, is defined by the energy stored in the cavity over the power flows outside of the cavity in one resonant period,

$$Q_{ext} = \frac{\omega U}{P_{ext}}. \tag{3.7}$$

Meanwhile, the coupling coefficient is defined by

$$\beta = \frac{Q_0}{Q_{ext}} = \frac{P_{ext}}{P_c}. \tag{3.8}$$

Then the loaded Q_L is defined as

$$Q_L = \frac{\omega U}{P_c + P_{ext}}, \tag{3.9}$$

which can be transformed into

$$Q_L = \frac{Q_0}{1+\beta} \quad \text{or} \quad \frac{1}{Q_L} = \frac{1}{Q_0} + \frac{1}{Q_{ext}}. \tag{3.10}$$

When a cavity mode discharges its power to an external load, its field amplitude will decay exponentially with time, the 1/e time being known as the decaying time which can be shown to be

$$\tau_d = \frac{2Q_L}{\omega_{mnp}}. \tag{3.11}$$

There is another way to characterise this property of the RF cavity. When an RF power generator or electron beam excite the electromagnetic fields of a cavity resonant mode to a steady state, a time period called filling time is needed for the cavity fields to reach $(1-1/e)$ of the steady state value. This filling time is equal to the decaying time.

3.2.3. Shunt impedance

The shunt impedance is the figure of merit to characterise the strength of the interaction between the cavity mode and a particle of unit charge. For the accelerating mode, it represents the power efficiency of the cavity to accelerate particles, therefore it becomes the key parameter for cavity optimization. The shunt impedance R_a of an accelerating mode is defined as

$$R_a = \frac{V_c^2}{P_c}. \tag{3.12}$$

The accelerating voltage V_c contains an electric field time dependent change which comes from the time period when the electron passes through the RF cavity with a speed close to light. Therefore the shunt impedance has included a transit time factor into its definition which is

$$T = \frac{V_c}{V_0} = \frac{\int_{-g/2}^{+g/2} E_Z(0,z)\cos(\omega \cdot z/c)dz}{\int_{-g/2}^{+g/2} E_Z(0,z)dz}. \tag{3.13}$$

Then, the shunt impedance can be rewritten as

$$R_a = \frac{V_0^2 T^2}{P_c}. \tag{3.14}$$

For the TM_{010} mode in the pill box cavity, the transit time factor and shunt impedance can be expressed analytically as

$$T = \frac{\sin(\omega_{010} g / 2c)}{\omega_{010} g / 2c} \tag{3.15}$$

and

$$R_a = \frac{Z_0^2 T^2 g^2}{\pi R_s J_1^2(2.405) r_c (r_c + g)}. \tag{3.16}$$

Equations (3.15) and (3.16) show that the length of the accelerating gap is a parameter to optimize for achieving high shunt impedance. For a normal conducting cavity a typical T value is in the range of 0.8 to 0.9.

Another important cavity figure of merit is defined through shunt impedance over quality factor. It is called the normalized impedance, usually written as

$$\frac{R}{Q} = \frac{V_c^2}{\omega U}. \tag{3.17}$$

This R/Q, also termed geometric shunt impedance, has nothing to do with cavity wall materials and depends primarily on the geometry of the cavity structure. For the TM_{010} mode of pill box cavity, R/Q can be derived as

$$\frac{R}{Q} = \frac{2Z_0 T^2}{2.405\pi J_1^2(2.405)} \frac{g}{r_c}. \tag{3.18}$$

3.3. Shunt Equivalent Circuit

The main characteristics of cavity resonant modes can be described by RCL shunt resonant circuits • and the cavity couplers can be represented by lossless transformers. For the fundamental mode, the shunt resonant circuit can represent not only the cavity resonant behavior but also its detuned property when viewed from the input waveguide. Figure 3.3 shows a cavity equivalent circuit with inductance L, capacitance C, and shunt conductance G_c. This cavity couples with a transmission line with a conductance of G_0 via an ideal transformer $1:n_c$.

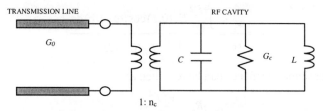

Figure 3.3 Equivalent circuit of an RF cavity

From circuit theory, one can easily find the equivalence of cavity and shunt resonant circuit parameters which are

$$\omega_c = \frac{1}{\sqrt{LC}}, \quad Q_0 = \frac{\omega_c C}{G_c} = \frac{1}{\omega_c L G_c}, \tag{3.19}$$

$$R_a = \frac{2}{G_c}, \quad \frac{R}{Q} = 2\sqrt{\frac{L}{C}} = \frac{2}{\omega_c C} = 2\omega_c L, \tag{3.20}$$

$$Q_{ext} = \frac{\omega_c n_c^2 C}{G_0}, \quad \beta = \frac{1}{n_c^2}\frac{G_0}{G_c}. \tag{3.21}$$

When a cavity is detuned from its resonance, its unloaded impedance is of complex form and can be expressed as

$$\tilde{Z}_c = \frac{1}{G_c + j\omega C + \dfrac{1}{j\omega L}} = \frac{1}{G_c(1 + j2Q_0\dfrac{\omega - \omega_c}{\omega_c})}, \tag{3.22}$$

which is often written as

$$\tilde{Z}_c = \frac{1}{G_c}\cos\psi_0 e^{j\psi_0}, \tag{3.23}$$

$$\tan\psi_0 = -2Q_0\frac{\omega - \omega_c}{\omega_c}, \tag{3.24}$$

where ψ_0 is the detuning angle of the unloaded cavity. Figure 3.4 shows the variation of the unloaded cavity voltage magnitude and the detuning angle with excited frequency. When a cavity is excited at extremely small coupling condition, its unloaded quality factor Q_0 can be obtained by measuring the frequency shift at 3dB power level or at detuning phase angle of $\pi/4$, i.e. setting $2Q_0(\omega - \omega_c)/\omega_c = 1$.

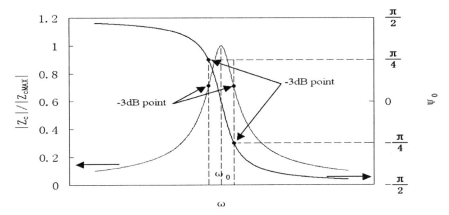

Figure 3.4 Cavity impedance magnitude and detuning angle

4. Steady State Beam Loading

4.1. *RF Component of Beam Spectra*

While traveling through an RF cavity, the electron beam not only gains energy from the cavity accelerating field, but also simultaneously excites fundamental mode and HOM fields as a current generator. These beam induced fields will react on the beam, decelerating or disturbing the moving electrons. As the beam current increases, the beam-cavity interactions cannot be neglected and have to be carefully controlled. The interaction of the beam with the fundamental mode impedance produces the beam loading to the RF power system, and the interaction with HOM impedance not only generates HOM power, but also possibly excites beam instabilities.

In light source storage rings, the electron beam can have various bunch patterns, such as single bunch, evenly and oddly distributed uniform bunches, and oddly distributed hybrid bunches, to meet the users' demands and to get longer lifetime as well as to avoid beam instabilities. In terms of the buildup of the beam-excited fields, the interactions between the bunch train and the cavity modes can be classified basically into two types, namely single bunch and multi bunch types. In the single bunch type, the field of a cavity mode excited by the former bunch disappears before the next bunch arrives in the cavity, whereas, in the multi bunch type, the field excited by the former bunch still remains in the cavity when the next bunch comes in. The fields induced by these successive bunches add as vectors and reach a steady value. Besides being a heavier beam loading to the accelerating mode, the multi bunch type of beam-cavity

428

interactions can build up strong beam induced HOM fields which convert the RF power into HOM power via the beam and affect the beam stability as well. Here we focus on the discussion of the fundamental beam loading issue and do not examine any further the related HOM issues of which the reader can find corresponding references elsewhere [2].

There are different ways to calculate the multi bunch induced voltage in RF cavities, either by vector superposition of the induced voltage of the successive passing bunches [9] or by current spectra method [10] which we will use, in the following, to calculate beam loading effects. Let us examine a circulating beam consisting of symmetrically distributed uniform bunches, its current being written as,

$$I(t) = \sum_{k=-\infty}^{+\infty} I_0(t - kT_b), \qquad (4.1)$$

where $I_0(t)$ is the current distribution with time, and T_b is the bunch time interval which equals the multiple of the RF accelerating field variation period $2\pi / \omega_{RF}$, for single bunch pattern $T_b = T_0 = 2\pi / \omega_{RF}$. The bunch charge and average beam current are

$$q = \int_{-\infty}^{+\infty} I_0(t)dt, \; I_b = \frac{q}{T_b}. \qquad (4.2)$$

By Fourier transformation, we can get

$$I(\omega) = \frac{2\pi}{T_b} \sum_{p=-\infty}^{+\infty} I_0(\omega)\delta(\omega - p\omega_b), \qquad (4.3)$$

where

$$I_0(\omega) = \int_{-\infty}^{+\infty} I_0(t)e^{j\omega t}dt, \quad \omega_b = \frac{2\pi}{T_b}. \qquad (4.4)$$

This shows that the beam can be considered as a current source with discrete frequency spectrum lines. In time domain, it can be expressed as

$$I(t) = I_b + 2\sum_{p=1}^{\infty} I_p \cos(p\omega_b t) , \qquad (4.5)$$

where

$$I_p = \frac{I_0(p\omega_b)}{T_b} \qquad (4.6)$$

For a bunch with a Gaussian distribution,

$$I_0(t) = \frac{q}{\sqrt{2\pi}\sigma_t} e^{-\frac{t^2}{2\sigma_t^2}}, \; I_0(\omega) = qe^{-\frac{1}{2}(\omega\sigma_t)^2}, \; I_p = \frac{q}{T_b} e^{-\frac{1}{2}(p\omega_b\sigma_t)^2}. \; (4.7)$$

The RF component of the beam current drives the fundamental mode of RF cavities, and produces beam loading effect on the RF system. This current component can be written as

$$i_b(t) = 2I_p \cos(\omega_{RF} t) . \tag{4.8}$$

When electrons pass through a cavity and interact with cavity modes, there exists a bunch form factor which describes the interaction strength. This bunch form factor comes from the vector sum of the fields induced by each electron within the bunch, and equals I_p over I_b. For short bunches, the bunch can be viewed as a point charge, and its beam current can be approximately written as

$$i_b(t) = 2I_b \cos(\omega_{RF} t) . \tag{4.9}$$

4.2. RF System with Beam Loading

For simplicity, a storage ring RF system consisting of an RF power source, a cavity and a circulator or an isolator inserted between the RF source and the cavity is examined in the following, and the corresponding results can be easily applied to the RF system with multi cavities. As shown in Figure 4.1 [9], the cavity fundamental mode is modeled with a resonance RCL circuit as stated in section 3.3, the power source with a circulator is treated as an RF current generator with frequency of ω_{RF} and the beam is simulated with a current generator with the same frequency as the RF current generator. The maximum generator power, i.e. available RF power, can be obtained by matching the cavity impedance to the generator impedance. Following the convention used in [9] • the generator current and the available RF power can be expressed as

$$i_g = I_g \cos \omega_{RF} t , \quad P_g = \frac{1}{8} \frac{I_g^2}{\beta G_c} \tag{4.10}$$

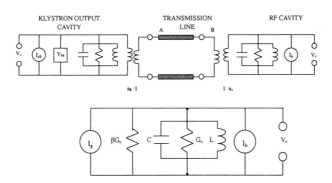

Figure 4.1 Equivalent circuits of an RF system

When the RF cavity is on resonance, it can be treated as a resistive load with an impedance of $1/G_c$. The generator and beam induced fundamental voltages are

430

$$V_{gr} = \frac{I_g}{G_c(1+\beta)} = \frac{2\sqrt{\beta}}{1+\beta}\sqrt{R_aP_g} \ , \tag{4.11}$$

$$V_{br} = \frac{2I_p}{G_c(1+\beta)} = \frac{I_bR_a}{1+\beta} . \tag{4.12}$$

When the RF cavity is excited at a frequency away from its resonant frequency ω_c, it appears as a load with complex impedance. This loaded impedance can be written as

$$\tilde{Z}_L = \frac{\cos\psi e^{j\psi}}{G_c(1+\beta)} \ , \tag{4.13}$$

$$\tan\psi = -2Q_L\frac{\omega-\omega_c}{\omega_c} . \tag{4.14}$$

When beam current and generator current are in phase, the generator and beam induced voltages can be written as

$$\tilde{V}_g = I_g\tilde{Z}_L = V_{gr}\cos\psi e^{j\psi} \ , \tag{4.15}$$

$$\tilde{V}_b = 2I_p\tilde{Z}_L = V_{br}\cos\psi e^{j\psi} . \tag{4.16}$$

In a practical RF system, the accelerating voltage is the vector sum of the RF generator and the electron beam induced voltages, while the beam current and generator current will automatically keep a proper phase difference which ensures that the electron bunch is accelerated at the synchronous phase of the RF cavity voltage. If we take the beam current as a relative phase reference and follow the synchronous phase defined in equation (2.9), we can build a phasor graph as shown in Figure 4.2.

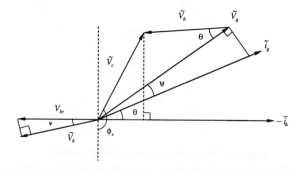

Figure 4.2 Phasor graph of the steady state beam loading

From the above phasor graph, we can find the accelerating voltage

$$V_a = V_c\sin\phi_s = V_{gr}\cos\psi\cos(\theta+\psi) - V_{br}\cos^2\psi \ , \tag{4.17}$$

$$-V_c\cos\phi_s = V_{gr}\cos\psi\sin(\theta+\psi) - V_{br}\cos\psi\sin\psi . \tag{4.18}$$

From equations (4.17) and (4.18), the generator power for establishing the required accelerating voltage in beam loaded RF cavities can be derived as,

$$P_g = \frac{V_c^2}{R_a} \cdot \frac{(1+\beta)^2}{4\beta} \cdot \frac{1}{\cos^2\psi} \left\{ \left[\sin\phi_s + \frac{I_b R_a}{V_c(1+\beta)}\cos^2\psi \right]^2 + \left[-\cos\phi_s + \frac{I_b R_a}{V_c(1+\beta)}\cos\psi\sin\psi \right]^2 \right\}$$

(4.19)

When the generator power is fed to an RF cavity, a part of the power, P_r, will be reflected back if the impedance of the cavity does not match that of the transmission line. From the equation $P_g = P_c + P_b + P_r$, one can get

$$P_r = \frac{V_c^2}{R_a} \cdot \frac{(1+\beta)^2}{4\beta} \cdot \frac{1}{\cos^2\psi} \left\{ \left[\sin\phi_s + \frac{I_b R_a}{V_c(1+\beta)}\cos^2\psi \right]^2 \right.$$

$$\left. + \left[-\cos\phi_s + \frac{I_b R_a}{V_c(1+\beta)}\cos\psi\sin\psi \right]^2 \right\} - \frac{V_c^2}{R_a} - I_b V_c \sin\phi_s$$

(4.20)

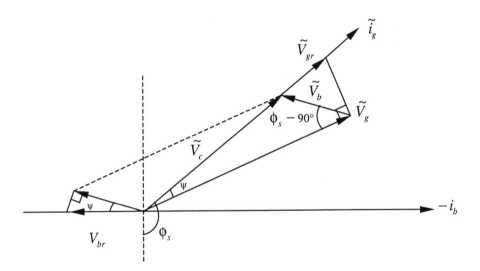

Figure 4.3 Phasor graph of the beam loading at optimum cavity tuning

As the electron beam is accelerated around the synchronous phase for phase focusing, it will present to the RF system not only a resistive, but also a reactive load. For minimizing the reflected power from the cavity due to its reactive impedance, the RF cavity needs to be detuned from its resonance to cancel the reactive part of beam load. This is achieved by keeping the generator current and the accelerating voltage in phase. The related phasor graph is shown in Figure 4.3, the frequency shift and the required generator power in the tuning condition can be expressed as

$$\tan \psi_m = \frac{I_b R}{V_c (1+\beta)} \cos \phi_s , \qquad (4.21)$$

$$P_g (\psi_m) = \frac{(1+\beta)^2}{4\beta} \cdot \frac{\left(V_c + V_{br} \sin \phi_s\right)^2}{R} , \qquad (4.22)$$

$$\frac{\Delta f}{f} = -\frac{1}{2} \cdot \frac{R}{Q} \cdot \frac{I_b \cos \phi_s}{V_c} . \qquad (4.23)$$

The cavity reflected power can be further minimized to zero by matching the coupling between cavity and RF power transmission line. The optimum coupling is

$$\beta_m = 1 + \frac{I_b R \sin \phi_s}{V_c} = 1 + \frac{P_b}{P_c} , \qquad (4.24)$$

thus the minimum generator power is

$$P_g (\psi_m, \beta_m) = P_c + P_b . \qquad (4.25)$$

When the generator current and accelerating voltage keep an offset angle ψ_L all the time, the optimized cavity detuning is given by

$$\frac{\Delta f}{f} = -\frac{1}{2} \cdot \frac{R}{Q} \cdot \frac{I_b \cos \phi_s}{V_c} - \frac{\tan \psi_L}{2Q_L} (1 + \frac{I_b R_a}{V_c (1+\beta)} \sin \phi_s) . \qquad (4.26)$$

4.3. Robinson Instability

Robinson instability is the most important beam cavity interaction effect in the fundamental RF system for storage rings. It was first studied by K. Robinson in 1956 [11]. The Robinson criteria include two instability limitations, impedance limitation and high current limitation, and can be expressed as

$$0 < -\sin 2\psi < -\frac{2V_c \cos \phi_s}{I_b R /(1+\beta)} . \qquad (4.27)$$

There are different ways to obtain these two limitations, the detailed derivation methods can be found in [9, 11~14]. The physical mechanism behind the Robinson criteria can be explained intuitively. As shown in Figure 4.4 [14] and Figure 4.5, the longitudinal focusing from the cavity impedance and the generator induced voltage determine the Robinson limitations. When a rigid bunch interacts with an RF cavity resonating at a frequency lower than the RF frequency in an electron storage ring, the bunch with higher than synchronous energy will have lower revolution frequency and will deposit more energy to the cavity mode. In contrast, a bunch with lower than synchronous energy will have a higher revolution frequency and will lose less energy in the cavity. This physical process gives a damping effect to the bunch synchrotron oscillation and sets the above impedance limitation. Luckily the cavity detuning for beam loading compensation in equations (4.23) or (4.26) is in the same direction to produce Robinson damping effect, normally, an offset Robinson detuning of the RF cavity with a few RF degrees or a narrow band beam feedback is adopted to damp the zero-mode dipole oscillation in practical RF systems.

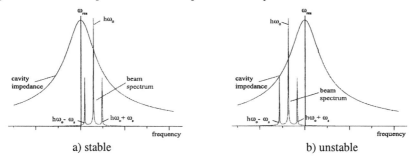

a) stable   b) unstable

Figure 4.4 Intuitive illustration of impedance limitation [14]

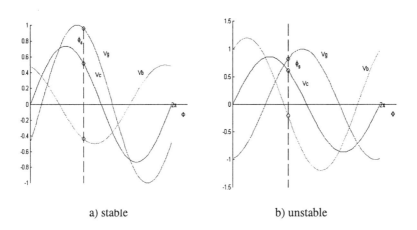

a) stable   b) unstable

Figure 4.5 Intuitive illustration of beam current limitation

In the high beam current case, the beam induced voltage has a big impact on the accelerating voltage. As a result of the synchronous acceleration, the generator induced voltage may be automatically shifted to defocus the rigid bunch (see Figure 4.5), and this yields the high beam current limitation. When the RF cavity runs in an optimized tuning condition, its Robinson criteria are automatically satisfied. Moreover the high current limitation can be transformed to a simpler form,

$$P_b < (1+\beta)P_c \qquad (4.28)$$

This indicates that the high current limitation of the SC cavity system is in marginal due to its extreme low cavity dissipated power, therefore the detuning technique or RF feedback is normally applied to ensure the system stability at high current operation.

5. Fundamental RF System

5.1. *Key Components of RF System*

In light source storage rings, RF systems have several key functions [15], which include (1) delivering energy to electron beam for compensating its energy loss due to synchrotron radiation and its interactions with beam chamber impedance, or ramping the electron beam to higher energy; (2) establishing RF voltage to capture and focus electrons into bunches; (3) controlling the energy acceptance and therefore the beam parameters, such as bunch length, beam lifetime, etc.; and (4) providing damping effects to the electron motions by synchrotron radiation and RF acceleration.

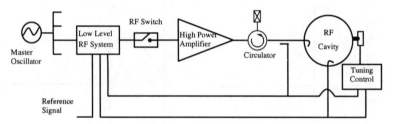

Figure 5.1 Schematic diagram of an RF system

Usually, an RF system comprises accelerating cavities, RF power amplifiers, high power transmission lines and low level RF control loops. Figure 5.1 shows a schematic diagram of the simplest RF system, and as an example, Figure 5.2 shows a 3D view of SLS RF system and its key components [16].

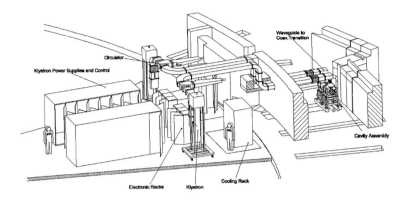

Figure 5.2 SLS RF System [16]

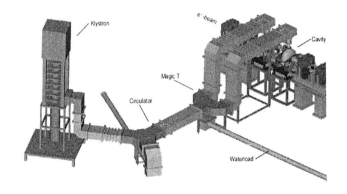

Figure 5.3 3D view of the ANKA RF plant [17]

Table 5.1 Parameters and features of Light Source Storage Ring RF systems [18]

Storage Ring	Ring Energy (GeV)	Current (mA)	Frequency (MHz)	Energy loss per turn (keV)	No. of Cavity	Type of Cavity	Power tube	No. of amplifiers
ALS	1.9	400	500	92 (BM) 129 (total)	2	1-cell	klystron	1
ANKA	2.5	200/400	500	662	4	1-cell	klystron	2
APS	7	100	353	5450 (BM)	16	1-cell	klystron	4
BESSY II	1.7	200	500	410	4	1-cell	klystron	4
ELETTRA	2	300	500	258 (BM) 320 (total)	4	1-cell not re-entrant	klystron	4
ESRF	6	200	352	4600 (BM) 6100 (total)	4	5-cell	klystron	2
HLS	0.8	300	204	16.3 (BM)	1	$\lambda/4$	tetrode	1
NSLS VUV	0.75	850	53	14.7	1	$\lambda/4$	tetrode	1
NSLS X-ray	2.5	300	53	504	3	$\lambda/4$	tetrode	4
PF	2.5	450	500	480	4	1-cell	klystron	4
PLS	2	300	500	225 (BM) 356 (total)	4	1-cell	klystron	4
SLS	2.4	400	500	600	4	1-cell	klystron	4
SPEAR3	3	200/500	476	580	4	1-cell	klystron	2
SPring-8	8	100	508	9200 (BM) 12900 (total)	32	1-cell	klystron	4
SRS	2	1000	500	254 (BM)	4	1-cell	klystron	1
TLS	1.5	240	500	91	2	1-cell	klystron	2

In light source storage rings, normally one or more cavities that are installed in one or more straight sections are providing the required RF voltage and beam power. For transferring the RF power from the high power amplifiers to the beam, the high power RF sources and RF cavities can be grouped into RF systems in various ways. The simplest one is one RF power amplifier energizing one RF cavity, e.g. the RF systems in BESSY-II, ELETTRA, HLS, PLS, PF, SLS and TLS. This scheme has more flexible low level controls and better redundancy. Another commonly adopted scheme, as shown in Figure 5.3, is one klystron powering several cavities, e.g. the RF systems in ALS, APS, ANKA, ESRF, SRS, SPEAR3 and Spring-8. This is a cheaper solution particularly for high energy light sources. The few exceptions are feeding one cavity with the output power from the combination of several amplifiers. The choice of the grouping way depends on the required RF power and the available commercial RF amplifiers, including quality, price and long term availability of the products. As a summary, Table 5.1 shows the technical features of a number of light source RF systems [18].

5.2. RF Cavities for Synchrotron Radiation Light Sources

Various types of RF cavities have been used in SR light source storage rings, these cavities work in different frequency ranges, typically around 50MHz, 200MHz, 350MHz and 500MHz. Technically, there are two kinds of RF cavities employed in SR light sources which are normal conducting and superconducting cavities. As the cases in TLS, CLS, Diamond and SOLEIL, shown in Table 5.2, superconducting cavities are being adopted for newly constructing and upgrading light source facilities due to their salient features, such as higher accelerating gradient, higher shunt impedance (about factor of 10^5 compared with the normal conducting cavity) and heavily damped HOM. Figure 5.4 and Figure 5.5 show the CESR type [19] and SOLEIL type [20] superconducting cavities. The superconducting cavity related technologies progressed rapidly in the past decade, thanks to the development of high energy particle colliders like CESR, KEKB and LHC, and now become mature although they are still complicated to construct and operate. The details on superconducting cavities can be found in the dedicated lecture of this school [21] and in the reference elsewhere [22].

Table 5.2 Superconducting cavities used in light sources [18]

Cavity	Geometry	Frequency (MHz)	R/Q (Ω)	$V_{c,\,max}$ (MV)
CESR Type	single cell	~500	89	3
SOLEIL Type	2 single cell	~352	90	4.8

438

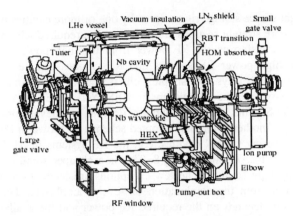

Figure 5.4 CESR type superconducting cavity [19]

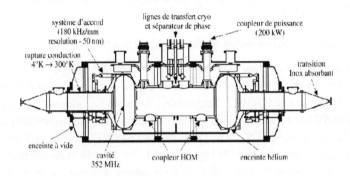

Figure 5.5 SOLEIL type superconducting cavity [20]

Normal conducting cavities, on the other hand, are still the dominant ones, various forms of which are now under operation and construction. In terms of accelerating fields, these cavities can be classified into the ones with acceleration based on TM_{010} mode and the ones based on TEM modes. The TEM mode cavity is a reasonable choice at low RF frequencies like 50MHz and 100MHz or even 200MHz for getting smaller physical size of cavity to fit the available space. Figure 5.6 shows the assembly of the NSLS new $\lambda/4$ cavity [23]. This TEM mode coaxial structure cavity is about $\lambda_{RF}/4$ long in length and is often known as $\lambda/4$ cavity. The combination of two $\lambda/4$ cavities can form a cavity with two gaps which is usually known as $\lambda/2$ cavity. The loading capacitance at the accelerating gap is usually optimized by compromising the cavity shunt impedance and the cavity's physical size. Normally, the shunt impedance of $\lambda/4$ cavities is about a few MΩ. As shown in Table 5.3, ALADDIN, MAX-III, NSLS and SSLS cavities are examples.

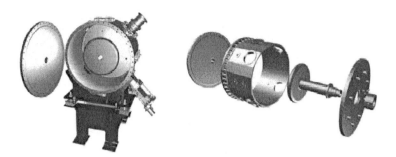

Figure 5.6 A 3D view of NSLS $\lambda/4$ cavity [23]

Table 5.3 Single cell TEM mode cavities used in light sources [18]

Facility	Shape	Frequency (MHz)	R (MΩ)	V_c (kV)
ALADDIN	$\lambda/4$	50.582	1.6	120
MAX-III	$\lambda/4$	99.956	3.2	200
NSLS X-ray	$\lambda/4$	52.887	2	250
SSLS	$\lambda/4$	55.517	0.58	120

The RF cavities based on TM_{010} mode have evolved from the pill box cavity, most of them are optimized to re-entrant shape with round corners for increasing shunt impedance via improving transient time factor, and to bell shape for increasing the accelerating gradient via reducing the ratios of surface fields to accelerating field. The typical shunt impedance of a single cell cavity is around 7~8 MΩ, which depends on the detailed design and manufacture technique. As shown in Table 5.4, ALS, ANKA, APS, BESSY-II, DELTA, ELETTRA, HLS, PF, PLS, SPring-8, SLS, SRS and TLS all use single cell cavities [18].

As sketched in Figure 5.7, the single cell PEP-II cavity with rectangular HOM damping waveguides [24], adopted in SPEAR3, represents the recent normal conducting upgrades. In the similar way, a new normal conducting cavity with circular HOM damping waveguides is being developed for light sources at BESSY-II under a European collaboration project, and it will be tested with beam first at DELTA [18]. Moreover, the normal conducting cavity with SiC beam pipe HOM dampers has been developed and operated at KEK PF [18]. A number of single cell resonators can cascade along the axis and couple together to form a multicell cavity [15]. Such kind of multicell cavities has been used in a few light sources like ESRF.

Table 5.4 Single Cell TM_{010} Mode Normal Conducting Cavities [18]

Facility	Cavity shape	Frequency (MHz)	R (MΩ)	$V_{c, max}$ (kV)
ALS	Nose-cone	499.654	10	750
APS	Nose-cone	351.927	11.2	1000
ANKA	Bell-shaped	499.652	6.8	600
BESSY II	Pill-box	499.654	5.8	350
DELTA(EC)	Nose-cone	499.654	>8	850
ELETTRA	Bell-shaped	499.654	6.8	600
HLS (new)	Nose-cone	204	7	250
PF (new)	Nose-cone	500.1	7.7	500
PLS	Nose-cone	500.08	8.6	400
SLS	Bell-shaped	499.65	6.8	650
SPEAR3	Nose-cone	476.337	7.6	1000
SPring-8	Bell-shaped	508.58	6.7	~530
SRS	Nose-cone	499.71	7.2	500
TLS (old)	Pill-box	495.666	6	550

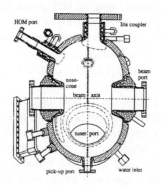
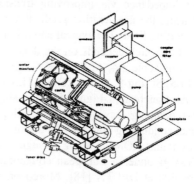

Figure 5.7 Single cell PEP-II cavity [24]

The tuner is one of the indispensable RF cavity elements, for it tunes the cavity fundamental mode to compensate its resonant frequency change due to beam loading variation, temperature fluctuation and detuning phase offsetting. There are mainly two ways to tune the cavity. One is using a plunger to disturb the high magnetic field in cavity, as shown in Figure 5.8, most of normal conducting cavities use this kind of tuners. The other is deforming the cavity longitudinally to disturb the high electric field, super conducting cavities and

ELETTRA type normal conducting cavity use this kind of tuners. Usually the fundamental mode tuning range is about several hundreds of kHz to several MHz. In addition, the tuner disturbs also the cavity HOM, therefore a second tuner or the cavity cooling water temperature regulation is sometimes used to tune the dangerous HOM in a direction away from beam spectrum lines.

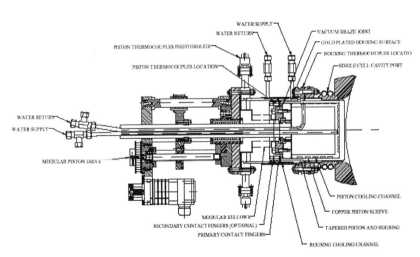

Figure 5.8 Plunger tuner of the APS RF cavity [25, 26]

442

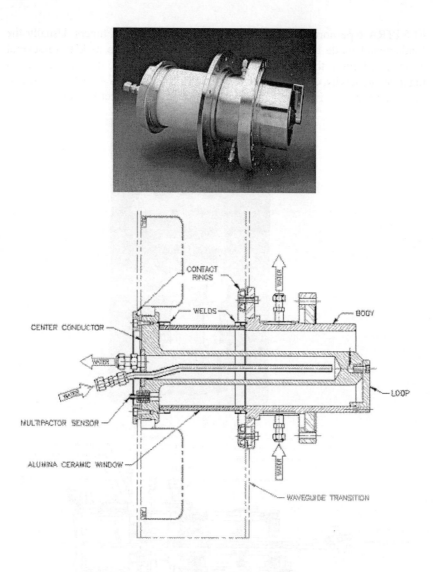

Figure 5.9 Input power coupler of the APS cavity [26]

Couplers are the devices to couple the cavity with the RF power source and external loads. There are three kinds of couplers installed on cavities, i.e. input power coupler for feeding the RF power, signal coupler (pick-up) for monitoring the amplitude and phase of the RF fields in cavities and HOM coupler for reducing HOM impedance. Aperture coupling, loop coupling and probe coupling are used in these input power couplers, and some couplers are capable to handle an RF power of up to 300kW. The loop and aperture input couplers are used for normal conducting cavities and the probe and aperture ones for superconducting cavities. Figure 5.9 shows an input power coupler of the APS RF cavity. Both loop and probe signal couplers are used in cavities. The HOM couplers appear in various forms, e.g. LC type, wave guide type and beam pipe type. For heavy HOM damping, the wave guide and beam pipe HOM couplers are used for both normal and superconducting single cell cavities.

5.3. High Power RF Sources

The high RF power source consists of power supplies and RF power amplifiers. Klystron and tetrode are the most commonly used high power RF amplifiers. In light sources, tetrodes with the output power levels up to 100kW work in the frequency range from 50MHz to ~200MHz, and klystrons, with output powers of 60kW, 180kW, 250kW, 300kW and 1MW, operate at frequencies about 350MHz and 500MHz. In addition, the 500MHz inductive output tube (IOT) is being used in light sources [27], and the 350MHz solid state power amplifier has been chosen as the RF power source for the SOLEIL booster [28]. A detailed technical introduction to these power amplifiers can be found elsewhere [29~31].

The high power RF transmission system [32] is used for transmitting the RF power from the high power RF source to the cavity or another load. It consists mainly of a transmission line, bender, circulator, phase shifter, power splitter (Magic-T) and RF load as well as directional coupler.

Rectangular waveguide and coaxial line are commonly used as power transmission line in the practical RF system for light sources. The coaxial line is used for low frequency or low power RF systems, and typically limited to 60~75kW at 500MHz. The coaxial transmission line, normally working in TEM mode with characteristic impedance of 50Ω, uses copper for the inner conductor and copper or aluminum for the outer conductor. The aluminum outer conductor is cheaper and has less weight, but it increases the transmission losses by about 10%. The coaxial line has no cutoff frequency, and its dimension is determined by power handling capacity and breakdown field strength. Examples can be found in all the RF systems with frequencies lower than 300MHz and 500MHz RF system with RF power less than 100kW, such as SSLS, NSLS, PLS ELETTRA and BESSY II [18].

Table 5.5 Standard Waveguide Characteristics [32]

Waveguide Designation	Inside Dimension (inches)	Operating Range TE_{10} mode(MHz)	Cutoff Freq. (MHz)	Cutoff Wave length (cm)
WR2300	23.000×11.500	320-490	256	116.84
WR2100	21.000×10.500	350-530	281	106.68
WR1800	18.000×9.000	410-625	328	91.44
WR1500	15.000×7.500	490-750	393	76.22

A rectangular wave guide is the only choice for a practical transmission line working at frequency higher than 350MHz and RF power larger than 100kW. Its dominant operating mode is TE_{10}. As shown in Table 5.5 [32], the dimension of the standard wave guides is chosen with a 2 to 1 transverse aspect ratio, this makes the cutoff frequency f_c of its first higher order mode TE_{20} twice the cutoff frequency of TE_{10}. Then the operating frequency is commonly chosen in the range from 1.25 f_c to 1.9 f_c, where the higher order mode cannot be excited and propagated. The wave guide has high power handling capacity, for instance, the maximum handling power of the standard aluminum wave guide WR1800 at 500MHz is about 1.9 times that of a 14-inch coaxial line [29] while keeping almost the same attenuation constant.

Circulator and Magic-T are the two indispensable elements in high power RF transmission system. The circulator in a light source RF system is typically a three-port isolator placed right after the klystron output port and used to get the voltage standing wave ratio (VSWR) less than 1.2 at the port where the klystron is feeding RF power into the transmission line. The Magic-T is a kind of power divider used for one klystron feeding two or more cavities. Its typical characteristics within 10% bandwidth are ±0.1dB power balance between the two co-linear arms, less than 0.1dB insertion loss, 30dB minimum isolation and VSWR less than 1.1.

In addition, directional couplers in transmission line are the necessary elements for measuring the phases and amplitudes of the input and reflected RF fields from the RF source to the cavities, and these RF signals are used as feedback signals for low level RF loops to control the RF system performance.

5.4. Low level RF Control System

Low level RF system is an indispensable part of the RF system to keep the light source performance optimized [17, 33]. The low level RF system consists mainly of a drive chain, tuning loop, voltage control loop and phase control loop, which regulate the resonant frequency, the RF field phase and amplitude of accelerating cavities to the required values during various operational states, such as injection, ramping and storage. For example, these loops keep the cavity system on tune, the cavity voltage constant and the RF voltage vectors of different cavities in phase during the beam operation of storage rings.

As shown in Figure 5.10, the drive chain is an RF signal distributing and processing circuit, where the initial RF signal from a master generator is split, regulated in phase and amplitude, and then amplified to drive klystrons. There is a station phase shifter in the drive chain for the RF system with two or more klystrons.

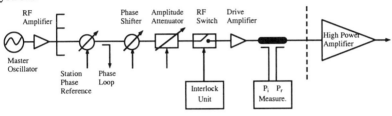

Figure 5.10 Schematic diagram of the drive chain

The tuning loop keeps the cavity system on tune by controlling the cavity tuner via a phase error signal detected from feeder transmission line and the cavity itself, as shown in Figure 5.11. Keeping the tuning loop working is essential to compensate the beam loading and temperature detuning effects on the cavity and then to minimize the RF power reflected from the RF cavity. Normally the detuning amount falls in a range of a few tens of kHz during the full current operation. The tuning speed of cavities is dominated by their mechanical tuner with the typical value of about a few hundreds of Hz per second which is fast enough to handle the beam injection rate and temperature

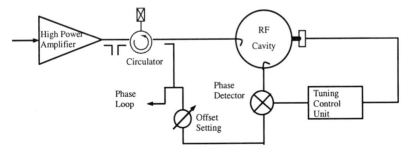

Figure 5.11 Schematic diagram of the cavity tuning loop

fluctuation. The dynamic phase error represents the tuning precision, which is about ±1°.

The amplitude loop keeps the cavity voltage variation within 1% while the beam loading decays with time during the operation. As shown in Figure 5.12, the RF voltage is picked up from the cavity signal coupler and compared with a reference signal from the operational setting. Their error signal is amplified and used to regulate the variable attenuator in the drive chain for adjusting the input power to cavity. For a klystron based RF system, there is another option to regulate the cavity voltage which is to modulate the anode voltage. Normally the amplitude loop has 3dB bandwidth up to a few of kHz which must be insensitive to synchrotron frequency.

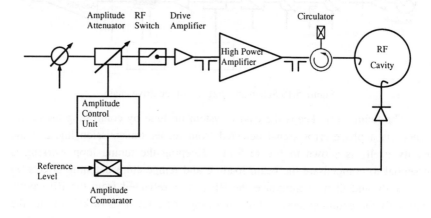

Figure 5.12 Schematic diagram of the cavity amplitude loop

The phase loop keeps the phase variation of the cavity voltage within ±1° and ensures all the cavities operating in phase during beam operations. As shown in Figure 5.13, the error signal represents the phase deviation of the RF voltage monitored by a pickup at the cavity or at the feeder waveguide close to the cavity with respect to the reference signal. This error signal is amplified and used to regulate the phase shifter in the drive chain. Normally, the phase loop is designed for controlling the phase change up to ±30° and its 3dB bandwidth is up to a few kHz which again must be away from synchrotron frequency.

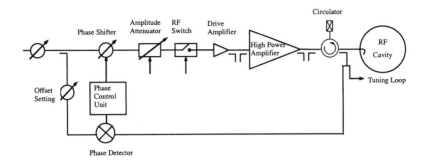

Figure 5.13 Schematic diagram of the phase loop

In addition, there are some RF systems equipped with auto gain control and phase lock loops around klystrons for getting more precise and more stable low level controls. Moreover, several other low level controls to enhance the RF system performance are developed and used in a few light sources. The direct RF feedback [34] is used for raising the threshold current of high intensity Robinson instability which is necessary particularly for the RF system based on SC RF cavities. Zero-mode beam feedback [34] is a synchrotron oscillation damping loop used to suppress the zero-mode longitudinal instability, which may be induced by noise in the RF system or the HOM in RF cavities. This loop samples the RF signal filtered from the beam signal detected by a beam position monitor (button BPM or stripline), and picks up the synchrotron oscillation signal by comparing this RF signal with a reference signal in a phase detector. This error signal, after properly amplifying and phasing, is used to drive a phase shifter in the drive chain for stabilizing the synchrotron oscillation. The RF phase modulation [35] is a complementary method to suppress longitudinal oscillation and to increase beam lifetime which is realized by applying a small amplitude modulation on the RF voltage with a frequency of 2 times the synchrotron frequency.

Finally, RF interlocks are indispensable parts of any RF system. They must be provided for watching the cavity vacuum, cooling water or liquid helium, abnormal temperature raise in the RF system, maximum input and maximum reflected power (or VSWR) as well as personal safety.

6. Harmonic RF System

6.1. *Harmonic RF acceleration*

The main role of harmonic acceleration is to control bunch length [36~39], either lengthening or shortening, and to provide Landau damping to suppress the longitudinal coupled bunch instability, thereby reducing the bunch energy spread induced. Most of the harmonic RF systems are used to improve the Touschek life time which is the dominant performance limitation in low energy or low emittance light sources. In this kind of harmonic RF system, the high harmonic RF voltage is used to cancel the longitudinal focusing and then lengthen the bunch, thus the reduced charge density makes the larger angle intra-beam (Touschek) effect weaker. However, there exists another case where the Touschek life time is determined by the bunch length modulated by longitudinal coupled bunch oscillations. The bunch lengthening RF cavity improves the energy spread deteriorated by the above bunch oscillation and results in higher light source undulator brightness besides the longer beam lifetime [40]. A few of the high harmonic RF systems are used to shorten the bunch particularly for free electron laser operation [39], where the high harmonic voltage is used to provide extra longitudinal focusing. In addition, there are some other harmonic RF systems dedicated for providing Landau damping or Robinson damping via proper detuning. For the RF system with fundamental and high harmonic cavities, the electron bunch experienced RF voltage can be expressed as

$$V(t) = V_c \sin(\omega_{RF} t + \phi_s) + V_{nc} \sin(n\omega_{RF} t + n\phi_n), \quad (6.1)$$

$$U_0 = eV_c \sin\phi_s + eV_{nc} \sin n\phi_n, \quad (6.2)$$

where V_{nc} is the harmonic cavity voltage, n is an integer ratio of high harmonic RF frequency to fundamental RF frequency. From equation (2.10) and (2.11)• one obtains

$$\begin{aligned}
\frac{d^2\Delta\phi}{dt^2} = -\frac{\alpha e\omega_0}{2\pi E_0}[&(V_c \cos\phi_s + nV_{nc} \cos n\phi_n)\Delta\phi \\
&-\frac{1}{2}(V_c \cos\phi_s + n^2 V_{nc} \cos n\phi_n)\Delta\phi^2 \\
&+\frac{1}{6}(V_c \cos\phi_s + n^3 V_{nc} \cos n\phi_n)\Delta\phi^3] + \cdots
\end{aligned} \quad (6.3)$$

As shown in Figure 6.1[38], to cancel the longitudinal focusing for lengthening the bunch, one can get the following conditions [37]

$$V_c \cos\phi_s + nV_{nc} \cos n\phi_n = 0 \quad (6.4)$$

$$V_c \sin\phi_s + n^2 V_{nc} \sin n\phi_n = 0 \quad (6.5)$$

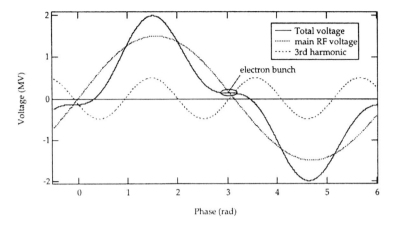

Figure 6.1 Main and harmonic voltages optimized for bunch lengthening

which yield the condition

$$\frac{V_{nc}}{V_c} = \frac{1}{n^2}\sqrt{n^2 \cos^2\phi_s + \sin^2\phi_s} \qquad (6.6)$$

$$\phi_n = \frac{1}{n}\tan^{-1}\left(\frac{1}{n}\tan\phi_s\right) \qquad (6.7)$$

Then, equation (6.3) becomes

$$\frac{d^2\Delta\phi}{dt^2} = -\frac{\alpha e\omega_0}{12\pi E_0}(V_c\cos\phi_s + k^3 V_{nc}\cos\phi_n)\Delta\phi^3 + \cdots \quad . \qquad (6.8)$$

For an electron performing synchrotron oscillation with phase amplitude ϕ_{max}, the synchrotron frequency can be derived from (6.8) and is given by

$$\omega_{ns} = \omega_s \frac{2\pi^{3/2}}{\Gamma^2(1/4)}\left[\frac{n^2-1}{6}\right]^{1/2}\phi_{max} \quad . \qquad (6.9)$$

The above equation indicates that the synchrotron frequency of a charged particle accelerated in the storage ring with harmonic RF cavity is dependent on its own phase oscillation amplitude. This effect is termed as the Landau damping and this harmonic cavity therefore called Landau cavity. From the fundamental and harmonic RF focusing force and Haissinski equation, the bunch length in the ideal condition stated in (6.4) and (6.5) can be derived as [36]

$$\sigma_{nl} = \frac{2\sqrt{\pi}}{\Gamma(1/4)}\left(\frac{3}{n^2-1}\right)^{1/4}\left(\frac{\alpha_c\omega_{rf}}{\omega_s}\frac{\sigma_E}{E_0}\right)^{1/2}\frac{c}{\omega_{rf}} \quad . \qquad (6.10)$$

To further enhance the longitudinal focusing with harmonic RF voltage, equation (6.3) can be linearized as

$$\frac{d^2\Delta\phi}{dt^2} - \frac{\alpha e \omega_0}{2\pi E_0}(V_c \cos\phi_s + nV_{nc} \cos n\phi_n)\Delta\phi,$$ (6.11)

and its synchrotron frequency can be written as

$$\omega_{ns} = \omega_s \left(1 + \frac{nV_{nc} \cos n\phi_n}{V_c \cos\phi_s}\right)^{1/2}.$$ (6.12)

Thus the bunch length can be expressed as

$$\sigma_{nl} = \sigma_l \left(1 + \frac{nV_{nc} \cos n\phi_n}{V_c \cos\phi_s}\right)^{-1/2}.$$ (6.13)

In the optimum phase condition, where $\cos\phi_n$ is -1, the synchrotron frequency and the corresponding bunch length are

$$\omega_{ns} = \omega_s \left(1 - \frac{nV_{nc}}{V_c \cos\phi_s}\right)^{1/2},$$ (6.14)

and

$$\sigma_{nl} = \sigma_l \left(1 - \frac{nV_{nc}}{V_c \cos\phi_s}\right)^{-1/2}.$$ (6.15)

6.2. Active and passive harmonic RF System

The harmonic RF system is widely used to lengthen the bunch in light source storage rings. It is operated in both active and passive ways [37, 38]. The active harmonic RF system with dedicated harmonic RF power source is the ideal one to control the bunch length over the whole beam current range. The passive harmonic system powered by the beam current itself is even simpler and cheaper and can also effectively control the bunch length and provide the Landau damping. For a harmonic cavity with shunt impedance R_n and detuning angle ψ_n, the beam (with bunch factor F_n) induced harmonic voltage is given by

$$\tilde{V}_{nc} = I_b F_n R_n \cos\psi_n e^{j\psi_n},$$ (6.16)

$$\tan\psi_n = -2Q_{nL}\frac{\Delta\omega_{nc}}{\omega_{nc}},$$ (6.17)

$$\psi_n = \frac{\pi}{2} + n\phi_n.$$ (6.18)

From equation (6.6) and (6.7), the optimum detuning and the required impedance of the harmonic cavity for bunch lengthening are given by

$$\psi_n = \frac{\pi}{2} + \tan^{-1}\left(\frac{1}{n}\tan\phi_s\right), \qquad (6.19)$$

$$R_n = \frac{V_c}{n^2}\frac{\sqrt{n^2\cos^2\phi_s + \sin^2\phi_s}}{I_b F_n \cos\psi_n}. \qquad (6.20)$$

Table 6.1 Harmonic Cavities used in light sources [18]

Facility	Purpose	Operation mode	Cavity type	Frequency (MHz)	ω_h/ω_R F
ALADDIN(O)	Lengthening	Passive	NC	~202	4
ALS (O)	Lengthening	Passive	NC	~1500	3
BESSY-II (O)	Lengthening	Passive	NC	~1500	3
BESSY-II(D)	Lengthening	Passive	SC	~1500	3
ELETTRA(C)	Lengthening	Passive	SC	~1500	3
MAX-II(O)	Lengthening	Passive	NC	~1500	3
NSLS-VUV(O)	Lengthening	Active	NC	~211	4
SLS (C)	Lengthening	Passive	SC	1500	3
Super-ACO(O)	Shortening	Active	NC	~500	5
UVSOR(O)	Lengthening	Passive	NC	~270	3

O: operation, C: commissioning, D: Development

As shown in Table 6.1, many light sources employ the high harmonic cavities to improve the beam lifetime and the energy spread. Most of the harmonic cavities are operated passively, and for taking the benefit of high shunt impedance, a few superconducting high harmonic cavities are under development and commissioning at BESSY-II [41], ELETTRA and SLS [42]. The passive Landau cavity must be detuned to Robinson unstable direction, therefore its tuning angle should be close to $\pi/2$ to avoid the strong instability excitation even considering the Robinson damping from main RF cavities. In practical operations, the high harmonic RF system has improved the beam lifetime by a factor of 1.5 to 2 which is very effective and favorable to synchrotron radiation users.

References

1. A. Chao, *Physics of Collective Beam Instabilities in High Energy Accelerators.* John Wiley & Sons, Inc., New York, (1993)
2. K. Akai, Proc. of the Asian Accelerator School, Beijing, 118-149 (1999)

452

3. D. A. Edwards and M. J. Syphers, *An Introduction to the Physics of High Energy Accelerators*. John Wiley & Sons, Inc., New York, 18-28 (1993)
4. K. Wille, *The Physics of Particle Accelerators*. Oxford University Press, Oxford, 152-184 (2000)
5. M. Sands, *The Physics of Electron Storage Rings*. SLAC-121 (1970)
6. J. LeDuff, Nucl. Instr. Meth in Phys. Res. A239, 83-101 (1985)
7. D. M. Dykes and D. S. G. Higgins, Proc. of PAC95, 1762-1764 (1995)
8. T. Wangler, *RF Linear Accelerators*. John Wiley & Sons, Inc., New York, 30-31 (1998)
9. P. B. Wilson, AIP Conf. Proc. 87, 450-581 (1982)
10. A. Hofmann, Proc. of CERN Accelerator School, CERN 95-06, (1995)
11. K. Robinson, *Radio Frequency Acceleration II*. C.E.A-11, (1956) and CEAL-1010 (1964)
12. F. Pederson, IEEE, NS-22, No.3, 1906-1909 (1975)
13. R. Garoby, CERN/PS 91-33, (1991)
14. Wiedemann, *Particle Accelerator Physics II*. Springer-Verlag Berlin, 163-228 (1999)
15. D. J. Thompson and D. M. Dykes, *RF Systems, in Synchrotron Radiation Sources: A Promer*. Ed. H. Winick, World Scientific, 87-118 (1994).
16. M. E. Busse-Grawitz, P. Marchand and W. Tron, Proc. of PAC99, New York, 986-988 (1999)
17. D. Einfeld, F. Perez, A. Fabris and et al, Proc. of PAC99, New York, 809-811 (1999)
18. http://accelconf.web.cern.ch/accelconf/jacow/proceedings.htm, Joint Accel. Conf. Proceedings
19. S. Belomestnykh, P. Barnes, E. Chojnacki and et al, Proc. of PAC99, New York, 980-982 (1999)
20. J. Jacob, A. Mosnier, J.-M. Filoh and et al, Proc. of EPAC02, Paris, 269-271 (2002)
21. H. Padamsee, these proceedings
22. H. Padamsee, J. Knobloch and T. Hays, *RF Superconductivity for Accelerators*. John Wiley & Sons, Inc., New York, (1998)
23. J. Keane, P. Mortazavi, M. Thomas and et al, Proc. of PAC99, New York, 1028-1030 (1999)
24. R. Rimmer, M. Allen, J. Saba and et al, Proc of PAC95, 1729-1731 (1995)
25. J. Jones, D. Bromberek, Y. Kang and et al, Proc. of PAC97, Chicago, 2929-2931 (1997)
26. http//www.aps.anl.gov/
27. A. Fabris, P. Carievich, C. Pasotti and et al, Proc. of EPAC02, Paris, 2142-2144 (2002)
28. T. Ruan, Proc. of 5th ESLS RF Meeting, Berlin, (2001)
29. R.G. Carter, Proc. of CERN Accelerator School, CERN 92-03, 269-300 (1992)

30. S. Isagawa, Proc. of the Joint US-CERN-Japan International School, 674-715 (1996)

31. G. Ya. Kurkin, Proc. of the Joint US-CERN-Japan International School, 742-767 (1996)

32. R.K. Cooper, Proc. of CERN Accelerator School, CERN 92-03, 245-268 (1992)

33. A. Fabris, M. Boccilai and et al, Proc. of EPAC96, Sitges, 1917-1919 (1996)

34. SPEAR3 Design Report (1999)

35. S. Sakanaka, M. Izawa, T. Mituhashi and et al, Proc. of EPAC2000, Vienna, 690-692 (2000)

36. A. Hofmann and S. Meyers, Proc. of 11th Conf. on High Energy Accelerators, Geneva, 610-614 (1980)

37. R. Biscardi, S. L Kramer and G. Ramirez, Nulc. Instr. Meth. In Phys. Res. A366, 26-30 (1995)

38. J. Byrd, J. Jacob, M. Georggson and et al, Proc. of PAC01, Chicago, 380-384 (2001)

39. G. Flynn, M.-E Couprie, M. Billardon and et al, Proc. of EPAC98, Stockholm, 954-956 (1998)

40. A. Andersson, M. Eriksson and M. Georggson, Proc. of EPAC98, Stockholm, 273-275 (1998)

41. P. vom Stein, W. Anderson, S. Belomestnykh and et al, Proc. of PAC01, Chicago, 1175-1177 (2001)

42. S. Chel, P. Marchand, M. Svandrlik and et al, Proc. of EPAC02, Paris, 2217-2219 (2002)

VACUUM SYSTEM

J. R. CHEN

National Synchrotron Radiation Research Center
101 Hsin-ann Road,
Science-based Industrial Park, Hsinchu 30077, Taiwan
E-mail: jrchen@srrc.gov.tw

1. Introduction

The vacuum system in an accelerator functions mainly to provide a clear path for the particle beam (to reduce the interactions between the particle beam and the gas molecules or other particles) and to provide a clean environment for critical components (to maintain their high performance). Generally, the vacuum system in an accelerator is comprised of vacuum chambers, pumps, gauges, valves, mechanical and electrical feedthroughs, the related control units, and several subsidiary components (such as cooling and heating units).

2. Pressure and vacuum units

Although the word *vacuum* means "empty" in some contexts, vacuum here refers to a condition of a given space with a gas pressure of under 1 atmosphere. Several pressure units, Pa, bar, torr, are commonly used to describe a vacuum.

$Pa = Newton/m^2$ *(SI unit), 1 Newton = 1 kg-m-sec^{-2}*
Bar = 10^6 dyne/cm^2, 1 dyne =1 g-cm-sec^{-2}
mb = milli-bar, 10^{-3} bar, 10^3 dyne/cm^2
Torr = mm-Hg (at 0 °C)

1 torr = 1.333 mb = 133.3 Pa $\leq$ 1.316 × 10^{-3} atm
1 Pa = 10^{-2} mb $\leq$ 7.5 × 10^{-3} torr $\leq$ 9.869 × 10^{-6} atm
1 atm $\leq$ 760 torr $\leq$ 1,013 mb $\leq$ 1.013 × 10^5 Pa

In general applications, various pressure ranges are used.

Low Vacuum: 760 – 25 torr
Medium Vacuum: 25 – 10^{-3} torr
High Vacuum (HV): 10^{-3} – 10^{-6} torr

Very High Vacuum: $10^{-6} - 10^{-9}$ torr
Ultra High Vacuum (UHV): $10^{-9} - 10^{-12}$ torr
Extreme High Vacuum (XHV): $< 10^{-12}$ torr

Gas pressure can be scaled to "molecular number density" using the gas formula,

$$P = nkT$$

Where, P: pressure
  n: molecule density
  k: Boltzmann constant, k= 1.3805 x 10^{-16} erg-°K^{-1}

Approximately 3.2 million gas molecules fill a volume of 1cm³ at 10^{-10} torr (at room temperature).

3.   Design considerations for an accelerator vacuum system

3.1. *Beam-related considerations*

3.1.1 *Beam interactions and beam lifetime*

The lifetime of the circulating beam in a storage ring is primarily limited by two effects, the Touschek effect and the residual gas effect. The interactions among particles in a single bunch (Touschek effect) and interactions between the beam-particle and the residual gas (residual gas effect) scatter the beam-particles out of their stable trajectory, and cause the beam current to decay.

Fewer interactions between the beam and the particles (of any kind) in the beam duct are required to increase the lifetime of the beam. Furthermore, a lower number density of gas molecules (or particles) reduces the blow up of the beam size. Several particles, such as gas molecules, micro dust, photoelectrons, and other secondary particles (ions and electrons) fill the vacuum duct. Fewer particles in the space cause fewer interactions between the circulating beam and the particles.

a)   Beam lifetime

The lifetime of the electron beam in an electron (or positron) storage ring can be described by

$$\tau^{-1} = \tau_T^{-1} + \tau_{RGS}^{-1} + \tau_{trap}^{-1}$$

where τ is the beam lifetime, τ_T is the Touschek lifetime, τ_{RGS} is the residual gas lifetime, and τ_{trap} is the lifetime due to particle trapping. In general, $\tau_T < \tau_{RGS}$. Both τ_{RGS} and τ_{trap} are determined by the vacuum system. Both factors depend on the surface conditions of the vacuum chamber and the environmental conditions during its assembly.

The residual gas lifetime is determined by three factors: the bremsstrahlung- (inelastic-) scattering lifetime, the nuclear- (elastic-) scattering lifetime and the electron-electron-scattering lifetime.

$$\tau_{RGS}^{-1} = \tau_{BS}^{-1} + \tau_{NS}^{-1} + \tau_{ee}^{-1}$$

where τ_{BS} is the bremsstrahlung-scattering lifetime, τ_{NS} is the nuclear-scattering lifetime, and τ_{ee} is the electron-electron-scattering lifetime. Typically, $\tau_{NS} < \tau_{BS} << \tau_{ee}$.

Ion trapping (electron ring), electron cloud (positron ring), and dust trapping are three trapping effects. The residual gas molecules are ionized by the stored beam; the ionized gas molecules can then be trapped in the potential well formed by the beam (so-called ion trapping). The slow motion of the trapped ion beam is such that the stored beam easily hits the trapped ions and is scattered. Irradiating the beam duct with synchrotron light generates many photoelectrons. These electrons can be clustered (electron cloud) by the potential of the stored positron beam. Beam emittance can increase rapidly due to the scattering between the electron cloud and the stored beam.

When a vacuum chamber is being assembled, micro dust can easily be introduced into it. This micro dust can be ionized by scattered synchrotron light or electrons. Since the micro dust is charged, it is easily trapped by the potential of the stored beam. The stored beam is likely to hit the trapped dust because the dust is very heavy and moves slowly. The intensity of the stored beam can be greatly reduced when it hits trapped micro dust.

b) Residual gas lifetime

Vacuum pressure is the most important factor in the design of a vacuum system. Formulas which correlate the residual gas lifetime with the gas pressure are essential for ensuring a reasonable pressure. Three formulas are commonly used to calculate the three components of the residual gas lifetime.

i. *Bremsstrahlung-scattering lifetime*

$$\tau_{BS}^{-1} = c \, \sigma_{BS} \, N = c(\rho/X_0)W$$

where, X_0: radiation length of the residual gas $(g - cm^{-2})$
ρ: density of the residual gas $(g - cm^{-3})$
c: velocity of light $(3x10^{10}cm\text{-}sec^{-1})$

$$W = \frac{4}{3} \ln(\gamma/\Delta\gamma) - \frac{5}{6}, \text{ probability that energy loss} > \Delta\gamma, \ \gamma = E_e/m_ec^2$$

$\rho = MP/24500 \times 760$, at room temperature
M: mass of the residual gas $(a.m.u.)$
P: pressure $(torr)$

If $\Delta\gamma/\gamma = 1\%$, then

$$\tau_{BS}^{-1} = 8539 \, MP/X_0 \, sec^{-1} = 3.1 \times 10^7 \, MP/X_0 \ hr^{-1}$$

where, $M/X_0 = \Sigma_i (M/X_0)_i$

	H	C	O	CH$_4$	H$_2$O	CO	Ar	CO$_2$
M	1	12	16	16	18	28	40	44
X$_0$	58	42.5	34.2	45.5	35.9	37.3	19.4	36.1

ii. *Nuclear-scattering lifetime*

$$\tau_{NS}^{-1} = [c_1(E^2A_0^2/P\beta_0)(1/<\beta>)]_x^{-1} + [c_1(E^2A_0^2/P\beta_0)(1/<\beta>)]_y^{-1}$$

where, C_1 : 1.0×10^{-7} hr- GeV $^{-2}$- nTorr^{-1}
E : electron energy (GeV)
P : pressure $(nTorr)$
A_0: limiting aperture $(min.[vacuum \ chamber, \ dynamic \ aperture])$
β_0 : Betatron function at the limiting aperture
$<\beta> = \int_{ring} \beta ds/L$, average betatron function

iii. Electron-electron-scattering lifetime

$$\tau_{ee}^{-1} = c\sigma_{ee} N$$

where $\sigma_{ee} = 5.0 \times 10^{-25} (Z/\gamma)(\gamma/\Delta\gamma) (cm^2)$: electron-electron scattering cross section

Z: atomic number of the residual gas
$N = 3.2 \times 10^{16} P$ (molec./cm³)
P: pressure (torr)

3.1.2 Beam stability

The vacuum pressure affects beam lifetime and beam size. The structure of the vacuum chamber also affects the properties of the beam, such as the stability of the beam's orbit, the beam's impedance, and the beam's response to the pulse magnetic/electric field.

The vacuum system should be mechanically as stable as possible. Any vibration or thermal expansion of the vacuum chambers can induce a movement of the magnets or the beam position monitors (BPMs). Movements of the magnets or the BPMs can change the orbit of the beam.

The cross section of the vacuum duct should be as smooth as possible. The wake field generated by an abrupt change of the beam duct's cross section increases the beam impedance and degrades its stability. Beam energy is transferred to the chamber and heats up its components.

The material and thickness of the vacuum chamber should allow the external electric/magnetic field (dc, ac or pulse field) to be applied to the beam. An improperly designed vacuum chamber may cause eddy current, field distortion, poor frequency response, and/or heating problems.

3.2. Vacuum system considerations

When designing the vacuum system in an accelerator, many vacuum system-related (vacuum engineering-related) issues must be addressed.

3.2.1 Basic vacuum issues

Two basic factors, pressure and thermal load, are the most important design criteria governing the vacuum system of an accelerator. They may not be violated.

The system must be leak tight to ensure a low pressure, and the outgassing rate of the vacuum material must be low. Surface-treatment techniques are often applied to achieve this low outgassing rate. Furthermore, the type of pumps and the pumping configurations must be carefully arranged to produce a low pressure with a cost-effective pumping speed.

Materials with high thermal conductivity, such as aluminum and copper, are widely used for making vacuum chambers and thermal absorbers. A design that depends on a grazing incidence angle between the incident beam (such as high intensity synchrotron light) and the chamber surface (or absorber) is always chosen to reduce the heat density on the surface of the chamber or the absorber, and to prevent meltdown. In the direct cooling configuration, cooling channels are built in the substrate to be cooled. The configuration is much better than the indirect cooling configuration in which the cooling channel is not the same piece of material as that of the substrate. In the latter case, the quality of the thermal contact between cooling channel and the substrate to be cooled is not easy to control. For an extremely high thermal load, a material with a low atomic number and high thermal conductivity, such as beryllium, is often adopted. Then, the thermal power density is reduced by the lower attenuation (less absorption) of the intensity of the incident beam.

3.2.2 System operation issues

During construction, installation, and operation, several more criteria must be met to ensure a high-quality vacuum system and/or to eliminate unnecessary costs. This section considers three major factors, vacuum protection, resistance to environmental conditions, and precise mechanical structure.

a) *Protecting a vacuum system in case of an accident*

The three layers of architecture - vacuum devices, the utility system and the interlock system, must all be tightly connected. Reliable vacuum devices are a basic requirement in protecting a vacuum system from an accident. Devices with self-protection, such as the over-load protection of an ion pump, the over-pressure protection of an ionization gauge, and the over-temperature protection of a turbomolecular pump, are preferred. As well as the vacuum devices, the utility systems supplied to the devices (such as cooling water, compressed air, and electricity), must be as reliable as (or even more reliable than) the vacuum devices. A reliable interlock system is also crucial. Hard wire protection logic with redundant sensors is always used for crucial devices.

b) *Reducing the impact of severe environmental conditions*

Temperature, radiation, humidity and dust are four major environmental factors that impact an accelerator's vacuum system and its components. The dimensions and performance of several vacuum components are sensitive to temperature. Temperature should be maintained within a certain range to ensure the performance of the vacuum components and the vacuum system. The high radiation field in an accelerator often causes problems of radioactivity and radiation damage. Although the problem of radioactivity is less severe in an electron/positron machine than in a proton or heavy-particle machine, the vacuum materials must also be carefully chosen to reduce the impact of radiation. Materials highly resistant to radiation are required, particularly in some cables and insulators. High humidity can reduce the quality of welding. Improper insulation in a high-humidity environment can increase pumping time and degrade vacuum performance. The surface of some critical components, such as bellows and brazed parts, should be isolated (for example, under a coating) from humid air or any condensation of water vapor, to prevent corrosions in the high radiation field. As stated in Section 3.3.1.a, micro dust is harmful to the performance of the beam. Fabrication and installation under clean room conditions are useful to avoid micro dust getting into the vacuum chamber.

c) *Maintaining a precise mechanical structure (even after baking)*

Careful dimension control during machining and welding is the basic requirement for ensuring a precise mechanical structure. Rigid fixed points are often designed at the locations of beam position monitors, heavy components, or some critical components, as reference points. Minimizing the force on the vacuum chamber or critical components is also a basic requirement. At least one bellow must be designed between two rigid fixed points, so that displacement of the rigid fixed reference points can be greatly reduced. Flexible supports constrained in all directions, except the direction of motion, are used to support the moving parts. Springs with suitable spring constants effectively reduce the force exerted by weighty components on the vacuum chamber. The force of some components during baking can differ significantly from that without baking. In such a case, the design includes pre-displacement to create an optimal force condition during baking. Thermal masks and cooling channels must be appropriately designed so that the temperature distribution is uniform and the deformation of the vacuum chamber is minimal.

4. *Gas desorption*

Gas desorption should be reduced to ensure a good vacuum. The following formula relates pressure, P, outgassing rate, Q, and pumping speed, S, in a vacuum system

$$P = Q / S$$

Reducing the outgassing rate is much more effective than increasing pumping speed in reducing pressure because the increase in pumping speed, S, is limited by the size or cost of the pumps, whereas the outgassing rate, Q, can be easily decreased by orders of magnitudes after proper treatment, such as baking or some cleaning procedures.

Two general types of desorption mechanisms may occur - thermal desorption and stimulated (or induced) desorption. Gas desorption excited by thermal energy is called thermal desorption. It is called stimulated desorption if the energy of energetic particles, such as electrons, photons or ions is involved.

4.1 *Thermal desorption*

Thermal desorption is the most common type of gas desorption in a vacuum system. The desorption rate (or outgassing rate), Q_{th}, is a function of the adsorption energy, E_d, and the thermal energy, kT

$$Q_{th} \sim \exp(-E_d/kT)$$

where E_d the binding energy of the gas desorbed on the substrate
k the Boltzmann constant ($8.6 \times 10^{-5} eV\text{-}K^{-1}$)
T the temperature (°K).

The thermal outgassing rate is controlled mainly by surface desorption and the diffusion process. A less contaminated surface and less contaminated outer surface layers enable a lower thermal outgassing rate to be achieved. Generally, thermal outgassing can be effectively reduced by chemical cleaning and/or in-situ baking. Water vapor is the dominant residual gas in a vacuum system before baking. A baking process with enough temperature and uniformity can reduce water vapor dramatically. The amounts of hydrocarbons and some carbides can also be reduced after baking if the baking thermal energy is sufficiently high. Hydrogen is the major outgas during baking because it is in infinite supply inside the bulk material. A barrier formed on the surface can obstruct the diffusion of hydrogen through the surface. Elastomers and materials with high vapor pressure are not recommended for use in accelerators which require an UHV system.

4.2 Photon-stimulated desorption (PSD)

Photon-stimulated desorption is the most severe outgassing problem of an electron/positron storage ring. Large numbers of photoelectrons are released when intense synchrotron light is shining upon the surface of the chamber. Gas molecules can be desorbed from the chamber's surface when photoelectrons leave or return to the surface. The following describes the desorption process of gas molecules; related parameters and the formula are also presented.

e-beam → Synchrotron Radiation→ Photo-electron→Gas molecules
$$I \quad \rightarrow \quad d/dt\,(d^2N(\varepsilon)/dId\varepsilon) \quad \rightarrow \quad Y(\varepsilon)F(\theta) \quad \rightarrow \quad 2\eta$$

$$Q_{psd} = I \times \int \frac{d}{dt} \frac{d^2N(\varepsilon)}{dId\varepsilon} Y(\varepsilon)\, F(\theta)\, d\varepsilon \times 2\eta$$

where, I beam current (mA)
  $Y(\varepsilon)$ photoelectron yield (No. of electrons / No. of photons)
    for aluminum, $Y(\varepsilon) \le 2.61\, \varepsilon^{-0.94}$, $10eV< \varepsilon$ $1560eV$
      $Y(\varepsilon) \le 441.9\, \varepsilon^{-1.13}$, $1560eV< \varepsilon < 10keV$
  $\varepsilon = h\nu$, photon energy
  η desorption coefficient (No. of molecules / No. of electrons)
  $F(\theta) \le \sin^{-1/2}\theta$
  θ angle between the surface and the incident beam

From the formula, the PSD rate, Q_{psd}, is lower when the chamber surface has a lower photoelectric yield (smaller Y), when it is cleaner (smaller η), and when the incidence angle is closer to the surface normal ($F(\theta)$ is minimum at $\theta=90°$). To simplify the calculation, the energy spectrum of synchrotron light can be approximated as,

$$\frac{d}{dt} \frac{d^2N(\varepsilon)}{dId\varepsilon} \le 1.51 \times 10^{14} \rho I\, E^2\, (\varepsilon/\varepsilon_c)^{-2/3}, \text{ for } \varepsilon \quad \varepsilon_c$$
$$\le 0 \quad \text{for } \varepsilon > \varepsilon_c$$

Therefore,

$$Q_{psd} \le 8.6 \times 10^{17}\, I\, E\, \varepsilon_c^{-1/3}\, Y(\varepsilon_c)\, F(\theta)\, 2\eta$$

where, E electron beam energy (GeV)

ε_c critical photon energy = $2.21 \times 10^3 \ I \ E^3/\rho$

ρ bending radius (m)

for aluminum, $Y(\varepsilon_c) = (0.41 - 1.66\varepsilon_c^{-0.6})$, $h\nu$ $1560eV$

$Y(\varepsilon_c) = (1 - 216.2\varepsilon_c^{-0.6})$, $h\nu > 1560eV$

4.3 *Ion-beam-induced desorption*

Gas burst in a proton ring is associated not only with problems of lifetime reduction and emittance growth, but also with some unusual instabilities. A special gas desorption phenomenon, pressure bump, has been observed in proton storage rings. When a circulating beam ionizes gas molecules, the ions (may be repelled by the beam potential (approximately 100 V per A of beam current). The energetic ions bombard the chamber wall and induce gas desorption. Many electrons can be emitted from the wall's surface. Furthermore, the multipactoring, due to the resonant interactions between the circulating beam and the secondary electrons, causes many more secondary emissions. Electrons trapped within a circulating proton beam can also neutralize the beam, leading to an unwanted tune-shift.

The ion-induced-desorption rate, Q_{iid}, can be given by the formula,

$$Q_{iid} = 103 \ P\sigma\eta \ \frac{I}{e}$$

where, P pressure
I beam current
σ ionization cross section
η desorption coefficient
e electron charge
$E_{ion} \leq 60 \ I \ \log(R/r)$, ion (residual gas ion) energy
R radius of vacuum chamber
r radius of proton beam

The above formula implies that the desorption rate, Q_{iid}, is proportional to the pressure, P. Balancing the gas flow rate in a slice dx of a chamber,

$$A \ (dP/dt) \ dx = (q_{th}dx + PI\sigma\eta \ dx) - S_0P \ dx$$

where, A chamber surface area
q_{th} thermal outgassing rate

464

I   beam current
σ   ionization cross section
η   desorption coefficient
S_0 distributed pumping speed

an equivalent pressure can be derived

$$P = q_{th} / (S_0 - I\sigma\eta)$$

The blow up resonance condition is met when a critical beam current, I_c, is reached

$$I_c = \frac{S_0}{\sigma\eta}$$

Increasing the distributed pumping speed, reducing the ionization cross section, and reducing the desorption coefficient effectively increase the critical current.

5. Pumping configuration and pressure distribution

The beam duct of an accelerator is always narrow and available locations for pumps are always limited. Several factors must be considered in using pumps effectively. They must be chosen to optimize the pressure distribution which will give a better beam performance. This section introduces the conductance-limited effective pumping speed, pumping configuration, the calculation of the pressure distribution and the considerations in choosing or operating pumps.

5.1. Conductance and effective pumping speed

Throughput is the amount of gas flow at a known pressure and temperature that passes a plane in a known period. Throughput (or outgassing rate) can be described as the product of pressure and pumping speed. If no desorption or absorption occurs along the path, then the throughput at a certain position should equal the outgassing rate of the gas source

$$Q = P'_{ch} S'_{ch} = P_p S_p = C (P'_{ch} - P_p)$$

where, P'_{ch} the gas pressure at chamber
       S'_{ch} the effective pumping speed at chamber
       P_p the gas pressure at pump
       S_p the pumping speed of pump

The throughput Q can also be determined by multiplying the pressure difference and a parameter by the conductance, C,

$$Q = C (P'_{ch} - P_p)$$

C is the conductance of the tube (unit: l/s). It is a function of the geometry and is independent of pressure in the molecular flow regime. A narrower or longer tube has a lower conductance.

Rearranging the above two formulae yields

$$1/S'_{ch} = 1/S_p + 1/C$$

As the conductance of a narrow and long tube is low, a large pump is useless because it is the conductance limiting the effective pumping speed.

5.2. Pumping configuration

The conductance of the beam duct in an accelerator is always small so special pumping configurations are required to meet the strict low pressure requirements, particularly in a machine with extensive gas desorption. Two pumping configurations, distributed pumping and localized pumping, are commonly used in storage ring vacuum systems.

5.2.1 Distributed pumping

Many small pumps are preferably located as close to the gas sources as possible to maintain the pumping speed of a large pump when pumping through a conductance-limited tube. This method effectively reduces the pressure in the case of a distributed gas source, for example, PSD in an electron storage ring that could be distributed continuously along the beam duct. Distributed ion pumps are popularly implemented in the bending magnet chambers using the magnetic field of a dipole magnet. They can easily evacuate photo-desorbed gases released in the bending chamber region. Getter strips, for example strips of non-evaporable getters (NEG), are used in the locations with no magnetic field, for example, in chambers in the straight sections. Recently, a technique of coating the beam duct with NEG material has been developed. In such a case, the coated chamber itself is also a well - distributed pump.

5.2.2 Localized (enhanced) pumping

As well as the distributed pumping mentioned above, another pumping configuration, localized (or enhanced) pumping, is also widely used. The basic concept is the opposite of that applied by distributed pumping in a synchrotron light source: localized pumping uses photon-masks to stop photons at certain positions so that PSD-sources are discretely distributed. A pump with a high pumping speed is installed beneath (or very close to) the PSD source. The conductance between the pump and the gas source can thus be much less constrained. This pumping configuration offers some advantages, such as a lower thermal load on the beam duct (if an independent photon-mask is used), fewer gas sources and less photoelectrons in the beam duct (farther away from the circulating beam).

5.3. Pressure distribution

The pressure in an accelerator is determined by several factors, such as the conductance of the vacuum chamber, the outgassing rate, the location and pumping speeds of the pumps, and others. When the distributions of the pumping speed, the outgassing rate and the conductance are known, the pressure distribution along the vacuum duct can be calculated according to the continuity equation for gas flow,

$$S_i P_i = Q_i + C_i(P_{i-1} - P_i) + C_{i+1}(P_{i+1} - P_i)$$

where, S_i (S_{i+1}) the pumping speeds at position i (i+1)

P_i (P_{i+1}) the pressure at position i (i+1)

C_i the conductance of the beam duct between positions i-1 and i

C_{i+1} the conductance of the beam duct between positions i and i+1

Q_i the outgassing rate at position i

Accordingly, under equilibrium conditions, the gas throughput evacuated by the pump at position i , $S_i P_i$, is equal to the outgassing rate at position i , Q_i, plus the gas flow into position i from position i-1 , $C_i(P_{i-1} - P_i)$, plus the gas flow from position i+1, $C_{i+1}(P_{i+1} - P_i)$. In an electron storage ring, the outgassing rate is governed mainly by photon-stimulated desorption, which can be estimated using the formula in Section 4.2. The dimensions of beam duct determine the conductance. With respect to the vacuum pressure point, a beam duct with a larger diameter is preferred, if the dimensions are tolerable in relation to the bore radius of the magnet.

A computer program helps to simulate the distribution of pressure in a vacuum system. Particularly in the design stage, several tests can be attempted before the optimized final design is achieved. Usually, locations for installing pumps are quite limited. Therefore, the design goal is to achieve a cost effective distribution of pumping speeds.

6. Vacuum components and reliability

This section introduces the vacuum components commonly used in accelerators. However, descriptions of mechanisms and the structures of the components which can be found in many vacuum textbooks are not provided. Only materials related to the operation and performance of accelerators are discussed.

6.1. Vacuum chamber material

Pressure and temperature are two of the most important fundamental considerations in choosing vacuum chamber material. In some cases, short lifetime radioactivity is also required to shorten the waiting time for maintenance. Although mass is not an important factor in choosing the vacuum chamber material of an accelerator, a light material is easier to handle during installation and maintenance.

To meet the low pressure requirement in an UHV system, the material from which the vacuum chamber is constructed must have few defects in the bulk to prevent virtual or real leaks. Low vapor pressure, low outgassing rate and bakeability are also required to help to maintain a clean vacuum. With respect to machining cost and quality point of view, the vacuum chamber material must be easy to machine and easy to weld.

High thermal conductivity of the vacuum chamber material is a basic requirement to protect against damage from over-heating. Heat density is commonly reduced using a grazing incidence design. In cases of extremely high thermal density, different thinking is adopted that does not attempt to stop the high-intensity beam over a short distance. In such a case, a material with low atomic number, such as beryllium, is a good candidate. In some cases, the dimensions of the chamber must be insensitive to variations in temperature. A low thermal expansion coefficient is the most important requirement of the chamber material in such a case.

6.2. *Surface treatments*

The design, treatment and operation of the UHV system in an electron storage ring are important. This section considers some treatment methods of vacuum components, such as chemical cleaning, baking, glow discharge cleaning and beam self cleaning. Surface treatments are performed to eliminate surface contamination and thus reduce surface outgassing rate.

6.2.1. *Chemical cleaning*

Dust, oil, oxides, carbides, and other matter are attached to the surface of a vacuum chamber after conventional oil-lubricated machining. Chemical cleaning is one of the most convenient and effective ways to eliminate the contamination from the chamber surface. Chemical cleaning processes are different for various materials or requirements. An aluminum alloy chamber can be easily machined in an ethyl alcohol vapor environment to prevent oil contamination and then form a clean oxide layer. In such a case, the surface needs to be cleaned only by light degreasing before vacuum system is set up. The same degreasing process can also be applied to an aluminum vacuum chamber fabricated using a special extrusion method (in a clean atmosphere of Ar + 10% O_2), to ensure that the inner surface is sufficiently clean. In some other cases, the chamber must be cleaned by a thorough chemical cleaning process. A cleaning process suggested by the LEP vacuum group at CERN is commonly used to clean aluminum vacuum chambers

1. Immersion in NaOH (concentration $45gl^{-1}$) at 45°C for one to two minutes.
2. Rinsing in demineralized water.
3. Immersion in an acid bath that contains HNO_3 (concentration 50% by volume) and HF (concentration 3% by volume).
4. Rinsing in demineralized water.
5. Drying.

Ultrasonic agitation in demineralized water is often performed between processes 4 and 5. The cleaning processes of aluminum, mentioned above, may only be performed before welding and coating. After welding or coating, the aluminum components should not be cleaned by the above chemical process.

The following chemical cleaning processes are popular for stainless steel components.

1. Rinsing in alkaline cleaner.

2. Rinsing in tap water (better with bubble agitation).
3. Pickling in hydrofluoric - nitric acid solution (HF 3 %, HNO_3 50%); the immersion time is determined by the extent of contamination.
4. Rinsing in tap water (better with bubble agitation).
5. Ultrasonic agitation in demineralized water.
6. Drying.

The cleaning of titanium material is similar to that of stainless steel. However, it takes less time than that for stainless steel, because the chemical activity of titanium is stronger. The titanium film coated on the wall of the Ti-sublimation pump or the ion pump should be scraped off before chemical cleaning.

Special care must be taken with the chemical solvent to protect the environment. Several new processes have been developed to meet the requirement of using a light chemical solvent.

Vacuum components must be wrapped in clean aluminum foil after they have been cleaned to prevent recontamination on the cleaned surface. During assembly, all components should be handled with clean and lint-free gloves, which should not be reused after they have been contaminated.

6.2.2. Baking, discharge cleaning and beam-self cleaning

Most of the vacuum system in a storage ring is designed to be bakeable. The baking must satisfy the requirement of uniform temperature distribution to increase the heating efficiency (in reducing outgassing rate) and prevent serious distortion. For aluminum alloy, the baking temperature is ~150°C; for stainless steel, it is ~250°C. Thermal insulators on the baking parts should be well wrapped to save electricity during baking. During baking, turbo molecular pumps are always used to evacuate the released gas from the vacuum system. The ionization gauges, ion pumps and titanium sublimation pumps must be degassed when the temperature begins to be reduced after bakeout.

High temperature degassing at ~950°C is normally applied to stainless steel components. During the high temperature degassing, not only can contamination on the surface be removed but also the solved gas can be eliminated from the bulk. The pre-bake process to high temperature in air or in oxygen will generate an oxide barrier on the surface. It is useful for obstructing the diffusion of hydrogen (from the bulk) through the surface.

The Ar glow discharge cleaning process has been adopted in some electron storage rings. It is effective in increasing beam lifetime in the beginning of the commissioning of the machine. However, the re-emission of Ar is a problem in limiting the beam lifetime later. Discharge of other gases, including He and H_2, is also effective in surface cleaning, and is less affected by the re-emission problem (because their atomic numbers are smaller and thus their cross section of interaction between the gas and the circulating beam is less).

A storage ring with a full-energy injector can be cleaned by intensive synchrotron light, in the so-called beam-self cleaning mode. It is a regular cleaning process used in many rings to clean the vacuum system after new beam ducts were installed. An accumulated dose of several tens of A-hours is required to achieve a reasonable beam lifetime of several hours.

6.3. *Valves*

Four kinds of valves, the gate valve, the angle valve, the variable leak valve and the fast closing valve, are popular for use in accelerator vacuum systems. The vacuum duct of an accelerator should be divided into sections by gate valves to facilitate maintenance and enable a vacuum system to adapt to new equipment. The sector gate valves must be equipped with RF bridges to reduce the RF impedance. A gate valve is also sometimes positioned between titanium-sublimation-pump (TSP) and the main chamber, if the Ti-filament (or Ti-ball) must be frequently replaced. A gate valve or an angle valve is often used with a turbo-molecular pump to isolate the vacuum system from the atmosphere in case of an accident that would otherwise cause the turbo-molecular pump to fail. Variable leak valves are required to control the quality (including, for example, low humidity) and the flow rate of the venting gas, if venting is required. A fast-closing valve (or shutter) is a standard device at the front-end of a beam line. The seal of a fast-closing-valve is need not be absolutely tight. The reaction time must, however, be sufficiently short to stop the shock wave in the case of a rush of air caused by an accident.

Leak tightness is the most basic requirement of a valve. If UHV conditions must to be maintained, then all-metal valves are the best choice. Where less radiation is present, O-ring valves which can be baked to a reasonable temperature (say 200 °C) are also good candidates. Response time, type of actuation (manual or pneumatic), mechanical reliability and lifetime are other important considerations in selecting valves.

References

1. J. Kouptsidis and A. G. Mathewson, DESY report, DESY 76/49, 1976.
2. H. Wiedemann, *"Coulomb scattering and vacuum chamber aperture,"* SSRL-ACD-NOTE, Dec.13,1983.
3. A.G. Mathewson et al., KEK report, KEK-78-9, (1978).
4. N.B. Mistry, *"Ultrahigh vacuum systems for storage ring accelerators,"* Cornell report, CLNS-86/720, 1986.
5. D.C. Chen et al., J. of Vac. Soc. of ROC 1(1), 24(1987).
6. A.G. Mathewson, presented at the Xth Italian National Congress on Vacuum Science and Technology, Stresa, Italy, 12-17 Oct. 1987.

RFQ DESIGN AND PERFORMANCE

JIAXUN FANG

Institute of Heavy Ion Physics,
Peking University, 100871, Beijing, PR China
E-mail: jxfang@pku.edu.cn

In the area of ion accelerators with low energy, the RFQ linear accelerator becomes more and more important. It combines the advantages of a strong beam current with an accelerating energy in the MeV range which cannot be met by the popular Van de Graaff or Cockcroft-Walton accelerators simultaneously. Further, it also has a small size and a high efficiency, and it is convenient to use. Therefore, it developed very quickly and is used in many fields. In this lecture, its principle, beam dynamics, resonator cavity, development, and applications are described concisely.

1. Introduction

In the area of ion accelerators with low energy, the Cockcroft-Walton type and the electrostatic type (Van de Graaff and Tandem) are the main traditional accelerating facilities. The Cockcroft-Walton accelerator has a strong beam current on the mA level, but its accelerating energy usually is a few 100 keV, well below 1 MeV. Electrostatic accelerators can reach the energy level of MeV, but its beam current is usually below 10 μA. However, this situation has been changed since the Radio Frequency Quadrupole (RFQ) accelerator appeared in the 1980's.

In 1970 Kapchinskii and Teplyakov proposed a new linear ion accelerator — RFQ [1]. In 1980, a prototype of a 4-vane RFQ was successfully built and tested in LANL (Figure 1). Its cavity length was only 111 cm with a diameter of 15 cm, and it accelerated a proton beam from 100 keV to 640 keV with a beam current of 30 mA and a transmission efficiency of 87% [2]. It demonstrated the advantages of an RFQ such as:

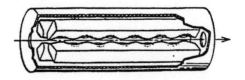

Figure 1 4-vane RFQ

- Strong beam current as high as $10^1 - 10^2$ mA
- Accelerating energy possibly reaching MeV per nucleon
- Small size
- High transmission efficiency over 80% or 90%
- Low injector energy directly from ion source at ground potential
- Integrating more functions in an RF cavity without external elements: not only effective accelerating and strong focussing, but also beam bunching and transverse matching
- Operating and controlling simple and convenient.

Due to the above merits, the proton and light ion RFQ and its applications developed rapidly since 1980. Then, the heavy ion RFQ and high power RFQ succeeded. In 1989 a BEAR (Beam Experiment Aboard Rocket) RFQ with 1 m length was carried by the rocket Alice to test a 1 MeV H⁻ beam in the air. In 2000, the LEDA (Low Energy Demonstration Accelerator) RFQ was built and tested at LANL. A DC beam of 100 mA was accelerated to 6.7 MeV in a cavity 8 m long [3]. Its average beam power was as high as 670 kW! It obviously showed the outstanding features of the RFQ. Due to this favorable performance, in particular, the combination of MeV accelerating energy with a high beam current of 10 – 100 mA which can not be achieved simulaneously by neither Cockcroft – Walton, nor Van de Graaff, nor the tandem accelerator, RFQs replace these accelerators widely in the low energy region, not only as powerful injectors of ion linacs and cyclic accelerators for high energy physics, spallation neutron source, accelerator-driven-system, and medical therapy etc., but also on its own for the beam to be used directly.

2. Principle of the RFQ

In conventional ion linac such as Wideroe, Alvarez, Spiral, Split-ring, QWR, etc., the basic accelerating structures are drift tubes. Ions are accelerated in the gaps between drift tubes, and focused by magnetic quadrupole lenses in the drift tubes. They also need a buncher in front of the linac to bunch the DC beam into a pulsed beam for the longitudinal matching as well as a transverse matching to the RF linac.

In the RFQ, the basic structure are 4 electrodes which generate an RF electric field in its channel to accelerate and focus ion beams simultaneously. The RF field can also play the role of transverse matching and longitudinal bunching functions according to the dynamics design. The RFQ ingeniously integrates the above four basic functions in one compact RF cavity.

474

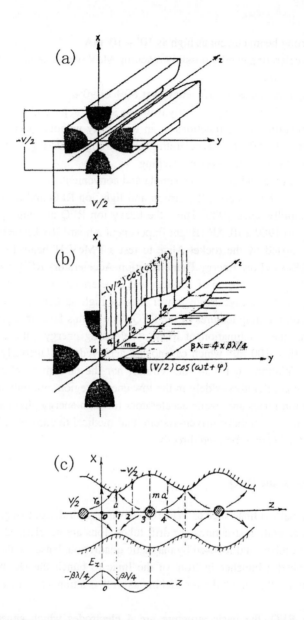

Figure 2 Illustration of the RFQ principle (a) electrostatic quadrupole lens
(b) electrode profile of the RFQ (c) electric field in the x-z plane of the RFQ

The RFQ concept can be easily understood from the static electric
quadrupole lenses which are popularly used as focusing elements in the beam
transport of accelerators and have only transverse electric field for focusing as

shown in Figure 2 (a). To make the electric quadrupole lenses having longitudinal electric field components for accelerating ions, the surfaces of the lenses can be modulated periodically along the beam axis. Then, there are periodic accelerating and decelerating electric fields around the beam axis. To realize a continuous acceleration, an RF voltage must be applied to the electrodes instead of a DC voltage and the modulation period can be made equal to $\beta \lambda$ as in linear ion accelerators with drift tubes where β and λ are relative velocity v/c and RF wavelength in free space, respectively. Then, the ions can be accelerated synchronously and focused transversely as in the conventional linear ion accelerators. This analysis is demonstrated in Figure 2 (b) which shows the partial profile of the RFQ electrodes from the corresponding Figure 2 (a). Figure 2 (c) shows the profile of the electrodes in the X-Z plane as well as the electric field lines and the E_z distribution along the axis.

To study the field distribution and beam dynamics in more detail, the calculation will be performed in the following.

To calculate the electric field around the axis the electric potential can be expressed as:

$$\psi(r,\theta,z,t) = U(r,\theta,z)\cos(\omega t + \phi) \tag{1}$$

Considering the symmetry of electrodes, the spatial part in general is

$$U(r,\theta,z) = \sum_{l=0}^{\infty} F_{ol} r^{2(2l+1)} \cos 2(2l+1)\theta$$
$$+ \sum_{n=1}^{\infty}\sum_{l=0}^{\infty} A_{nl} I_l(knr)\cos 2l\theta \sin knz \tag{2}$$

where $k = 2\pi/\lambda$.

For simplicity, usually only the first two terms are taken as:

$$U(r,\theta,z) = -\frac{V}{2}[F_{00}r^2\cos 2\theta + A_{10}I_0(kr)\sin kz] \tag{3}$$

$$A_{10} = A = \frac{m^2-1}{m^2 I_0(ka) + I_0(mka)} \quad \text{------ accelerating efficiency} \tag{4}$$

$$F_{00} = F = 1 - AI_0(ka) \quad \text{------ focusing efficiency} \tag{5}$$

m ------ modulating efficiency

I_0 ------ modified Bessel function

From U(r, θ, z), the electric field components can be obtained as

$$E_z = \frac{kAV}{2} I_0(kr)\cos kz \tag{6}$$

$$E_r = \frac{FV}{a^2} r\cos 2\theta + \frac{kAV}{2} I_1(kr)\sin(kz) \tag{7}$$

$$E_\theta = -\frac{FV}{a^2} r\sin 2\theta \tag{8}$$

From (4)-(7) and Figure 2(c), it can be seen that A and F denote the strengths of accelerating and focusing, respectively, and can be adjusted by modulating m. When m is increased, E_z and A will be larger whereas E_r and F will be smaller. It is the same the other way round.

In the x-z plane, the equation of the electrode tip shown in Figures 2 (b) and 2(c) can be derived by setting $\theta = 0$ and U= -V/2 in equation U(r, θ , z)

$$F(x/a)^2 + A\, I_0(kx)\, \sin kz = 1 \tag{9}$$

Similarly, the equation of electrode tip in y-z plane shown in Fig. 2 (b) is

$$F(y/a)^2 - A\, I_0(ky)\sin kz = 1 \tag{10}$$

The x and y coordinates shown in Figure 2 (b) can be calculated as

at z =	0	$\beta \lambda /4$	$\beta \lambda /2$	$3\beta \lambda /4$	$\beta \lambda$
x =	r_0	a	r_0	ma	r_0
y =	r_0	ma	r_0	a	r_0

where $r_0 = a/\sqrt{F}$ is the characteristic radius.

Usually, an RFQ can be divided into many cells with a length equal to $\beta \lambda /2$, and the distance from axis to tip from a to ma or from ma to a.

3. Dynamics and design

3.1 Longitudinal motion

From the above analysis and the expressions of the electric field, it is clear that the longitudinal motion in RFQs can be treated as in general linacs. Therefore, the energy gain per cell is

$$\Delta W_s = qe \int_{-L/2}^{L/2} E_z dz = qeE_0 TL\cos\varphi_s \tag{11}$$

where the average electric field is $E_0 = AV/L$ $\qquad$ (12)

the transit-time factor $T = \pi/4$ $\qquad$ (13)

and the cell length L= $\beta \lambda /2$. $\qquad$ (14)

In an RFQ the equation of longitudinal motion can be derived as

$$\frac{d}{dz}(W - W_s) = qeE_0T(\cos\varphi - \cos\varphi_s) \qquad (15)$$

$$\frac{d}{dz}(\varphi - \varphi_s) = -\frac{2\pi}{\varepsilon_0\beta_s^3\gamma_s^3\lambda}(W - W_s) \qquad (16)$$

or

$$\frac{d}{dz}\left[\beta_s^3\gamma_s^3\frac{d}{dz}(\varphi - \varphi_s)\right] = -\frac{2\pi}{\lambda}\cdot\frac{qeE_0T}{\varepsilon_0}(\cos\varphi - \cos\varphi_s) \qquad (17)$$

The equations are of the same form as in general ion linacs. Therefore, all the discussions and results of longitudinal and phase motions in general ion linacs can be applied to RFQs as well. But it should be pointed out that the energy gain and longitudinal motion can be adjusted locally by changing the local m.

3.2 Transverse motion

It is an obvious advantage that an RFQ features strong transverse focusing by its own RF electric field which is quite different from other ion linacs with magnetic quadrupoles. It makes the RFQ simpler and smaller.

Based on the expression of the electric radial component and the approximation

$$I_1(kr) \approx \frac{1}{2}kr \qquad (18)$$

the equation of transverse motion can be derived as

$$\frac{d^2x}{dz^2} = \frac{qe}{\varepsilon_0\gamma\beta^2}\left[\frac{FV}{a^2}\cos(\omega t + \varphi) + \frac{k^2AV}{4}\sin kz\cos(\omega t + \varphi)\right]x \qquad (19)$$

On the right hand side, the first term corresponds to the focusing force of the RF quadrupole field. The second term means the defocusing force caused by the accelerating field. It varies with both time and coordinate. Then take its average in one cell

$$\frac{d^2x}{dz^2} = \frac{qe}{\varepsilon_0\beta^2\gamma}\left[\frac{FV}{a^2}\cos(\omega t + \varphi) - \frac{k^2AV}{8}\sin\varphi_s\right]x \qquad (20)$$

and, similarly,

$$\frac{d^2 y}{dz^2} = \frac{qe}{\varepsilon_o \beta^2 \gamma} \left[-\frac{FV}{a^2} \cos(\omega t + \varphi) - \frac{k^2 AV}{8} \sin \varphi_s \right] y \tag{21}$$

With the coordinate transformation

$$\tau = \frac{z}{\beta\lambda} = \frac{\omega t + \varphi}{2\pi} \tag{22}$$

equation (20) becomes

$$\frac{d^2 x}{d\tau^2} + (\Delta - B\cos 2\pi\tau)x = 0 \tag{23}$$

where

$$B = \frac{qe\lambda^2 FV}{\varepsilon_o \gamma a^2} \tag{24}$$

and

$$\Delta = \frac{qe\pi^2 AV \sin \varphi_s}{2\varepsilon_o \beta^2 \gamma} \tag{25}$$

B and Δ also denote the focusing force of the RF quadrupole field and the defocusing force of the RF accelerating field, respectively.

Equation (23) is a Mathieu equation. The transverse motion can be stable if the parameters are chosen in a certain region.

3.3 Beam dynamics design

The beam dynamics design of an RFQ is performed in terms of a given ion species (charge states and mass number), input energy, beam current and emittances, operating frequency, inter-electrode voltage, through choosing proper dynamics parameters, to reach the requirement of output energy, beam current and emittances.

The determination of operating frequency (f) and inter-electrode voltage (V) is combined with not only beam dynamics, but also the RFQ cavity and RF transmitter parameters. To reach enough focusing force and initial cell length, a heavy ion RFQ (low ratio of charge states to mass number Q/A) needs a low operating frequency. An RFQ for protons and light ions with high Q/A can operate at higher frequency. However, low operating frequencies will lead to a longer and larger RFQ cavity compared with high frequencies for a given output energy per nucleon.

Usually, higher inter-electrode voltages are beneficial to reach higher energy gain, shorter cavity and better performance in terms of beam dynamics. However, they need more RF power and face the danger of RF sparking. There is a popular criterion for RF sparking known as Kilpatrick criterion [4]

$$f = 1.643 E_S \exp[-8.5/ E_S] \qquad (26)$$

where E_s (MV/m) is the sustained maximum surface RF field in the cavity at frequency f (MHz).

For the 4-vane RFQ,

$$E_S = kV/r_0, \qquad k \approx 1.25 - 1.55 \qquad (27)$$

The maximum E_s exists at $a = r_0$ (see Figure 2) where the distance between adjacent vanes is minimum. From equation (26), the E_s is called a unit of 1 Kilpatrick. In practice, RF sparking is a complicated problem. It depends not only on the frequency f, but also on the RF pulse duration time (duty factor) and the surface conditions. For safety, the designed E_S is taken as ≤ 1.8 Kilpatrick.

As mentioned above, the beam dynamics design is about choosing the proper parameters to meet the goals of RFQ dynamics subject to certain given conditions. From the previous discussion of beam dynamics in longitudinal and transverse motions, it can be seen that these conditions allow only 3 independent parameters to be chosen which are usually taken as the synchronous phase Φ_s, the channel aperture a and focusing strength B. Therefore, the design is concentrated on the choice of these independent parameters.

Generally, according to the dynamics functions, an RFQ can be divided into 4 sections, namely, the Radial Matching Section (RM), the Shaper Section, the Gentle Bunching Section (GB), and the Accelerating Section. Figure 3 shows an example to illustrate the design of RFQ dynamics parameters.

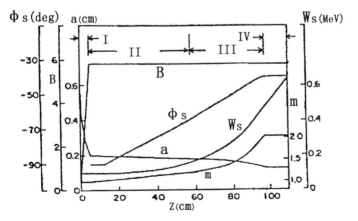

I — Radial Matching Section (RM)   II — Shaper Section
III — Gentle Bunching Section (GB)   IV — Accelerating Section

Figure 3 An Example of RFQ Dynamics Parameters

480

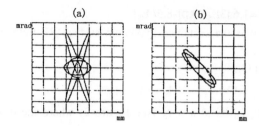

(a)   (b)

mrad   mrad

mm   mm

Figure 4 Illustration of RM function (a) RFQ acceptance without RM (b) RFQ
acceptance with RM

(1) Radial Matching Section

It is at the entrance of an RFQ extending over several cells. Its shape is like
a trumpet shrinking from the large aperture a down to normal r_0 without
modulation (m=1). The focusing strength B increases from 0 to a normal value
(Figure 3). For an RF quadrupole field, its acceptance at the entrance is time-
varying (Figure 4a). However, the input beam is a dc one. They cannot be
matched. After setting an RM section, the acceptance becomes nearly time-
independent (Figure 4b). The radial matching can be greatly improved.

(2) Shaper and Gentle Bunching Section

In an RFQ and other ion RF linacs, the accelerated beams are pulsed. But
the input beam is a dc one. It can not be captured completely. In other linacs, this
problem has been solved by setting an RF buncher upstream of the linac. It
bunches the dc beam into a pulsed one and matches it to the longitudinal
acceptance of the linac. But this needs an additional buncher and more space for
an ion drift section. This problem is solved ingeniously through adiabatic
bunching in the RFQ's own RF field. This bunching is realized by gradually
changing the synchronous phase from –90° to the one in the Accelerating Section
(usually about –30°, see Figure 3). In the GB section, the parameters are chosen
such as to not only keep bunching, but also to keep a nearly constant charge
density for reducing the effect of space charge. The Shaper Section upstream
GB is to offer the needed initial values for GB. The Shaper and GB Sections
make up for the RFQ's very high capture efficiency ($\geqslant$90%) while maintaining
a very compact structure. It is a unique feature of RFQs as compared with other
ion linacs.

(3) Accelerating Section

After well matching radially and longitudinally, the ion beam enters the
Accelerating Section. Here, Φ_s ($\sim$ –30°), m ($\sim$ 2) and B are kept nearly
constant (Figure 3) for very high accelerating efficiency and to reach the final
accelerating energy.

All the above procedures of dynamics design and calculations for an RFQ can be performed by a very popular program called PARMTEQ (Phase and Radial Motion in Transverse Electric Quadrupoles) developed by LANL [5]. After setting the needed parameters into input files, which has been discussed above, the program will simulate the motion of the ions. All the required final results such as output transverse and longitudinal phase diagrams, beam envelopes and losses, and the distribution of parameters as in Figure 3 are delivered.

The PARMTEQ program has played an important role in the development of the RFQ. However, its field calculation is based on the potential that takes only the first two terms in equation (3) in account. In the development of high power RFQs and heavy ion RFQs, the influence of higher multipoles and harmonic terms in equation (2) should be considered. Thus, a new version of PARMTEQM has been put forward by LANL. It takes the potential up to the first 8 terms in equation (2), and can be used for a kind of electrodes with two dimensional cutting for easy machining. PARMTEQM's output can also be used for the numerical milling of electrodes.

4. RFQ resonator cavity

An RFQ resonator cavity contains 4 electrodes as shown in Figure 2 (b) and excites the electric RF quadrupole field around the axis as described in equations (6) to (8) at the designed parameters. However, due to many reasons, the practical fields and parameters will deviate from the designed ones. But the deviations should be within the allowable values. Also there are many factors to disturb the field and parameters. A good RFQ structure should have enough stability against the disturbances.

Due to the skin effect, the RF current will result in heat losses on the surfaces of the electrodes, on the inner wall and other parts inside the cavity. In general, the voltage between electrodes can be as high as 100-200 kV. It will lead to heavy heat losses up to $\sim 10^2$ kW. These losses will consume much RF power from the RF transmitter. An efficient RFQ structure should create the designed voltage V with the least RF power P needed. To measure the RF efficiency of an RFQ structure, the shunt-impedance Z_S and the specific shunt-impedance ρ are defined as

$$Z_S = V^2/P \tag{28}$$

$$\rho = V^2/PL = Z_S/L \tag{29}$$

where L is the length of the cavity.

The units in equations (28) and (29) are usually taken as

$$Z_S — k\,\Omega, \qquad \rho — k\,\Omega\,m, \qquad V — kV, \qquad W — kW, \qquad L — m$$

Besides the above needed RF power for heat losses, another part of the RF power will be converted to the energy of the accelerated ion beams. So, the RF transmitter should deliver enough RF power for these two parts. In general, an RFQ needs RF power of $\backsim 10^{2}$ kW totally, and the thermal power is evacuated by water cooling. Note that the above values of V and W are peak values. To save RF power, an RFQ usually is designed to operate in pulsed mode. Therefore, a duty factor (d.f.) is defined as the ratio of the pulsed RF duration time in an RF cycle to the RF period time. Then the average RF power equals to the peak RF power multiplied with d.f.. Except high power RFQs, general RFQs have a d.f. as low as 10^{-4} — 10^{-2} and need not much average RF power. But for the high power RFQ which was developed recently, d.f. can reach 10^{-1}— 100% (c.w.). The high d.f. leads not only to a high average RF power, but also to serious sparking and water cooling problems.

Presently, there are several types of RFQ cavities. But basically they can be divided into two kinds, namely 4-vane and 4-rod RFQ cavities. The first one is well suited to operate at high frequency ($\backsim$200—450 MHz) with high RF efficiency to accelerate protons and light ions. The second is better suited to work at low frequency ($\backsim$20—200 MHz) with high RF efficiency for heavy ions. Both will be discussed in the following.

4.1. 4- vane RFQ cavity

The first 4-vane RFQ cavity was built at LANL in 1980. It was formed by attaching 4 vanes as 4 electrodes onto the inner wall of a cylindrical resonating cavity symmetrically (Figure 1). In the cavity, the TE_{210}-like mode can be excited to provide the radio frequency quadrupole electric transverse field in the central channel. Through the modulation of vanes, the longitudinal field component is achieved that is needed for acceleration (Figure2b and 2c). The calculation of an RFQ cavity can be performed by the two-dimensional SUPERFISH program as a symmetrical resonator cavity.

The 4-vane RFQ is widely used for protons and light ions. Especially above 200 MHz, it is the only suitable RFQ structure, up to now, that has a small size and high RF efficiency. For example, the 4-vane 425 MHz cavity of the BEAR RFQ has a diameter of 18 cm, a length of 1 m and weighs 55 kg. It accelerates H^{-} ions to 1 MeV with a beam current of 30 mA [6].

The 4-vane RFQ cavity is characterized by its field balance in the 4 quadrants. The field flatness is very sensitive to the symmetry, the mechanical tolerance, local disturbances, and the temperature. Any field unbalance in the 4 quadrants translates into the existence of the dipole mode TE_{110} that will lead to beam losses. This is a serious problem for a 4-vane cavity due to the very week coupling among the 4 quadrants. The solution of this problem is to strengthen the coupling using a Vane Coupling Ring (VCR), PISL, etc. It also means to increase the gap between the TE_{210} mode and the TE_{110} mode. In a 4-vane RFQ, the tuners are very useful to tune the operating frequency and to improve the field balance and flatness.

4.2 4-Rod RFQ Cavity

The 4-rod RFQ cavity was developed at the University of Frankfurt. Figure 5 shows the structure of the 4-rod RFQ with 202 MHz for H^- ions [7]. Figure 6 is an ISR RFQ with 26 MHz for O^+ ions the structure of which belongs to the 4-rod type of RFQ for heavy ions [8]. Their common features are that the two pairs of

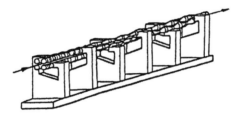

Figure 5 202 MHz 4-Rod RFQ

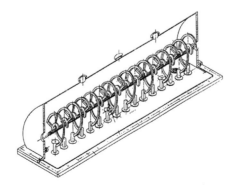

Figure 6 26 MHz ISR RFQ

484

RFQ electrodes are supported by a series of straight or arm stems alternately. From the RF point of view, the stems play a role as λ/4 resonant lines loaded with capacitance. Their behavior is illustrated in Figure 7. One end of the resonant line is at ground potential, and the other is resonating at high RF

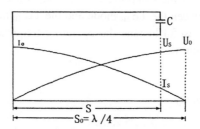

Figure 7 Illustration of a λ/4 resonant line loaded with capacitance.

potential U_0. However, due to its capacitive load C, the potential decreases from U_0 down to U_s while the length needs to decrease from S_0 down to S. In the 4-rod RFQ, the corresponding capacitance arises from the coupling between the two pairs of electrodes and the stems that are fixed to a common base plate at ground potential.

The RF features of the 4-rod RFQ cavity can be well simulated and understood by a lumped equivalent circuit as shown in Figure 8. In the circuit,

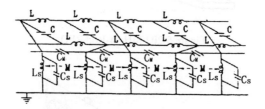

Figure 8 Lumped equivalent circuit of a 4-rod RFQ

the two pairs of electrodes are simulated by two chains of inductance L with coupling capacitance C. The stems are simulated by parallel resonant circuits with C_s and L_s having mutual inductances M and coupling capacitance C_m between adjacent stems. On the basis of this equivalent circuit for the 4-rod RFQ, the RF mode, frequency, field distribution and influence of disturbances can be illustrated very well [9].

Due to the complicated form and asymmetry of the 4-rod RFQ structure, its calculation and design should use codes like 3D MAFIA, equivalent circuit and cold model measurement.

The 4-rod RFQ structure is characterized by RF resonant lines. Therefore, its performance is not so sensitive to the cavity tank and other factors as in the case of the 4-vane RFQ. Furthermore, it does not suffer from the serious problem of dipole mode, and the gap between TE_{210} mode and TE_{110} mode is large enough. So, it is more stable to operate and has relaxed tolerances for mechanics and machining. Its frequency can be easily varied by using a movable plate to change the length of the support stems. Accordingly, a variable energy RFQ has been built in this way [10]. This method is also used to change the local field strength in order to reach flatness conveniently. Due to its simpler structure and small size, the cost of construction is also lower than that of the 4-vane RFQ. The 4-rod structure can be operated at low frequency for heavy ion RFQs. But too low a frequency ($\leqslant 20$ MHz) will lead to very long stems and insufficient rigidity. When the operating frequency exceeds 200 MHz, the stems become too short and will reduce RF efficiency. This limits the useful frequency range. Finally, larger electrodes that may be desirable for effective water cooling will lead to a heavy capacitance load. So, achieving a high duty factor with the 4-rod RFQ is a problem to be solved. Table 1 gives a concise parameter list of the completed HERA 4-vane and 4-rod RFQs [7].

Table 1 HERA 4 Vane and 4 Rod RFQ

		4-Vane	4-Rod
Ions		H⁻	H⁻
F	(MHz)	202	202
W	(keV)	750	750
I	(mA)	20 (max.54)	20 (max. 36)
V	(kV)	70.5	70.5
Length	(cm)	118	118
Diameter	(cm)	40	25
Q		11200	3500
Z_s	(k Ω)	57	63
P (at 20mA)	(kW)	102	94
Weight	(Kg)	600	100
Duty factor	(10^{-4})	2.5	2.5

Besides ISR RFQ and ISAC RFQ mentioned before [11], the Split Coaxial RFQ (SCRFQ) structure [12] and the Interdigital (IH) RFQ structure can also be analyzed on the basis of resonant lines [13]. They are also suitable to accelerate heavy ions.

5. Development and Application

The first successful test of a 4-vane RFQ, in 1980, spawned a rapid development of 4-vane proton and light ion RFQs [14]. They have become popular as injectors instead of bulky Cockroft-Walton accelerators, and are also widely used as low energy accelerators with strong beam current.

Then, the heavy ion RFQ was developed. It promoted the development of low frequency structures like the 4-rod RFQ structures. Heavy ion RFQs are not only used as injectors, but as ion implanters as well.

In recent years, the high power proton RFQs are strongly promoted by projects like the Spallation Neutron Source (SNS), the Accelerator-Driven-System (ADS), etc. [15-21]. The leading RFQ is LEDA. Usually, a high power RFQ has a high duty factor. The study and construction of high power RFQs will lead to an important development of the strong beam dynamics, high power cavity, and the related advanced technology in the field of RFQs.

By now, the study of the superconducting RFQ (SC RFQ) cavity has also achieved progress [22]. The success of SC RFQs will open up a new area , especially for RFQs with high power beam and c.w. operation.

Through the development of more than 20 years, the RFQ is now playing a more important role in the area of low energy accelerators, and is widely used in many fields. In the following tables, some RFQs and their applications are summarized.

Table 2 Proton and Light Ion RFQs (Completed)

Project/ Institution	Ion	F MHz	Type	W MeV	L m	Aim
LANL	P	425	4V	0.64	1.11	First Test 4V RFQ
BEAR	H⁻	425	4V	1.0	1.0	Test on Rocket
BNL	H	200	4V	0.75	1.5	Injector
AccSys	P/ H⁻	200	4V	2.0	1.4	Injector
AccSys	P/ H⁻	425	4V	2.0	1.6	Medical Use
AccSys	d	425	4V	1.0	1.35	Neutron Generator
AccSys	d	425	4V	2.0	2.3	Neutron Generator

Table 3 Heavy Ion RFQ (Completed)

Project/ Institution	Ion	F MHz	Type	W MeV/u	L m	Aim
INS	Li^+	100	4V	0.8	7.25	Injector
INS	Xe^{4+}	25.5	SCR	0.045	2.1	Injector
IKF	Xe^{21+}	80-110	4R	0.1- 0.2	1.45	Injector
GSI	U2+	27.15	4R	0.02	4.0	Injector
GSI	U^{4+}	36	IH	0.12	9.2	Injector
Heidelberg	Li^+, Be^+	108	4R	0.5	6.0	Injector
Saclay	N^{6+}	200	4V	0.183	2.3	Injector
CERN	O^{8+}	202	4V	0.14	0.86	Injector
CERN	U^{35+}	101	4V	0.1	1.4	Injector
ISAC	$\geq 1/30$	35	SPR	0.2-1.5	8.0	Injector
LBL	S^{4+}	200	4V	0.2	2.25	Injector
PKU	O^+	26	ISR	0.062	2.5	Implant
LEON	M^{50+}	85	4R	0.05	2.0	Cluster
GSI HLI	U^{28+}	108	4R	0.3	2.85	Implant
SIMAZU	B^+, N^+, P^+	70	4V	0.9	2.1	Implant
HITACHI	N^+, P^+	17.3	L - C	0.043	2.3	Implant
GSI	Kr^+	13.5	SCR	0.045	9.4	Implant

Table 4 High Power RFQ

Project/ Institut.	Ion	F MHz	Type	D.F %	W MeV	I mA	Aim	Status
FMIT	D	80	4V	100	2.0	100	Material	Compl.
CRNL	P	267	4V	100	1.27	75	Test	Compl.
CWDD	D	352	4V	100	2.0	80	Test	Compl
LEDA	P	352	4V	100	6.7	100	APT	Compl
JAERI	P	201	4V	10	2.0	100	SNS	Compl
JHP	H^-	432	4V	0.4	3.0	36	JHP	Compl
ISIS	P	202	4R	2.5	0.67	50	SNS	Compl
IPHI	P	352	4V	100	5.0	30	ADS	Constr
TRASCO	P	352	4V	100	5.0	30	ADS	Constr
KOMAC	H^- P	350	4V	100	3.0	20	Mult-use	Constr
IHEP	P	352	4V	6.0	2.5	50	ADS	Study
LBNL	H^-	402	4V	6.2	2.5	56	SNS	Test

488

References

1. I.M. Kapchinskii, V.A. Teplyakov, Prib.Tekh. Eksp. 2 (1970)
2. K.R.Crandall et al, Linac 79, 205 (1979)
3. L. Young, PAC 97 (1997)
4. W.D. Kilpatrick, LBNL, Rep. UCRL-2321(1953)
5. R.H.Stokes et al. IEEE NS-26,3469 (1979)
6. D. Schrage, L.M. Young et al, LA-UR-88-3042(1988)
7. A. Schempp et al, Linac 88, (1988)
8. C.E.Chen, J.X. Fang et al. EPAC 2000 (2000)
9. J.X. Fang, A. Schempp, EPAC 1992, 1331 (1992)
10. A. Schempp et al. PAC 95 (1995)
11. R. Poirier et al. Linac 98 (1998)
12. S. Arai et al. Linac 96,575 (1996)
13. U. Ratzinger et al. EPAC 96 (1996)
14. H. Klein, PAC 88,(1983)
15. J. Staples, Linac 94, 755 (1994)
16. J.M. Han, Y.S. Cho, B.H. Choi et al. EPAC 2000, 812 (2000)
17. Y. Yamazak et al. EPAC 2000, 286 (2000)
18. A. Pisent, M. Comunian et al. EPAC2000,857(2000)
19. J.M. Han, Y.S. Cho, B.H.Choi et al. EPAC2000, 812(2000)
20. S. Fu , S. Fang et al, Linac 2002 (2002)
21. H.V. Smith et al. EPAC2000, 969 (2000)
22. G. Bisoffi, V. Andreev et al. EPAC 2000,324 (2000)

INSERTION DEVICES: WIGGLERS AND UNDULATORS

C. S. HWANG

National Synchrotron Radiation Research Center (NSRRC),
101 Hsin-Ann Road, Hsinchu Scienced-Based Industry Park,
Hsinchu 30077, Taiwan
E-mail: cshwang@srrc.gov.tw

The main features of the magnet design and performance of insertion devices will be presented. The material includes synchrotron radiation features such as the radiation power and photon flux, magnet design concepts, shimming techniques, and performance optimization. The characteristics of various insertion devices will be discussed. Field measurement methods and systems are also discussed here.

1. Introduction

Insertion devices, including wigglers and undulators, are magnetic devices that produce a periodic field variation. They are all placed in straight sections of a storage ring. When insertion devices operate in a regime in which interference effects can be neglected, the resulting spectrum at higher photon energies is smooth and similar to that of a bending magnet [1]. Moreover, the intensity of radiation can increase significantly with the number of poles, and if the magnetic field is increased as well radiation with a higher critical energy would be generated. Therefore, the flux density will be proportional to the periodic number. When the periodic magnet is used in a regime in which interference effects are coherent, the device is called an "undulator". The flux density from an undulator is proportional to the square of the number of periods. Recently, third-generation synchrotrons are designed with many straight sections to maximize the use of such devices and reduce electron beam emittance to enhance the brightness of the photon beams, particularly from undulators. In the case of electrons or positrons traveling with an energy of γmc^2 at nearly the speed of light on a curved path, spontaneous emission falls within a narrow cone whose axis is the direction of motion of the electron. The power of the emitted radiation is proportional to the square of the electron energy and to the square of the magnetic field strength. Increasing the number of insertion devices in synchrotron radiation facilities can provide radiation with enhanced features over those of radiation from bending magnets. The main features of the radiation from insertion devices are that (1) they can have a higher photon energy, (2) higher flux and brightness, and (3) various polarization characteristics for the experiments.

Periodic magnets of this type are also at the heart of devices that generate coherent radiation, called free-electron lasers (FELs) [2]. The first undulators to be installed in a storage ring were at the VEPP-3 ring at INP, Novosibirsk, in 1979. Superconducting wigglers are currently operational in many synchrotron radiation facilities including SRS (UK), DCI and ESRF (France), UVSOR and Photon Factory (Japan), NSLS Xray Ring and CAMD (USA), SRRC (Taiwan), BESSY II (Germany). However, superconducting in-vacuum undulators (ANKA,

489

Germany, SSLS, Singapore) are under development. The trend of insertion devices in the near future will be toward superconducting magnets. Another reason for installing insertion devices in electron rings is to controllably change the properties of the electron beam, including the damping times, the energy spread, and the emittance.

2. Basic features of radiation from standard insertion devices [1 - 5]

Figure 1 reveals that an electron beam that travels in a curved path through a bending magnet at nearly the speed of light emits photons into a narrow cone with natural emission angle $\cong \gamma^{-1}$. Figure 2 shows that an electron beam that travels through a wiggler or undulator in the mid-plane along a spatially periodic sinusoidal field at nearly the speed of light emits photons into a narrow cone with natural emission angle $\cong k\gamma^{-1}$ where k is the deflection parameter.

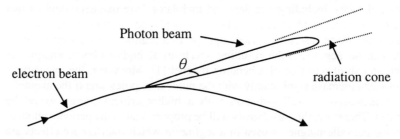

Figure 1. Synchrotron radiation emitted from an electron beam that is bent along a curved path.

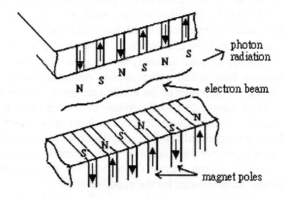

Figure 2. Synchrotron radiation emitted from an electron beam that is bent by a spatially periodic sinusoidal field in an insertion device.

Figure 3 compares the radiation patterns of various types of magnets. The dipole magnet yields a wide band of synchrotron radiation in the horizontal plane and a narrow cone in the vertical plane. The horizontal radiation cone of a wiggler is about $k\gamma^{-1}$ wide and the vertical cone is the same as that of the dipole magnet. However, the radiation cone of the undulator in horizontal and vertical planes is all close to γ^{-1}. Here, a deflection parameter of $k \cong 1$ is defined for the undulator.

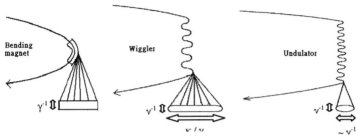

Figure 3. Synchrotron radiation emitted from (a) bending magnet, (b) wiggler, (c) undulator.

2.1 *Motion of electrons in insertion devices* [1]

The theory and characteristics of synchrotron radiation have been developed in detail from first principles and are reviewed elsewhere. Figures 4 and 5 show typical on-axis photon spectra from a bend magnet, a wiggler, and an undulator along with the spectral brightness of IDs at various facilities. The equation of motion of a relativistic electron that moves with an average velocity υ_z perpendicular to a sinusoidal wiggler field of magnitude $B_y = B_0 \cos k_u z$ and period $\lambda_u \equiv 2\pi / k_u$ is derived in the following. The velocity and trajectory [2] are mutually perpendicular to υ_z and $\vec{B}$, as indicated by the Lorentz force equation,

$$\vec{F} = \frac{d\vec{p}}{dt} = d(\gamma m \vec{v}) = e(\vec{E} + \vec{v} \times \vec{B}).\tag{2.1.1}$$

Accordingly, considering only the magnetic field and putting $B_y = B_0 \cos k_u z$ and period $\lambda_u \equiv 2\pi / k_u$ into Eq. (2.1.1) yields the trajectory on the transverse axis as in Eq. (2.1.2)

$$\frac{v_x}{c} = -\frac{K}{\gamma} \sin k_u z \quad \text{and} \quad x = \frac{K}{\gamma k_u} \cos k_u z,\tag{2.1.2}$$

where $\gamma = 1957E[GeV]$ and the deflection parameter $K = eB_0/k_u mc = 0.934$ $B_0[T]\lambda_u[cm]$. The horizontal motion of the electron will cause the electron velocity along the z-axis to vary as well. However, the total speed remains unchanged as

$$v_x^2 + v_z^2 = v^2 = (cons\tan t).\tag{2.1.3}$$

The peak angular electron deflection and the trajectory amplitude are K / γ and $(K / \gamma) \cdot (\lambda_u / 2\pi)$, respectively. For a wiggler with K>>1, the on-axis radiation photon flux is only twice per wiggle period. For undulators with $K \approx 1$, the on-axis radiation photons flux is the square of the wiggle period.

2.2 Radiation from wiggler magnet [1]

In a wiggler, the deflection parameter K is large (typically $K \geq 10$) and photon radiation from different poles of the electron trajectory is enhanced incoherently. The angular density of flux is then given by 2N (N is the number of magnet periods) times the formula for bending magnets of Eq. (2.2.1). The angular distribution of radiation emitted by electrons that are moving through a bending magnet on a circular trajectory in a horizontal plane is

$$\frac{d^2\bar{B}(w)}{d\theta d\psi} = \frac{3\alpha\gamma^2}{4\pi^2} \frac{I}{e} \frac{\Delta w}{w} \left(\frac{\varepsilon}{\varepsilon_c}\right)^2 \left(1+\gamma^2\psi^2\right)^2 \left[K_{2/3}^2(\xi) + \frac{\gamma^2\psi^2}{1+\gamma^2\psi^2} K_{1/3}^2(\xi)\right] \quad (2.2.1)$$

where ε and ε_c are the photon energy and the photon critical energy, respectively, θ and ψ are the observation angles in the horizontal and vertical directions, respectively, α is the fine-structure constant, I is the beam current, e is the electron charge, the subscripted Ks are modified Bessel functions of the second kind, and ξ is defined as $\xi \equiv \left(\varepsilon/2\varepsilon_c\right)\left(1+\gamma^2\psi^2\right)^{3/2}$. Radiation from a bending magnet is linearly polarized when observed in the mid-plane ($\Psi=0$). Out of this plane ($\Psi \neq 0$), the polarization is elliptical and can be decomposed into its horizontal and vertical components [2], [5]. The two terms in the last bracket of Eq. (2.2.1) refer to the horizontally and vertically polarized radiation. In the mid-plane, the second term vanishes and the polarization is purely linear. Off the mid-plane ($\Psi \neq 0$), both terms contribute and the polarization is elliptical.

The energy spectrum is smooth and broadband, peaking near and then falling off exponentially above the critical energy, $\varepsilon_c \equiv \hbar w_c = 3\hbar c\gamma^3/2\rho$. Half of the total integral power is radiated above and half below the critical energy ε_c or the critical wavelength λ_c, which in practical units is,

$$\varepsilon_c[\text{KeV}] = 0.665 E^2[\text{GeV}]B[\text{T}] \quad (2.2.2)$$

or
$$\lambda_c[A] = 18.64 / B[\text{T}]E^2[\text{GeV}]. \quad (2.2.3)$$

However, in a wiggler, the radiation is from the periodic structure and the interference is incoherent. The spectral flux as seen by an on-axis observer, at a photon energy of about ε_c, is 2N times that of an equivalent bending magnet. The harmonic peaks are spaced so closely that they blur together.

Frequently, experimental end stations share the radiation fan with a horizontal opening angle of $2K/\gamma$. At nonzero horizontal observation angle θ, the electron's effective radius of curvature ρ at the point at which the electron's trajectory is tangent to the direction of observation is larger and the off-axis critical energy will decrease to

$$\varepsilon_c(\theta) = \varepsilon_c(0)\sqrt{1 - (\theta\gamma/K)^2} .$$ (2.2.4)

Consequently, off the mid-plane ($\Psi \neq 0$), circularly polarized components from successive bends cancel and the radiation remains partially linearly polarized and partially unpolarized. The calculation of the brightness for wiggler radiation should take into account the depth-of-field effects to the apparent source size from different poles. The results will be presented in chapter 2.4.

2.3 Radiation from undulator magnet [1,2]

When K is moderate ($\approx$ or ≤ 1) in an undulator, the radiation from different periods interferes constructively, producing sharp peaks at harmonics of the resonant frequency, depending on the electron energy, the undulation period, the field strength and the observation position. Even when K is not small (≤ 10), sharp peaks are produced when the effects of electron-beam emittance can be ignored and low harmonics are considered. The photon wavelength is obtained by a Lorentz transformation of the undulation period into the beam frame followed by a relativistic Doppler shift back into the laboratory frame. The velocity used in the Lorentz transformation and the Doppler shift is the longitudinal electron velocity, which is less than the full electron velocity because the electron follows a curved path through the undulator. The photon energy of the fundamental (n=1) radiation peak is

$$\varepsilon_{n=1}(\theta,\psi) = \frac{2\gamma^2}{\lambda_u\left[1 + \dfrac{K^2}{2} + \gamma^2(\theta^2 + \psi^2)\right]}$$ (2.3.1)

This energy value holds on-axis ($\theta=\Psi=0$), and the fundamental energy and wavelength in practical units become

$$\varepsilon_1[\text{KeV}] = \frac{0.950E_e^2[\text{GeV}]}{(1 + K^2/2)\lambda_u[\text{cm}]}$$ (2.3.2)

$$\lambda_1[\overset{o}{A}] = \frac{13.06\lambda_u[\text{cm}](1 + K^2/2)}{E_e^2[\text{GeV}]}$$ (2.3.3)

The relative bandwidth at the nth harmonic is

$$\Delta\lambda/\lambda \cong \Delta w/w \cong 1/nN \, (n = 1,2,3,....).$$ (2.3.4)

As K increases, the spectral peak spacing at the various harmonic energies decreases and the peaks merge eventually. In this way, the insertion device changes from the undulator to the wiggler regime. Additionally, the resonant frequency on different observation angle blurs the radiation peaks for any angle-integrated sampling, leading to the superposition of a peaked spectrum on top of a broadband continuum. Meanwhile, when K increases, the peaks become buried in the continuum region. Therefore, the angular distribution of the radiation intensity of the nth harmonic on-axis is

$$\frac{d^2 \bar{B}_n}{d\theta d\psi}\bigg|_{\theta=\Psi=0} = \frac{I}{e}\frac{\Delta\varepsilon}{\varepsilon}\frac{\alpha N^2\gamma^2 K^2 n^2}{\left(1+K^2/2\right)^2}\left\{J_{\frac{n-1}{2}}\left[\frac{nK^2}{4\left(1+K^2/2\right)}\right]-J_{\frac{n+1}{2}}\left[\frac{nK^2}{4\left(1+K^2/2\right)}\right]\right\}^2 \quad (2.3.5)$$

where the J's are Bessel functions and n is the harmonic number. Usually, the central cone is of interest. Integrated over the central cone, the flux in practical units [photons/s/0.1% bandwidth] becomes approximately

$$\bar{B}_n = 1.431\cdot 10^{14}\frac{NI[A]K^2 n}{1+K^2/2}\left\{J_{\frac{n-1}{2}}\left[\frac{nK^2}{4\left(1+K^2/2\right)}\right]-J_{\frac{n+1}{2}}\left[\frac{nK^2}{4\left(1+K^2/2\right)}\right]\right\}^2 \quad (2.3.6)$$

The brightness characterizes the effective size of a radiative source, and is given in units of flux per phase space volume. The brightness is obtained in practical units of photon/s/mm^2/mrad^2/0.1% bandwidth.

$$\beta_n(\theta=0,\Psi=0) = \frac{B_n}{\left(2\pi\right)^2\cdot\sigma_{Tx}\sigma_{Ty}\sigma_{Tx'}\sigma_{Ty'}} \quad (2.3.7)$$

where $\sigma_{Tx}=\sqrt{\sigma_x^2+\sigma_P^2}$, $\sigma_{Ty}=\sqrt{\sigma_y^2+\sigma_P^2}$, $\sigma_{Tx'}=\sqrt{\sigma_{x'}^2+\sigma_{P'}^2}$, $\sigma_{Ty'}=\sqrt{\sigma_{y'}^2+\sigma_{P'}^2}$, σ_x (σ_y) and $\sigma_{x'}$ ($\sigma_{y'}$) are the beam size and divergence in the x (y)-direction. σ_p and $\sigma_{p'}$ are the extended-source size and divergence limits for single-electron radiation with the quantities $\varepsilon_x=\sigma_x\sigma_{x'}$ and $\varepsilon_y=\sigma_y\sigma_{y'}$. The diffraction-limited source size (rms) corresponding to the angular divergence $\sigma_{P'}$ is $\sigma_P\cdot\sigma_{P'}=\frac{\lambda_p}{4\pi}$ and $\sigma_P=\frac{1}{4\pi}\sqrt{\lambda_p L}$, λ_p being the photon wavelength and λ_u the length of the undulator period. Thus, a small emittance is essential to achieving an undulator radiation source with high brightness. Temporarily, the radiation maintains its phase over a distance $l_c=\lambda(\lambda/\Delta\lambda)\cong nN\lambda$, the coherence length.

2.4 Summary of the radiation features of bending magnets and insertion devices:

- Bending magnet

Brightness:

$$\beta = \overline{B} \Bigg/ 2\pi \left[\left(\sigma_x^2 + D^2 \sigma_\varepsilon^2 + \sigma_r^2 \right) \left(\sigma_y^2 + \sigma_r^2 + \frac{\varepsilon_y^2 + \gamma_y \sigma_r^2}{\sigma_\varphi^2} \right) \right]^{1/2} \qquad (2.4.1)$$

where γ_y is the third betatron function in the y direction, ε_y is the beam emittance, D is the dispersion function in the x direction, σ_ε is the energy spread, and σ_φ is the angular width-effective rms half-angle.

Critical energy:

$$E_c[keV] = 0.665 E^2[GeV] B[T] \qquad (2.4.2)$$

Total radiated power:

$$P_{tot}[kW] = 88.5 \cdot I[A] \cdot \frac{E^4[GeV]}{\rho[m]} \qquad (2.4.3)$$

where ρ is the bending radius.

- Wiggler magnet

Brightness:

$$\beta = \overline{B} \Bigg/ \sum_{\pm} \sum_{n=\left[-\frac{N}{2}\right]}^{\left|\frac{N}{2}\right|} \frac{1}{2\pi} \frac{\exp\left[-\frac{1}{2}\left(\frac{x_0^2}{\sigma_x^2 + Z_{n\pm}^2 \sigma_{x'}^2} \right) \right]}{\left[\left(\sigma_x^2 + Z_{n\pm}^2 \sigma_{x'}^2 \right)\left(\frac{\varepsilon_y^2}{\sigma_\psi^2} + \sigma_y^2 + Z_{n\pm}^2 \sigma_{y'}^2 \right) \right]^{1/2}} \qquad (2.4.4)$$

$Z_{n\pm} = \lambda_u \left(n \pm \frac{1}{4} \right)$ and $x_0 = \frac{k}{\gamma} \frac{\lambda_u}{2\pi}$ where λ_u is the wiggler period. The calculation of the brightness of wiggler radiation must take into account the depth-of-field effects which determine the contribution of the various poles to the apparent source size. The length is L=$n\lambda_u$.

Critical energy:

$$E_c[keV] = 0.665 E^2[GeV] B[T] \qquad (2.4.5)$$

Total radiated power:

$$P_{tot}[kW] = 0.633 \cdot E^2[GeV] B_0^2[T] L[m] I[A] \qquad (2.4.6)$$

- Undulator magnet [6]

Brightness and polarization ratio:

$$\beta_n = \overline{B}_n \big/ (2\pi)^2 \sigma_{Tx}\sigma_{Ty}\sigma_{Tx'}\sigma_{Ty'} \tag{2.4.7}$$

The angular distribution of the nth harmonic is concentrated in a narrow cone whose half-width is given by,

$$\sigma_{p'} = \sqrt{\frac{\lambda_p}{L}} = \frac{1}{\gamma}\sqrt{\frac{\left(1 + \frac{K^2}{2}\right)}{2Nn}} \tag{2.4.8}$$

and the radiation photon energy is

$$\varepsilon_n[keV] = \frac{0.95 E^2[GeV] \cdot n}{\lambda_u \left(1 + \frac{K^2}{2} + \gamma^2\left(\theta^2 + \psi^2\right)\right)} \tag{2.4.9}$$

The total spectral flux is $\overline{B}_n$ (Photons/s/0.1%BW) and the brilliance generated on the n-odd harmonic on-axis is β_n (Photons/s/mm^2/mrad^2/0.1%BW). The radiation by a single electron in the nth harmonic spectrum is characterized by the four Stokes parameters $S=(S_0(n),\ S_1(n),\ S_2(n),\ S_3(n))$.

$$\overline{B}_n = 1.431 \times 10^{14} nNI \left(\frac{S_0(n)}{1 + \frac{1}{2}\left(K_x^2 + K_y^2\right)} \right) \tag{2.4.10}$$

where

$$S_0(n) = V^2 + H_{//}^2 + H_{\perp}^2 \qquad S_1(n) = -V^2 + H_{//}^2 + H_{\perp}^2$$

$$S_2(n) = 2V\ H_{//} \qquad\qquad S_3(n) = 2V\ H_{\perp}$$

with

$$V = K_y J, \qquad H_{//} = K_x J \cos\phi, \qquad H_{\perp} = K_x J \sin\phi. \tag{2.4.11}$$

and $\quad J = J_{(n+1)/2}(nK) - J_{(n-1)/2}(nK),$ $\hspace{3cm}$ (2.4.12)

$$K = \frac{\sqrt{K_x^4 + K_y^4 + 2K_x^2 K_y^2 \cos(2\phi)}}{4\left(1 + \frac{1}{2}\left(K_x^2 + K_y^2\right)\right)}. \tag{2.4.13}$$

where $J_n(nK)$ represents the n-th integer order Bessel function as a function of the variable nK, ϕ is the phase difference between the horizontal and vertical sinusoidal field.

- The r.m.s. photon beam sizes and divergences are approximately given by

$$\sigma_{Tx} = \sqrt{\sigma_x^2 + \sigma_P^2} \quad \text{and} \quad \sigma_{Ty} = \sqrt{\sigma_y^2 + \sigma_P^2}$$

$$\sigma_{Tx'} = \sqrt{\sigma_{x'}^2 + \sigma_{P'}^2} \quad \text{and} \quad \sigma_{Ty'} = \sqrt{\sigma_{y'}^2 + \sigma_{P'}^2}.$$

where σ_x (σ_y) and $\sigma_{x'}$ ($\sigma_{y'}$) represent the electron horizontal (vertical) beam sizes and divergences, and

$$\sigma_{P'} = \sqrt{\frac{\lambda_P}{L}} = \sqrt{\frac{1+(K_x^2+K_y^2)/2}{2nN\gamma^2}}, \qquad \sigma_P = \frac{1}{4\pi}\sqrt{\lambda_P L}. \qquad (2.4.14)$$

- In a helical undulator, the n-th harmonic energy of the helical radiation spectrum on-axis is,

$$\varepsilon_n[KeV] = \frac{0.95\ n\ E^2[GeV]}{\lambda_u[cm]\left(1 + \frac{1}{2}\left(K_x^2 + K_y^2\right)\right)}$$

- The polarization ratios $P_i(n)$ of the radiation are defined as the difference between the fluxes polarized in two orthogonal directions, divided by the total polarization flux $P_i(n)= S_i(n)/S_0(n), i=1,2,3$. If the radiation is fully polarized, then $p = \sqrt{\sum_i p_i^2} = 1$

- The criterion for optimizing the flux and polarization ratio in different phasing positions on elliptical undulator devices should depend on the maximum value of $S_3^2 Fn$ ($S_2^2 Fn$) to define the merit flux.

- When the field strength B_x and B_y are equal, then the spectrum only comprises the fundamental ($n=1$) for which the circular polarization rate is $p_3=1$.

- Total power:

$$P_{tot}[kW] = 0.633 \cdot E^2[GeV]B_o^2[T]L[m]I[A]$$

- Figures 4 and 5 show examples of photon flux and brilliance distribution.

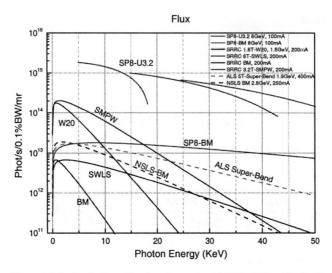

Figure 4 Calculated photon flux for storage rings SRRC, ALS and Spring8.

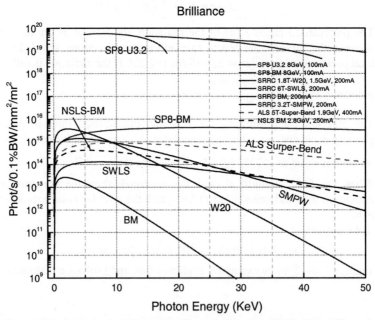

Figure 5 Calculated photon brilliance for different storage rings SRRC, ALS and Spring8.

3. Design criteria for magnets in insertion devices

The main parameters of the insertion devices as determined by the required characteristics of the radiation must be specified before the magnet can be designed. A 3D calculation code, like TOSCA or RADIA, should be used to design such magnets. The design criteria are described below.

A hybrid structure with a vanadium permendur pole and Nd-Fe-B magnets was chosen for the insertion devices. The period-to-gap ratio λ_u/g is always larger than 2 to obtain high field and the field errors of the hybrid structure are much less sensitive to the magnetization errors of the individual magnet blocks. However, the nonlinear relative permeability of the iron pole is such that the variations in the first field integral and in the tune shift of the hybrid structure are much larger than those of the pure magnet structure, if the phase of the magnet array needs to be changed. Therefore, the pure structure with Nd-Fe-B magnets was used for phase change undulator.

For the hybrid structure (Figure 6), the wedged-poles have a larger cross section at the pole tip to prevent excessive pole saturation. The chamfers are used to reduce local saturation and demagnetizing fields. Although the vertical recess lessens the field strength, the recess with shims can be used to minimize the on-axis field strength variation and maximize the field-tuning range. Magnet overhanging on both vertical and transverse axes is used to weaken the 3-D leakage flux from the sides of the pole and to reduce the roll-off effect. Many small magnet blocks of different thickness were mounted on the end poles to minimize the integral dipole strength and reduce the displacement of the trajectory. The longitudinal distances between each of the end poles, the pole height and the pole tilts are all adjustable to correct the first and second field integral as well as the multipole components. Two rows of trim magnet blocks with different sizes are present at both ends for shimming the multipole field. One row is for the normal field and the other is for the skew field component. The design concept of the end pole design can also be used for other insertion devices.

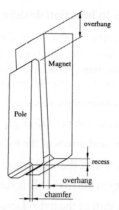

Figure 6 Schematic drawing of the hybrid structure design.

The hybrid insertion devices (Figure 6) utilize a permanent magnet to generate flux which directly enters high permeability, soft-iron pole pieces. This structure will generate a higher field for insertion devices. Figure 7 shows the basic design for the hybrid structure of the insertion devices [1]. In general, the ratio of the obtainable magnetic flux density versus the gap-to-period ratio for the hybrid structure and pure structure (only permanent magnet without iron piece) is very important in designing a magnet. Thus, if the ratio is smaller than 0.5, then the hybrid structure is better. The insertion devices with the pure structure utilize permanent magnets without iron to generate flux. Such a structure increases the field of insertion devices. Figure 8 shows the basic design of the insertion devices with the pure structure.

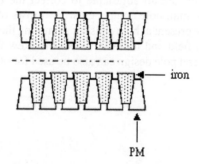

Fig. 7 Hybrid insertion device field strength enhancement technique (wedged poles).

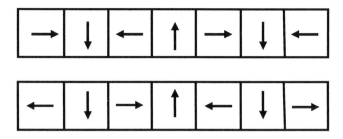

Fig. 8 Schematic drawing of the pure structure.

The design of the end poles is very important in optimizing the first (net change in angle) and second (net change in position) field integral. The relationship between the field integrals and the angle ($\phi(z)$) or position ($\delta(z)$) is defined as [2]

$$\phi(z) = \frac{e}{\gamma mc} \int_{-\infty}^{z} B_y(z)dz \qquad (3.1.1)$$

$$\delta(z) = \frac{e}{\gamma mc} \int_{-\infty}^{z} dz \int_{-\infty}^{z'} B_y(z')dz' \qquad (3.1.2)$$

Commonly, the angle and position at the entrance and exit of an insertion device should be zero. Accordingly, the end poles can be optimized to meet these conditions. The various end pole designs and the arrangement and dimensions of the poles are shown in Figure 9 [2]. For example, the poles are arranged in a sequence, +1/4, -3/4, +1, -1, +3/4, +1/4, and the field distribution is shown in Figure 10 [2]. However, the magnet code (such as RADIA) can be used to design and optimize the dimensions of the end pole. In general, an even pole design can more easily yield a wider, flat integral field range. However, the effective pole is smaller than that of the odd pole design. If the odd pole design is used, then the maximum number of effective poles can be obtained. However, obtaining a wider, flat integral field range is difficult.

502

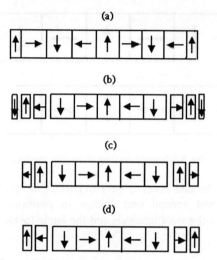

Figure 9   Design of various end poles for the pure-permanent magnet structure, (a) half dimension magnet piece at end pole, (b) different dimension block: magnetic piece with a constant gap between the magnet at the end pole region, (c) and (d) different dimension block: magnetic piece with different gaps between the magnet at the end pole region. (From reference [2])

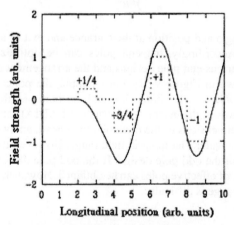

Figure 10   Sequence of magnetic poles (dotted line) resulting in no offset between the electron trajectory (solid line) and the magnet central axis (From reference [2])

4.   Magnetic structure of various insertion devices

Many magnetic array schemes, as shown in Figure 11 [2], produce elliptical polarization. They can also produce linear polarization. The HELIOS device is a planar helical undulator, and is depicted in Figure 11(a) [7]. The upper array and the lower array only produce the horizontal (B_x) and the vertical field (B_y), respectively. Therefore, the disadvantages of the HELIOS device are as follows:

(1) The roll-off range is shorter than that of the conventional ID, (2) second-order deflection (asymmetric field) occurs, (3) no linear polarization is available at $0°$, $360°$, and (4) the field is weak. Figure 11 (b) [8] presents a modification of the HELIOS device in which both arrays produce pure helical fields with equal field strength that exceeds that of the original HELIOS device without having its disadvantages. However, the helical field is fixed and no linear polarization is available. Figure 11(c) shows JAERI's proposed APPLE-II planar device [6, 9, 10, 11]. This device can produce any kind of polarization by changing the relative distance among the array of four magnets. It yields a higher field strength in terms of both B_y and B_x. However, the roll-off of B_y and B_x is much shorter than that of the conventional planar undulator. Finally, a structure, shown in Figure 11(d) [12], consisting of six magnetic arrays, was proposed by Kitamura of Spring-8. In that case, the upper and lower central arrays generate a vertical field while the outer four arrays generate a horizontal field. The devices have the same advantages as APPLE-II, the roll-off range is wider, but the B_x field is weaker.

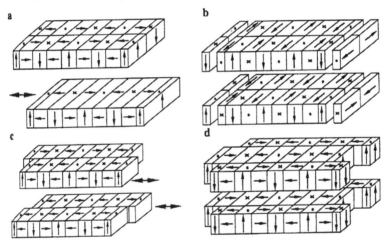

Figure 11 Various arrangements of planar permanent magnets. (a) ESRF's HELIOS planar device yields elliptically polarized light, (b) Modification of a HELIOS planar device produces gap-independent, circularly polarized light, (c) Block orientations for JAERI's proposed APPLE-II planar device and SSRL's elliptically polarizing undulator, and arbitrarily orientated linear polarization, (d) SPring-8 with arrays of six magnets for the elliptically polarizing undulator, and arbitrarily orientated linear polarization (from reference [2]).

In addition, some important undulators have been proposed. H. Kitamura of the Spring-8 team constructed and installed the in-vacuum undulator (Figure 12) in the Spring-8 storage ring [13]. Consequently, the vacuum chamber for the electron beam was not specially limited. Therefore, the magnetic gap was as small as possible.

Herbert O. Moser and R. Rossmanith of ANKA [14] proposed the Superconducting In-vacuo Miniundulator (Figure 13). The associated array of magnet is the same as the hybrid magnet structure except that a superconducting wire replaces the permanent magnet. The superconducting in-vacuo undulator is not spatially limited either, just like the in-vacuum undulator. In this case, the magnetic flux density is greater than that of the other type of undulator. A 1.33 T field can be obtained for a superconductive undulator with a periodic length of 14 mm and a pole width of 2 mm at a magnet gap of 5 mm and a current density of 1 kA/mm^2.

A staggered undulator, shown in Figure 14 [15,16], is also developed. A superconducting solenoid-derived field is used to magnetize iron poles, called staggered poles, to produce the vertical or horizontal field for linear polarization. It can also be designed as a helical undulator. The structure and operational reliability of the staggered undulator are better than those of the superconducting in-vacuo undulator. However, the field is a little weaker than that of the superconducting in-vacuo undulator. Meanwhile, the staggered undulator has a strong field in the longitudinal direction.

Finally, two coaxial solenoids separated by half the coil-pitch and carrying current in opposite directions generate a circularly polarized transverse field with no longitudinal field component. The schematic drawing is shown in Figure 15.

Figure 12 In-vacuum undulator, proposed by H. Kitamura of the Spring-8 team.

Figure 13 Superconducting in-vacuo miniundulator proposed by Herbert O. Moser and R. Rossmanith of ANKA (from reference [14]).

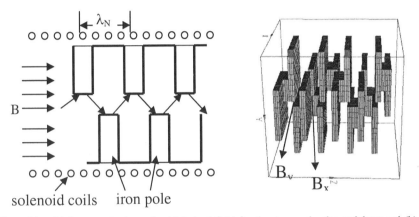

solenoid coils   iron pole

Figure 14   (a) Superconducting solenoid-derived field for the staggered poles undulator and (b) superconducting solenoid-derived – field for helical staggered undulator.

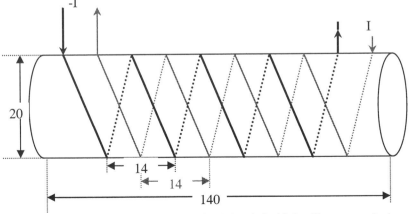

Figure 15 A bifilar helical undulator produces circularly polarized light with a superconducting wire (Unit is mm)

5. Measuring the field of an insertion device

Stretched-wire [17,18], Hall probe [19], Helmholtz coil [20] and the pulsed-wire [21,22] methods have been used to measure and characterize the field quality of insertion devices. The stretched-wire (long-loop-flip coil) method is for measuring quickly the first and second field integral. The 3-D Hall probe measurement method is mainly for shimming the trajectory and spectrum. The Helmholtz coil is used to measure the magnet dipole moments of the magnet block. The pulsed-wire method is for quickly mapping the field and measuring the distribution of the first and second field integral to simulate the angles and positions of the electron beams.

5.1 Stretched- wire system

A stretched-wire system was developed to measure and characterize the first and second magnetic field integrals of insertion devices and the harmonic field components of a lattice magnet. Figure 17 schematically depicts the stretched-wire measurement method [17,18]. This simple wire support structure was built with reliably high precision, and a high-speed automatic measurement system for measuring the transverse field integrals of various kinds of magnets. The measurement precision of the system is 1.5 $G \cdot cm$ (70 $G \cdot cm^2$) for the first (second) field integral of the insertion devices and 0.01% in the analysis of the harmonic field components of the lattice magnets. The scanning time for a complete measurement is five minutes for the first and second field integral distributions, and four minutes for the harmonic field components.

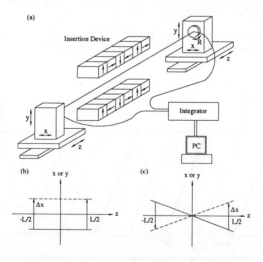

Figure 17 (a) System diagram of the open-loop wire for the first and second field integral measurements, (b) moving method for measuring the first field integral, and (c) moving method for measuring the second field integral (From reference [17]).

5.2 *Hall probe system with three-orthogonal axes*

An "on the fly" system, using three-orthogonal Hall probes [19], was used to map the elliptically polarizing undulator (EPU). The design concept depends on a reliable measurement to perform a synchronized fast measurement for the three magnetic field components. The uncertainty errors in the center position and the angule error, and the planar Hall effect are such that the relative orthogonal angular error of the three Hall probes should be readjusted and calibrate to within 0.1°. Then, the relative shift in the center position along the vertical and horizontal transverse axes between the three Hall probes can be precisely calibrated to within 100 μm. The positional shift on the two transverse axes and the planar Hall effect can be calibrated and readjusted in the Adjustable Phase Undulator (APU) or EPU magnet, individually. The calibration data were input into the software to correct the field strength error as much as possible. Consequently, this system can obtain a real field distribution with a small field modification. An rms precision of 5 G-cm for the three field components can be achieved in mapping the multipole field and the maximum integral field error between the Hall probe and stretch wire is about ±20 G-cm.

5.3 *Highly automatic Helmholtz system*

The performance of hybrid undulators and wigglers depends on the quality of the field induced from the permanent NdFeB magnet blocks. A variation of the magnetic moment on the easy axis from block to block leads to a variation in the magnetic potential from pole to pole. This field variation affects the quality of photon spectrum. The two minor magnetic moment components can simultaneously induce an undesirable quadrupole and skew dipole field components, thereby affecting the dynamics of the electron beam. Accordingly, arranging the magnet block by the block sorting can either reduce the maximum field deviation of each pole on the axis of the insertion device or reduce the quadrupole and skew dipole strengths to a minimum. After the blocks are sorted, the field quality of the insertion device is five times greater than without block sorting in the case of a random block assembly, subsequently leading to an improved field distribution for an assembled magnet.

This is a highly automatic measurement system [20] that can reduce the time required for measurement and prevent human error. It has been used to measure a large number of blocks. The magnet block is installed on a block holder whose center is located at the center of symmetry of the coil-pair (shown in Figure 18). An induced magnetic flux can be integrated when the block holder is rotated by 360° around the vertical axis of the coil-pair. In calculating the three magnetic moments, the magnet block orientation must be changed in two steps: firstly the block is rotated by 180° around the coil-pair axis and then by 90° around the third orthogonal axis. From these measurements, the three components of the magnetic moment can be obtained using a relatively simple mathematical algorithm.

A reference magnet is used to test the stability and reproducibility of the measurement system over long periods. The experimental results demonstrate a precision of 0.04% for this system's easy component. The rate of measurement is 40 blocks per hour, but the speed can be improved if necessary. The coil-pair geometric constant. of the system has been calibrated using the voltage-field reciprocity principle.

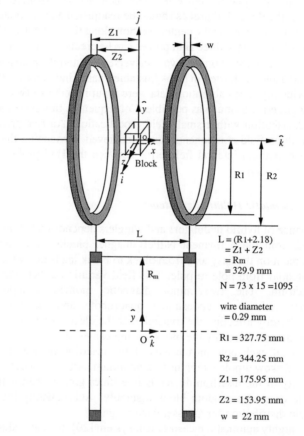

$L = (R1+2.18)$
$= Z1 + Z2$
$= Rm$
$= 329.9$ mm

$N = 73 \times 15 = 1095$

wire diameter
$= 0.29$ mm

$R1 = 327.75$ mm

$R2 = 344.25$ mm

$Z1 = 175.95$ mm

$Z2 = 153.95$ mm

$w = 22$ mm

Figure 18 Geometry and structure of the Helmholtz coil pair, in the magnet coordinate system (From reference [20]).

5.4 *Pulsed-wire system*

The pulsed-wire method for measuring magnetic field has been developed by several groups and applied to measure wigglers and undulators. The benefits of this method are evident in the following situations.

1. Making in-situ measurements of wiggler magnetic fields, first integral fields, and second integral fields, to monitor the field error, calculate the radiation spectrum and cancel the wiggler steering errors.

2. Solving the dynamic behaviors of interest.
3. Easily making a point measurement of the mini-gap undulator.

A pulsed-wire system [21,22] is required to measure the small-gap superconducting multiple wiggler. Figure 19 schematically depicts the pulsed-wire system. For measuring the first integral field, a digital function generator and a power transistor are used to generate a current pulse of 0.2 ms and 0.5 Ampere. The photo-coupler H21A1 is used to detect the horizontal displacement of the Be-Cu wire, due to the interaction between the wire current and the magnetic field. The wire imperfection, stiffness, and tension stress will limit the precision and accuracy of the pulsed-wire system. The stiffness of the wire causes dispersive behaviour. However, some spurious signals distort the original field signals. Applying a high mechanical tension to the wire and using thin wires are recommended to reduce the effect of wire stiffness. Moreover, the uniformity of the wire is of interest. These tail signals will also in turn distort the measured waveform of the ending poles. A thick Be-Cu wire with a diameter of 0.25mm diameter is then used to average out the imperfections, though the thick wire has a significant stiffness effect. However, with the reduction of the contribution of the tail signals, other spurious signals can be attacked, including that from the effect of the dispersion of acoustic wave due to the finite stiffness of the wire. Until now, this method of measurement has not been used to measure an undulator. Therefore, the method must be improved to meet the requirement for precision in characterizing insertion devices.

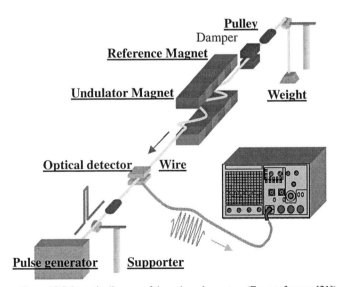

Figure 19 Schematic diagram of the pulse-wire system (From reference [21]).

510

Acknowledgement

The author would like to thank the SRRC people for supporting this article. Especially, it is a pleasure to thank the magnet group people in SRRC, C. H. Chang, T. C. Fan, F. Y. Lin, H. H. Chen, and M. H. Huang, to work together for various helpful discussions.

References

[1] Synchrotron Radiation Sources- A Primer, World Scientific Publishing Co. Pte.

[2] CERN Accelerator School, Synchrotron radiation and free electron lasers, edited by S. Turner, CERN 98-04, 3 August 1998.

[3] Neil Marks, CERN Accelerator School, Fifth general accelerator physics course, edited by S. Turner, CERN 94-01, 26 January,1994, Vol. II.

[4] CERN Accelerator School, Fifth advanced accelerator physics course, edited by S. Turner, CERN 95-06, 22 November,1995, Vol. II.

[5] Handbook of accelerator physics and engineering, World Scientific Publishing Co. Pte. Ltd., edited by Alexander Wu Chao and Maury Tigner (1998).

[6] C. S. Hwang, Shuting Yeh, 1999, January, "Various polarization features of a variably polarized undulator with different phasing modes", *Nucl. Instrum. Meth. A 420*, (1999) 29-38.

[7] P. Elleaume, Nucl. Instr. And Meth. A 306 (1994) 83.

[8] B. Diviacco and R. P. Walker, Nucl. Instr. And Meth. A 292 (1990) 517.

[9] S. Sasaki, Nucl. Instr. And Meth. A 304 (1991) 719.

[10] C. S. Hwang, C.H. Chang, T.C. Fan, F.Y. Lin, Ch. Wang, Shuting Yeh , H.P. Chang, K.T. Hsu, L.H. Chang, P. K. Tseng, T. M. Uen, 1997, September, " *Performance and characteristics of a prototype symmetric hybrid adjustable phase undulator*", Nucl. Instrum. Meth. in Physics A399, (1997) 463-476

[11]R. Carr, Nucl. Instr. And Meth. A 360 (1994) 431.

[12] X. Marechal et al., Rev. Sci. Instr. 66 (1991) 391.

[13] H. Kitamura, J. Synchrotron Rad. 5, 184 (1998).

[14] Andreas Geisler, Achim Hobl, Detlef Krischel, Herbert Moser, Robert Rossmanith, Michael Schillo, presented at the ASC2002 conference (2002).

[15] Y. C. Huang, et al., Nucl. Instr. And Meth. A 341 (1994) 431.

[16] C. H. Chang, C. S. Hwang, Ch. Wang, T. C. Fan, G. H. Luo, Presented at XIV Russian Synchrotron Radiation Conference (2002).

[17] C. S. Hwang, C.H. Hong , F.Y. Lin and S.L. Yang, *Nucl. Instrum. Meth. A* **467** (2001) 194-197.

[18]D. Zangrando, R. P. Walker, *Nucl. Instrum. Meth. A* **376** (1996) 275-282.

[19] C. S. Hwang, T.C. Fan, F.Y. Lin, Shuting Yeh , C.H. Chang, H.H. Chen, P. K. Tseng, J. Synchrotron Rad. 5, 1998, 471-474.

[20] C. S. Hwang, Shuting Yeh, P. K. Tseng, and T. M. Uen, Rev. Sci. Instrum. 67(4), 1996 1741-1747.

[21]T. C. Fan, C. S. Hwang, C. H. Chang, Ian C. Hsu, "A systematic study on pulse wire system for magnetic field measurement on long undulator with high field", Rev. Sci. Instrum., 73, 1430 (2002).

[22] R. W. Warren, Nucl. Instr. and Meth., A272 (1988) 257.

MEDICAL AND INDUSTRIAL APPLICATIONS OF ELECTRON ACCELERATORS*

YUZHENG LIN

Tsinghua University, Beijing 100084, China
E-mail: linyz@mail.tsinghua.edu.cn

1. Medical Applications of Electron Accelerators

§1. 1 Introduction

1.1.1 *The need for medical accelerators [1,2]*

Cancer is a major health problem throughout the world. Cancer seriously threatens human health and life.

Cancer is the first or second cause of death in many countries, including China, United States, Europe, and Japan, accounting for one-fourth of all deaths. It is the leading cause of death in the 35~54 age group.

Cures for cancer nowadays are achieved for about 45% of all cancer patients using the currently available therapeutic strategies: surgery, radiation therapy and chemotherapy. Surgery is the most successful therapy which contributes 22% to the overall cure rate. Radiation therapy contributes 12% to the cure rate alone and 6% in combination with surgery. Chemotherapy and the other remaining modalities account for the last 5% of the total cure rate. So radiation therapy plays a crucial role in the treatment and cure of malignancies.

In Japan, the number of new cancer patients was about 500,000 in 2000.

The Chinese Hygiene Ministry estimated that the number of new cancer patients per year in China was more than 2 million in 2001.

The U.S.A 1986 Blue Book recommended one treatment machine (accelerator) for every 250 patients.

According to this recommendation, more than 24,000 medical accelerators would be required worldwide. But so far only 7,000~8,000 medical machines have been put into use.

There is a serious need to develop medical accelerators and to study radiotherapy for cancer treatment.

---

* This work is supported by the China National Natural Science Foundation, 10135040.

512

1.1.2 *Chronology of Development [1,2]*

The history of radiotherapy can be traced back to 1895. Only a few months after the discovery by W. K. Roentgen in 1895 of a type of radiation, later called the roentgen ray after its discoverer, or X-radiation, the first attempts were made in Germany and the U.S.A to use this radiation in the treatment for malignant tumors in patients.

1. *The Era of X-ray Tube Machines*

Later, X-ray tube machines with 150~400KV served as radiotherapy machines. Unfortunately, from the medical point of view, X-rays generated by these machines were fairly soft; and the depth-dose curves were particularly disadvantageous since the maximum dose could only be delivered at the skin surface and then would rapidly decrease with tissue depth. (see Fig.1.1).

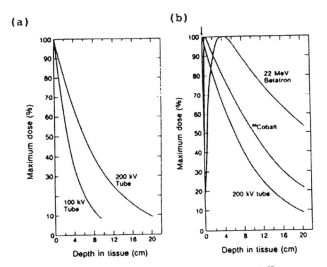

Fig.1.1 Depth-dose curve for beam from (a) 200KV X-ray; (b) ^{60}Co gamma radiation and (c) 22MeV betatron X-rays.

2. *The Era of ^{60}Co*

In the early half of the 1950s, the era of cobalt sources for radiotherapy began. Cobalt-60, emitting gamma rays with energies of 1.17 MeV and 1.33 MeV, made it possible to obtain far better depth-dose curves than those of the X-ray tube machines.

3. *The Era of Betatrons*

A dramatic increase in the photon radiation energy was made possible by the development of the betatron. The first patient was irradiated in 1949 with X-rays generated by a 20MeV electron betatron which was installed in Urbana (U.S.A). Betatrons played a significant role in the development of radiation therapy during the 1950s and 1960s.

The first Chinese domestic medical betatron was made in 1974, with electron energy of 25 MeV.

The betatron's major disadvantages are its weight, the relatively low intensity of the X-ray beam, as well as the small treatment field area. After 1970, the use of the betatron has shown a gradual decline.

4. *The Era of RF Linear Accelerators*

The advances made during World War II in the construction of magnetrons for radar techniques have made it possible to use microwave generators for electron acceleration in radiation therapy. In England, the first traveling-wave linear accelerator was demonstrated by D.W. Fry in 1946. The first electron linear accelerator (linac) was subsequently installed at Hammersmith Hospital, England, in 1952.

In 1962 in the U.S.A, Varian Associates introduced a fully rotational isocentric 6 MeV bent beam linac. This company also applied the side coupled standing wave accelerator structure, which was invented at Los Alamos National Laboratory, introducing a fully rotational isocentric 4MeV in-line linac in 1970. Since then, the field of medical electron linac has developed rapidly. The era of RF electron linacs began and RF linacs have taken a dominant place on the world market of medical accelerators. At present the world's total number of RF linacs used in radiation therapy exceeds 7,000~8,000. The number of these machines increases at a rate of approximately 10% per year.

The first Chinese domestic RF electron medical linac was developed successfully in 1977. In 2001, about 640 medical linacs were installed on mainland China. Among them, 300 sets were home-made.

In the treatment for cancer, the use of X-rays and electron beams, generated from X-ray tube machines, ^{60}Co machines, betatrons and RF electron linacs, is considered conventional radiotherapy. Other particle beams, such as protons, neutrons, and ions, as well as heavy ions have also been used to treat cancer. Most of these are in the research stage or test therapy. These are considered unconventional radiotherapy. But in recent years, proton therapy is becoming a conventional therapy.

Table 1.1. gives a chronology of the development of radiotherapy and nuclear medicine, with special consideration of accelerator techniques [2].

Table 1.1. Chronology of the development of radiotherapy and related techniques.

Year	Event
1895	K. Roentgen discovers X-rays
1913	W.E. Coolidge develops the vacuum X-ray tube
1931	E.O. Lawrence develops a cyclotron
1937	1MV Air-insulated Van de Graaff accelerator installed at Huntington Memorial Hospital (Boston)
1938-1943	R. Stone treats 250 patients with fast neutrons at the University of California
1939	First medical cyclotron, Crocker (U.S.A)
1940	1.25 MeV pressure Van de Graaff accelerator at Massachusetts General Hospital
1946	Medical radioisotopes commercially available (U.S.A)
1946	Wilson suggests medical use of protons and heavy ions
1946	20MeV electron beam for cancer treatment developed using a betatron (Urbana,U.S.A)
1949	First patient treated with 20 MV X- rays from a betatron at the University of Illinois (Urbana, U.S.A)
1951	First European radiotherapy betatron put in operation at the Cantonal Hospital in Zurich (Switzerland)
1951	Cassen develops a mechnical radioisotope scanner
1952	First three cobalt units produced and installed at Canadian facilities in Saskatoon and Ottawa (Canada), and at Oak Ridge (U.S.A)
1952	First medical 8 MeV RF linear accelerator installed at Hammersmith Hospital in London
1955	Medical cyclotron MRC installed at Hammersmith Hospital in London
1956	First American 6 MeV linear accelerator to generate a 4MeV beam put in operation at Stanford (U.S.A);
1956	First medical application of a proton beam in Sweden
1957	Anger demonstrates his scinticamera
1958	First scanning electron beam (5 to 50 MeV) from a linear accelerator developed for cancer therapy (Chicago, U.S.A)
1962	Isocentric RF linear accelerator installed in the U.S.A
1965	Resumption of neutron therapy studies at Hammersmith Hospital in London
1973	Positron camera (PET)

1976	First irradiation with an ion beam in an RF linear accelerator LAMPF (U.S.A)
1982	First cyclotron used for neutron rotation therapy in the U.S.A
1990	First superconducting cyclotron for neutron therapy (Harper, U.S.A)
1990	First hospital-- based proton synchrotron used for radiotherapy (Loma Linda, U.S.A)

1.1.3 Fundamental principles of radiation therapy [1,2]

1. *Fundamental principles*

Radiation therapy is mainly based on the ionizing effect of radiation beams in cancer tissue. Although radiation dos not selectively target cancer cells, the selective destruction of a tumor is achieved to some extent by the following facts:

(1) Considerable efforts are made to concentrate the dose in the tumor tissue and so spare the surrounding normal tissue as much as possible;
(2) Some tumor cells are more radiosensitive than normal cells;
(3) The reparability of malignant tissues is thought to be less efficient than that of normal tissue.

So the strategy of radiotherapy can be presented briefly as follows:

(1) Deliver a sufficiently high dose to eradicate the tumor;
(2) Minimize the dose to the normal tissues, particularly those in critical organs.

According to this strategy, the following requirements should be satisfied:

(1) The characteristic of the depth dose plotted as a function of the tissue thickness should be almost rectangular; i.e., it should have a flat maximum, a quite rapid disappearance of the trailing edge, and a rather steep rise. Fig.1.2 (a) shows such an idealized characteristic, necessary for treating tumor T situated as in Fig.1.2 (b); the real shape of the characteristic for some particles is seen in Fig.1.2 (c) [1].
(2) The energy of the particles should ensure that their range extends to the tumor; easy energy regulation is required so as to vary the range.
(3) The beam intensity should be relatively high so that the irradiation times can be short and can be varied over a wide range.

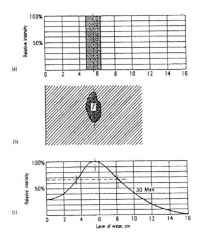

Fig.1.2 Characteristic of depth dose: (a) idealized; (b) position of tumor; (c) the real characteristic for some particles.

(4) The beam intensity in the cross-section perpendicular to the axis should be uniform; the beam should be easily collimated for precise definition of the irradiated field.

(5) Moreover, easy maneuverability of the beam is needed so that it can be used for both stationary and rotary therapy.

2. Unit of Absorbed Dose

The unit of the dose absorbed in tissue is the Gray (Gy), the dose at which 1 Joule of Energy is absorbed in each kilogram of substance.

$$1Gy=1 \ Joule/kg=100cGy$$

Usually, the median treatment for a cure is more like 25-35 fractions of 200 cGy in 25 to 35 days for a total of 5,000 to 7,000 cGy.

1.1.4 Conventional Accelerator Therapy

At present, radiotherapy with X-rays (photon beams) or electron beams is called conventional or routine radiotherapy.

1. Photon Therapy

When the accelerated beam hits the high Z target, X-rays can be generated.

Fig.1.3 shows the curve of the depth dose as function of the thickness of the water layer for conventional 200KV X-rays, ^{60}Co, γ-rays and X-rays with energies ranging from 5 to 35MeV [1].

518

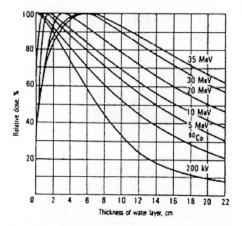

Fig.1.3 Depth dose as function of thickness of water layer for 200KV X-rays, 60Co, γ -rays, and betatron X-rays with energies of 5,10,20,30 and 35MeV.

With X-ray energy, the relative absorption in various tissues is different (see Fig.1.4). This is due to different physics effects. In the 01~1.0MV range, the main physics effect is the photon-electron effect, $\propto Z^3$; in the 1MV~10MV range, the main physics effect is the Compton scatter effect, $\propto Z$. The third physics effect can occur for ranges from 1.022 MV, but its main effect is for energies greater than 10 MV [2].

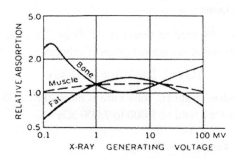

Fig.1.4 Relative absorption of ionizing radiation as a function of energy for muscle, bone and fat tissues.

An essential development in photon therapy was the use of a multi-energy (or dual photon) modality (mixed photon beam therapy). In a single treatment, the patient is irradiated with beams of photons with two energies; for example, one at 6MV and the other at 22 MV (Fig.1.5). By a selection of weighting of the two photon beams, a depth-dose characteristic is obtained for any energy between 6MV and 22MV [2].

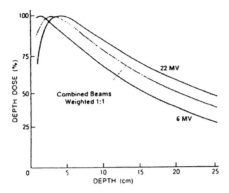

Fig.1.5 Dose distributions for a patient treated with6MV, 22MV and a mixed beam (weighted 1:1).

The disadvantages of photon radiotherapy are that it can not concentrate the dose in the tumor tissue and there is a high dose in the exit side of the patient's body [1] (Fig. 1.6, Fig.1.3).

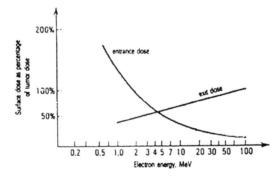

Fig.1.6 Ratio of entrance (surface) dose and exit dose to depth dose at 10cm, as a function of X-ray energy.

In 1962, Varian Associates developed an isocentric-mounted medical accelerator. Since then rotational therapy or arc therapy could be employed (Fig.1.7).

2. Electron Therapy

Radiotherapy can be carried out by direct use of accelerated electrons. Compared with photon therapy, relative low energy electron therapy has an obvious advantage. In its depth-dose curve, the beams have a flat peak and a fast drop, and the therapeutic range depends on electron beam energy (Fig.1.8). But for higher beam energies the shape of this curve becomes similar to that for photon beams [2] (Fig.1.8 and Fig.1.3).

520

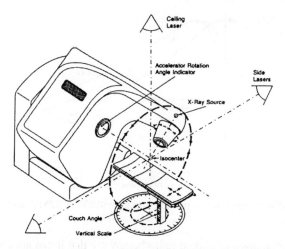

Fig.1.7 Location of the isocenter, gantry rotation axis, treatment couch and positioning lasers on a 6MeV linac [3].

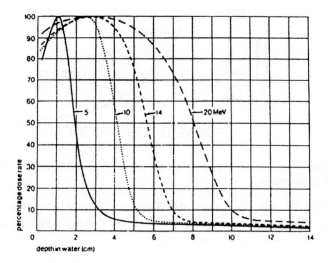

Fig.1.8 Depth-dose distribution in a water phantom for 5 MeV, 10MeV, and 14MeV electron beams.

Of all patients receiving radiation therapy, the proportion of patients treated with electrons is estimated to be between 18% and 25%.

Like photon therapy, electron treatment also employs electrons with various energies.

1.1. 5 *Unconventional Accelerator Therapy [2]*

To date, X-rays (photons) and electron beam therapies still play an important role in radiotherapy. But they have obvious shortcomings. It is necessary to find new types of ionizing radiation.

An ideal depth-dose should have a rectangle-like shape; i.e. most of the energy should be delivered to the target area (tumor) corresponding to the tumor tissue, while the energy absorbed by the intervening normal tissue around the tumor should be close to zero. Unfortunately, X-rays, as well as electron beams used in photon and electron therapies, fall short of this ideal (Fig.1.3 and Fig.1.8). However, using heavy particles much better results can be obtained. The depth-dose curves for 187MeV protons and 190MeV deuterons are shown in Fig.1.9 In both cases, there is a sharp maximum near the end of the particle trajectory (Bragg peak) [1].

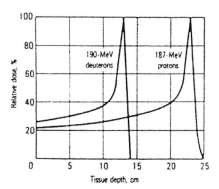

Fig.1.9 Depth dose in water for 190MeV deaterons and 187MeV protons.

The depth-dose curve for irradiation of a mediastinum tumor with ^{60}Co, γ-rays and a proton beam are given in Fig.1.10 [1].

This means that we can use the physical properties of new types of radiation beams to improve the efficiency of radiotherapy.

Irradiation by conventional radiotherapy, expressed in units called Grays, determines a physical dose rather than a biological one since biological interactions depend not only on the value of ionization but also on the ionization distribution in the tissue. A densely ionizing track interacts more strongly with a cell than a few less dense tracks with the same concentration of ions. This concentration is proportional to the amount of energy dissipated (E) by the particle along the unit path (1). The amount is defined as Linear Energy Transfer

522

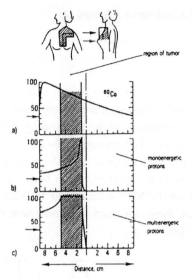

Fig.1.10 Irradiation of a mediastinum tumor with a proton beam: (a) depth-dose for ^{60}Co, γ-rays; (b) depth-dose for monoenergetic protons; (c) depth-dose for multienergetic protons.

(LET), which is commonly expressed by the energy loss in units of KeV per $1 \mu m$.

$$LET = \frac{dE}{dl}\left(KeV/\mu m\right)$$

Unfortunately, X-rays as well as electron beams are low LET beams. For example, a 200KV X-ray radiation has a LET equal to $3 KeV/\mu m$. Fast neutrons are an example of high LET beams. Heavy ions, from helium (z=2) to uranium (z=92) accelerated to 1~16000MeV per nucleon, have LET values ranging from 10 to 10000 $KeV/\mu m$.

Radiation with a LET value less than 10 $KeV/\mu m$ is considered low LET radiation.

Radiations with different LET values have different degrees of biological effectiveness.

The differences in the effect of radiations of various types for the same physical dose are accounted for by introducing a Coefficient of Relative Biological Effectiveness (RBE). RBE is a parameter which describes the biological effectiveness of the different radiations to the same tissue. It is defined as a ratio of the needed dose of the discussed radiation to the needed dose of 200KV X-rays in order to get the same biological effectiveness.

$$RBE = \frac{D_{200KVX-ray}}{D_{discussed-radiation}}$$

Sometimes in radiation protection, this ratio is also called quality factor Q.

Radiation with a relative high LET value has a relative high RBE value. But RBE reaches a maximum at LET values equal to $100{\sim}120\ KeV/\mu m$ (Fig.1.11).

In radiotherapy, another important fact to be considered is that tumors, especially their central parts, contain more oxygen-poor tissue. And this oxygen-poor tissue is not sensitive to X-rays, electron beams, and other some radiations. As a result of the oxygen effect, the oxygen-poor central parts of tumors are more resistant to irradiation than the surrounding healthy tissue. But the oxygen effect differs in different kinds of radiations.

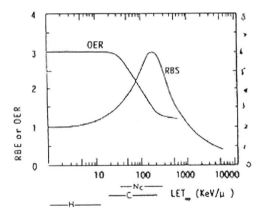

Fig. 1.11 Dependence of RBE and OER on LET [2].

A value of the Oxygen Enhancement Ratio (OER) is used to express the ratio of the needed doses in an anaerobic (oxygen-poor) and in a well-oxygenated medium to induce the same biological effect:

$$OER = \frac{D_{anaerobic-tissues}}{D_{well-oxygenated\ tissues}}$$

524

The OER value depends on the LET value to a considerable degree (Fig.1.11). From Fig.1.11, one can conclude that in order to diminish the oxygen enhancement effect and to increase the biological effectiveness, high LET radiations should be used. So it is desirable to find and use low OER radiation beams. Heavy ions like neon, alpha, argon and carbon as well as pi-mesons (poins) and neutrons are examples of this kind of low OER radiation.

As previously mentioned, the unconventional accelerator therapies can contain the following three promising approaches to improve the efficiency of radiotherapy:

Employ particle beams which improve only the physical selectivity of the irradiation, i.e. the dose distribution (e.g. proton beams or helium ion beams);

Employ the high-LET radiations which produce different types of biological effects, and which aim at improving the differential effect between tumor and normal tissues (e.g. fast neutrons);

Combine the above two approaches to seek a high physical selectivity with high-LET radiation (e.g. heavy ions.)

Fig.1.12 gives the depth-dose distributions for neutron and various heavy particle beams. In Fig.1.12, the depth dose is normalized at 10cm [2].

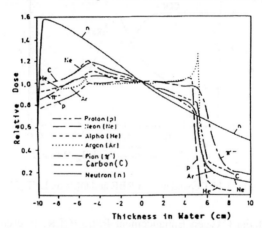

Fig.1.12 Depth-dose distributions for neutron and various heavy particle beams.

From Fig.1.12, all the particles exhibit a characteristic plateau. As for pions, they have the advantage of being scattered only slightly by the tissue at the beginning of the range where they lose energy mostly by ionization. At the end of the range they are captured by the nuclei of carbon, oxygen and hydrogen atoms in the tumor tissue. Then they quickly become absorbed and the excited nuclei thus formed either emit alpha particles, protons, heavy ions or fast neutrons, or disintegrate into heavy fragments. As a result of these transformations, a relatively high dose of radiation at the end of the pion range can be effectively used in therapy.

1.1. 6 Accelerators for Unconventional Therapy [2]

1. *Chronology of Development*

(1) Neutron therapy [5]

Differing from low-LET radiation in the radiobiological effect, neutron has a higher RBE, less repair of radiation injury to cell, less variation of radioresistance through every stage of the cell cycle and less radioresistance to hypoxic cells, i.e., a lower Oxygen Enhancement Ratio (OER) than low-LET radiation.

The earliest attempts at employing neutrons in radiation therapy were made over 50 years ago. In 1938, R. S. Stone of the University of California used fast neutrons from a beryllium target bombarded with 8MeV deuterons. But the result was unsatisfactory due to serious complications. In the early 1960s, thorough radiobiological studies were begun using the 16MeV cyclotron at Hammersmith Hospital (England). They finally led to the resumption of fast neutron therapy in 1966-1969. By 1997 over 15,000 patients worldwide had been treated with fast neutrons.

(2) Proton therapy [10]

The possibility of utilizing high energy protons in radiation therapy was first considered by Wilson in 1946. The first irradiations, however, were started in 1957 at the synchrocyclotron in Uppsala (Sweden).

There is an increasing number of projects which aim at using protons to treat many other tumor types, and larger proportions of patients. One of the most impressive is the Loma Linda project at Los Angeles (U.S.A).

By 2001, the number of patients treated with proton beams had reached about 30,000.

(3) Meson therapy

In the early 1960s, P. H. Fowler and D. H. Perkin proposed that mesons could be used in radiation therapy. It was only in the early 1970s that the basis for employing mesons in radiotherapy was created when high intensity medium energy accelerators, the so-called meson factories, were put in operation. The first patient irradiations with pions began in 1974 in Los Alamos (U.S.A).

(4) Heavy ion therapy

It was also P. H. Fowler who, in 1946, considered the possibility of using heavy ions in radiation therapy. The first treatments of this kind began in 1957 in Berkeley (U.S.A) with the use of helium ions. In recent years, neon, silicon and argon ions have been used at this facility for radiotherapy.

By 2001, it was estimated that the number of patients irradiated with proton and heavy ions as well as meson had exceeded 34,440. But except for protons, other kinds of heavy particle therapy are of an experimental nature, about 4,400 patients accepting heavy particle therapy (including pions). Proton radiotherapy is now becoming a conventional form of therapy.

2. Accelerators for Unconventional Therapy

(1) Accelerators for neutron therapy

i) Cyclotron (including isochronous cyclotron)

In a cyclotron with a neutron head, deuterons are accelerated to an energy of the order of approximately 10~15MeV and then used to bombard an appropriate target (e.g. beryllium) to produce a neutron beam. At a deuteron energy of 14 MeV, a current of $100(\mu A)$, and a distance of 125cm, a dose of 0.44 Gy is attained in a 20×20cm field.

ii) Neutron generator

In a neutron generator, deuterons are accelerated to 200~300KeV, then the deuteron beam of several mAs is converted on a tritium or deuteron target. The $T(d,n)^4He$ reaction is used to obtain a 14 MeV neutron beam with a flux of 7×10^{11} neutrons/sec; or in the $D(d,n)^3He$ reaction, 3MeV neutrons are produced with a flux of 10^{10} neutrons/sec.

The Haefely Neutron Therapy System produces a neutron dose rate of more than 0.15Gy/min for a treatment field of 13cm $\times$ 13cm at source-to-skin distance of 100cm.

iii) RF proton linac

Tens of MeV protons accelerated in a RF proton linac to bombard a beryllium target can also be used to treat tumors with the generated neutrons.

In IHEP, China, a 35MeV proton linac has been in part time use since 1991 to produce neutrons to treat tumors.

(2) Accelerators for proton therapy [14]

Synchrocyclotrons, isochronous cyclotrons and proton synchrotrons with beam energies of 70~250 MeV can be used as proton sources for proton therapy.

A beam intensity of less than $1\,nA$ is sufficient to produce a dose of 1Gy (100rad) in 10 min with an irradiation field of 10cm and an energy of 200MeV. In ordinary therapy, complete treatment requires doses of 60Gy on an area of about 25cm². With 115MeV protons, this dose can be produced with 0.8×10^{12}protons, which corresponds to $120\,nA/\text{sec}$.

The first generation of accelerators used in proton radiotherapy was designed for physics studies. Therefore, they had to be specially adapted to therapy, which often required quite complex technical procedures.

Accelerators of the second generation employed for proton therapy are those added to medical cyclotrons originally intended for the production of radionuclides and neutron therapy. Later the proton channel was put in operation.

In recent years, several dedicated proton accelerators have been developed to ensure that proton therapy will eventually become a routine method of treatment. Among them, the most famous proton synchrotron system is in the Loma Linda University Medical Center (U.S.A).

3. Accelerators for heavy ion therapy

Fig.1.13 gives the light ion ranges in therapeutic application vs. kinetic unit energy.

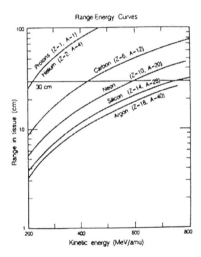

Fig.1.13 Light ion ranges in therapeutic application vs. kinetic unit energy.

The full penetration range of these ions should usually lie between 4 and 30cm for different tumors in different body positions. It should be noted that the above are energy unit values in Fig.1.13; that is, they are expressed as per single nucleon of an ion. Total ion energies must be correspondingly greater. For example, a neon ion accelerated to an energy of 470MeV/nucleon (20cm range) has a total energy of 9.4GeV. So an accelerator for heavy ion therapy is a high energy accelerator.

Up to 1992, the world's only facility devoted to heavy ion therapy was located at Lawrence Berkeley Laboratory (U.S.A). The facility, called Bevatron, was modified to accelerate heavy ions for nuclear science research and biomedical studies, which included radiation therapy.

In 1994, the Heavy Ion Medical Accelerator (HIMAC) in Japan set in operation for heavy ion therapy. Until June, 2001, 917 patients were treated. Carbon beams of 290, 350 and 400MeV/nucleon are used in this radiotherapy. HIMAC is the world's first dedicated light ion therapy facility.

4. Accelerators for pi-meson therapy

Negative pi-mesons (Pions), like other high-energy charged-particles, feature a characteristic depth-dose distribution curve (Fig.1.12). Since the pion rest-mass is 139MeV, a large fraction of this energy is absorbed in the region of mutual absorption in the form of short-range secondary charged-particles which cause a high density of ionization. As for their LET value, pions, together with such ions as He and C for which RBE is less than 2, are characterized by medium LET values.

Any accelerator to be used for pion therapy would require a medium energy and intensive beam current, because only one pion is generated in 10^6 interactions in a carbon target for the proton energy of 700 MeV. Until this kind of accelerator, called a meson factory, was set in operation, pion therapy was restricted to clinical use.

In the early 1970's, in the best known meson factories, LAMPF (U.S.A), TRIUMF (Canada), and SIN, experimental work on meson therapy began. They constructed meson therapy channels to treat patients.

According to the estimation as of July 2001, the total number of patients irradiated with pions was 1,100. To date, there is no information of the construction of any dedicated pion therapy facility.

§1.2 RF Electron Linear Accelerators for Conventional Therapy [4,6,15]

In most conventional therapies the RF linear accelerator serves as the radiation source. In some hospitals RF microtrons are also employed. During the 1950s to 1960s, the betatron was widely adopted. Only the RF linear accelerator will be discussed here.

1.2.1 *Clinical Requirements*

1. *Radiation Energy Range*

At present, in photon radiotherapy, the upper energy range of 15 to 20 MeV is considered optimal. With increased energy, the beam penumbra becomes wider as a result of a larger number of high energy secondary electrons, and unwanted neutron radiation appears. This effect has caused the manufacturers to set the maximum accelerator energy to 22-25 MeV. The low limitation of energy in photon therapy is determined by clinical requirements of the treatment of head and neck tumors or some lymphatic vessels for which the energies of 4-6 MeV are optimal.

Routine clinical applications of electron therapy encompass the same energy range: from 4 to about 25 MeV. When the electron energy is increased to above 18 MeV, the uniformity of the field is difficult to maintain.

2. *Dose Rate and Precision of the Dose Delivered*

The needed dose rate is about 100~500cGy/min.m for photon therapy, and 100~1000cGy/min.m for electron therapy.

As for the precision of the dose, in 1976, the International Commission on the Radiation Units and Measurement formulated the following recommendation in Report No.24 (ICRU 4-24): "If the eradication of the primary tumor is sought, the accuracy in the absorbed dose to the target volume should lie within $\pm 5\%$". The practical implementation of this recommendation in therapy is very difficult, because there are many error factors which influence the absorbed dose precision. That is why the requirements imposed on the accelerator itself should be as high as possible.

At present, the accuracy of the dose rate delivered from commercial medical linear accelerators can reach $\pm 2\%$.

3. *Radiation Field Size and Flatness*

In general, symmetry measured at a distance of 100 cm from the source is varied from 2×2cm to 25×25cm for electron therapies and from 2×2cm to 40×40cm for photon therapies. Field flatness for electron and photon therapies should be less than $\pm 3\%$ and $\pm 5\%$, respectively.

1.2. 2. Accelerator Fundamental System

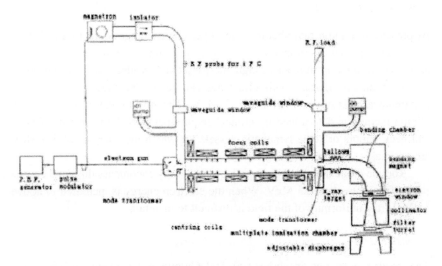

Fig.1.14 A Schematic of traveling-wave medical linac.

Fig.1.14 and Fig.1.15 give the schematic of a traveling-wave medical linac (linear accelerator) and a standing-wave medical linac.

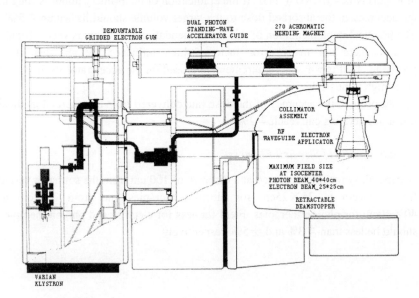

Fig.1.15 A Schematic of a Varian Clinic 18 standing-wave linac [3].

1. *Accelerating Waveguide*

From the mid-1950s to the end of 1960s, medical linacs employed a traveling-wave accelerating structure (Fig.1.16 a.b, Fig.1.17), the so-called disk-loaded accelerating structure [3].

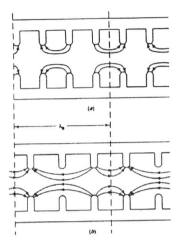

Fig.1.16 RF electric field in disk loaded.

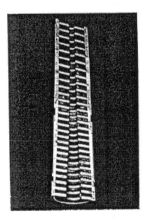

Fig.1.17 Cutaway view of a TW accelerator accelerating structure with buncher.

At the end of the 1970s, the standing wave (SW) accelerator structure was widely employed (Fig.1.18). This made it possible to increase the linear accelerating gradients up to the currently obtainable 12 to 18MeV m-1, which in turn, allowed a considerably shorter accelerating structure with comparable energies. In this way, the accelerator became more compact.

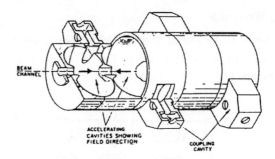

Fig.1.18 Perspective of interior of side coupled SW accelerator structure.

In order to change the accelerator's output energy, the RF power in the accelerating structure was usually varied.

In Varian Clinac 1800 and Clinac 2300 machines, an energy switch was introduced. Its schematic is shown in Fig.1.19. The accelerating structure, 1.4m long, is divided in two sections: 0.4m and 1.0m long. The shorter section mainly bunches electrons and the longer one is an accelerating section. Both sections are coupled together by an energy switch. By adjusting the position of the energy switch, the coupling between the buncher section and the accelerating section can be changed so the output energy can be varied and a good energy spectrum can be maintained at the same time (Fig.1.20).

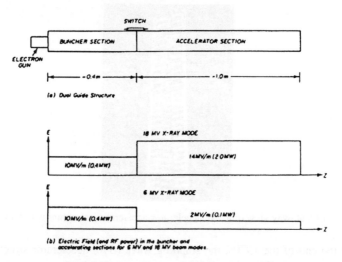

Fig.1.19 Varian Clinac 1800 accelerator with 6MV and 18MV energy switch.

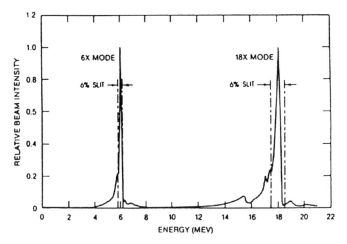

Fig.1.20 6MeV and 18MeV energy spectrum (Varian Clinac 1800) [3].

2. *RF Power Source and Transport System*

Accelerators used in radiotherapy are powered with magnetrons or klystrons, generally operating at the S-band of the frequencies of 2998 MHz or 2856 MHz or operating at the X-band of the frequency, 9300MHz. The operating peak power is about 1-5MW with 4 μs pulse width and 250pps.

If a klystron is employed, the klystron generator is housed in a separate stationary cabinet. RF power passes through a rotating RF joint to SW accelerating structure installed horizontally in the rotating gantry (Fig.1.15).

3. *Gantry*

There are 4 types of gantry structures:

(a) With an in-line treatment head; (b) with a horizontal accelerating structure and a 90° bending magnet: (c) with a horizontal accelerating structure and a 270° bending magnet; (d) with an in-line treatment head, without beam bending (Fig.1.21) [2].

4. *Beam Transport and Bending System*

(1) Beam focusing and steering

In general, the magnet field generated by focusing coils focuses the beam. The beam size is about $\phi 2mm$. Several pairs of steering coils are usually employed to control the beam position and direction so that there are small penumbras and symmetrical field distribution.

534

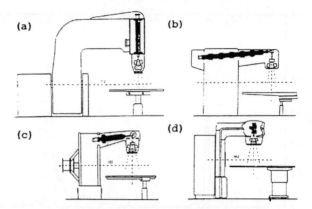

Fig.1.21 Gantry development of therapy linac.

(2) Bending system

In the early stage of medical electron linac development, 90° bending magnets were adopted. Although this kind of magnet featured all the advantages of simplicity of construction and small dimensions, it was not a monochromatic system. Later, 270° bending magnets systems were employed. Good achromatic properties can be achieved in a 270° magnet [2] (Fig.1.22).

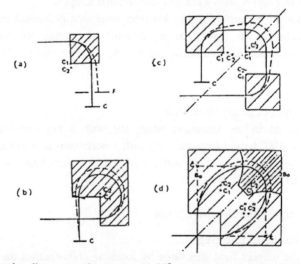

Fig.1.22 Beam bending magnetic systems (a) 90° magnet, (b) triple focal 270° magnet(c) multisectional 270° magnet, (d) pretzel-type 270° magnet.

5. Treatment Heads

Low energy (4 to 6 MeV), accelerators are employed exclusively for X-ray therapy; the beam of accelerated electrons is converted to bremsstrahlung radiation and is not used directly in electron therapy. The accelerating structure is mounted in the treatment head (Fig.1.23) without a bending magnet.

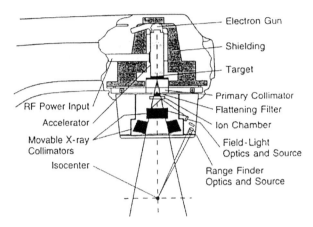

Fig.1.23 Schematic of a treatment head of Varian Clinac 6/100 accelerator for X-ray therapy.

Middle and high energy (10-20 MeV) accelerators use $90°$ or $270°$ bending magnets mounted in the head (Fig.1.24).

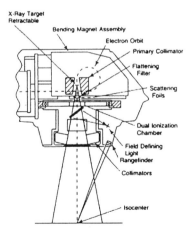

Fig.1.24 Clinac 18 treatment head [3].

Fig.1.25 shows the structure and components of a multi-energy machine with X-rays and electron rays [3].

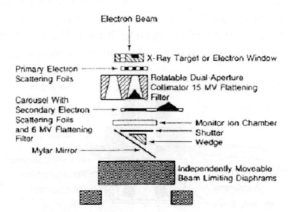

Fig.1.25 Philips SL25 treatment head.

(1) Fig.1.26 gives the X-ray (Bremsstrahlung) spectrum of 10MeV electrons bombarding a 3mm-thick High Z target. Unlike the characteristic spectrum, this curve is continuous. Although the energy of the primary electrons is 10MeV, after conversion, the number of photons with energies close to 10 MeV is relatively small. The equivalent energy is about one-third of the beam's primary energy.

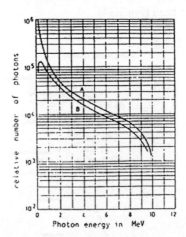

Fig.1.26 Bremsstrahlung spectrum from conversion of 10 MeV electrons(a) calculated curve, (b) spectrum obtained from 3mm thick tungsten target [2].

(2) Initially, targets were made from heavy metals, such as tungsten, gold, platinum and copper, in order to get a good depth dose distribution; in present-day designs, layer-type targets are employed. For example, a target may consist of a 2mm-thick gold layer, 6mm-thick copper layer, 2.5mm water layer (cooling) and 0.5 mm- thick stainless steel layer.

(3) Bremsstrahlung radiation is emitted within a relatively narrow forward cone around the direction of the primary electrons. The cone width becomes narrower with increasing energy. Fig.1.27 shows the distribution of bremsstrahlung radiation for a thick target bombarded by 8 to 30MeV primary electrons.

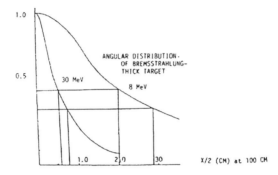

Fig.1.27 Angular distribution of bremsstrahlung for 8 and 30MeV electrons incident on a thick target.

To create satisfactory radiation field uniformity, a flattening filter is employed for X-ray treatment and scattering foils are used for electron therapy.

In some designs, the beam scanning method is applied to make a uniform radiation field.

(4) Collimators and applicators
(i) X-ray collimators
As already indicated in Fig.1.24 and Fig.1.25, the X-ray collimation diaphragms consist of two pairs (upper and lower) of movable collimation jaws. Fig.1.28 shows the main operation modes of a collimation diaphragm (symmetrical mode and asymmetrical mode).

In recent years, conformation therapy has been developed. The principle of conformation therapy, which is in fact a 3D therapy, is that the treatment field is divided into individual sections. Each section has a separate treatment plan (Fig.1.29). In one example, 64 independent collimating leaves (sections) comprise one pair of collimating jaws so as to encompass irregularly shaped tumors. This collimator is called a multi-leaf X-ray collimator. The multileaf collimator can be rapidly set under computer control and can be programmed to follow the changing tumor outline for dynamic therapy.

(ii) Electron applicators

In electron therapy, applicators are designed for shaping electron beam profiles after the beam passes through scattering foils (Fig.1.30).

Fig.1.31 shows a frequently used box-type applicator which is employed for collimation of beams whose cross-section is square, rectangular or circular. The basic set includes five applicators with apertures of 6×6cm, 10×10cm, 15×15cm, 20×20cm and 25×25cm. Some manufacturers supply electron applicators with an automatically controlled aperture which permits variable collimation of the beam.

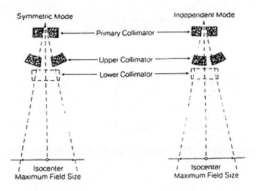

Fig.1.28 Symmetrical and independent movement of collimation jaws of an X-ray beam collimating diaphragm [2].

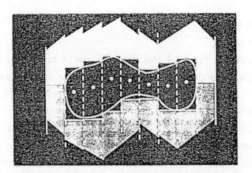

Fig.1.29 The principle of conformation therapy.

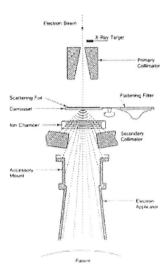

Fig.1.30 Beam subsystem and electron applicator for electron therapy [3].

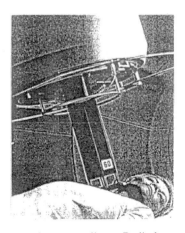

Fig.1.31 Box-type electron applicator Radiotherapy patient [2].

1.2.3. *The New Development of Conventional Therapy with Linac [4,6,7,8]*

Although linear accelerators have been used for radiotherapy for more than 40 years, the technology and the applications have been continued to develop. These new developments will be outlined below.

(1) Dynamic Therapy

(i) Rotation therapy

The simplest form of dynamic therapy is rotation therapy, otherwise described as arc therapy or proportional arc therapy. Rotation therapy requires that the gantry should rotate in a highly controllable way during irradiation. In particular, the dose monitoring and control system have to be linked to the gantry control system. In proportional arc therapy, the gantry rotation speed is proportional to the dose rate. Rotation therapy does not need to be continuous or cover the full 360° of a gantry rotation. It is often desirable to restrict the length of an arc to avoid irradiation of a critical organ such as the spinal cord.

(ii) Three-dimensional conformal therapy with intensity modulation

The term "conformal therapy" has been applied to techniques where particular attention is paid to 'conforming' the dose distribution to the target volume. In conformal radiotherapy, the high-dose volume is tailored to match the planning target volume (PTV), sparing organs-at-risk (OAR). All such techniques depend on three-dimensional identification of the target volume requiring state of the art imaging, three-dimensional treatment planning taking into account the properties of the radiation beams and the individual anatomy of each patient and in some cases, assessment of the expected biological effects of the proposed treatment in addition to the dose distribution. One major advance of these techniques has been in faster and more practical methods to tailor the geometrical shape of the beam to the projected area of the planning target volume in the direction of viewing of the beam.

Although the use of uniformly shaped beams reduces the volume of irradiated tissue it does not necessarily result in a uniform dose distribution. In the case of co-planar beams there is a further constraint in that the boundary of the irradiated volume in the plane of the beams is always convex, so it is not possible to create re-entrant targets that might be needed to wrap a treatment volume about a sensitive organ (OAR organs-at-risk). This problem is shown schematically in Figure 1.32 [6].

To produce concave targets, a reasonable first approximation to the required profile for each beam can be made by setting the intensity at each point to be proportional to the length of chord projected through the target volume from that point. Beam intensity modulation by use of dynamic control.

To produce concave targets, a reasonable first approximation to the required profile for each beam can be made by setting the intensity at each point to be proportional to the length of chord projected through the target volume from that point. Beam intensity modulation by use of dynamic control of an MLC (Multi-Leaf Collimator) has been demonstrated. The generation of modulated beams can be thought of as an extension of the generation of wedged fields by dynamic collimation although in this case the algorithms required calculating the collimator trajectories are a great deal more complicated. The problem is two-dimensional and, as the trajectories for adjacent leaves will not in general be the same, the motion of all the leaves must be calculated so that leaf collisions are avoided. Fig.1.33 illustrates this mentioned method. Each pair of leaves generates a slit field which is scanned across the field. As the slit moves, its width is varied so that points in the treatment plane are exposed for variable lengths of time and thus receive varying doses [6].

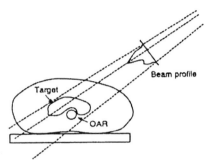

Fig.1.32 A re-entrant (concave) target volume closes to a sensitive organ at risk (OAR).

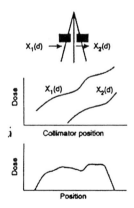

Fig.1.33 A method of beam modulation by dynamic collimation: scanning slit. The separation of the two trajectories at each point in the field determines the dose at that point.

An "intensity-modulated arc therapy" (IMAT) technique has been developed, which can be considered as the inverse of CT scanning; however the inverse analogy is not complete. In theory any arbitrary dose distribution could be generated by projecting therapy beams with the same function used for CT. Unfortunately, although the manipulation of negative numbers in a computer is easy, in the case of a therapy beam projection, the beam modulation is constrained by allowing only positive intensities. One approach to this problem has been developed, which makes use of a multivane intensity modulation system which consists of forty tungsten leaves arranged in two rows of twenty. Each leaf is pneumatically powered and can occlude a 1cm×1cm area of the beam, with a maximum size of 20cm×20cm. A rotation treatment is split into many segments of a few degrees each and the leaves are opened for the predetermined fraction of the time needed to irradiate each segment.

Three-dimensional conformal therapy with intensity modulation beams (IMB) makes treatment plans by forward treatment planning become quite impossible. For this reason the treatment-planning process has to be radically changed to meet the standard "given a prescription of desired outcomes, compute the best beam arrangement". This problem is solved by "inverse treatment planning" using a computer with human guidance rather than by human experience alone.

Fig. 1.34 shows the concept of intensity-modulated beams. Two (large sets of) such beams with 1D intensity- modulation are shown irradiating a 2D slice. The beams combine to create a high-dose treatment volume spanning the PTV which has a concave outline in which might lie OAR. Such uniform high-dose treatment volumes cannot be achieved using beams without intensity-modulation. Planning such treatment relies on tools such as the simulated annealing method to solve the inverse problem.

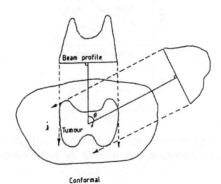

Fig.1.34 The concept of intensity-modulated beams [7].

(1) Stereotactic Radiosurgery

Stereotactic radiosurgery is a one-time application of a high dose of radiation to a stereotactically defined target volume (e.g., a brain tumor). Radiosurgery is a unique hybrid of surgery and radiotherapy, but it differs markedly from both conventional neurosurgery and conventional external-beam radiotherapy. Conventional neurosurgery seeks to resect the pathologic process physically. The goal of conventional radiotherapy is to eradicate or control the local disease process. Conventional radiotherapy is typically delivered in 10 to 60 treatments involving delivery of small doses of radiation once or twice daily through two to four static treatment fields. Differences in biological sensitivity and the repair capacity of normal and pathologic tissue are employed to injure the target lesion selectively. Radiosurgery differs radically from conventional neurosurgery in that it is generally an outpatient treatment with no incision and few acute complications. Radiosurgery also differs in several important respects from conventional radiation treatment. Radiosurgery usually relies on one high-dose treatment as opposed to the multiple low-dose fractions inherent in conventional treatment. Treatment delivery involves multiple stereotactically targeted, arced fields versus a limited number of conventionally simulated, static fields. Therefore, radiosurgery relies more on the extreme accuracy of the radiation delivery than on radiobiological differences in tissue sensitivity and repair capacity. The goal of radiosurgery is to deliver a high dose to the target and a minimal dose to normal tissue just a few millimeters away.

An ideal radiosurgery treatment plan would deliver 100% of the desired dose to the treatment target (e.g., a brain tumor) and none to the normal brain. This is not possible in reality, but the primary goal of radiosurgery treatment planning is to achieve a plan that conforms to the target as closely as possible. Another goal of dose planning is to adjust the dose gradient such that critical brain structures near the target receive the lowest possible dose of radiation. This can be accomplished through the use of multiple beams, and with a unique entrance and exit pathway, yet all directed at a single target. This also is the basis of modern radiotherapy treatment planning, which generally uses two to four fixed, coplanar radiation fields defined by collimators within the linac and further shaped by custom blocking. The concept of using multiple beams is extended by the radiosurgery treatment paradigm used for linac and gamma knife systems. Gamma knife systems use 201 separate cobalt sources, all aimed at one target. Fig.1.35 illustrates the change in dose concentrations by the addition of multiple radiation beams. Doses are represented as isodose lines (labeled) in a coronal plane for a single field (A), two fields (B), and radiosurgical treatment (nine arcs) (C). The radiosurgery paradigm is the equivalent of hundreds of radiation beams [4].

544

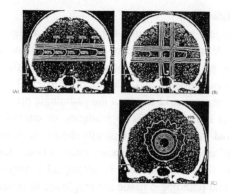

Fig.1.35 Illustration of the change in dose concentrations by the addition of multiple radiation beams.

Treatment planning generally begins by directing nine equally spaced arcs of radiation at the center of the target. Fig.1.36 gives an example of radiosurgery treatment planning at the University of Florida. Fig.1.36 (A)

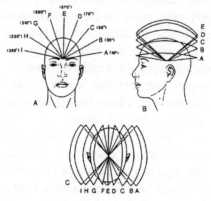

Fig.1.36 Radiosurgery treatment planning at the University of Florida begins with nine spaced arcs of radiation.

shows the coronal orientation of the nine arcs. Each arc span is 100 degrees, and each arc is spaced 20 degrees from its neighboring arcs. Treatment table angles (in degrees) to deliver each arc are noted in parentheses, and each arc is lettered A,B,...G,H,I; (B) Lateral view showing the arc orientational one; (C) Superior view of the nine arcs lettered to coincide with each table angle shown in (A) [4].

As regards the radiation source, linear accelerators (linacs) are the most common source of stereotatic radiosurgery. The most common machines used for stereotatic radiosurgery have maximum energies between 4MV to 15MV (In general, 6MV machines are used). For radiosurgery, the delivered X-ray beam should be refined so that it is suitable for patient treatment. More specifically, a radiotherapy beam must be homogeneous over defined areas that comprise the treatment field. To achieve this, the beam is both collimated to define edges or field sizes and filtered to make the beam flat and even over the treatment field area. In order to treat a variety of lesion sizes and shapes, the collimators are available in 2 to 5 mm increments from 5 to 40mm, as shown in Fig.1.38. The figure shows how the linac gantry rotates around the isocenter for stereotactic radiosurgery [4].

Fig.1.37 Tertiary radiosurgery collimators ranging in size from 5 to 40mm [4].

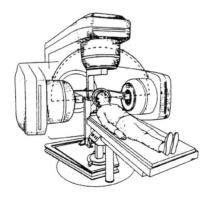

Fig.1.38 The linac gantry rotates in the indicated manner around the isocenter [4].

For accurate localization during stereotactic radiosurgery, an attachment of a stereotactic head ring is required (Fig.1.39) [4]. The rigidly attached ring allows spatially accurate information to be acquired from angiography, computed

546

tomography (CT), and magnetic resonance imaging (MRI). The images obtained with this ring establish a fixed relationship between the ring and the target lesion that are later translated during the treatment planning so that the treatment target is accurately placed at the precise isocenter of the radio delivery device.

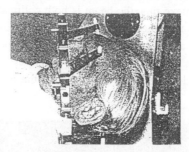

Fig.1.39 A Patient with a steeply sloping forehead who required long pins in the anterior posts for ring attachment.

Stereotactic Radiotherapy (SRT) is different from stereotactic Radiosurgery; SRT applies the principles of stereotactic radiosurgery to the fractioned treatments of radiotherapy. Although a single fraction of radiosurgery is of proven value for many types of intracranial disease, situations exist for which it cannot be used effectively. For example, a single fraction dose of >8Gy has an unacceptable risk of injuring sensitive structures (optic, neuron, etc) and is clearly too low to achieve long-term tumor control. The ideal treatment would incorporate conventional fractionation to spare the critical structure, but it would use stereotactic targeting and dosimetry techniques to reduce the volume of normal brain exposed to potentially harmful doses of radiation. This is stereotactic radiotherapy. For rigid skull fixation during repeated imaging and dosimetry procedures, two approaches have been studied: noninvasive stereotactic frames and frameless techniques.

1.2.4. The domestic RF electron linacs for conventional therapy in mainland China [12]

The urgent need for radiation therapy spurred the development of domestic electron linacs for medical application. From 1972 to 1977, four groups from the Shanghai, Beijing, Nanjing, and Sichuan areas started to develop medical linacs. First, Tsinghua University and Beijing Medical Equipment Institute (BNEI) completed an 8-10MeV traveling wave medical (BJ-10) linac in July, 1977 (Fig.1.40). Before long, HEPI of the High Energy Physics and Shanghai Medical Nuclear Instruments Factory (SMNIF) constructed another 8-10MeV TW medical machine (ZJ-10) (Fig.1.41).

Fig.1.40 BJ-10 TW medical linac.

Fig.1.41 ZJ-10 TW medical. linac

In 1984, BMEI developed the first 4MeV standing-wave (SW) electron medical linac. In 1990, Guangdong Weida Medical Apparatus Group Co., in cooperation with Beijing Vacuum & Electronics Research Institute (BVERI) and Tsinghua University developed the first 6 MeV SW medical linac.

Tsinghua University and BVERI successfully developed two 14 MeV SW accelerating wave-guides with 270° deflecting chamber in 1994 and 1995 respectively. These waveguides can provide two mode X-rays, 6MV and 15MV, and five mode E-rays, 6, 8, 12, 14 MeV. They were used to equip two medical machines (BJ-14, WDZ-14C) separately, and have been installed at two hospitals. Fig.1.42 shows a view of the BJ-14 SW medical linac installed at a hospital.

Fig.1.42 BJ-14 SW medical linac.

After more than 20 years of development, the total number of domestic RF linacs for conventional therapy has reached about 300 sets.

2. Radiation Processing Application of Electron Accelerators

§2.1 Introduction

Radiation processing application of accelerators began in the 1950s, and rapidly came into use in the early 1980s. The main reasons for the rapid growth of these applications of accelerators were the quick reduction of unit costs of both power and beam energy, the considerable advances made in radiation chemistry, giving reduction of required dosages, and the development of suitable commercial technologies and their introduction into practical operations .

The situation in China followed almost the same course. The research and development of radiation accelerators and their radiation processing applications started in the 1960s. At present there are about 70 radiation accelerators for radiation processing. Among them, 44 sets are home-made machines. By 2001, the total beam power capacity was about 3500KW.

Table 2.1 and Fig.2.1 gives the status of radiation processing facilities with beam power exceeding 5KW and the beam power capacities in mainland China [11,12].

Table.2.1 Summary of radiation processing facilities with beam power exceeding 5KW.

year	Before 1989	90	91	92	93	94	95	96	97	98	99	01
Sets of acc.	6	10	13	18	29	30	36	40	44	46	49	55
Beam power(KW)	204	225	400	650	935	955	1295	1436	1950	2690	2830	3300

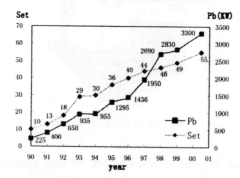

Fig.2.1 The increasing status of radiation processing accelerators and their beam power in mainland China.

§2.2 Principal Types of Processes [1]

Principal types of radiation processes include the following items:

1. Radiation-induced Cross-linkage of Polymers

This is undoubtedly one of the most important industrial applications of radiation treatment. Under the influence of ionizing radiation, additional chemical bonds are formed between the long chains of high molecular mass that make up polymers. As a result of the cross-linkage, structures with higher resistance to temperature and solvents are formed. They also display much better strength characteristics than those of the starting materials. Accelerators are used on a commercial scale most frequently for radiation-induced cross-linkage of polyethylene in the production of heat-skrinkable foil, cable insulation, and heat-shrinkable tubing.

2. Graft Copolymerization

Radiation treatment is one radiation technique used to modify the properties of fibers. It permits an appreciable improvement in the wrinkle resistance of fabrics and modification of their surface properties that are decisive in regard to their resistance to soiling and potting.

3. Radiation-induced Curing of Coatings

This process consists of the radiation-induced polymerization of lacquers or other protective coatings.

4. Radiation-induced Degradation

This process is the reverse of cross-linkage, which increases the molecular mass. Instead, it decreases the mass under the action of radiation. Degradation makes it possible to obtain materials with properties that are difficult to get by other methods.

Table 2.2 shows the principle types of radiation processing used on a commercial scale and lists the required doses and the theoretical processing yield in kilograms of the given material per accelerator kilowatt hour [1].

Table 2.2 Kinds of radiation processing and required doses.

Kind of processing	Dose required, kGy	Theoretical processing capacity, kg/kWh
Destruction of pests	0.25~1.0	3600~14000
Food preservation	1.~25.	144~3600
Cellulose depolymerization	5.~10.	360~720
Graft copolymerization	10. ~20.	180~360
Medical sterilization	20.0~30.0	120~180
Curing of coatings	20~50	72~180
Polymerization of emulsion	50~100	36~72
Vulcanization of silicones	50~150	24~72
Cross-linkage of ploymers	100~300	12~36
Vulcanization of rubber	100~300	12~36

§2.3 Processing Parameters, Irradiation Techniques [1]

1. Dose and Dose Rates

The dose unit in radiation processing is the Gray (Gy), the dose at which 1 Joule of energy is absorbed in each kilogram of substance. Hence, a dose with a practical value of 10kGy (1Mrd) corresponds to an absorption of 10kW-sec/kg.

The dose rate is expressed most often in kGy/sec or kGy/min in radiation processing.

2. Electron Range

The energy of accelerated electrons used for radiation processing usually lies in the range from 200KeV to 3MeV. At energies above 1MeV, the electron ranges in the material undergoing irradiation are approximately proportional to the energy of the electrons; the range of 1MeV electrons in water is about 0.55cm, while that of 2MeV electrons in water is about 1.1cm.

The electron range is inversely proportional to the density of the material being irradiated. Thus, a 100 μm layer of aluminum with a density of 2.7g/cm^3 is equivalent to a 270 μm layer of water.

3. Irradiation Methods

Fig.2.2, Fig.2.3 and Fig.2.4 give the irradiation methods of thin foil, wires and cables as well as liquid.

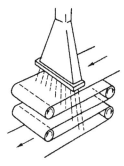

Fig.2.2 Two –sided irradiation of thin foil [1].

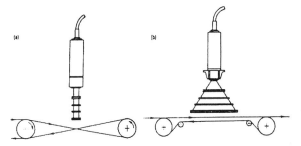

Fig.2.3 Two-sided irradiation of insulation of wires and cables [1]. (a) figure-of-eight method; (b) parallel-wire method.

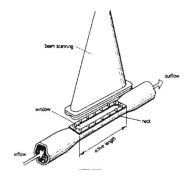

Fig.2.4 Arrangement for irradiation of liquid [1].

4. *Irradiation Efficiency (η)*

The total efficiency-the percentage ratio of useful energy to the total energy of the beam-depends on both the accelerator characteristics and the irradiation technique, as well as on a number of other factors determined by the kind of material, its geometric dimensions, and the packaging.

For example, in cross-linking of polyethylene rods, the efficiency is 60%~70%, while in cross-linking of wire (and cable) insulation it is 15%~50%.

5. Processing Capacity

When the true efficiency is taken into account, the capacity of an accelerator system for radiation processing can be found from

$$W = 3600 \frac{P}{D} \times \frac{\eta}{100}$$

Where W is the capacity (kg/h), P is the beam power (kW), D is the required dose (key), and η is the efficiency (%).

With a beam power of 7.7KW, a dose of 100kGy, and an efficiency of 45%, for example, an hourly capacity of 1250kg of irradiated material is attained.

§2.4. Examples of Applications [1]

1. Wire and Cable Processing

It is in this field that most extensive use is made of accelerators for radiation-induced cross-linking.

The wires and cables cross-linked by electron radiation possess very good electrical properties, high resistance to the heat of soldering irons, low agreeability, resistance to organic solvents and nonflammability.

When polyethylene is used as insulation for wires and cables, a standard increase in the softening temperature to about 135°C is achieved through radiation processing; for non-cross-linked polyethylene, this temperature is 95~105°C.

2. Production of Packaging Materials

Cross-linkage of polyethylene foil produces heat-shrinkable foil. Cross-linked polyethylene possesses elastic memory. Upon being heated, it remains elastic but is easily deformed; however, on cooling it retains the shape imparted to it. Foil drawn over an object to be packaged fits tightly on the object after heating and cooling.

3. *Curing Coatings*

The main advantages of electron curing of coatings are lower electricity consumption, and the absence of solvents released during drying. It is also important that the curing process is very rapid, taking only 0.1~1 sec. It is not accompanied by the release of heat, which is particularly important in the case of heat-sensitive materials such as wood, cardboard, and plastics.

4. *Radiation-induced Degradation*

Degradation of teflon waste by irradiation with accelerated electrons in the presence of air is carried out on a commercial scale. The resulting product is used in the production of aerosol lubricants.

5. *Semiconductor Device Radiation*

The irradiation of accelerated high energy electrons (5~14MeV) can induce the defect energy levels on the semiconductor, forming recombination centers to reduce the life time of minority carriers and to improve the electronic characteristics considerably. By this technique, the switching velocity of the thyristors and transistors are noticeably increased.

§2.5 Radiation-Processing Accelerators [2]

High-voltage electron accelerators and RF electron linacs are widely employed in radiation processing on a commercial scale. Among high-voltage accelerators, the most widely adopted accelerators are Dynamitrons and Insulation-Core Transformers.

1. *Dynamitrons*

Dynamitron is the trade name used for accelerators manufactured by the U.S Company Radiation Dynamics Inc. (RDI). The dynamitron operating principle is shown in Fig.2.5. Two large D-shaped electrodes supplied with power from an RF generator at ~100 kHz are mounted inside a pressure tank. The voltage from these electrodes, which constitutes a kind of antenna, induces the corresponding secondary voltages in separate segments of the received unit. The segments are coupled to rectifiers and rectified voltages for each segment are then added up to supply a HV electrode. As seen from Fig.2.5 the accelerating chamber is installed inside these segments. The tank is filled with an insulating gas, SF6.

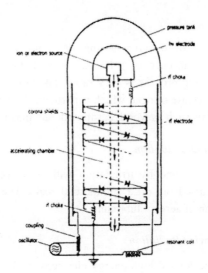

Fig.2.5 A schematic of Dynamitron-type accelerator [2].

Usually, the maximum accelerating voltage is 2~4 MeV, the beam power being several tens of KW.

2. *Insulating-core Transformers*

Conventional high-voltage transformers with a grounded magnetic core are employed in particle acceleration, mainly because the grounded core technique involves some insulation difficulties. In order to overcome this drawback, an insulating-core transformer was developed. Its design is schematically shown in Fig.2.6.

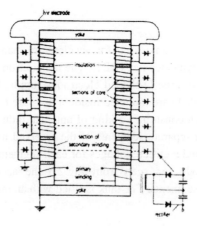

Fig.2.6 A schematic of an insulating-core transformer [2].

The core is divided into sections separated by a thin insulating space. The core is excited through primary winding in a three-phase system. Each secondary section is coupled to a rectifier operating as a voltage divider and is an independent 50 kV unit provided with a characteristic equipotential end-ring. The transformer current load capacity is dependent on the core diameter and on the total output voltage. The transformer constitutes a compact unit, housed in a pressure tank also filled with an insulating gas, SF6.

3. TW Electron RF Linear Accelerators

Traveling-wave (TW) electron linear accelerators (linacs) make use of the TW field to accelerate electrons. In order to accelerate electrons synchronouslys, a disk-loaded waveguide structure is employed to slow the phase velocity of the TW. Fig.2.7 shows a disk-loaded waveguide structure. The electromagnetic wave field in the waveguide is excited by a microwave power source (Magnetron or Klystron) through a coupler. The electrons are accelerated by this RF field along the disk-loaded waveguide in the axial direction with a velocity equal to the TW velocity.

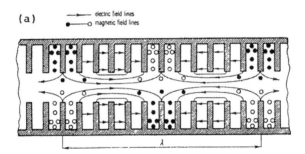

Fig.2.7 A disk load TW structure.

The advantage of the TW linac is high accelerated energy. Usually the 5~15MeV beam energy can be easily reached. But the beam current is not so high.

§2.6 Radiation Processing Accelerators in Mainland China [12]

Table 2.3 shows the main parameters of home-made radiation processing accelerators.

Table 2.3 Main parameters of home-made radiation accelerators.

Type		Energy(MeV)	Beam current(mA)
D.C. high voltage single accelerator	Insulating-core transformer	0.3	30
		0.6	40~50
		1.2	6~10
	Dynamitron	2.0	30
		2.5	20
		3.0	30
	Long filament type	0.2	20
Pulsed resonance accelerator	Single cavity standing wave linac	2.0	10
	TW linac	5.0	0.15
		14.0	~0.2
		4~12	0.45

The total number of home-made radiation processing accelerators in mainland China is about 44 sets.

§2.7 Main applications of radiation accelerators in mainland China [12]

Industrial production scale radiation processing which adopts accelerators as the radiation source concentrates mainly on heat-shrinkable (tube, film) materials, and wire and cable in mainland China. They occupy a dominant position in the beam power of the equipment and Annual Output Value (about 80~85%). Other applications such as coating curing (including paint, jewels, non-dry glue and printing) are not at the production scale level. For all these applications, the beam energy is below 3MeV. 5~14MeV traveling-wave electron linacs are employed to modify properties of semiconductor devices.

By 2001, the total radiation accelerators with beam powers over 1KW reached 70 sets. The annual output value of radiation processing was about 2 billion Chinese Yuan.

3. Radiography Application of Electron Accelerators

§3.1 Introduction

The history of radiography can be traced back to the event of Roentgen discovering the X-ray and his using the ray to radiate his wife's hand to make a film in 1895. The first commercial high-energy X-ray source for radiographic application was a 1 MeV resonant transformer in 1939. In the1940s~1950s, the betatron payed an important role in radiography. Since the 1960s, the main accelerator for radiography has become the electron linear accelerator (linac), especially the standing-wave linac.

The electron beam energy range of today's commercial radiography linacs is mainly from 4MeV to 16 MeV. The total number of machines is about 800~1000 sets worldwide. Radiography has become one of the classical applications of electron accelerators.

Accelerator radiography as one of the methods of non-destructive testing (NDT) has been mainly employed in the field of checking the flaws and welded joints of high-pressure vessels, main axles, large casting, atom reaction top lids, torpedoes, and missiles.

In recent years, linear accelerators as x-ray sources have also been used in industrial computed tomography since the 1980s.

Electron linear accelerators have also been employed as high energy X-ray sources for the inspection of customs containers since the 1990s.

§3.2 Generation of X-ray for radiography and its characters

When an accelerated electron hits a high-Z target, the electron decelerates or changes its moving direction. At that time, the electron emits electromagnetic radiation called bremsstrahlung radiation (BR). The emitted BR's energy spectrum is a continuous one. Fig.1.26 shows the BR spectrum of a 10 MeV electron beam hitting a 3mm-thick tungstom target.

The BR beam is emitted in a relatively narrow cone in about the same direction as the primary electrons. Angular distributions of the dose rate of the BR beam (X-ray) are plotted as examples in Fig.1.27 and Fig.3.1 [9].

The intensity of the emitted BR is dependent on the electron beam intensity and energy as well as on the conversion efficiency. Photon conversion efficiency is dependent on electron beam energy and the target material. Fig 3.2 shows the efficiency [2].

Fig.3.3 shows the measurement curves of the dose rate of the emitted BR at $0°$ and $90°$ directions via the initiated beam energy from the NCRP-51 report.

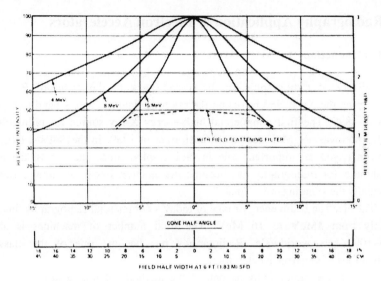

Fig.3.1 Angular distribution of BR for Linatron-type machines.

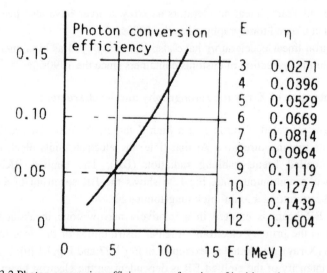

Fig.3.2 Photon conversion efficiency as a function of incident electron energy.

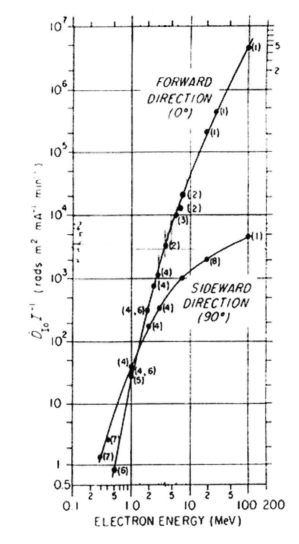

Fig.3.3 The relation between the dose rate of the emitted BR via the initiated electron beam energy.

The following formula can be used to calculate the dose rate (J_x)

$$J_x = 0.067 I_b \cdot E_e^{n} \qquad (3.1)$$

Where I_b is the electron beam current (μa); E_e electron energy (MeV); $n = 2.6 \sim 3$.

As shown in Fig. 1.26, the BR spectrum is a continuous one. The maximum X-ray energy is almost equal to the emitted electron energy but the most part of X-ray energy is less than the electron beam energy. One uses usually one-third of the maximum X-ray energy to represent the X-ray energy. But measuring the X-ray energy is a difficult issue. Usually one characterizes the X-ray energy by its penetrating power. And the penetrating power is defined most often in terms of the half-value layer (HVL), i.e. a layer that attenuates the intensity of the radiation by 50%. The Varian Company gives the HVL curve for steel as a function of the X-ray energy as shown in Fig.3.4.

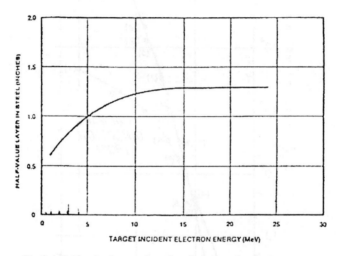

Fig.3.4 Half-value layer of steel as function of radiation energy.

Table 3.1 Typical broad beam HVL at characteristic Linton machine energies of Varian Company for various materials.

Material	HVL(cm)						
	1MeV	2MeV	4MeV	6MeV	8MeV	10MeV	16MeV
Tungsten	0.55	0.90	1.15	1.20	1.20	1.20	1.15
Lead	0.75	1.25	1.60	1.70	1.70	1.70	1.65
Steel	1.60	2.00	2.50	2.80	2.80	3.20	3.30
Aluminum	3.90	5.40	7.50	8.90	8.90	10.00	11.00
Concrete	4.50	6.20	8.60	10.20	10.20	11.50	12.70
Solid propellant	6.10	8.40	11.60	13.80	13.80	16.50	20.40

The penetrating power is usually expressed with the following formula:

$$I = I_o e^{-\mu x} \qquad (3.2)$$

where I_o --the initial intensity of the entrance X-ray beam

I --the transmitted radiation intensity

x --the path length of the transmitted object

μ --the linear attenuation coefficient, which is related to the material property and X-ray energy.

From (3.2), HVL can be connected to μ with the following rotation:

$$HVL = \frac{\ln 2}{\mu} \qquad (3.3)$$

Three physics effects make up the X-ray attenuation, which causes the reduction of X-ray intensity due to transmitting the object. The first effect is the "photoelectric absorption" process. When the energy of the X-ray is less than 0.1MV, this effect plays a major role. The second effect is the "Compton scattering" process. For photons in the beam with energies between 0.1 and 10 MeV, Compton scattering is the major attenuation process. The third process, "pair production', in which the photon is completely absorbed and an electron-positron particle pair is created, has a threshold energy of 1.02 MeV. This effect becomes significant when the X-ray energy is greater than 6MeV.

Fig.3.5 illustrates the principle of taking a radiography.

An accelerated electron beam (1) striking a target (2) produces a beam of X-rays (3). This beam penetrates through the object under examination (4) and hits a radiographic film (6), blackening it. The image obtained on the film consists of areas of various degrees of blackening, depending on the absorption of radiation by the various areas of the object. If the object contains a flaw (5) (e.g., air bubbles in a casting or in a weld), the X-rays will be absorbed less in this area and the film will thus be blackened more in that spot. This blackening or optical density is expressed by the logarithm (base ten) of the intensity ratio of the light incident on the film to the transmitted light.

The film holders used in accelerated radiography do not differ basically from those used in the conventional X-ray technique. Since the photosensitive emulsion of film has a relatively low absorption coefficient for X-rays, X-ray radiography is done with metal intensifying screens (7) in the form of foil applied to either side of the films (Fig.3.5) even with multiple-layer films [1]. These screens are made of lead or tantalum. The X-rays produce secondary

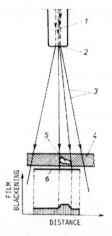

FILM BLACKENING

DISTANCE

Fig.3.5 Principle of taking radiographs (1) Accelerator and electron beam (2) target (3) X-ray beam (4) object under inspection (5) flaw (6) radiographic film (7) metal intensifying screens.

electrons in the lead foil and, striking the emulsion, these electrons enhance the effect of the irradiation.

This technique of radiographic inspection is usually called "Non Destructive Testing (NDT)", due to its ability to leave the inspected object unharmed while testing its interior.

§3.4 Application for radiography

The most relevant parameters in radiographic application are radiographic sensitivity, image quality, penetration capability or the thickness of the inspected object, and exposure time.

1. *Radiographic sensitivity*

Radiographic sensitivity is sometimes called "defect detectability", which refers to the percent ratio of the thickness of the minimum detectable defect to the thickness of the object examined. Thus, 1% detectability means that a 1 mm-high defect can be detected in a 100 mm –thick steel plate. Several reference standards for checking the sensitivity, such as penetrameters, have been adopted. Most commonly used are wire penetrameters. The most widespread adopted standard is an American standard-ASTM E142. For example,"ASTM 1-1T" means that a 1 mm steel wire behind a 100 mm steel plate can be inspected. The wire sensitivity is related to X-ray energy and the inspected material, as shown in Fig.3.6.

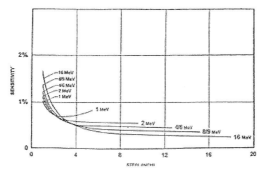

Fig.3.6 Wire sensitivity for steel: fine grained film, lead screens, low scatter.

2. *Penetration capability*

The HVL of X-rays with various energies has been shown in Table 3.1. From the data in the table, one can calculate the linear attenuation coefficient, μ, for the various materials.

3. *Exposure times*

The sensitometric properties of various commercial X-ray films are different, each having the different sensitometric speeds. So the correct exposure time should be established by exposing each film in a real radiography arrangement until it will produce a film darkening, when developed, of a specified density, for example, 2.0 on the H&D units. Each commercial X-ray radiography machine would give the responsive exposure cures.

Fig.3.7 shows the exposure curves for various X-ray energies and steel thicknesses made by the Varian Company (Linatron).

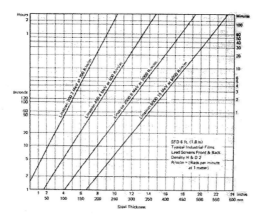

Fig.3.7 Exposure curves of Linatron machines.

§3.5 Accelerators for radiographic application

During the 1940s~1950s, Van de Graff generators with energies of 1 ~2MeV and betatrons with electron beam energies ranging from 15MeV to 31MeV were used as radiographic accelerators. Since the 1960s, RF electron linear accelerators (linacs) have rapidly come into use due to their high dose rate and compactness, and have taken the most important place among radiographic accelerators. As mentioned above, the amount of radiographic linacs has reached 800~1000 sets. One of the main manufacturers is the Varian Company in the U.S.A, which called its machines Linatrons. Other manufacturers are Siemens, Mitsubishi and D.V. Yefremov Institute (Russia). Table 3.2 shows the Linatron's performance characteristics.

Table 3.2   Linatron performance characteristics.

Model	Linatron200	Linatron400	Linatron3000A	Linatron6000	Linatron6000H
Max.beam energy(MeV)	2	4	9	15	15
X-ray output (cGy/min,at1m)	200	400	3000	6000	20000
Max.focus spot(mm)	2	2	2	3	~2.0
HVL(mm)	20.0	25.0	31.0	33.0	33.0

In the 1970s, Beijing Research Institute of Automation for Machine Building Industry, China (BRIAMI) mass-produced 15 MeV and 25MeV betatrons for radiography.

Since 1988, the China Institute of Atomic Energy, BRIAMI and Tsinghua Tongfang Co. Ltd have mass-produced about 10 sets of electron linacs for radiography. Fig.3.8 shows a view of a Chinese radiography workshop at a heavy industry company [12].

Fig.3.8 A workshop for radiography with 4 MeV linac.

§3.6 Application for industrial computed tomography

Since the first medical CT was successfully developed in 1971, CT has been widely used for clinical diagnoses. Although X-ray transmission computed tomography (CT) originally served a medical purpose, CT technology caused technologists to be interested in other applications. The medical CT uses the X-ray tube machine as a radiation source, which has less penetration capability. It is difficult to use this CT in industrial fields. However, a strong demand for evaluating industrial products with high energy X-ray CT has rapidly grown. In the early 1980s, a CT system based on the RF electron linac quickly came into use to examine crack and gas voids in solid rocket motors as well as defects in large high-accuracy machine components. Fig. 3.9 shows a block diagram of the high energy X-ray CT system using the electron linear accelerator as its radiation source.

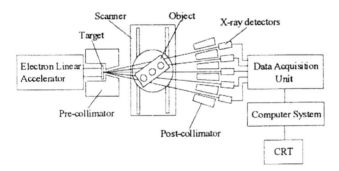

Fig.3.9 Block diagram of the high energy X-ray CT system [13].

In general, electron beam energy of linear accelerators is 4MeV~16MeV. Fig.3.9 shows what appears to be a third generation CT system, but is actually a second-generation CT. In order to reduce the scattered photons from adjacent detectors (crosstalk) due to the high energy X-rays' scattering effect between the adjacent detectors, the detectors must be shielded using tungsten spacers. The adjacent detectors can not be close together and each detector must be collimated. Then a continuous fan of X-rays is substituted for the pencil-like X-ray beams and each detector can only accept X-rays along a line extending from the X-ray target (Fig.3.9).The detectors must translate-rotate to get enough data to reconstruct the imaging. As shown in Fig.3.9, for CT imaging, a collimated X-ray beam passes through the tested object and is measured by an array of detectors, which is shielded by tungsten spacers to overcome the crosstalk. The object is rotated during the measurement process. After the detectors collect the data at one angle, they then continue to rotate to another angle to acquire more data. When a section of the object has been entirely scanned, a full image of the section can be formed. Fig.3.10 shows a CT image of an exhaust manifold, provided by the ARACOR Company, CA, and U.S.A.

Industrial computed tomography (CT) is very different from industrial digital radiography (DR). Fig.3.11 gives the comparison of DR and CT imaging.

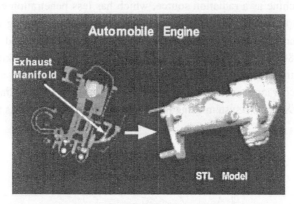

Fig.3.10 A CT image of an exhaust manifold.

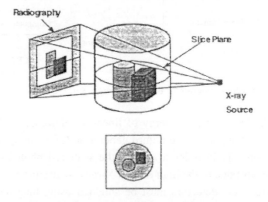

Fig.3.11 Comparison of DR and CT imaging.

DR only provides qualitative images which represent the relative X-ray attenuation along the various ray paths through the object; i.e., it only provides a projective image. By comparison, a CT image is free of any superimposed structure from outside the area of interest. This structural "noise", a disadvantage with DR, is absent with CT. Furthermore, CT provides quantitative data. It can measure internal dimensions and density distributions, resolving differences of approximately 0.1% with a high confidence level. These capabilities enable CT to characterize the internal structure of complex objects, and to detect, size, and precisely locate defects. But CT can only provide the image for one section. To get an image of different sections it is necessary to translate the object to acquire the data in another section.

§3.7. Application for inspecting large containers for customs [12]

The technology of DR can also be used to inspect large containers for customs. To do this, high energy X-rays scan the total container. There are two ways to accomplish this. One way is to fix the X-ray source, and move the container to let the X-ray penetrate it. Another one is to fix the container, and move the X-ray source. The former is called a fixed type container inspecting system. Its principle is shown in Fig.3.12

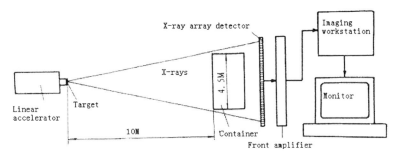

Fig.3.12 The principle of the fixed type container inspecting system.

In general, in order to get high penetration capability from a more compact system, the container inspecting system employs the electron linac as the X-ray source, with energies ranging from 2 MeV to 9MeV.

The large container inspecting system is a very powerful tool for customs against the smuggling of contraband goods, arms and drugs, as well as import duty avoidance. Tsinghua University, Beijing, China in cooperation with Tsinghua Tongfang Co. Ltd has developed three types of large container inspecting systems, including 9 sets of the fixed type, 15 sets of the relocatable type, 5 sets of the mobile type and one set of airport container inspecting systems. These systems have been installed at Tianjin, Dalian, Qingtao, and Shanghai Ports. In addition, five sets of container inspecting systems have been exported to Australia, Iran and the United Arab Emirates.

Fig.3.13 gives a picture of an S-band 9 MeV traveling-wave (TW) linac from the electron gun side for the fixed type system.

Fig.3.13 Shows a view of the building and working field for the fixed type system at the Tianjin seaport.

Fig.3.14 Building and work field for the fixed type inspecting system at the Tianjin seaport.

The relocatable type inspection system, which employs an S-band 6 MeV electron linac as its X-ray source, is shown in Fig.3.15

Fig.3.15 Relocatable type inspection system.

The mobile container inspection system, which employs an X-band 2.5 MeV electron linac as its X-ray source is given in Fig.3.16.

Fig.3.16 Mobile container inspection system.

Other companies such as the ARACOR Company, U.S. and the HAIMANN Company, Germany etc. also provide some kinds of container inspection systems with linacs as their X-ray source.

Reference

[1] W.H. Scharf. Particle Accelerators and Their Uses (part II) Harwood academic publishers, (1986)

[2] Waldemar H.Scharf, Biomedical Particle Accelerators AIP press. New York, (1993)

[3] C.J. Karzmark, C.S. Nunan, E. Tananbe, Medical Electron Accelerators, McGrawhill. INC, (1993).

[4] W.A. Fr.edman et al, Linac Radiosurgery-A Practical Guide, Springer-Verlag, (1998)

[5] R.S. Cabilic et al., Fast Neutrons and High-LET particals in Cancer Therapy, Spring-Verlag, (1988)

[6] D. Greene et al., Linear Accelerators for Radiation Therapy, Second Edition, Institute of Physics Publishing, (1997)

[7] S. Webb, the Physics of Conformal Radiotherapy-Advances in Technology, Institute of Physics Publishing, (1997)

[8] P. Metcalfe et al., the Physics of Radiotherapy X-rays from Linear Accelerators, Medical Physics, Publishing, (1997)

[9] W. H. Scharf, Particle Accelerators-Applications in Technology and Research, Research Studies Press Ltd., (1989)

[10] 郁庆长，罗正明等著，质子治疗技术基础，原子能出版社，（1999）

570

[11] Proceedings of the fourth industrial irradiation accelerator and their applications symposium, in China (1996)

[12] Yuzheng Lin, Applications of Low Energy Accelerators in China, APAC2001, Beijing, (2001)

[13] H. Miyai, et al.,Response of silicon detector for high energy X-ray computed tomography, IEEE Trans.NS.,Vol.41, NO.4,1994

[14] W. Wieszczycka and W. H. Scharf, proton radiotherapy accelerators, world scientific, (2001)

[15] J. V. Dyk, the modern technology of radiation oncology, medical physics publishing (1999)

HIGH GAIN FREE ELECTRON LASERS

Li Hua YU

725C, NSLS
Brookhaven National Laboratory
Upton, NY11973
E-mail: lhyu@bnl.gov

We describe the basic principle of the high gain free electron lasers (FEL) and the Self Amplified Spontaneous Emission (SASE). We discuss several deferent schemes of seeded single pass FEL to improve the temporal coherence. We also discuss the advantages and limitations of these schemes. We report the High Gain Harmonic Generation (HGHG) experiment in the infrared region, and recent DUVFEL experiment using HGHG method at the BNL.

1. The Basic Principle of an FEL

In 1976, John Madey and company demonstrated the free electron laser by an experiment in the infrared region [1]. This first free electron laser operated in a regime we now call small gain regime. The basic principle is shown in figure 1. In this device, an electron beam, collinear with an input laser, goes into a magnet with alternating magnetic field called undulator on its axis. A simple argument in the following will explain that if the electron beam energy is appropriate, the laser light will be amplified. Let us consider a reference frame moving in the same direction of the electron beam. In this frame, due to Doppler effect, the wavelength of the laser light will be longer, and due to the same reason, the wavelength of the undulator will be shorter because it is moving in the opposite direction of the reference frame. If we choose the speed of the reference frame appropriately, these two wavelengths will become equal. We then define the reference frame as a "resonance frame". In this frame, the two electro-magnetic waves, traveling in opposite directions, will form a standing wave, with node points where the field is minimum and peak points where the field is maximum, as shown in figure 1. Based on the basic electro-magnetic theory, we know that the electrons, oscillating in this standing wave, due to their oscillating speed and the alternating magnetic field in the standing wave, experience a "ponderomotive force", and tend to move into the node points. This "ponderomotive" force can be described by a sinusoidal potential well as shown in figure 1. If the electrons are not standing still in the resonance frame, but are moving in the forward direction slowly, as shown in the figure, then, like many initially uniformly distributed bicycles riding in a sinusoidal mountainous area, some of the electrons moving down slope speed up while others moving up slope slow down, as clearly indicated in the figure. Therefore, after a while, the electrons will be clustered near the valleys of the mountain (the node points of the ponderomotive

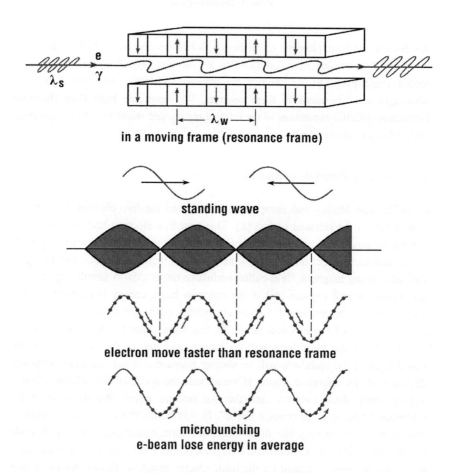

in a moving frame (resonance frame)

standing wave

electron move faster than resonance frame

microbunching
e-beam lose energy in average

Fig.1 The basic principle of FEL

potential well). We call this clustering "micro bunching" because in the laboratory frame the spacing of the clusters is equal to the optical wavelength of the input laser. However, because the electrons are moving forward in the resonance frame, the micro bunching is not exactly located in the node points of the standing wave but slightly to the right of those node points, as shown in the figure. Thus the picture is not symmetrical, and more electrons are moving up the slope than the electrons moving down the slope. It is clear that more electrons are losing energy than those that are gaining energy. Hence, on average, the electrons are losing energy when all the electrons are taken into account in the laboratory frame. Due to energy conservation, this lost energy must be compensated by the increase of the laser energy, thus the input laser is amplified at the exit of the undulator. This is the basic principle of the free electron laser (FEL).

It is clear that the concept of "micro-bunching" plays the crucial role in this process of amplification: because the radiation from the different micro bunches, separated by an integer number of optical wavelengths, are coherent with each others, as compared with the incoherent radiation from individual electrons when they are not micro bunched. This coherence greatly enhances the output radiation, just like the "super-radiation" in the atomic physics.

The first free electron laser worked in the "small gain regime", where after passing through the undulator, the laser is amplified by a few percent, as compared to the "high gain regime" developed later where the gain is in the order of magnitude of millions.

In the small gain regime, to achieve high power output, the FEL uses an "oscillator "configuration, as shown in figure 2. In this configuration, the electron beam consists of a train of micro pulses called "macro-pulse". As an example, these trains of pulses are provided by a linac. In figure 2, we give a typical set of parameters. The macro-pulses are separated by a few tens or hundreds of milliseconds, while the trains have a length of typically a few micro seconds, with each micro pulse a few pico-seconds long, separated by a few tens of nano-seconds. In the oscillator configuration, as shown in figure 2, the first electron micro pulse generates in the undulator a spontaneous radiation pulse, which is reflected back by a mirror to the right of the undulator. Then after traveling back through the undulator in the opposite direction, it is reflected by the mirror to the left of the undulator. When it is going through the undulator again, it is arranged that the second micro pulse of the macro pulse train happens to be overlapping with the radiation, thus the radiation is amplified. The radiation is reflected back by the right mirror again, and the process will repeat itself by the third micro pulse, etc. After many repetitions of this process, when a few tens or a few hundreds of the micro pulses of in the macro pulse pass through the undulator, the radiation is amplified by million times. Eventually, the

574

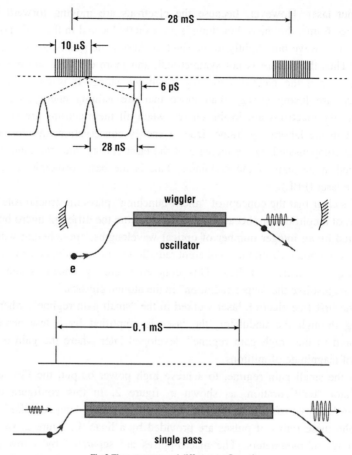

Fig.2 Time structure and different configurations

radiation intensity is built up so high, that within one pass of the micro pulses through the undulator, the energy losses of the electrons are so large, that their energy moves out of resonance with the input radiation wavelength, and the gain is reduced to be equal to the losses in the cavity formed by the two mirrors and the space between them. This is the basic principle of the oscillator configuration. Usually this configuration operates at the small gain regime.

2. High Gain FEL, SASE, and Gain length

One of the disadvantages of the oscillator configuration is that for wavelength shorter than Deep UV (below one hundred nanometers) there is no mirror with good reflectivity available, let alone within the X-ray region. Thus the concept of a new configuration of "high gain FEL" is developed in the early eighties. In an FEL, as explained before, the radiation produces micro-bunching in the electron beam, and in turn, the micro bunching amplifies the radiation because it is coherently in phase with the input radiation. This is a positive feedback mechanism, hence if the quality of the electron beam is very good, and the undulator is sufficiently long, there is an exponential growth of radiation.

When there is no input in the FEL, the spontaneous radiation produced at the beginning of the undulator is amplified exponentially. This is called the "Self Amplified Spontaneous Emission", abbreviated as SASE [2,3]. If the length of the undulator is long enough, and the SASE gain is so high that in a single pass of the electron beam through the undulator, the energy loss of the electrons is large enough to be out of resonance with the input radiation, the exponential growth stops before the end of the undulator. This is called the saturation. The main advantage of the SASE FEL is that there is no need for mirrors, making it possible to generate intense hard X-rays by FEL.

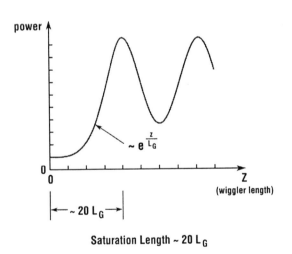

Saturation Length ~ 20 L$_G$

Fig. 3 Gain length and saturation

The distance in the undulator along which the radiation power increases by a factor e is called the power gain length L$_G$. Theory shows it takes 20 L$_G$ of the undulator for the SASE power to grow by a factor of about 10^8 and reach saturation (see figure 3). Thus whether it is possible to achieve hard X-ray FEL depends on whether it is possible to make 20 L$_G$ a realistic distance in practice. The advances on high gain FEL theory and high brightness electron beam in the middle to late half of eighties made this possible.

3. Relations Between the Gain Length and the Electron Beam Quality

The earlier works on high gain FEL were for the idealized cases: The theory was one dimensional, so the diffraction loss was neglected; the electrons were assumed to be all parallel to the axis of the undulator, i.e., the angular spread was neglected. Because of angular spread and energy spread, different electrons have different longitudinal velocity. In turn, this causes the micro bunching formed during the FEL process to disperse. As shown in figure 4, the electrons in resonance with the radiation and in phase with the wave front will remain in phase after one undulator period, but the electrons with different longitudinal velocity will be out of phase with the electrons in resonance, causing the micro bunching to disperse. The figure illustrates that within one period, one electron is

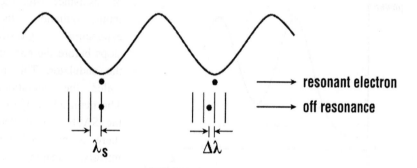

Fig. 4 Energy spread and angular spread degrade micro-bunching

out of phase with the resonant electron by $\Delta\lambda$. If the accumulated de-bunching in one gain length is comparable to a quarter of wavelength, the micro-bunching generated by the exponential growth can not compete with the de-bunching process, and the gain will be significantly reduced. Hence angular spread and energy spread are very important qualities of the electron beam.

To achieve high gain, we need to focus the electron beam in the undulator to maintain small beam size so that the current density is high. However, because the transverse phase space area remains constant, the stronger focusing will increase the angular spread. Hence this phase space area is an important measure of the quality of the electron beam. Roughly speaking, it is the product of the beam size and the angular spread of the electron beam, and in the accelerator terminology, it is defined as the emittance ε. During acceleration, the product of the emittance ε and energy γ remains constant, hence the quality of the electron beam is characterized by this product and defined as the normalized emittance $\varepsilon_n=\varepsilon\gamma$.

Theory shows that to prevent the angular spread from significantly reduce the gain given by the one-dimensional theory, emittance should be much smaller than the wavelength, thus to achieve hard X-ray, the emittance should be much smaller than an Angstrom. The quality of the electron beam, characterized by the normalized emittance, the energy spread, and the peak current, is crucial in determining how short a wavelength a high gain FEL can achieve.

Much work has been done to calculate the effect of diffraction, focusing, and electron beam quality on the gain of FEL in the mid-eighties [4]. The result can be summarized as a scaling function [5]. This work showed that the gain length can be calculated as a function of emittance, energy spread, and focusing strength in the undulator, in a scaled form. That is, all these parameters are expressed as their ratio to a scaling constant which is determined by the electron beam energy and current. The scaling function shows a favorable scaling to shorter wavelength: the relevant required parameters are scaled with a weak dependence, the square root of the wavelength. For a given high gain free electron laser at a longer wavelength λ, to reduce the wavelength and to keep the growth rate per undulator period the same, the electron energy should be increased as $\gamma \propto 1/\sqrt{\lambda}$, the electron beam current should be increased as $I_0 \propto 1/\sqrt{\lambda}$, the normalized emittance should be reduced as $\varepsilon_n \propto \sqrt{\lambda}$, and the relative energy spread σ_γ / γ should be kept constant. This relatively weak dependence on wavelength pointed the promising direction towards X-ray FEL development. The scaling function provides fast calculation of the gain length to determine the required electron beam quality for hard X-ray FEL.

Another important advance is the development in the mid-eighties of a high brightness electron source, the photo-cathode radio frequency electron gun [6]. In this device, a short laser pulse of a few pico-seconds is injected into a cathode inside a radio frequency (RF) cavity, the photo-electrons from the cathode are accelerated by the high RF field. Without the high field in the cavity, the space charge of the electron pulse would blow up the electron beam size, and hence increase the emittance. With the high field, the electrons are accelerated to relativistic speed in a very short time. The space charge effect is largely reduced once the electrons are highly relativistic, hence a much lower emittance can be achieved.

To increase the peak current, the electron pulse length can be reduced in a process called magnetic compression. For the sake of space, we will not elaborate about this process here, but only point out that because of phase space conservation, the product of electron pulse length and the energy spread remains constant during the compression. Thus the energy spread σ_γ is increased when the electron pulse length is reduced. This apparently is harmful for the FEL. However, according to the scaling gain function we mentioned before, when the wavelength λ is reduced, the scaling parameter is σ_γ / γ, rather than σ_γ itself. Since the energy γ should be increased as $\gamma \propto 1/\sqrt{\lambda}$, and the electron beam current should be increased as $I_0 \propto 1/\sqrt{\lambda}$, it is easy to see indeed σ_γ / γ can be maintained constant. Thus the compression happens to match the FEL gain scaling relation.

4. Proposals of X-FELs, the Experimental Confirmation of the SASE Theory and the Lack of Temporal Coherence in SASE

Since the advent of the photo-cathode RF gun technology, the electron beam quality has been steadily and dramatically improved. This advance, in the early nineties, combined with the advances in the high gain FEL theory, makes it possible to consider the development of an X-ray FEL. There were a series of workshops [7] organized to identify the crucial direction of development and the possible parameter range.

Following this period, SLAC proposed the LCLS project for an FEL at 1.5 Angstrom using the existing linac [8]. According to the proposal, the linac energy is 14 GeV and the required electron beam normalized emittance is 1.5 mm-mrad, with peak current of 3400 ampere. For this electron beam and an undulator period of 3 cm, the gain length is 5.6 m and the required total undulator length is 120 meter. The output radiation peak power is 10 GW, which is eight orders of magnitude higher than what the undulator radiation can produce at the present.

In Europe, DESY also proposed a TESLA VUV FEL project aiming at the wavelength of 60 Angstrom [9]. In this proposal, the accelerator is a 1 GeV super-conducting linac, the required e-beam has a peak current of 2500 ampere with a normalized emittance of 2 mm-mrad. For an undulator period of 2.73 cm, the gain length is 1.3 meter, and the total undulator length is 26 meter. The

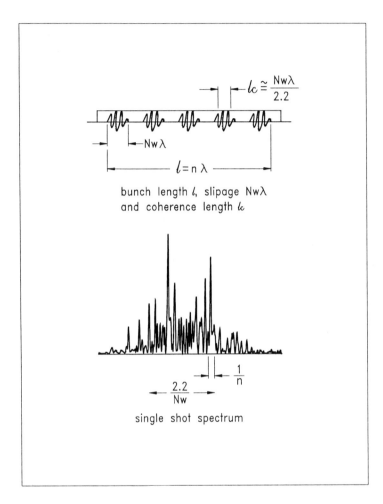

Fig. 5 SASE pulse and spectrum: showing
lack of temporal coherence

output power is expected to be 2 GW.

However, the output of the SASE scheme is not temporally coherent, i.e., the coherence length of the output pulse is much shorter than the pulse length itself. As shown in figure 5, the output pulse consists of a train of pulses, each of them originates from the amplified spontaneous radiation of one electron, since the phases of different electrons are random, the coherence length of the output train of pulses is determined by the length of each pulse. Since the wave front of the radiation of an electron moves ahead of the electron by one wavelength after each undulator period (see figure 4), at the end of the undulator it is ahead of the electron by $N_w\lambda$, where λ is the radiation wavelength. As a result, the coherence length of the train is at most equal to $N_w\lambda$, the length called slippage distance. Actually, theory showed that the pulses formed at the saturation have the coherence length less than the slippage distance by a factor of about 2.2. If the electron pulse length is $n\lambda$, which is much longer than the coherence length as shown in figure 5, i.e., if $n \gg N_w$, the output spectrum would consist of many spikes. Theory predicts that the number of spikes is equal to the ratio of the electron pulse length over the coherence length, i.e., the number of coherence length within the electron pulse. As shown in figure 5, the width of the spikes is $1/n$, and the width of the spectrum is $2.2/N_w$ at saturation. The bandwidth of the spontaneous radiation of an undulator is $1/N_w$, so the SASE bandwidth is even larger than the spontaneous radiation.

Related to this is the large shot to shot fluctuation. Since the phases of different electrons are random, the distribution of the spikes in the spectrum is also random. Therefore, if one uses a monochromator with bandwidth of $1/n$ to achieve full coherence, its spectral window may fall on or off one of the spikes, resulting in 100% intensity fluctuation.

Since 1997, a series of SASE experiments by different laboratories around the world confirmed the SASE theory, including the gain length, the saturation power, the bandwidth, the spikes in the spectrum, and the fluctuation, etc. Among them are,

- LANL, UCLA. Collaboration, SASE at 12 μm (1998)[10],
- BNL, SASE at 0.8 μm (1998) [11]
- APS (project LEUTL), SASE saturation at 0.4 μm (2000) [12]
- BNL, SLAC, UCLA collaboration (project VISA), SASE saturation at 0.8 μm (2001)[13]
- DESY: SASE saturation at 0.1 μm (2002)[14]
- BNL (project UVFEL), SASE at 0.266 μm (2002)[15]

The success of these experiments clearly showed the rapid progress towards shorter and shorter wavelength. The electron beam parameters from the photo-cathode RF electron gun are approaching the required value for the LCLS proposal [8]. For example, the beam parameters from the RF-gun at the Accelerator Test Facility (ATF) of the BNL during the VISA experiment were already close to the design parameters in LCLS, the charge per pulse was 0.5 nC, with normalized emittance of 0.8 mm-mrad [13]. In the mean time, much research work is being conducted to study how to preserve the quality of the electron beam during the electron beam transport and the previously mentioned magnetic compression process to increase peak current.

However, these experiments also confirmed the statistical properties of the SASE process: the spectrum with bandwidth larger than spontaneous radiation, the spikes in the spectrum, and the large intrinsic fluctuation. As an example, in figure 6, a single shot spectrum, measured during the 400 nm SASE of the UVFEL experiment at BNL in April 2002, shows the random spikes. Obviously, it is desirable to develop a method to generate temporally coherent FEL output.

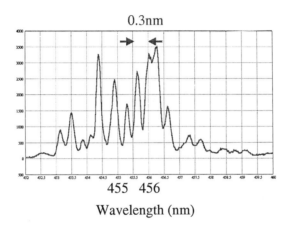

Fig. 6 SASE spectrum at 400 nm for a 4 ps electron pulse

5. The Approaches to Improve the Temporal Coherence and the Basic Principle of HGHG

One example of the different approaches to improve the temporal coherence is the "Two staged FEL" scheme proposed by DESY group [16]. As shown in figure 7, the SASE output from the first undulator is sent through a monochromator to reduce the bandwidth, and then sent into the second undulator as a seed, while the electron beam, after the exit of the first undulator, is by-passed through the monochromator and sent into the second undulator to amplify the narrow band seed. The output from the monochromator will have 100% fluctuation if the bandwidth is chosen to achieve full coherence, as we explained before. When the second undulator is sufficiently long, the fluctuation can be reduced, but to avoid the SASE from the second undulator from competing with the amplified seed, its length should not approach the saturation length, and

582

hence there is always a residual intrinsic fluctuation, even for a perfectly stable electron beam.

At BNL, in the late eighties, we have realized the lack of temporal coherence in the SASE output during our theoretical study, and began to develop

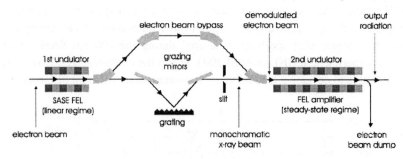

Fig. 7 Two staged FEL scheme (DESY)

a scheme called High Gain Harmonic Generation (HGHG) to achieve short wavelength and full coherence [17, 18].

In figure 8 we describe the basic principle of the HGHG FEL. At the point "a" on the top of figure 8, the initial time and energy phase space distribution, the electron density profile and the its density distribution are all uniform, as given in figure 8a. Then, a small energy modulation is imposed on the electron beam (see figure 8b) by its interaction with a seed laser in a short undulator (the modulator). The energy modulation is converted to a coherent spatial density modulation (see figure 8c) as the electron beam traverses a dispersion magnet (a three-dipole chicane, shown as the section between b and c on the top part of figure 8). A second undulator (the radiator), tuned to a higher harmonic of the seed frequency (ω), causes the micro-bunched electron beam to emit coherent radiation at the harmonic frequency ($n\omega$), followed by exponential amplification (see figure 8d) until saturation is achieved. The HGHG output radiation has a single phase determined by the seed laser and its spectral bandwidth is Fourier transform limited. In figure 9, we schematically show the initial coherent radiation and then the exponential growth and saturation in the radiator.

This process is a harmonic generation process, and the harmonic is amplified by high gain FEL so the output power is higher than the input power, hence the name of High Gain Harmonic Generation. The output of HGHG has nearly full temporal coherence, so its bandwidth is much narrower than SASE. Also, because it is not started from noise, it is much more stable. Another advantage is that the output pulse length is determined by the seed laser, thus it is possible to generate an output pulse length much shorter than the electron pulse length.

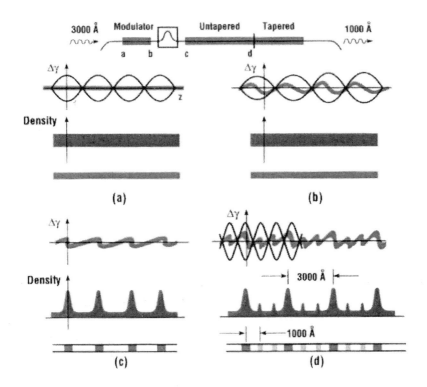

Fig. 8 The basic principle of HGHG

This process is a harmonic generation process, and the harmonic is amplified by high gain FEL so the output power is higher than the input power, hence the name of High Gain Harmonic Generation. The output of HGHG has nearly full temporal coherence, so its bandwidth is much narrower than SASE. Also, because it is not started from noise, it is much more stable. Another advantage is that the output pulse length is determined by the seed laser, thus it is possible to generate an output pulse length much shorter than the electron pulse length.

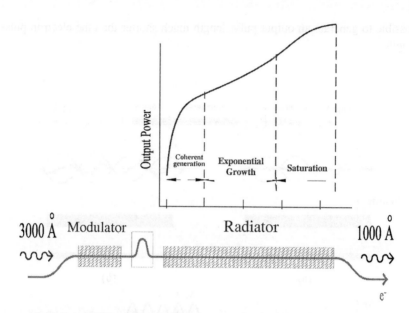

Fig. 9 Power (logarithmic scale) as function of distance in the radiator

The conventional method to generate harmonics (HG) from laser is to use the non-linearity of the crystals or gas, which needs high intensity input. Since the output of this conventional HG process is many orders of magnitude smaller than the input, it is impossible to cascade the HG process, i.e., to use the output of one HG stage as the input of next HG stage to reach much shorter wavelength. However, because the output power of the HGHG is the same order of or much larger than the input power, it becomes possible to cascade several HGHG stages to reach X-ray region [19].

These advantages make the HGHG an attractive direction of development towards X-ray FEL. Since at the time of late eighties when this concept was at its beginning the e-beam quality from the photo-cathode RF electron gun was still not good enough for X-ray FEL, it was our opinion that the development towards X-ray FEL should be step-wise, i.e., we should start from longer wavelength and gradually develop towards shorter wavelength. At BNL, in the early nineties, we proposed to develop a Deep UVFEL program, using the HGHG process to generate intense and coherent radiation at or below 100 nm, starting from a seed at 300 nm or shorter [17,18]. Before starting this DUVFEL program, we first planned to carry out a proof-of-principle experiment, converting 10 μm CO_2 laser into high power 5 μm output by HGHG.

6.   The HGHG experiments at the BNL

The first HGHG experiment was successfully accomplished in 1999 at the Accelerator Test Facility (ATF) of the BNL. The results are published in Science

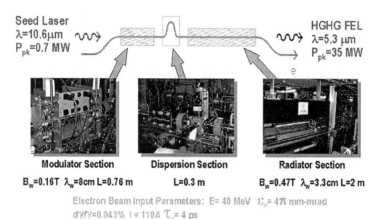

Fig. 10 The setup for the proof-of-principle experiment for HGHG

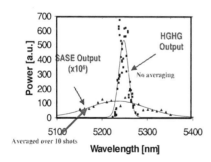

Fig. 11 Multi-shot spectrum of HGHG and SASE for the same electron beam condition

[20], and the detailed characterization of the output coherence properties is published in PRL [21]. A schematic diagram is shown in figure 10, together with the photos of the main components. The radiator is 2 m long, sufficiently long for the HGHG to reach saturation. In figure 11, we show the measured multi-shot spectrum of the HGHG as compared with the SASE output from the same undulator under the same electron beam condition. We need to multiply the SASE power spectrum by 10^6 in order to show it on the same scale as the HGHG spectrum. To achieve the same output power for the SASE, the undulator should be 3 times as long, i.e., 6 m long. Because the large fluctuation, each point for the SASE spectrum is the average of ten shots while each point for HGHG is a single shot, this shows the stability of the HGHG

586

output is much better than SASE. The fluctuation in HGHG output is due to the fluctuation in the electron beam parameters and the CO_2 laser power. Also the coherence length of the HGHG output was found approximately equal to the pulse length, i.e., it is nearly fully coherent. In short, this experiment confirmed the theory and the simulation. In 2000, we started the work on the DUVFEL experiment.

The DUVFEL experiment is carried out in the Source Development Lab (SDL) of the National Synchrotron Light Source (NSLS). In February of 2002,

SASE Signal at DUV-FEL February 2002

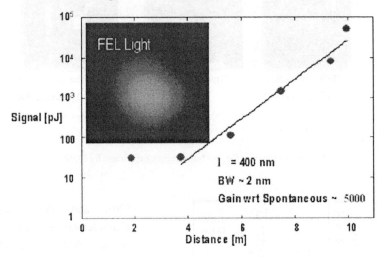

Gain length $L_G = 0.9$ m

Fig. 12 The output radiation profile and the measured power as function of distance in NISUS in the 400 nm SASE experiment of Feb. 2002

we achieved SASE at 400 nm in the radiator undulator named NISUS [15]. In figure 12, the radiation power measured along the NISUS undulator is plotted versus the distance in the NISUS, showing the exponential growth. This plot gives the gain length of 0.9 m. When compared with the gain length of 1.1 m based on the designed parameters, this shows that the system already satisfied the required condition for the HGHG experiment. In May we began to install the seed laser injection line, and in September we installed the 0.8 m modulator named the MINI undulator. At the end of the October, we achieved successful output of HGHG, generating 266 nm from an 800 nm Titanium Sapphire laser seed. In figure 13 we show the HGHG and the SASE single shot spectrum for the same electron beam condition. In order to show them on the same scale we need to multiply the SASE spectrum by 10^4. The fluctuation of the HGHG is 30%, as compared with the 100% of SASE for the same electron beam condition. The fluctuation of HGHG is due to the large electron beam fluctuation on the day of the measurement, while that of the SASE came from both the intrinsic fluctuation and the beam fluctuation, showing the significant improvement of the stability of the HGHG process.

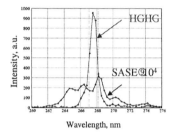

Fig. 13 The single shot spectrum of HGHG and SASE for the same electron beam condition in the 266 nm HGHG and SASE experiment on October 28, 2002.

The output reached 130 µJ with a sub-pico-second pulse length, showing deep saturation. The theory predicts a 3rd harmonic output at 88 nm to be 1% of the fundamental at 266 nm when saturation is achieved, so we expect about 1 µJ of 88 nm radiation in the output, which will become useful for many new types of experiments in chemistry. Hence our next step will be to measure and characterize this 88 nm output and start the user application of the DUVFEL. Actually, if we upgrade our electron beam energy from the present maximum of 200 MeV to 300 MeV, we can generate coherent radiation below 100 nm with more than 100 µJ per pulse, this will be an unprecedented high brightness radiation in this wavelength region which will open many new applications in different branches of chemistry.

In the meantime, we are considering a cascading scheme of a soft X-ray FEL program [19], as shown schematically in figure 14, as a part of the NSLS upgrade proposal. In this scheme, we start from a 266 nm seed, generate radiation at 53.2, 10.6, and 2.1 nm respectively through three HGHG stages, and finally amplify the 2.1 nm radiation to saturation at an output power of 1.7 GW.

One feature of this scheme is that it can produce output pulses as short as 20 fs, which will find many important applications in a wide range of fields. For the

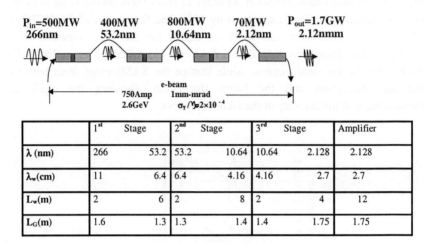

	1st Stage		2nd Stage		3rd Stage		Amplifier
λ (nm)	266	53.2	53.2	10.64	10.64	2.128	2.128
λ_w(cm)	11	6.4	6.4	4.16	4.16	2.7	2.7
L_w(m)	2	6	2	8	2	4	12
L_G(m)	1.6	1.3	1.3	1.4	1.4	1.75	1.75

L_{total}=36 m to reach 1.7 GW

Fig.14 The Schematic diagram for a soft-X-ray FEL
based on cascading HGHG concept

sake of space, we will not elaborate on the scheme here, and only point out that the input and output wavelengths of the three HGHG stages, their undulator periods, the gain lengths, and lengths of different sections are all given in the table, adding up to a total length of undulator of 36 meter. One of the technical challenges for this soft X-ray FEL is the time jitter between the laser pulses and the electron pulses. Since the electron pulse is about 500 fs and we need to use the different parts of this pulse several times during the cascading of several HGHG stages, we should reduce the time jitter to below 100 fs. At the present, the time jitter without feedback system is 500 fs. With a feedback system, the desired goal is possible, even though it is a significant challenge. Overcome this challenge will be one of the important steps towards our final goal: the temporally coherent X-ray FEL.

Acknowledgments

This work was carried out with the support of DOE Contract DEAC No. DE-AC02-98CH10886 and AFOSR/ONR MFEL Program #NMIPR01520375.

References

1. L. R. Elias, W.M. Fairbank, J.M.J. Madey, H.A. Schwettman and T. Smith, PRL 36, 717 (1976)
2. Y.S. Debenev, A.M. Kondratenko and E.L. Saldin, *Nucl. Instrum Methods Phys. Res* A**193**, 415 (1982).
3. R. Bonifacio, C. Pellegrini, and L. Narducci, *Opt. Commun.* **50**, 373 (1984).
4. The references on this subject can be found in, for example, G.T. Moore, Nucl, Instru. Meth. , A239, 19 (1985); E.T. Scharlemann, A.M. Sessler, and J.S. Wutele, PRL 54, 1925 (1985); J.M. Wang and L.H. Yu, *Nucl. Instrum Methods Phys. Res* A**250**, 484 (1986); K. J. Kim, *Nucl. Instrum Methods Phys. Res.* A**250**, 396 (1986); K.J. Kim, *Phys. Rev. Lett.* **57**, 1871 (1986); S. Krinsky and L.H. Yu, *Phys. Rev.* A**35**, 3406 (1987);
5. L.H. Yu, S. Krinsky, and R.L. Gluckstern, *Phys. Rev. Lett.* **64**, 3011 (1990). An earlier qualitative discussion on the scaling can be found in: C. Pellegrini, Nucl. Instru. Meth. A272, 364 (1988)
6. J.S. Fraser, et. al., Proc. 1987 IEEE Particle Accerator Conf. Cat. No. 87CH.2387-9 (March 1987) p.1705.

7. "Proceedings of the Workshop Prospect for a 1 Å Free-Electron-Laser", J.Gallardo, (Editor) Sag Harbor, New York, April 22-27, 1990, BNL 52273; "Workshop on Fourth Generation Light Sources", M. Cornacchia and H. Winick (Editors), SRRL report 92/02, Feb 24-27, 1992; "Report on the workshop Towards Short Wavelength Free Electron Lasers", I. Ben-Zvi and H. Winick (Editors), Brookhaven National Laboratory, May 21-22, 1993, BNL Report 49651; "Workshop on Scientific Applications of Short Wavelength Coherent Light Sources", W. Spicer, J. Arthur and H. Winick (Editors), SLAC Report 414, 1992; "Workshop on Scientific Applications of Coherent X-rays", J. Arthur, G Materlik and H. Winick (Editors), SLAC Report 437, 1994.
8. "Linac Coherent Light Source (LCLS) Design Study Report", The LCLS Design Study Group, Stanford Linear Accelerator Center (SLAC) Report No. SLAC-R-521, 1998.

9. "A VUV Free Electron Laser at the TESLA Test Facility at DESY: Conceptual Design Report", DESY Print, June 1995, TESLA-FEL 95-03
10. M. J. Hogan et al., *Phys. Rev. Lett.* **22**, 4867 (1998).
11. M. Babzien, et. al. , Phys. Rev. E, 57, 6093 (1998)
12. S.V. Milton et al., Phys. Rev. Lett., **85**, 988 (2000).
13. A. Murok, et. al., P.2748, "Proceedings of the 2001 Particle Accelerator Conference", Chicago; The ATF beam parameters can be found in , V. Yakimenko, et. Al., *Nucl. Instr. Methods Phys. Res.* A**483**, 277 (2002).
14. V. Ayvazyan, et. Al., PRL 88, 10, 104802(2002)
15. A. Doyuran, et. al., to be published in Nucl. Instru. Meth. (2002)
16. J. Feldhaus et al., Optics Communications 140 (1997) 341
17. I. Ben-Zvi, L.F. Di Mauro, S. Krinsky, M. White, L.H. Yu, *Nucl. Instr. Methods Phys. Res.* A**304**, 151 (1991).
18. L.H. Yu, *Phys. Rev* A**44**, 5178 (1991)
19. Juhao Wu, and Li Hua Yu, "High Gain Harmonic Generation x-ray FEL", 2716, Proceedings of PAC2001, Chicago (2001)
20. L.H. Yu et. al., Science, 289 (2000) 932
21. A.Doyuran et. al., Phys. Rev. Lett. 86, 5902 (2001)

PROTON THERAPY: ACCELERATOR ASPECTS AND PROCEDURES

HANS-UDO KLEIN, DETLEF KRISCHEL

ACCEL Instruments GmbH
Friedrich-Ebert-Str. 1
D-51429 Bergisch Gladbach
Germany
E-mail: klein@accel.de

Based on advanced accelerator technology proton therapy (PT) is going to become a powerful tool in cancer treatment. We describe a PT equipment based on a superconducting 250 MeV proton cyclotron. Superconductivity is used to provide advantageous design options for the essential components within the cyclotron. The main subsystems of the accelerator equipment from the proton source to the patient rooms with some so-called gantries are described as well as the principles of scanning the beam over a maximum field size.

1. Purpose of a proton therapy facility

A proton therapy (PT) facility is being developed and built with the aim of treating and, preferably, destroying tumors in the human body by means of proton radiation. More precisely the goal is to bring particle dose exactly into the tumor volume, exactly in dimension, in intensity, and in time. The principle is shown in Fig. 1. Protons penetrate human tissue with little interaction until a maximum of the particle energy i. e. of radiation dose is deposited within a relatively small volume, described as the Bragg peak. Varying the energy of the protons over a suitable range will move the Bragg peak in depth to become a Spread-Out Bragg Peak (SOBP), thus allowing to irradiate the full tumor volume.

The much more advantageous depth-dose behaviour of protons in contrast to that of photons is the basis to the already proven treatment successes of proton therapy to reach a much better conformity with the tumor and by far less irradiation of healthy tissue /1/.

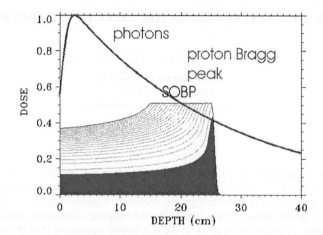

Figure 1. Dose distribution over depth in tissue for photons and protons [Courtesy: PSI /2/]

2. Design of the facility

As an example for PT equipment we go along the several subsystems of the Rinecker Proton Therapy Center (RPTC), actually being built in Munich, Germany. The layout of the accelerator configuration is given in Fig. 2. The building in which the RPTC is housed can be separated roughly into two parts: one part contains the plant for the creation of the high energy proton radiation as well as the area for its application (equipment section); the remaining area houses all diagnostic facilities, doctor's rooms, patient preparation rooms and administration.

The technical core of the Proton Therapy System (PTS) for RPTC is a superconductng cyclotron for the acceleration of protons to a kinetic energy of 250 MeV (Figure 3). This fixed extraction energy of the proton beams will be tuned to the necessary energy for irradiation by using a so-called energy selection system (ESS) that slows down the proton beams within graphite blocks and selects particles of the requested energy range by slit systems and an achromatic magnet selection. The conditioned proton beams are then distributed in sequence to the 5 treatment rooms through suitable beam transfer lines. Four of the five treatment rooms are equipped with so-called gantries which allow the proton beam to be directed to the patient from all angles. The fifth room is for the treatment of eye tumors. For other PT facilities also synchrotrons are used as the proton accelerator.

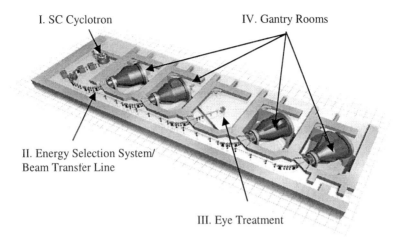

I. SC Cyclotron

IV. Gantry Rooms

II. Energy Selection System/
Beam Transfer Line

III. Eye Treatment

Fig. 2: Layout of the RPTC Proton Therapy Facility

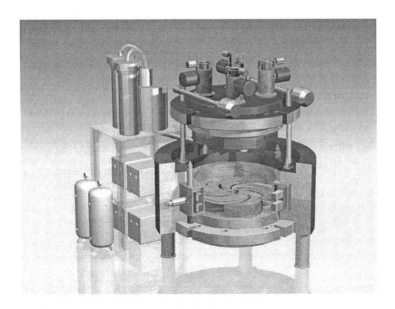

Fig. 3 CAD View of Superconducting 250 MeV Proton Cyclotron

3. Equipment for production, transfer and application of the beam

3.1. *The cyclotron*

A so-called Cold-Cathode-Source in the centre of the cyclotron serves as the proton source. The protons are accelerated over acceleration distances which are provided by four resonators, the so-called Dees. By a particular shape of the magnetic field, that guides the particles in spiral trajectories, one achieves that the protons cycle with the same frequency at all energies. This is the so-called isochronicity condition and the particles remain synchronous to the RF-acceleration field. Superconducting solenoidal coils are used for the creation of the magnetic field. They are encased within a cylindrical cryostat, which in turn is positioned within magnetic steel. The exact shape of the magnetic field is achieved by contouring of the magnetic steel. Pole tips seen from above the cyclotron exhibit a four-fold symmetry. One can detect two distinct areas: in the so-called hills the magnetic steel reaches almost to the mid-plane of the cyclotron and leaves a vertical gap for the beam; in the so-called valleys the magnetic steel recedes considerably and leaves room for the previously mentioned dees. The acceleration is done continuously to the final energy of 250 MeV. The accelerated particles are then deflected in two steps by two electrostatic deflectors. With the aid of an arrangement of so-called focussing bars they are guided through the extraction channel to the outside.

Main characteristics and features of the cyclotron are:
- Internal ion source
- High extraction efficiency of more than 80 % (in multi turn extraction mode), resulting in a moderate activation of the components
- Creation of the magnetic field with reliable, low-energy consuming super-conducting coils made of NbTi
- The iron yoke and all cyclotron components outside of the cryostat are <u>at room temperature</u>
- Compact construction (outer diameter of yoke rings is only about 3 m)
- Self-sustained cryosystem with automatic reliquefaction of the helium. No handling of cryogenic liquids is necessary
- Automatic operation via a control system
- The modular construction allows quick and simple access by means of an automatic lifting system of the upper iron yoke. All connections to the inside of the iron yoke are shaped in such a way that no joints have to be removed before opening of the yoke

- The superconducting coils allow a higher field strength as well as a bigger gap than normal conducting copper coils and therefore a higher extraction efficiency, more generous spatial conditions and more reliable operation
- The superconductive coils operate under constant ambient conditions, no 'warming-up' phase for the magnet is necessary as is unavoidable for large normal conducting magnet coils
- The cryosystem and the cryostat remain cold even during maintenance, resulting in simple and fast maintenance procedures
- Automatic operation via the control system, no personnel is required during normal operation.

3.2. Degrader and Energy-Selection-System

Degrader and Energy-Selection-System (ESS) serve for the selection of the particle energy requested to irradiate at a certain depth in tissue. To achieve that, the energy of the proton beam is reduced from 250 MeV to values between 70 and 250 MeV by the introduction of material (carbon) blocks of various thickness within the so called degrader. Due to scattering effects the degrader not only lowers the energy, but also influences other properties of the beam. The beam optics before that is designed in such a way that the increase of beam emittance is minimised, resulting in a maximisation of the useable part of the beam that is created within the cyclotron.

The beam that leaves the degrader is not mono-energetic. The ESS, consisting of dipole magnets, quadrupole magnets and a system of beam slits, reduces the energy distribution to a quasi mono-energetic beam. The selection of energy is through an adjustable vertical slit between the deflection magnets.

For the control of the beam characteristics a sufficient number of measuring devices is used, checking the beam position, the current and the profile, at selected locations within the conditioning and guide system of the beam. The precise shape of this measurement system is determined in the design phase of the degrader sub-system.

The ESS contains a Faraday cup for the control of the proton beam intensity. A kicker magnet directs the beam onto that Faraday-Cup. This only happens during fractions of a second during the changing of the beam energy. This Faraday cup can also be inserted into the path of the beam during emergency stops. The degrader is located within an evacuated tank.

3.3. *The beam transfer lines*

The beam transfer lines direct the beam into a particular treatment room. Figure 4 shows an example of a beam transfer line for similar purposes. The beam tubes are under complete vacuum.

Choosing a treatment room is simply done by the energisation of the respective dipoles. The beam optics is adjusted in such a way that the beam is rotationally symmetric at the entrance to the gantry. This makes the gantry beam optics independent of the angular location of the gantry.

The safety concept includes beam-stops before each treatment room. An inadvertently arriving beam will thus be totally blocked. Each stopper remains in the beam path until the control system of the selected treatment room reports beam-readiness and a request for beam is made.

A series of control magnets allows, together with monitoring equipment for the beam position, a fine tuning of the horizontal and vertical position of the beam. The proton beam is guided either into one of the four gantry rooms or into the room with fixed horizontal beam guidance.

Figure 4. Example of a typical proton/ light ion beam transfer line, COSY FZ Julich

3.4. *Beamlines and Beam control in the gantry room*

In the gantry rooms the gantry optics provide a beam with various diameters und round cross section at a well defined reference point, the iso-centre. There the beam is guided, via the gantry structure and with the aid of dipole and quadrupole magnets, onto any part of the patient by rotation of the gantry. The gantry rotation can be controlled in steps of 0.1 degree and a maximal deviation from the position of the rotating centre to the isocentre of 1 mm. In Figure 5 several options for the possible layout of a gantry are given. The beam optics allow the tuning of the beam profile, necessary for the therapy. The present design makes it possible to create a smallest beam of 3 mm (1 σ) radius at the location of treatment.

Eccentric compact gantry (A)

New centric compact (B)

Long throw gantry (C)

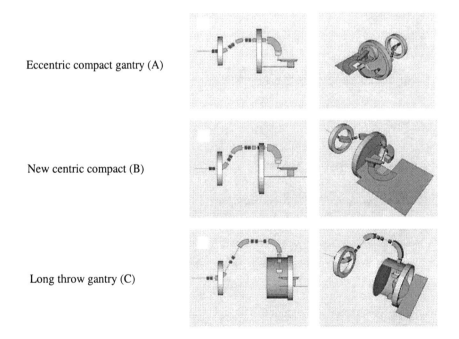

Figure 5. Gantry Options [Courtesy: PSI /2/]

Behind the last bending magnet of the gantry, which directs the beam perpendicular to the gantry axis, is the gantry nozzle with the scanning system. The purpose of the scanning system is to distribute the focussed beam, in a controlled way, over a fixed area. It consists of two orthogonal scanner magnets and is capable of irradiating an area of 30 x 40 cm, a field size sufficient for almost all tumors. Figure 6 provides a sketch of the nozzle elements.

The delivered dose is verified for each beam position, with the aid of space-resolving detectors within the nozzle, by comparing it with the pre-determined irradiation pattern. For high precision irradiation of small areas the nozzle may be fitted with a fixation mechanism for shutters and collimators.

The beam path is evacuated up to behind the scanner magnets. When it leaves the thin exit window the beam travels the short distance to the patient through air. For special irradiation requirements it may pass collimators or shutters where it can be partially absorbed. In general, however, the scanning method does not include scattering or absorbing elements within the beam path.

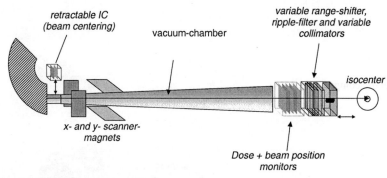

Fig. 6: Layout of a Scanning Nozzle

3.5. *Beam guidance in the room with horizontal beam*

In the room with a horizontal fixed beam line only small areas in the head region (almost exclusively eyes) are irradiated. When in therapy position, a passive scattering system guarantees a homogenous proton dose over a diameter of about 40 mm. In order to keep the beam position stable and to create a homogenous source the beam in the horizontal-beam-nozzle is being focussed onto a scattering foil made of tantalum. Through that an angular distribution with a half-width of about 12 mrad is created. The non-central parts of the proton beam are stopped by carbon collimators. A time-averaged uniform distribution of the depth dosage is being created by a rotating modulator wheel. The horizontal-beam-nozzle is not evacuated. The beam leaves the vacuum tube through an exit window behind the last quadrupole magnet. During irradiation the proton beam is stopped in the irradiated tissue.

4. Integrated Control System

All the subsystems and operations as described above must be carefully steered and controlled by an integrated control system. Figure 7 gives the general architecture of such a system.

During a Treatment Planning (TP), independently from the operations but consistent with the available parameters and features of the equipment, e. g. irradiation angles of the gantry, the medical doctor must define the required dose distribution over the tumor volume. The TP will provide the required input for the Treatment Control in the respective gantry room for the individual patient. Beam parameters are controlled by the Accelerator and Beam transport Control System. Beam must not be applied before the Scheduler is requesting and allowing for it. Safety regulations are imposed independently over all hardware systems, mostly hard wired, including a fast interlock system.

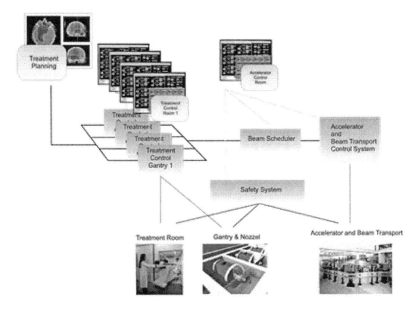

Figure 7. Integrated Control System for a Proton Therapy facility

References

/1/ Suit, Herman, The Gray Lecture 2001, Coming Technical Advances in radiation Oncology

600

/2/ PSI - Team and PSI web site,
http://radmed.web.psi.ch/asm/gantry/gantry_master.html

INTRODUCTION TO SYNCHROTRON RADIATION APPLICATIONS

H.O. MOSER, O. WILHELMI, P. YANG

Singapore Synchrotron Light Source
National University of Singapore
5 Research Link
Singapore 117603
E-mail: moser@nus.edu.sg

Synchrotron radiation as a tool for materials and process characterization and modification has a huge range of applications covering many disciplines from chemistry to engineering, environmental science, life sciences, materials science, and to physics. Within the limited scope of this book only a few very common applications are presented.

1. Introduction

When synchrotron radiation impinges on matter it interacts with the electric charges and the magnetic momenta. The interaction can be elastic or inelastic. As a result, radiation is scattered and/or absorbed, excitations may be created, molecules ionized or dissociated, and particles released.

In the case of elastic interaction, the scattering of radiation includes phenomena described as interference, diffraction, refraction, all depending on the geometric structure of the material, i.e., the spatial distribution of the electric charges and magnetic momenta.

In the case of inelastic interaction, the incoming radiation may be partly or completely absorbed and eventually lead to the creation of fluorescence radiation, the release of photoelectrons, or the creation of excited states and of quasiparticles such as plasmons and polaritons. Furthermore, if the radiation dose is high enough, radiation damage may ensue, including the photodissociation of molecules, the scission or crosslinking of polymer chains, and the creation of free radicals. The inelastic interaction may also only modify the energy of the incident photons when they interact with the quasiparticles in a solid such as phonons.

All these effects are exploited in a panoply of methods to characterize materials, and of processes to modify materials. Some of them are workhorses that can be used at almost any facility. Others are more sophisticated and require the specific performance of a given facility so that they can be found only in a few places.

In this lecture, we select examples of applications that are available at SSLS and at the majority of other sources, too. For reviews of more sophisticated state-of-the-art methods and their applications the reader may refer to the websites of sources worldwide as given in the survey D1.2.

2. Beamlines and experiments

Although there is a huge variety of beamlines and experimental stations that reflects the great many of methods and ways to implement them, some underlying principles governing the set-up of beamlines can be described.

The basic objective for a beamline is to match the source to the sample, more precisely, the photon beam emittance to the sample acceptance, to perform the required spectral selection if any, and to couple the sample environment physically to the source point in the UHV vacuum of the storage ring. Moreover, it has to fulfill safety functions in terms of personnel radiation safety and operational safety.

Ideally, the volume of the source phase space should be a subset of the sample acceptance phase space volume as defined by considerations such as maximum tolerable angular range for diffraction or for small angle scattering experiments or the size of the sample. In reality, it is rarely the case that all photons offered by the source can be channeled onto the sample. Quite often, the progress made in reducing the source phase space volume at constant photon flux is traded off against reductions in sample size.

As illustrated in Figure 1, the optical components to actually shape the phase space volume are mirrors and monochromators. The basic functions of mirrors in a beamline are to collect as much as possible of the source flux, to relay the photon beam collected through slits and other apertures along the beamline, and

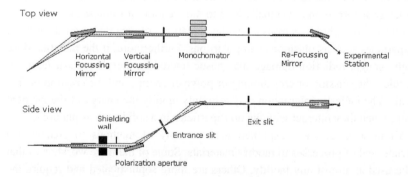

Figure 1: Optical schematic of a soft X-ray beamline equipped with 4 interchangeable spherical gratings in the monochromator (SINS beamline at SSLS). Overall length from source to sample is about 21 m.

to deliver an either focused or collimated beam on the sample. Slits or other apertures only passively cut out a certain region of the phase space, thereby reducing photon flux. Furthermore, monochromators primarily cover the function of spectral selection, but frequently have focusing properties as well.

Depending on the spectral range, the optical elements look and are used differently. In the hard X-ray range above a few keV, mirrors work in grazing incidence to exploit total reflection for good reflectivity. Monochromators use Bragg reflection from suitable crystals, mostly Si. In the soft X-ray range below, and extending to a few tens of eV, mirrors still work in grazing incidence, but monochromators use gratings, in grazing incidence as well. In the Ultraviolet and Visible, near-to-normal incidence can be used for mirrors as well as for gratings. Finally, in the Infrared, near-to-normal incidence is used for mirrors whereas the spectral selection is performed using Fourier Transform Interferometers that are basically Michelson interferometers.

3. Internal structure of materials – PCI

Using X-rays to look through objects is Wilhelm Conrad Roentgen's discovery, made in 1895, that won him the Nobel Prize in 1901. In his first pictures, the contrast is mainly formed through various absorptions of materials such as bones, muscles, soft tissue, and the wedding ring of Mrs. Roentgen (Figure 2).

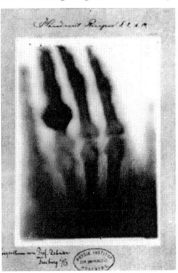

Figure 2. Roentgen discovered X-rays on November 8, 1895, in Würzburg and took this photo of the hand of his wife, Anna Bertha Roentgen, on December 22, 1895.

However, X-rays are not only absorbed, they are scattered and may show diffraction and refraction phenomena that are ultimately caused by phase shifts

604

among waves, hence the name phase contrast imaging (PCI). As phases depend on the spatial variations of the refractive index that translate into electron density and further into atomic density and atomic charge, phase contrast imaging gives insight into variations of atomic density and composition. In fact, all contrast mechanisms contribute to the imaging simultaneously as illustrated in Figure 3. It

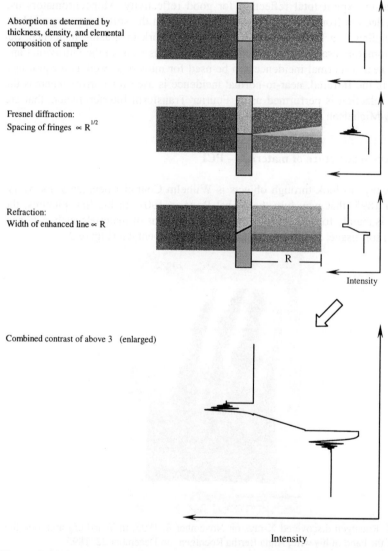

Figure 3. Overview of contrast formation mechanisms in phase contrast imaging – absorption, Fresnel diffraction, refraction, and their superposition.

is only a matter of the specific sample and the experimental set up which one will prevail. At SSLS' PCI beamline, the distance from source to sample is 16.71 m and from sample to detector between 2 and 3 cm, respectively.

For soft matter, a low divergence incident beam, and an optimized distance between sample and detector, refractive contrast will dominate. Figures 4 and 5 show examples of hydrogen bubbles during electroplating[1] and the head of an ant.

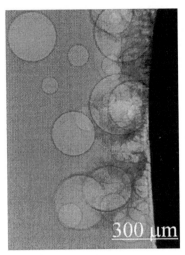

Figure 4. Hydrogen bubbles formed during electroplating of zinc on copper (courtesy of Nature).

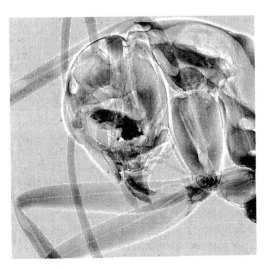

Figure 5. Head of an ant taken at SSLS (frame size is about 3 mm).

Features such as the dark ring delineating the bubbles in the electrolyte or the vessels in the legs of the ant as well as the white lines at the edges of the ant's legs that are caused by total reflection are characteristic and can be explained on the basis of Snell's law. Figure 6 shows a schematic for the case of the hollow sphere.

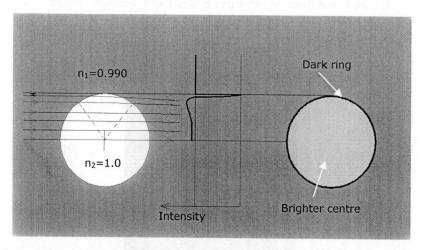

Figure 6. Schematic explanation of the dark ring at the edge of a bubble. Even incident rays hitting the bubble close to the equator are still deflected by refraction, no ray can reach the dark ring. The bubble picture on the right-hand side of the illustration is rotated by 90°.

The interest of phase contrast imaging lies in the visualization of the internal unknown structure, more precisely, the distribution of refractive index, in objects that are opaque to visible light, and in the time-resolved monitoring when such structures undergo changes.

4. Atomic distances in crystalline materials – XRD

4.1. Laue diffraction

Measuring the distance between atoms in a crystal was Max von Laue's famous discovery in 1912 that won him the Nobel Prize in 1914. As we are already familiar with, X-rays, on their way through a crystal, are scattered by the electrons. When the atoms are regularly spaced as in a crystal, interference phenomena become visible, i.e., X-rays are only diffracted into specific directions in which the scattered waves constructively interfere.

As illustrated in Figure 7, the mathematical condition for constructive interference at a 3D regular arrangement of atoms was cast by von Laue into three equations that bear his name

$$\bar{a}_1 \cdot (\bar{s} - \bar{s}_0) = h\lambda$$
$$\bar{a}_2 \cdot (\bar{s} - \bar{s}_0) = k\lambda \qquad (1)$$
$$\bar{a}_3 \cdot (\bar{s} - \bar{s}_0) = l\lambda$$

Here $\bar{s}, \bar{s}_0, \bar{a}_1, \bar{a}_2, \bar{a}_3$ are unit vectors of the scattered and incident light and the primitive vectors subtending the lattice, respectively, λ is the wavelength, and h, k, l are integers. The three vectors $\bar{a}_1, \bar{a}_2, \bar{a}_3$ define an affine coordinate system

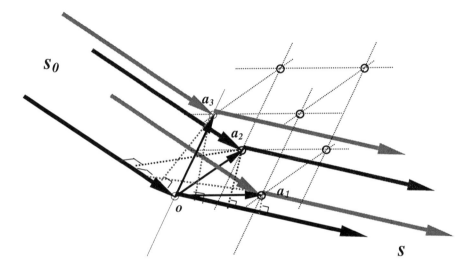

Figure 7. Schematic of Laue diffraction. For constructive interference, the optical path difference must be an integer multiple of the wavelength λ in all three directions given by $\bar{a}_1, \bar{a}_2, \bar{a}_3$, thus satisfying Eqs. (1).

which describes the crystal lattice in real space. Their reciprocal vectors are $\bar{b}_1, \bar{b}_2, \bar{b}_3$ as given by Eqs. (2).

$$\bar{b}_1 = \frac{\bar{a}_2 \times \bar{a}_3}{\bar{a}_1 \cdot \bar{a}_2 \times \bar{a}_3}$$
$$\bar{b}_2 = \frac{\bar{a}_3 \times \bar{a}_1}{\bar{a}_1 \cdot \bar{a}_2 \times \bar{a}_3} \qquad (2)$$
$$\bar{b}_3 = \frac{\bar{a}_1 \times \bar{a}_2}{\bar{a}_1 \cdot \bar{a}_2 \times \bar{a}_3}$$

The reader may check that $\vec{a}_i \cdot \vec{b}_j = \delta_{ij}, i, j = 1,2,3$ where δ_{ij} is the Kronecker symbol.

The reason for introducing the reciprocal vectors is that they allow us to readily find a solution for the scattering vector $\vec{s} - \vec{s}_0$. Expressing it as a linear combination of the reciprocal basis vectors

$$\vec{s} - \vec{s}_0 = \lambda(h\vec{b}_1 + k\vec{b}_2 + l\vec{b}_3) \tag{3}$$

we can recover von Laue's equations when we multiply $\vec{s} - \vec{s}_0$ with each of the basis vectors successively. The important content of equation (3) is that every vector from one lattice point to another in the reciprocal lattice, namely, $h\vec{b}_1 + k\vec{b}_2 + l\vec{b}_3 = \vec{G}$, is a possible scattering vector. Together, the incident vector $\vec{s}_0$ and the scattered vector $\vec{s}$ describe a sphere. Those points of the reciprocal lattice that lie on the sphere are then possible Laue reflexes. The sphere is called Ewald sphere (Figure 8). Drawing the Ewald sphere in the reciprocal lattice offers an elegant way to find the diffracted reflexes that is rather difficult otherwise.

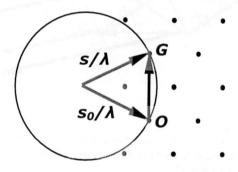

Figure 8. Ewald sphere in reciprocal lattice. The radius of the sphere is $1/\lambda$, the origin of the reciprocal lattice is at **O**. For all reciprocal lattice points falling on the sphere the scattering vector equals a reciprocal vector, i.e., $(\vec{s} - \vec{s}_0)/\lambda = \vec{G}$, and a diffracted ray is possible.

Figure 9 shows the original diffractogram as taken by W. Friedrich and P. Knipping who did the experimental work for von Laue, and modern Laue pattern pictures. Obviously, Laue pattern indicate directly the crystal symmetry with respect to the axis of the incoming photon beam.

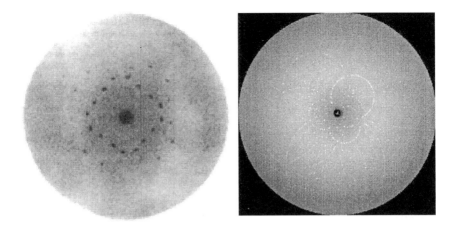

Figure 9. Laue diffractograms. Left: one of the first pattern obtained by Friedrich and Knipping from a CuSO$_4$ crystal[2]. Right: modern Laue pattern from a photo-active yellow protein[3].

Today, Laue diffraction is widely used for macromolecular structure determination, a technique that allows finding the positions of the order of between 10^4 to 10^5 atoms in a large protein molecule or a virus with a resolution down to about 0.7 Å. Laue pattern are usually recorded with light covering a certain bandpass of 1-2 Å width.

4.2. *Bragg reflection*

Instead of considering all atoms as individual scatterers and evaluating the interference of all scattered waves as it is done in the von Laue description, atoms can be thought to be grouped into planes, so-called lattice planes, and the interaction with the incoming X-rays described as a reflection. The interference condition is then given as a reflection leading to Bragg's law as

$$2d \sin \vartheta_B = m\lambda \qquad (4)$$

wherein λ is the wavelength, d the distance between the equally oriented and spaced lattice planes, ϑ_B the Bragg angle enclosed between the incident rays and the lattice planes, and m an integer (Figure 10). This relationship won father and son W.H. Bragg and L.W. Bragg the Nobel prize in physics in 1915.

Von Laue's and Bragg's descriptions are equivalent as we are showing now. Referring to the Ewald's sphere sketch (Figure 8 and 11) we see that the scattering may be interpreted as reflection of the incident ray at a plane perpendicular to vector $\vec{G}$ (Figure 10). Thus, the scalar product between a vector describing that plane and $\vec{G}$ must be zero

$$(\vec{s} - \vec{G}/2) \cdot \vec{G} = 0 \qquad (5)$$

Equation (5) translates into

$$2\vec{s} \cdot \vec{G} = G^2$$

The mutual distance of planes parallel to $\vec{s} - \vec{G}/2$ through adjacent reciprocal lattice points in real space is $1/G$. We then obtain

$$2 \cdot \frac{1}{m\lambda} \cdot \frac{1}{d} \cdot \cos(\frac{\pi}{2} + \vartheta_B) = \frac{1}{d^2}$$

and retrieve, with $\cos(\frac{\pi}{2} + \vartheta_B) = \sin\vartheta_B$, Bragg's law of equation (4).

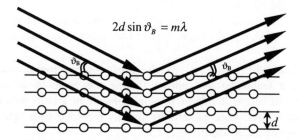

Figure 10. Bragg description of X-ray diffraction as reflection at lattice planes. λ is the wavelength, d the distance between the equally oriented and spaced lattice planes, ϑ_B the angle enclosed between the incident rays and the lattice planes, and m an integer.

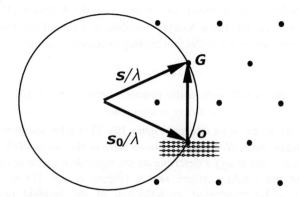

Figure 11. From Ewald to Bragg in reciprocal space. The reciprocal vector G is perpendicular to the Bragg planes. Its length is the inverse of the spacing ($|G|=1/d$)

4.3. *Powder diffraction*

A third way to exploit diffraction for structure determination is the so-called powder diffraction from polycrystalline materials which is due to Peter Debye and Paul Scherrer (1916). As illustrated in Figure 12, in a powder, a great many

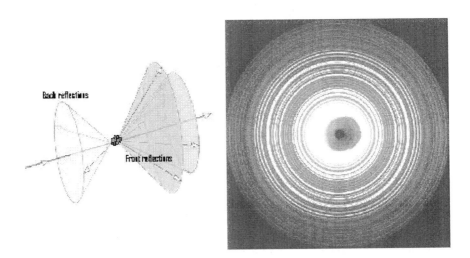

Figure 12. Debye-Scherrer method for powder diffraction. When X-rays are scattered by a powder of poly-crystalline grains their Bragg reflexes describe cones around the incident photon beam. Left: schematic of the method[4]; Right: pattern from olivine powder[5]

individual grains of a specific material are randomly oriented which makes their Bragg reflexes describe cones around the incident photon beam axis and circles in a detector plane perpendicular to the photon beam axis.

Powder diffraction is the workhorse for the characterization of phases in a wide variety of materials in physics, chemistry, metallurgy, geology, bio-science, and, in particular, in engineering and material science. It enables even *ab initio* structure determination, mainly thanks to the Rietveld refinement method[6].

5.   Short-range order and valency – XAFS

When a photon is absorbed by an atom it can release an electron from a bound state into the continuum if its energy is large enough. The characteristic energy levels of the electrons in an atom translate into so-called absorption edges, i.e., the minimum photon energies beyond which a photon can release an electron from a certain energy level. A typical X-ray absorption spectrum showing the K edges – the electron is released from the K shell – of carbon, nitrogen and oxygen is illustrated in Figure 13.

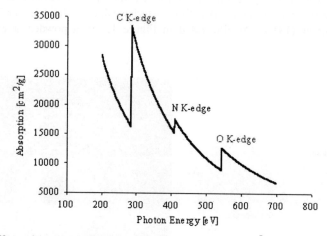

Figure 13. X-ray absorption edges of carbon, nitrogen, and oxygen[7].

When the absorption edges are measured with a higher resolution the X-ray absorption fine structure (XAFS) becomes visible (Figure 14). Usually, two regions are distinguished, one within about ±10 eV from the edge and another extending to about 500 eV beyond the edge. In the immediate neighbourhood of the edge the absorption of a photon does not lift the electron in the continuum, yet, but in higher lying bound states the energy of which is influenced by the

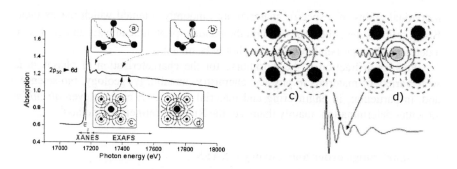

Figure 14. XAFS spectrum showing XANES and EXAFS region of U L_{III}-edge in $CaUO_4$. Insets a) and b) relate the multi-scattering and single scattering of the photoelectrons to corresponding spectral features whereas c) and d) illustrate the destructive and constructive interference of the outgoing with the back-scattered photoelectron, respectively. Schematics c) and d) are enlarged on the right-hand side. XANES: X-ray Absorption Near Edge Structure; EXAFS: Extended X-ray Absorption Fine Structure[8].

surroundings. The spectroscopy in this range below and close to the absorption edge is called X-ray Absorption Near Edge Structure (XANES). Figures 15 and 16 show examples for Ti and As compounds. Although theoretical approaches exist[9] XANES is frequently used in a fingerprinting mode, i.e., comparing the signal of the unknown sample with known cases.

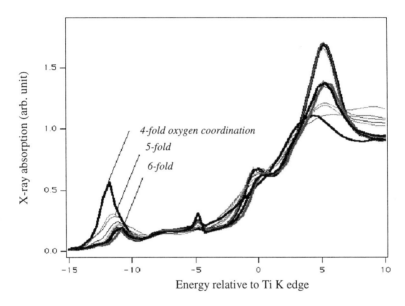

Figure 15. XANES at the Ti K-edge for an amorphous film of Ba, Sr, and Ti grown by metal organic chemical vapour deposition (MOCVD). As the ratio of Ba/Ti changes from 1/1 to 2/1the oxygen coordination around the Ti goes from 6-fold (octahedral) for $BaTiO_3$ over 5-fold in between to 4-fold (tetrahedral) forming amorphous Ba_2TiO_4 [10].

If the energy of the photon is high enough to release an electron an outgoing electron wave with wave number $k = \sqrt{2m_e E}/\hbar$ is created. E is the kinetic energy of the electron and is given by the photon energy minus the electron binding energy. This wave is scattered at the surrounding atoms and as the scattered waves interfere with the original wave, either constructively or destructively, the absorption curve exhibits a slight oscillation when the photon energy is varied. Upon Fourier transforming the absorption curve the radial distribution function of nearest-neighbour shells is obtained. This is called Extended X-ray Absorption Fine Structure (EXAFS). For details on EXAFS see references[11] (Figure 17).

614

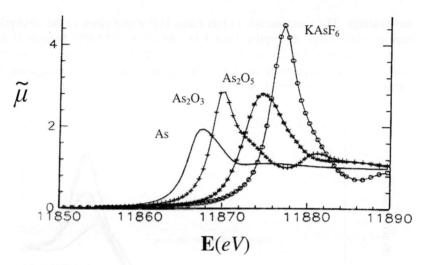

Figure 16. XANES showing the K-edges in As (—), As_2O_3 (+), As_2O_5 (*) and in $KAsF_6$ (°). The edge is shifted to higher energy with increasing valency of As[12]

XAFS is an inelastic method that can deliver its full power only with synchrotron radiation. It allows complete chemical speciation of dirty samples in

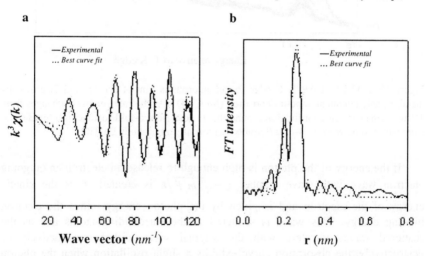

Figure 17. EXAFS spectrum of Pt/carbon nano-tube and the radial distribution function. a) k^3-weighted EXAFS oscillation $k^3\chi(k)$ for 2 wt% Pt/carbon nano-tube in wave vector (k) space; b) Fourier transform of the $k^3\chi(k)$ to distance (r)-space. Solid lines: experimental. Dashed lines: best curve fits obtained with Pt-Pt coordination number of 5.4, bond distance of 0.255 nm and Debye-Waller factor of 106 pm^2, with the assumption of face-centered-cubic packing and spherical cluster geometry[13].

air, quasi *in situ*. Speciation means determination of the element, its valency, and the phase of the sample. It also delivers the radial distribution function around the atoms of a selected element as a description of the short-range order. The range of applications of XAFS is very wide and includes catalysis [14], environmental analysis[15], chemistry[16], polymer engineering[17], surfaces and thin films[18].

6. Electronic and magnetic properties of materials – PES, XMCD

The huge variety of phenomena related to the electrical, optical, and magnetic properties of materials such as metals, semiconductors, and superconductors, is reflected in the electrons' spin-dependent energy-momentum distribution in the periodic potential of the crystalline solid[19]. While the comprehensive description of these phenomena makes use of concepts such as band structure, energy gap, and Fermi surface that were introduced about 70 years ago, it is only a few years that synchrotron-radiation based methods can visualize them directly. This was achieved by means of photoelectron spectroscopy (PES), in particular, when it is implemented as angle-resolved PES (ARPES), and magnetic dichroism in its two forms as either circular or linear dichroism (XMCD, XMLD).

As is well known about a free electron gas enclosed in a cube of edge L the kinetic energy ε_k of the electrons is related to their momenta by

$$\varepsilon_k = \frac{\hbar^2}{2m}(k_x^2 + k_y^2 + k_z^2) \tag{6}$$

The wave functions are sinusoidal and vanish on the wall of the cube. The components of the wave vector satisfy $k_i = \pm \frac{2\pi}{L} n_i$ where i stands for x, y, z and n_i is an integer with $n_i \geq 0$. The Fermi surface that, by definition, encloses all occupied energy states of the Fermi electron gas in the ground state as a whole and at absolute zero is a sphere in momentum space.

In a crystalline solid there is an additional constraint from the spatial periodicity of the electrostatic potential as caused by the ion cores. As a consequence there are no solutions for the wave function around integer multiples of the basis vectors of the reciprocal lattice, and, more so, there is a range of forbidden energy levels around these points in reciprocal space. The allowed and forbidden energy ranges are called bands and gaps, respectively. The Fermi surface is no longer a sphere, but becomes distorted.

The method of choice for measuring Fermi surfaces and band structures is angle-resolved photoelectron spectroscopy (ARPES). A typical experimental set up for ARPES is depicted in Figure 18. A tunable photon beam hits the sample and releases photoelectrons. Their yield is measured as a function of their kinetic energy and angle to the surface normal. The angle is related to their momentum parallel to the surface. The kinetic energy E_k, photon energy $\hbar\omega$, and the energy

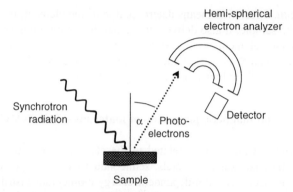

Figure 18. Schematic set-up of an ARPES experiment

required to release the electron, also called binding energy, E_b are connected via

$$E_k = \hbar\omega - E_b$$

Measurement of the photoelectron yield versus E_k and $k_\parallel$ for a given $\hbar\omega$ delivers E_b that is related to the energy E_i of the initial state of the electron and the ground state E_g via $E_b = E_g - E_i$. In this way, energy bands and Fermi surfaces can be directly mapped (Figure 19).

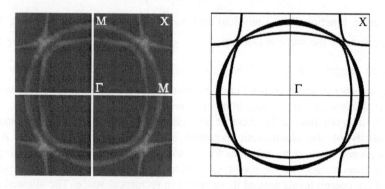

Figure 19. Study of the Fermi surface of the superconductor Sr_2RuO_4. Intensity map of an ARPES experiment (left) together with corresponding theoretical results (right)[20]

Pioneered by G. Schütz et al[21] XMCD is used for measuring and visualizing magnetization in materials that are important for magnetic data storage such as d-band transition metals and their compounds. Here, we focus on XMCD relying upon the reviews by J. Stöhr et al[22][23] and Sacchi and Vogel[24]. This method exploits the different absorption of either right-hand or left-hand circularly polarized photons (RCP, LCP) when, for example, p shell electrons in transition metals are excited into the valence d band in the presence of a magnetization. As illustrated in Figure 20 the clue is the difference in the density of states in the

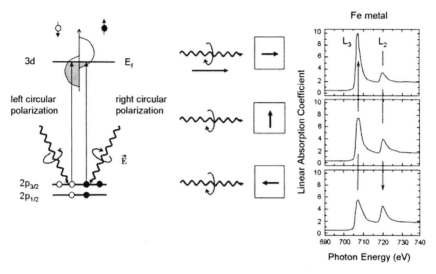

Figure 20. The schematic on the left shows the split of the d-valence band in a magnetic metal into spin-up and spin-down states. Since spin flips are forbidden in X-ray absorption, the measured resonance intensity directly reflects the number of d-band states of a given spin. In the experimental results at the right, the polarization of the X-rays was kept constant and the direction of the magnetic field was changed instead as indicated in the middle (right–hand circularly polarized photon and magnetization of the sample relative to the direction of photon propagation)[22].

valence band between spin-up and spin-down electrons and the dependence of the electron spin orientation on the polarization of the photon and on the L_3 or L_2 edge. LCP photons create more spin-up electrons at the L_3 edge and less at the L_2 edge than RCP photons and vice versa for RCP photons. As illustrated in Figure 21 the effect can be used as a contrast mechanism for photoemission electron microscopy to visualize areas of different magnetization on surfaces. The resolution is already good enough to show the single bits of a magnetically stored record.

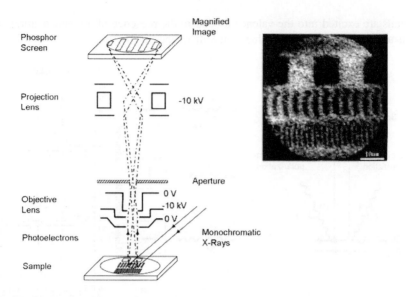

Figure 21. XMCD microscopy of a data-storage disk. Circularly polarized synchrotron radiation illuminates the sample and the photoelectrons are visualized by a PEEM (Photo Emission Electron Microscope). The polarization dependent photoelectron yield images the magnetic domains on the sample. The tunable energy of the synchrotron radiation makes the analysis element specific[22,25].

7. Dynamics of molecules – Infrared spectro- and microscopy

Synchrotron infrared radiation is a preferred tool for identifying and measuring molecules and for investigating low energy electron excitations in condensed matter[26]. Compared to the common globar source in standard laboratory Fourier-Transform Interferometers both, flux and brilliance, of the synchrotron IR can be substantially higher, even more so if the source point is positioned in the edge region of the magnetic dipole field where the magnetic field is about half the maximum value and its slope maximum. Figure 22 shows simulated data of the planned ISMI beamline at SSLS[27].

The enhanced flux and brilliance widen the range of accessible spectroscopic applications and enable IR microscopy. A very strongly benefiting field is life sciences. Taking microscopic images of living cells and tissues in a small bandwidth around a wavelength corresponding to an interesting spectral feature, e.g., a vibrational band, and comparing images made at a few different wavelengths, is used for diagnosis purposes on medical issues. This technique is called IR spectromicroscopy. Spectral features are themselves correlated to specific molecular groups of interest such as proteins, phosphates, amides, carbonate, and so on.

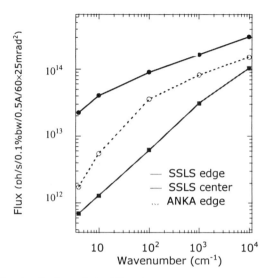

Figure 22. Simulated flux of the ISMI beamline edge source point compared with a source point in the center of the arc and with ANKA edge source point (left).

Results on the composition of a rat retina[28] as well as on dioxin-induced changes in human liver cells[29] are shown in Figures 23 and 24. Besides life sciences

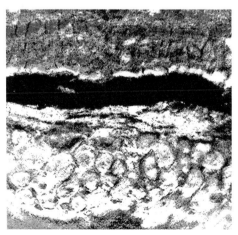

Figure 23. Cross section of a rat retina (image width ~500 µm) viewed with an infrared microscope/spectrometer. Imaging molecular chemistry of individual retina layers is made possible by combining microscopy and infrared spectroscopy (courtesy of Science).

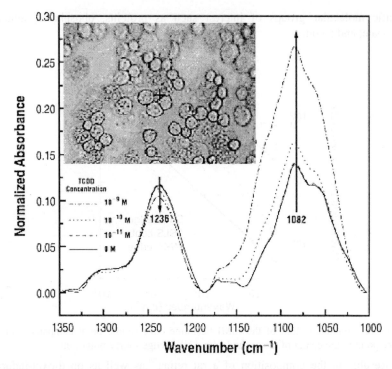

Figure 24. Dioxin-induced changes in human liver cells. FTIR spectra near phosphate bands show differences for cells treated with varying concentrations of TCCD (graph). The inset shows HepG2 cells (human liver tumor) as seen through an IR microscope. Spot size for IR spectromicroscopy: 10 μm.

there are many applications in other fields that have also benefited from IR spectro/microscopy including cosmetics, geochemistry, surface science, molecular physics, and solid state physics.

8. Micro- and nano-manufacturing – DXRL

Deep X-ray lithography combined with electroplating and plastic molding as replication processes has strongly impacted micro- and nanomanufacturing since about 20 years when the LIGA process was pioneered by Becker et al.[30].

Figure 25 illustrates the basic process steps including mask making, substrate preparation, irradiation, development, electroplating, and hot embossing. As shown, the LIGA process starts with the batch type process characteristic for integrated circuit manufacturing and merges into the classical high-volume manufacturing methods like plastic molding.

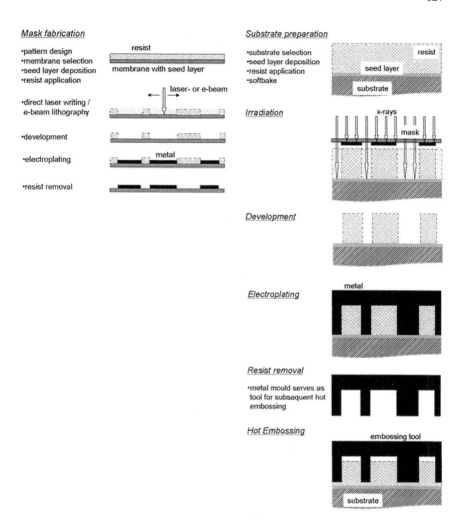

Figure 25. Schematic of LIGA process from mask making (top left) over substrate preparation and irradiation to hot embossing (right column).

Exploitation of the LIGA process has led to a large variety of mechanical, electromechanical, fluidic, and optical microstructures that are characterized by a high aspect ratio and a broad choice of materials including metals like Ni, Cu, Au, NiFe, NiCo, and many plastics suitable for transfer-injection or hot embossing.

For an X-ray wavelength of less than 1 nm and structure details of larger than 1 μm DXRL may be looked at as essentially a shadow-casting exposure technique that can be used to generate deep structures (up to more than 1 mm) owing to the small divergence angle of synchrotron radiation.

While the single exposure process is like an extrusion of the 2D mask pattern, thus producing cylindrical structures of almost arbitrary cross section, the potential to create 3D structures is greatly enhanced when inclined irradiation combined with rotation and wobbling are considered[31]. Figure 26 illustrates the simple inclined irradiation. This technique has been used to produce photonic band-gap structures[32] (Figure 27).

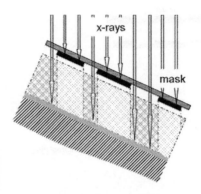

Figure 26. Inclined irradiation

Figure 27. Photonic band gap structures produced by inclined irradiations with rotation in between[32].

Figure 28 shows a separation nozzle structure that was the original reason to develop the LIGA process. More examples of microgears, micromotors, and a turbine-driven microcutter for plaque removal in arteries are depicted in Figure 29 (see ref. [33, 34]).

One of the approaches to go from the micrometer to the nanometer range is explained in Figure 30. It is called super-resolution process and makes use of the diffraction effects that become more visible when the mask structures go down to a few 100 nm size, as well as of the dose-dependent solubility of resist in a developer[35]. As illustrated, the actual dose profile can be tailored to be a sharp

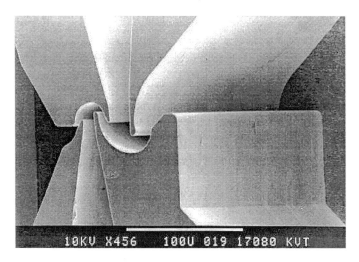

Figure 28. Separation nozzle as developed by FZK.

Figure 29. Gears, motors (IMM), and cutters (IMT/FZK) manufactured with the LIGA process.

peak within the opening of the clear mask feature. Setting the resist development level to a relatively high dose level, resist features can be made substantially smaller than the mask features. In this way, 380 nm holes on an Au mask were reduced to 180 nm holes in PMMA resist.

Finally, the diagnostic means offered by phase contrast imaging can also be applied to micro-manufacturing. Figure 31 shows microstructures made from SU-8 negative resist on top of a Si wafer as a substrate. In contrast to SEM pictures PCI allows to look "down" the whole hollow cylinder along its axis and to see imperfections of wall thickness and formation of skins. The technique is being developed at SSLS as a valuable diagnostic tool for the improvement and control of micro- and nano-manufacturing processes.

624

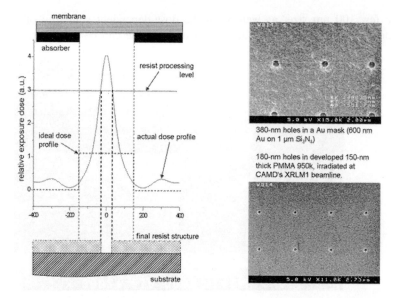

380-nm holes in a Au mask (600 nm Au on 1 µm Si₃N₄)

180-nm holes in developed 150-nm thick PMMA 950k, irradiated at CAMD's XRLM1 beamline.

Figure 30: The schematic on the left hand side outlines the principle of the super-resolution process. The curve is the simulated exposure dose in PMMA behind a 300 nm mask feature in 80 µm distance. By selection of an appropriate resist processing level a 4× feature size reduction can be achieved. The right hand side shows micrographs of first experimental results. The 380 nm holes of a mask were reduced to 180 nm holes in resist.

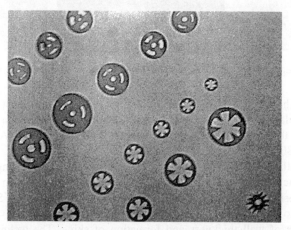

Figure 31: Transmission phase contrast images of 250 µm high SU-8 resist structures on a 525 µm silicon wafer. The largest cylindrical structures have an outer diameter of 150 µm, the smallest ones of 50 µm. PCI allows to investigate the inside of hollow high aspect ratio microstructures.

9. Conclusion

The applications outlined refer to the determination of the microscopic imaging of biological and technological matter, the long-range and short-range order in crystalline and amorphous matter to sub-atomic resolution, the measurement of the valency states in samples quasi *in situ*, the determination of band structures and Fermi surfaces in solids, the measurement of atomic magnetization and its use for microscopic imaging, and the manufacturing of micro and nanostructures. Although they cover a wide range of topics and methods, these applications represent but a small fraction of what is possible with synchrotron radiation overall.

Acknowledgments

Work supported by NUS Core Support 2002/03. The authors are grateful to Mr. Zheng Hongwei for help with some of the figures.

References

1. W.L. Tsai, P.C. Hsu, Y. Hwu, C.H. Chen, L.W. Chang, J.H. Je, H.M. Lin, A. Groso, G. Margaritondo, Nature, vol 417, 9 May 2002, p. 139; Y. Hwu, Wen-Li Tsai, A. Groso, G. Margaritondo and Jung Ho Je, J. Phys. D:Appl. Phys. 35(2002)R105 – R120.
2. http://detserv1.dl.ac.uk/Herald/images/cuso4.gif.
3. http://www.fys.ku.dk/~axel/.
4. http://www.matter.org.uk/diffraction/x-ray/powder_method.htm.
5. http://iucr.sdsc.edu/iucr-top/cif/mmcif/ndb/cbf/.
6. H.M. Rietveld, Acta Cryst. 22(1967)15, and J. Appl. Cryst. 2(1969)65.
7. http://unicorn.mcmaster.ca/stxm-intro/STXM_poly1.JPG.
8. P.Behrens, Trends in Analyt. Chem. 11(1992)218,
http://www.fz-rossendorf.de/FWR/XAS/CaUO4.gif,
http://www.fz-rossendorf.de/FWR/XAS/XAS_1.html,
http://detserv1.dl.ac.uk/Herald/xray_review_spectroscopy.htm.
9. Takashi Fujikawa, *Multiple Scattering Approach,* in *X-ray Absorption Fine Structure for Catalysts and Surfaces,* Series on Synchrotron Radiation Techniques and Applications, Vol. 2, pp. 77, Yasuhiro Iwasawa, ed., World Scientific, 1996.
10. http://www.ceramics.nist.gov/programs/thinfilms/exafsamorphous.
11. D.E. Sayers, E.A. Stern and F.W. Lytle, Phys. Rev. Lett. 27(1971)1204; D.C. Koningsberger and R. Prins, *X-ray Absorption: Principles, Applications, Techniques of EXAFS, SEXAFS and XANES*, Wiley, New York, 1988.
12. B. Lengeler, Adv. Mat. 2(1990)125.

13. http://www.nature.com/nature/journal/v412/n6843/extref/412169aa.html.

14. Yasuhiro Iwasawa, ed., X-ray Absorption Fine Structure for Catalysts and Surfaces, Series on Synchrotron Radiation Techniques and Applications, Vol. 2, World Scientific, 1996.

15. Molecular Environmental Science: Speciation, Reactivity, and Mobility of Environmental Contaminants, SLAC-R-95-477, 1995; Molecular Environmental Science and Synchrotron Radiation Facilities, SLAC-R-97-517; D.B. Hunter, P.M. Bertsch, K.M. Kemner, and S.B. Clark, *Distribution and Chemical Speciation of Metals and Metalloids in Biota Collected from Contaminated Environments by Spatially Resolved XRF, XANES, and EXAFS*, J. de Physique IV France 7(1997)C2-767.

16. Tsun-Kong Sham, ed., *Chemical Applications of Synchrotron Radiation, Part I:Dynamics and VUV Spectroscopy, Part II: X-ray Applications*, Advanced Series in Physical Chemistry, Vol. 12 A and B, World Scientific, 2002.

17. Harald Ade and Stephen Urquhart, in *Chemical Applications of Synchrotron Radiation, Part I:Dynamics and VUV Spectroscopy*, Advanced Series in Physical Chemistry, Vol. 12 A, pp. 285, Tsun-Kong Sham, ed., World Scientific, 2002.

18. Steve M. Heald, D.T. Jiang, in *Chemical Applications of Synchrotron Radiation, Part II: X-ray Applications*, Advanced Series in Physical Chemistry, Vol. 12 B, pp. 761, Tsun-Kong Sham, ed., World Scientific, 2002.

19. Charles Kittel, Introduction to Solid State Physics, Wiley.

20. A. Damascelli et al., Phys. Rev. Lett. 85 (2000) 5194, and http://www.physics.ubc.ca/~damascel/andrea_intro.pdf .

21. G. Schütz et al., Phys. Rev. Lett. 58(1987)737.

22. J. Stöhr, et al, Surface Review and Letters 5(6)(1998)1297.

23. J. Stöhr and Y. Wu, *New Directions in Research with Third-Generation Soft X-Ray Synchrotron Radiation Sources*, NATO ASI Series, Series E: Applied Sciences, Vol. 254, 221-250 (Eds. A. Schlachter and F.J. Wuilleumier, 1994).

24. Maurizio Sacchi and Jan Vogel, *Magnetism and Synchrotron Radiation*, Lecture notes in physics, Vol. 565, E. Beaurepaire, F. Scheurer, G. Krill, J.-P. Kappler, eds., Springer 2001

25. A. Bienenstock and A.L. Robinson, *Impact of Synchrotron Radiation on Materials Science*, http://www.slac.stanford.edu/pubs/beamline/25/2/25-2-bienenstock.pdf

26. P. Calvani and P. Roy, eds., *Infrared Synchrotron Radiation*, Editrice Compositori, Bologna, 1998

27. Y.-L. Mathis, unpublished, 2001

28. David L. Wetzel and Steven M. LeVine, Science 285(5431)(1999)1224

29. H.-Y.N. Holman et al., ALSNews, 160 (2000)

30. E.W. Becker, W. Ehrfeld, P. Hagmann, A. Maner, D. Münchmeier, Microelectronic Engineering 4(1986)35-56

31. H.O. Moser, W. Ehrfeld, M. Lacher, H. Lehr, *Fabrication of Three-dimensional Microdevices from Metals, Plastics, and Ceramics*, Proceedings 1st Japanese-French Congress on Mechatronics, Besançon, Oct. 20-22, 1992, Institut des Microtechniques de Franche-Comté; W. Bacher, P. Bley, H.O. Moser, *Potential of LIGA technology for optoelectronic interconnects*, in Optoelectronic Interconnects and Packaging, SPIE Critical Reviews of Optical Science and Technology, Vol. CR62, pp. 442-460, 1996.

32. G. Feiertag et al., Appl. Phys. Lett. 71(11)(1997)1441.

33. http://hikwww1.fzk.de/imt/liga/.

34. http://www.imm-mainz.de.

35. J.R. Kong, O. Wilhelmi, H.O. Moser, *Gap optimisation for proximity x-ray lithography using the super-resolution process*, ICMAT 2003, Symposium G, 7-12 December 2003, Singapore.